CAMBRIDGE TRACTS IN MATHEMATICS

General Editors

B. BOLLOBAS, F. KIRWAN, P. SARNAK, C.T.C. WALL

129 Gaussian Hilbert Spaces

Svante Janson
Uppsala University

Gaussian Hilbert Spaces

CAMBRIDGE UNIVERSITY PRESS
Cambridge, New York, Melbourne, Madrid, Cape Town, Singapore, São Paulo

Cambridge University Press
The Edinburgh Building, Cambridge CB2 8RU, UK

Published in the United States of America by Cambridge University Press, New York

www.cambridge.org
Information on this title: www.cambridge.org/9780521561280

First published 1997
This digitally printed version 2008

A catalogue record for this publication is available from the British Library

ISBN 978-0-521-56128-0 hardback
ISBN 978-0-521-05720-2 paperback

Contents

Introduction vii

Chapter 1. Gaussian spaces 1
1. Preliminaries 1
2. Gaussian random variables 3
3. Gaussian Hilbert spaces 4
4. Complex Gaussian spaces 12
5. Feynman diagrams 15

Chapter 2. Wiener chaos 17

Chapter 3. Wick products 23
1. Wick products of Gaussian variables 24
2. Wick exponentials 32
3. General Wick products 35

Chapter 4. Tensor products and Fock space 42
1. Spaces 42
2. Operators 45
3. Absolute continuity 53

Chapter 5. Hypercontractivity 57
1. Real hypercontractivity 57
2. Complex hypercontractivity 67
3. Multipliers 72

Chapter 6. Variables with finite chaos decompositions 78

Chapter 7. Stochastic integration 86
1. Brownian motion and Itô integrals 86
2. Stochastic integrals on general measure spaces 95
3. Skorohod integrals 105
4. Complex Gaussian stochastic measures 110

Chapter 8. Gaussian stochastic processes 117
1. The covariance function 117
2. The pseudometric 119
3. Continuity 120
4. The Cameron–Martin space 120

Chapter 9. Conditioning 127
1. Introduction 127
2. Conditional expectations 128
3. Conditional distributions 132
4. Example: Random potential in a network 133

Chapter 10. Pairs of Gaussian subspaces 142
1. Projections and singular numbers 142
2. Measures of dependence 143

Chapter 11. Limit theorems for generalized U-statistics 150
1. U-statistics 150
2. Asymmetric statistics 157
3. Random graphs 165

Chapter 12. Applications to operator theory 184

Chapter 13. Some operators from quantum physics 197
1. The Fourier–Wiener transform 197
2. Annihilation and creation operators 200
3. Position and momentum operators 210

Chapter 14. The Cameron–Martin shift 216
1. The Cameron–Martin shift 216
2. Shifts and the Cameron–Martin space 223

Chapter 15. Malliavin calculus 228
1. The directional derivative 229
2. The gradient operator 236
3. Higher derivatives 239
4. Absolute continuity 240
5. Sobolev spaces 245
6. The L^2 case 257
7. Dense subspaces 261
8. The Meyer inequalities 266
9. The divergence operator 274
10. Smoothness 281
11. A new look at the Skorohod integral 283

Chapter 16. Transforms 286
1. The T- and S-transforms 286
2. More general Wick products 293
3. Fock space operators 298
4. Stochastic integration 300
5. The Cameron–Martin shift 304
6. Malliavin calculus 304

Appendices 309
A. The monotone class theorem 309
B. Stochastic processes 311
C. Banach-space-valued functions and random variables 312
D. Polarization 314
E. Tensor products 315
F. Reproducing Hilbert Spaces 323
G. Analytic functions in Banach spaces 326
H. Hilbert–Schmidt operators and singular numbers 327

References 329

Index of notation 335

Index 337

Introduction

A Gaussian Hilbert space is a (complete) linear space of random variables with (centred) Gaussian distributions. This simple notion combines probability theory and Hilbert space theory into a rich and powerful structure, and Gaussian Hilbert spaces and connected notions such as the Wiener chaos decomposition and Wick products appear in several areas of probability theory and its applications, for example in stochastic processes and fields, stochastic integration, quantum field theory and limit theory for various statistics. There are also applications to non-probabilistic analysis, for example Banach space geometry and partial differential equations.

Although there are many references dealing with such applications where Gaussian spaces are treated and used, see for example Hida and Hitsuda (1976), Hida, Kuo, Potthoff and Streit (1993), Holden, Øksendal, Ubøe and Zhang (1996), Ibragimov and Rozanov (1970), Kahane (1985), Kuo (1996), Major (1981), Malliavin (1993, 1997), Meyer (1993), Neveu (1968), Nualart (1995, 1997+), Obata (1994), Pisier (1989), Simon (1974, 1979a), Watanabe (1984), there seems to be a shortage of works dealing with the basic properties of Gaussian spaces in general, without connecting them to a particular application. (One exception is the paper by Dobrushin and Minlos (1977).) This book is an attempt to fill the gap by providing a collection of the most important definitions and results for general Gaussian spaces, together with some applications to special Gaussian spaces. For further results and applications, see the references above.

In our point of view, the situation is similar to that of Hilbert spaces in functional analysis. All spaces (Hilbert or Gaussian, respectively) of the same dimension are isomorphic, so from an abstract point of view it suffices to study one of them. Nevertheless, there are many different concrete examples of such spaces that arise in different contexts, and the importance (and perhaps beauty) of the general theorems is revealed by their interpretations for the various examples. Hence, on the one hand, it seems best to develop the theory of Gaussian spaces abstractly, formulating definitions and results in terms of the intrinsic structure only. On the other hand, different Gaussian spaces may then be used to illustrate and apply the general results.

We take a probabilistic point of view, regarding the basic objects as random variables. In accordance with tradition in probability theory, we will therefore assume these variables to be defined on some probability space, but we will not make any further assumptions on this probability space, and we will usually not consider it explicitly. (In some examples, however, we use specific probability spaces for special purposes.)

The core of the book consists of Chapters 1–5, where the central parts of the theory are developed. Gaussian Hilbert spaces, and some related notions, are defined in Chapter 1, where also many examples (both simple and less simple) are given. Moreover, we introduce Feynman diagrams as a convenient bookkeeping method in moment calculations.

Every Gaussian Hilbert space induces an orthogonal decomposition, known as the (Wiener) chaos decomposition, of the corresponding L^2-space of all square integrable random variables that are measurable with respect to the σ-field generated by the Gaussian Hilbert space. This decomposition is introduced and studied in Chapter 2.

The chaos decomposition further forms the basis for the definition of Wick products in Chapter 3. This chapter also includes the definition of Wick exponentials, which form a simple family of random variables that is useful on many occasions.

It is shown in Chapter 4 that the Wiener chaos decomposition and the Wick products can be regarded as a concrete realization of the symmetric tensor products of the Gaussian Hilbert space. This point of view leads to new results; in particular, a contractive linear map of one Gaussian Hilbert space into another extends in a canonical way to a linear contraction between the corresponding L^2-spaces. This mapping is further shown to be a contraction on L^p for every $p \geq 1$. An important example of a mapping that is obtained as such an extension is the Mehler transform.

Chapter 5 is a continuation of Chapter 4, treating hypercontractivity, i.e. the property that the operators defined in Chapter 4 under suitable conditions are contractions from one L^p-space into another with a different exponent. This chapter is perhaps more technical than the preceding ones; on the other hand, it contains powerful theorems that have rather deep applications.

The remaining chapters present various applications or further developments. They build upon the first chapters, but are to a large extent independent of each other, and the reader may choose rather freely among them according to his or her interests. The main purpose of these chapters is to show how the general theory is used in different contexts. We do not intend to give complete coverage of the topics studied there; on the contrary, we concentrate on results that are directly related to the main theme of this book, and we try to avoid going too deep. The reader who has a special interest in some of these applications will certainly need other sources for further results, but we hope that the present book may serve as a useful introduction or complement.

Chapter 6 studies the distribution of random variables with a finite chaos decomposition; in particular, variables with no terms beyond the second order are studied in detail. (Such variables occur frequently in limit theorems of the type studied in Chapter 11.)

Chapter 7 contains some important applications to stochastic integration, beginning with the usual Itô integral with respect to Brownian motion and then introducing some extensions; in particular the Gaussian stochastic integral over general measure spaces and the Skorohod integral.

In Chapter 8 we study a few aspects of Gaussian stochastic processes, emphasizing the Hilbert space geometrical point of view. We further define and study the Cameron–Martin space, which is a Hilbert space of (deterministic) functions on the index set associated to a Gaussian stochastic process.

In Chapter 9 we give some simple results on conditioning in Gaussian spaces. The results are applied in an example treating a random Gaussian potential in a network.

Chapter 10 studies pairs of subspaces of a Gaussian Hilbert space. The relation between two subspaces is described by a projection operator and a corresponding sequence of numbers, and it is shown how various measures of dependence between the two subspaces can be estimated using these numbers.

In Chapter 11 we give applications to results on the asymptotic distributions of U-statistics and related random variables, some of which appear in the study of random graphs. It is noteworthy that in this chapter, unlike the preceding ones, the Gaussian Hilbert spaces are not present in the problem from the beginning; they are introduced as a convenient technical tool for the solution.

Chapter 12 contains applications to operator theory related to Grothendieck's theorem. In particular, both an upper and a lower bound to Grothendieck's constant are given. Also in this chapter, the Gaussian Hilbert spaces are introduced as a technical tool.

In the remaining chapters we return to the study of general Gaussian Hilbert spaces. In Chapter 13 we define and study the annihilation, creation, position and momentum operators that are important in quantum mechanics. (No physics is assumed or explained; our treatment is purely mathematical.) In connection with this we also present Wick's original definition of the Wick product, using 'Wick ordering', and show how it relates to the Wick product for Gaussian Hilbert spaces.

In Chapter 14 we study an operation on random variables generalizing the shift of a Brownian motion studied by Cameron and Martin (1944).

Chapter 15 contains an introduction to Malliavin calculus, based on results in Chapters 13 and 14. In particular, we give a detailed treatment of Gaussian Sobolev spaces and the Meyer inequalities, and results on existence and smoothness of densities. We also give another interpretation of the Skorohod integral.

Finally, Chapter 16 introduces some transforms that map random variables to continuous functions on a suitable space (typically the Gaussian Hilbert space itself). These transforms are useful in several contexts, but we have chosen to put them in the last chapter and merely give some examples of their use rather than introducing and using them earlier. One application of them is a more general definition of the Wick product than the one given in Chapter 3. Another application is a new definition and extension of the Skorohod integral.

Some background material from probability theory and functional analysis is presented in Appendices A–H. (Thus, for example, 'Theorem A.1' and 'equation (C.1)' refer to the appendices.)

A graduate course could be based on the first five chapters together with a selection of material from later chapters. Suitable prerequisites for this book are a standard course or two in probability theory, some integration theory and some functional analysis, especially elementary Hilbert space theory.

The selection of material and methods presented here is, of course, partly a matter of taste. For example, when deriving the basic properties in the first chapters, we often prefer to use Feynman diagrams and simple combinatorics rather than generating functions, and we avoid using properties of Hermite polynomials that have to be verified by other methods (instead we show that many of these properties follow from the probabilistic results).

The results in this book are collected from many different sources, and although some proofs are our own, the original references are seldom given and absence of references does not imply that the results are new.

I would like to thank the many other mathematicians who have contributed with helpful discussions, references and corrections; in particular I would like to mention Pontus Andersson, Persi Diaconis, Allan Gut, Sten Kaijser, Paul Malliavin, Bernt Øksendal, Gunnar Peters, Jim Propp and Philip Protter.

The manuscript has been typed by Lisbeth Juuso, Zsuzsanna Kristófi, Eira Tersmeden and myself; I thank the three first mentioned. I also thank Cambridge University Press for helpful proofreading.

Most of this book has been written in Uppsala; parts of the work have also been done during visits to the Mittag-Leffler Institute in Djursholm and the Institute for Mathematics and its Applications in Minneapolis. The work on this book was supported by the Göran Gustafsson Foundation for Research in Natural Sciences and Medicine.

My daughter Sofie was born while I was completing this book. Any remaining inconsistencies or errors are entirely due to her distracting influence.

Uppsala, January 1997,

Svante Janson

I

Gaussian spaces

1. Preliminaries

We recall some standard notions from probability theory, integration theory and functional analysis, at the same time fixing some basic notation.

A real or complex *random variable* is a measurable function on a probability space $(\Omega, \mathcal{F}, \mathrm{P})$. We usually allow complex variables without further comment; in many cases there will be no difference between the real and complex cases, and we then allow both possibilities. (Complex variables are important in some applications but not used in others.)

We will occasionally also consider random variables with values in some other space E, for example vector-valued variables, but we will then always say so explicitly. (We will only be interested in cases when E is a topological space equipped with its Borel σ-field, which we denote by $\mathcal{B}$.)

We identify random variables that are equal a.s.; hence we usually write $X = Y$ rather than the longer $X = Y$ a.s., but these expressions are equivalent. In fact, many constructions will give us variables that are uniquely defined only a.s. (Formally, random variables may be defined as equivalence classes of measurable functions, but we will not stress this point of view.)

The underlying probability space $(\Omega, \mathcal{F}, \mathrm{P})$ is completely arbitrary and is often not even mentioned. Of course, it is understood that whenever we talk about joint distributions or sums of random variables, they have to be defined on the same probability space. In particular, a linear space of random variables (the main theme of this book) is a linear space of measurable functions on some probability space. We will tacitly use $(\Omega, \mathcal{F}, \mathrm{P})$ to denote the underlying probability space of any random variable or space of random variables under discussion, as long as there is no danger of confusion.

Given a set A of random variables on some probability space $(\Omega, \mathcal{F}, \mathrm{P})$, we let $\mathcal{F}(A)$ denote the σ-field generated by the variables in A; this is the smallest σ-field such that all variables in A are measurable. Clearly, $\mathcal{F}(A) \subseteq \mathcal{F}$.

The *distribution* of a random variable X with values in some space E is the probability measure on E induced by $X\colon (\Omega, \mathcal{F}, \mathrm{P}) \to E$. We write $X \overset{\mathrm{d}}{=} Y$ if X and Y have the same distribution, and denote convergence in distribution (i.e. convergence of the corresponding distributions in the usual sense) by $\overset{\mathrm{d}}{\to}$.

The L^p-norm of a random variable X is defined by

$$\|X\|_p = (\mathrm{E}\,|X|^p)^{1/p}$$

for $0 < p < \infty$, for $p = \infty$ this is modified to $\|X\|_\infty = \operatorname{ess\,sup} |X|$. (This is really a norm only when $1 \le p \le \infty$, since the triangle inequality fails for $0 < p < 1$.) For $0 < p \le \infty$, $L^p = L^p(\Omega, \mathcal{F}, \mathrm{P})$ is the linear space of all random variables X defined on $(\Omega, \mathcal{F}, \mathrm{P})$ such that $\|X\|_p < \infty$.

We will mainly use L^2, which is a Hilbert space with the inner product $\langle X, Y \rangle = \mathrm{E}(X\overline{Y})$; thus $\|X\|_2^2 = \langle X, X \rangle$. Note that if X and Y are real random variables with expectations 0, then $\|X\|_2^2 = \operatorname{Var}(X)$ and $\langle X, Y \rangle = \operatorname{Cov}(X, Y)$. In particular, two such variables are orthogonal if and only if they are uncorrelated. (Convergence in L^2 of random variables is traditionally known as *convergence in mean square.*)

L^p is a Banach space for $1 \le p \le \infty$. If $1 \le p < \infty$, the dual space of L^p equals $L^{p'}$, where p' is the conjugate exponent defined by $1/p + 1/p' = 1$.

We further use $L^0 = L^0(\Omega, \mathcal{F}, \mathrm{P})$ to denote the space of all random variables on $(\Omega, \mathcal{F}, \mathrm{P})$, equipped with the topology of convergence in probability. This is a complete metric topological vector space, where the metric may be chosen as $\mathrm{E}(|X - Y| \wedge 1)$; for further topological properties of this space see for example Dunford and Schwartz (1958, Section IV.11, where the space is denoted by TM).

Convergence in probability is denoted $\xrightarrow{\mathrm{p}}$.

Recall that (for a probability space) the L^p-spaces decrease as p increases; if $0 \le p \le q \le \infty$, then $L^p \supseteq L^q$. More precisely,

$$\|X\|_p \le \|X\|_q, \qquad 0 < p \le q \le \infty,$$

which is known as *Lyapounov's inequality*. Furthermore, L^q is a dense subspace of L^p when $0 \le p \le q \le \infty$, because truncations of any random variable $X \in L^p$ converge to X in L^p.

As stated above, we allow complex random variables. We usually make no distinction in the notation between the real and complex cases (assuming both are valid); for example, L^p stands for both the real spaces of real variables and the corresponding complex spaces of complex variables. When it is necessary to distinguish between spaces of real and complex random variables, we write $L^p_{\mathbb{R}}$ and $L^p_{\mathbb{C}}$, respectively. We use these subscripts in the same way also for various spaces of functions on $\mathbb{R}$ or on other sets.

A set in a topological vector space is *total* if the family of (finite) linear combinations of elements of the set is dense in the space. For a Banach space (or more generally a locally convex space), this is equivalent to requiring that every continuous linear functional that vanishes on the set vanishes identically.

We let $\mathbf{1}[S]$ denote the indicator function of a statement S, which is defined to be 1 if the statement holds and 0 otherwise. Similarly, if A is a set, $\mathbf{1}_A(x) = \mathbf{1}[x \in A]$ is the function that equals 1 when $x \in A$ and 0 when $x \notin A$. (This is known as the indicator function of A, and outside probability theory also as the characteristic function of A.)

2. Gaussian random variables

In this book a *Gaussian* or *normal* random variable is a real-valued random variable with characteristic function $\exp(i\mu t - \frac{1}{2}\sigma^2 t^2)$ for some $\mu \in (-\infty, \infty)$ and $\sigma^2 \geq 0$. Note that we include the degenerate case $\sigma^2 = 0$, when the variable a.s. equals μ. The variable is *centred* or *symmetric* if $\mu = 0$, and *standard* if $\mu = 0$ and $\sigma^2 = 1$. We denote the distribution of a Gaussian variable by $\mathrm{N}(\mu, \sigma^2)$, with μ and σ^2 as above.

We assume that the reader is familiar with the basic properties of normal distributions, for example that μ is the mean and σ^2 the variance of the variable, that all moments are finite, and that if $\sigma^2 > 0$, then the normal distribution has the density $(2\pi\sigma^2)^{-1/2}\exp\bigl(-(x-\mu)^2/2\sigma^2\bigr)$, $-\infty < x < \infty$; moreover, if $X \sim \mathrm{N}(\mu, \sigma^2)$, then $\mathrm{E}\, e^{zX} = \exp\bigl(\mu z + \frac{1}{2}\sigma^2 z^2\bigr)$ for any complex number z.

A finite number of random variables $\xi_1, \dots, \xi_n$ are said to have a *joint normal distribution* if $\sum_1^n t_i \xi_i$ has a normal distribution for any real numbers $t_1, \dots, t_n$. We say that an infinite set of random variables has a joint normal distribution if every finite subset has.

The joint characteristic function of a finite set $\xi_1, \dots, \xi_n$ of jointly normal variables is $\mathrm{E}\exp(i\sum_j t_j \xi_j) = \exp\bigl(i\sum_j t_j \mathrm{E}\,\xi_j - \frac{1}{2}\sum_{j,k} t_j t_k \operatorname{Cov}(\xi_j, \xi_k)\bigr)$. Hence the distribution of a finite set of jointly normal variables is determined by their means and covariances; this extends to infinite sets by a standard argument, see Example A.3.

In particular, two or more jointly normal variables are independent if (and only if) their covariances vanish. For centred variables, this is equivalent to the variables being orthogonal.

The Gaussian variables considered in this book will almost always be centred. This is not very restrictive, since any Gaussian variable can be written as the sum of a centred Gaussian variable and a constant.

We will occasionally also use complex or vector-valued Gaussian variables. We say that the random vector $\xi = (\xi_1, \dots, \xi_n)$ has a normal (or Gaussian) distribution in $\mathbb{R}^n$ if $\xi_1, \dots, \xi_n$ have a joint normal distribution, i.e. if $\xi \cdot t$ is normal for every vector $t \in \mathbb{R}^n$. (This notion can be extended to random variables in infinite-dimensional vector spaces, see Example 1.13 below.) Equivalently, a random vector ξ in $\mathbb{R}^n$ is normal if and only if its characteristic function equals $\exp(i\mu \cdot t - \frac{1}{2}t'\Sigma t)$ for some vector $\mu \in \mathbb{R}^n$ (the mean) and some semi-definite matrix Σ (the covariance matrix).

We similarly say that a complex random variable ζ is Gaussian if its real and imaginary parts $\operatorname{Re}\zeta$ and $\operatorname{Im}\zeta$ have a joint normal distribution. We return to further comments on complex Gaussian variables in Section 4.

Note that although general random variables are allowed to be complex in this book, Gaussian variables are always real unless we explicitly consider complex (or vector-valued) Gaussian variables.

REMARK 1.1. Although these definitions are standard, the reader should note that some authors prefer slight variations; for example, some define Gaussian random variables to be symmetric. We find it less confusing (although somewhat tedious) instead to repeat 'symmetric' or 'centred' whenever necessary. Also, we use the expressions 'Gaussian variable' and 'normal variable' as synonyms, without attaching any significance to the choice of one or the other.

3. Gaussian Hilbert spaces

DEFINITION 1.2. A *Gaussian linear space* is a real linear space of random variables, defined on some probability space $(\Omega, \mathcal{F}, \mathrm{P})$, such that each variable in the space is centred Gaussian. Obviously, a Gaussian linear space is a linear subspace of $L^2_{\mathbb{R}}(\Omega, \mathcal{F}, \mathrm{P})$, and we use the norm and inner product of L^2 on it. A *Gaussian Hilbert space* is a Gaussian linear space which is complete, i.e., a closed subspace of $L^2_{\mathbb{R}}(\Omega, \mathcal{F}, \mathrm{P})$ consisting of centred Gaussian random variables.

Note that we require all random variables in a Gaussian space to be *real*. We will give some comments on linear spaces of complex Gaussian variables in the next section, but we will almost exclusively study the real case. On the other hand, we will often study complex random variables (Gaussian or non-Gaussian) constructed from the real variables in a Gaussian space; this can usually be done without any complications, and it is important for many applications.

Note also that we require the variables in a Gaussian space to be *centred* Gaussian; this is important for the properties developed in the sequel and spaces consisting of general Gaussian variables seem to be of much less use. If one for some reason is given a linear space V of general (real) Gaussian variables, one can always study the Gaussian space $G = \{\xi - \mathrm{E}\,\xi : \xi \in V\}$, and use the inclusion $V \subseteq G \oplus \mathbb{R}$ to transfer results to V.

A Gaussian linear space can always be completed to a Gaussian Hilbert space.

THEOREM 1.3. *If $G \subset L^2_{\mathbb{R}}(\Omega, \mathcal{F}, \mathrm{P})$ is a Gaussian linear space, then its closure $\overline{G}$ in L^2 is a Gaussian Hilbert space.*

PROOF. Suppose that $\xi \in \overline{G}$. We have to show that ξ has a centred normal distribution.

There exists a sequence $\xi_n \in G$ such that $\xi_n \to \xi$ in L^2. Let $\sigma^2 = \|\xi\|_2^2 = \mathrm{E}\,\xi^2$ and $\sigma_n^2 = \|\xi_n\|_2^2$. Then $\sigma_n^2 \to \sigma^2$ as $n \to \infty$. Convergence in L^2 implies convergence in distribution, and since $\xi_n \sim \mathrm{N}(0, \sigma_n^2) \xrightarrow{\mathrm{d}} \mathrm{N}(0, \sigma^2)$ as $n \to \infty$, we obtain $\xi \sim \mathrm{N}(0, \sigma^2)$. □

In view of this result, there is no real loss of generality in considering only Gaussian Hilbert spaces, and we will do so in much of the sequel. Nevertheless, incomplete Gaussian spaces appear naturally, as in several of the examples below, and it may sometimes be of interest to consider them as they are instead of immediately completing them.

Since Gaussian variables have moments of all orders, a Gaussian linear space is a subspace of L^p for every finite p.

THEOREM 1.4. *The L^p-norms, $0 < p < \infty$, are all proportional on a Gaussian linear space G. Hence all L^p-topologies coincide on G; they also coincide with the topology of convergence in probability.*

Moreover, the closure of G in any L^p, $0 \leq p < \infty$, equals the Gaussian Hilbert space $\overline{G}$. In particular, a Gaussian Hilbert space is a closed subspace of every $L^p_{\mathbb{R}}$, $0 \leq p < \infty$.

PROOF. A standard computation shows that if ξ is a centred Gaussian variable, then

$$\|\xi\|_p = (\mathrm{E}\,|\xi|^p)^{1/p} = \kappa(p)\|\xi\|_2, \qquad 0 < p < \infty, \tag{1.1}$$

where $\kappa(p) = \sqrt{2}\big(\Gamma(\frac{p+1}{2})/\sqrt{\pi}\big)^{1/p}$. Hence, a Cauchy sequence in one L^p-norm in G is also a Cauchy sequence in any other L^p-norm, which shows that the closures in the two topologies coincide.

Moreover, suppose that $(\xi_n)_1^\infty$ is a sequence in G such that $\xi_n \xrightarrow{\mathrm{p}} \xi$ as $n \to \infty$ for some random variable ξ. Then $\xi_m - \xi_n \xrightarrow{\mathrm{p}} 0$ as $m, n \to \infty$, and since the variables $\xi_m - \xi_n \in G$ are Gaussian, this implies $\|\xi_m - \xi_n\|_2 \to 0$. Consequently, $(\xi_n)_1^\infty$ is a Cauchy sequence in L^2. Thus the sequence converges in L^2 to some limit, which has to be ξ, and the remaining assertions follow. □

We have required that each variable in a Gaussian space has a normal distribution. It follows that the variables furthermore are jointly normal.

THEOREM 1.5. *Any set of random variables in a Gaussian linear space has a joint normal distribution.*

PROOF. By definition, it suffices to consider a finite set. Thus, let $\xi_1, \dots, \xi_n$ belong to a Gaussian linear space G. If $t_1, \dots, t_n$ are arbitrary real numbers, then $\sum_1^n t_i\xi_i \in G$, and thus $\sum_1^n t_i\xi_i$ has a normal distribution. This implies that $\xi_1, \dots, \xi_n$ has a joint normal distribution. □

We give some important examples of Gaussian spaces.

EXAMPLE 1.6. Let ξ be any non-degenerate, normal variable with mean zero. Then $\{t\xi : t \in \mathbb{R}\}$ is a one-dimensional Gaussian Hilbert space.

EXAMPLE 1.7. Let $\xi_1, \dots, \xi_n$ have a joint normal distribution with mean zero. Then their linear span $\{\sum_1^n t_i\xi_i : t_i \in \mathbb{R}\}$ is a finite-dimensional Gaussian Hilbert space. Conversely, by Theorem 1.5, every finite-dimensional Gaussian space is of this type.

EXAMPLE 1.8. More generally, if $\{\xi_\alpha\}$ is any set of centred jointly normal variables, then the linear span of $\{\xi_\alpha\}$ is a Gaussian linear space, and, by

Theorem 1.3, the closed linear span of $\{\xi_\alpha\}$ in $L^2_{\mathbb{R}}$ is a Gaussian Hilbert space. These spaces are called *the Gaussian linear space spanned by* $\{\xi_\alpha\}$ and *the Gaussian Hilbert space spanned by* $\{\xi_\alpha\}$, respectively.

EXAMPLE 1.9. Let $\{\xi_\alpha\}$ by any set (finite or infinite, possibly uncountable) of independent standard normal random variables. By the preceding example, their closed linear span

$$\Big\{\sum_\alpha a_\alpha \xi_\alpha : \sum_\alpha a_\alpha^2 < \infty\Big\}$$

is a Gaussian Hilbert space.

We observe that every Gaussian Hilbert space is of this type. In fact, let $\{\xi_\alpha\}$ be any orthonormal basis in the space. The variables ξ_α are uncorrelated, and thus independent, standard normal variables, and their closed linear span equals the given space. From an abstract point of view, all Gaussian Hilbert spaces of the same dimension are thus the same. Nevertheless, different concrete examples of Gaussian Hilbert spaces are useful for different applications, and we continue with some further examples.

EXAMPLE 1.10. Let B_t, $0 \le t < \infty$, be a standard Brownian motion. By Example 1.8, the closed linear span of $\{B_t\}_{t\ge 0}$ is a Gaussian Hilbert space, which we denote by $H(B)$. As will be seen in detail in Chapter 7, this space has a simple representation in terms of stochastic integrals, viz.

$$H(B) = \Big\{\int_0^\infty f(t)\,dB_t\Big\},$$

where f ranges over the set of (deterministic) functions in $L^2_{\mathbb{R}}([0,\infty), dt)$.

EXAMPLE 1.11. Similarly, if $(X_t)_{t\in T}$ is any Gaussian stochastic process in discrete or continuous time (T is a suitable subset of $\mathbb{R}$), and if (for simplicity) $\mathrm{E}\,X_t = 0$ for all t, then the closed linear span of $\{X_t\}$ is a Gaussian Hilbert space. By taking the closed linear span of e.g. $\{X_t\}_{t\le t_0}$ or $\{X_t\}_{t\ge t_0}$ instead, we obtain closed Gaussian subspaces of this space.

EXAMPLE 1.12. More generally, we may, for any set T, define a Gaussian stochastic process indexed by T to be a family X_t, $t \in T$, of jointly normal random variables. (This simple definition is sufficient for our purposes; as is well-known, it is not well suited for studying pathwise properties such as sample path continuity when T is uncountable; see Appendix B.) Again, if the process is centred, i.e. $\mathrm{E}\,X_t = 0$ for all t, then the closed linear span of $\{X_t\}_{t\in T}$ is a Gaussian Hilbert space. Hence a centred Gaussian stochastic process indexed by T is the same as a function from T into some Gaussian Hilbert space.

We study this example in some detail in Chapter 8.

EXAMPLE 1.13. Let $\mathcal{X}$ be a real Banach space, or more generally, a locally convex topological vector space. A Borel probability measure μ on $\mathcal{X}$

is said to be Gaussian if each continuous linear functional $x^* \in \mathcal{X}^*$, regarded as a random variable defined on the probability space $(\mathcal{X}, \mathcal{B}, \mu)$, is Gaussian. If furthermore μ is symmetric, this means that $\mathcal{X}^*$ is a Gaussian space, which may be completed to a Gaussian Hilbert space contained in $L^2(\mathcal{X}, \mathcal{B}, \mu)$. (Some elements x^* may be 0 μ-a.e., in which case the Gaussian space is really a quotient space of $\mathcal{X}^*$.)

Similarly, a random variable ξ with values in $\mathcal{X}$ is said to be Gaussian if the real-valued random variable $\langle x^*, \xi \rangle$ is Gaussian for every $x^* \in \mathcal{X}^*$, or equivalently, if the distribution of ξ is a Gaussian measure on $\mathcal{X}$. If ξ is also symmetric, then the set $\{\langle x^*, \xi \rangle : x^* \in \mathcal{X}^*\}$ is a Gaussian linear space; if μ is the distribution of ξ, this space is naturally isomorphic to the Gaussian space just constructed by considering $\mathcal{X}^*$ as a space of random variables defined on $(\mathcal{X}, \mathcal{B}, \mu)$.

EXAMPLE 1.14. A simple special case of Example 1.13 is obtained by taking $\mathcal{X}$ to be a finite-dimensional Euclidean space $\mathbb{R}^d$. The Gaussian measures on $\mathbb{R}^d$ are just the d-dimensional Gaussian distributions, i.e. the distributions of d-dimensional Gaussian random vectors.

Given a centred Gaussian measure on $\mathbb{R}^d$, the Gaussian Hilbert space constructed in Example 1.13 is just the space of all linear functionals on $\mathbb{R}^d$. This space is d-dimensional, provided the measure is not singular.

An important example is the *standard d-dimensional Gaussian measure*, given by the density

$$(2\pi)^{-d/2} e^{-|x|^2/2};$$

this is just the product of d standard Gaussian measures on $\mathbb{R}$, and is therefore the distribution of $(\xi_1, \dots, \xi_d)$ with ξ_i independent standard normal.

EXAMPLE 1.15. Consider the probability space $(\mathbb{R}, \mathcal{B}, \gamma)$, where γ is the standard Gaussian measure $d\gamma = \frac{1}{\sqrt{2\pi}} e^{-x^2/2} dx$. Then $\xi = x$ is a standard normal random variable, and $H = \{tx : t \in \mathbb{R}\}$ is a one-dimensional Gaussian Hilbert space. (This is a special case of Example 1.6; it is also the simplest but perhaps most important case of Example 1.14.)

This example is an important bridge between the probabilistic theory of Gaussian Hilbert spaces and real analysis; we will several times obtain interesting results in real analysis by interpreting general theorems for this particular Gaussian Hilbert space.

EXAMPLE 1.16. (Rather long and technical, and may be skipped at the first reading.) An example of central importance in quantum field theory and elsewhere is obtained by specializing Example 1.13 to the space $S'(\mathbb{R}^d)$ of real tempered distributions, the dual of the space $S(\mathbb{R}^d)$ of real rapidly decreasing smooth functions (we omit the subscript $\mathbb{R}$ on S in this example).

Let φ denote the canonical embedding of $S(\mathbb{R}^d)$ into its bidual. (In fact, the space is reflexive so φ is an isomorphism.) Explicitly, if $f \in S(\mathbb{R}^d)$, then $\varphi(f)$ is the linear functional $u \mapsto u(f)$ defined on $S'(\mathbb{R}^d)$.

Suppose that μ is a symmetric Gaussian probability measure on $S'(\mathbb{R}^d)$. Then, by Example 1.13, $\varphi(f) \in L^2(S'(\mathbb{R}^d), \mu)$ is a symmetric Gaussian random variable for every $f \in S(\mathbb{R}^d)$. Let H be the closure in $L^2(\mu)$ of $\{\varphi(f) : f \in S(\mathbb{R}^d)\}$; thus H is a Gaussian Hilbert space.

Define the symmetric semi-definite bilinear form $\mathcal{E}$ on $S(\mathbb{R}^d)$ by

$$\mathcal{E}(f,g) = \langle \varphi(f), \varphi(g) \rangle_H = \mathrm{E}\, \varphi(f)\varphi(g).$$

If $f_n \to f$ in $S(\mathbb{R}^d)$, then $\varphi(f_n) \to \varphi(f)$ everywhere on $S'(\mathbb{R}^d)$, and thus also in probability and by Theorem 1.4 in L^2. Hence φ is a continuous map of $S(\mathbb{R}^d)$ into H and $\mathcal{E}$ is continuous.

If we, for simplicity, assume that $\mathcal{E}$ is non-degenerate, then $(S(\mathbb{R}^d), \mathcal{E})$ is a pre-Hilbert space and may be completed to a Hilbert space $S_{\mathcal{E}}$; φ then extends to an isometry of $S_{\mathcal{E}}$ onto the Gaussian Hilbert space H.

We have here regarded μ as given, but we may as well start from the bilinear form $\mathcal{E}$ instead, since it follows from Minlos's theorem (Gelfand and Vilenkin 1961, Chapter IV) that any continuous semi-definite symmetric bilinear form on $S(\mathbb{R}^d)$ corresponds to a (unique) Gaussian measure on $S'(\mathbb{R}^d)$ as above.

The most important special case is obtained by taking $\mathcal{E}(f,g) = \int fg\,dx$, the usual inner product in $L^2_{\mathbb{R}}(\mathbb{R}^d)$. Then the completion $S_{\mathcal{E}}$ equals $L^2_{\mathbb{R}}(\mathbb{R}^d)$, and the construction above yields a Gaussian measure μ on $S'(\mathbb{R}^d)$ and a Gaussian Hilbert space $H \subset L^2(\mu)$ such that $\varphi\colon L^2_{\mathbb{R}}(\mathbb{R}^d) \to H$ is an isometry. This Gaussian measure μ is called the *white noise measure* on $S'(\mathbb{R}^d)$. The study of the white noise measure and the corresponding Gaussian Hilbert space has developed into a whole theory, the *white noise calculus*, see Hida, Kuo, Potthoff and Streit (1993), Kuo (1996) and Obata (1994).

Another special case, known as the *free Euclidean Bose field in d dimensions* (Simon 1974, Hida, Kuo, Potthoff and Streit 1993, Glimm and Jaffe 1981, Chapter 6), is obtained by taking $\mathcal{E}(f,g) = \int f(1-\Delta)^{-1}g = (2\pi)^{-d}\int(1+|y|^2)^{-1}\hat{f}(y)\hat{g}(y)dy$, with $\hat{f}(y) = \int e^{-ix\cdot y}f(x)\,dx$. In this case $S_{\mathcal{E}}$ is the Sobolev space $\mathcal{H}_{-1}(\mathbb{R}^d)$.

EXAMPLE 1.17. Let us finally note that a closed subspace of a Gaussian Hilbert space is always a Gaussian Hilbert space. (See Example 1.11 for an important instance.)

We also introduce two further, closely related, definitions. Recall that a *linear isometry* of a Hilbert space into another is a linear map that preserves the norm, or, equivalently, the inner product. Linear isometries that are onto are also called *isomorphisms* (but note that this term has a different meaning for Banach spaces) or (in particular for complex spaces) *unitary* operators.

DEFINITION 1.18. A *Gaussian Hilbert space indexed by a (real) Hilbert space H* is a Gaussian Hilbert space G together with a specific linear isometry $h \mapsto \xi_h$ of H onto G.

On a formal level, this definition adds rather little, and it will not be much used in this book where the emphasis is on the Gaussian space itself. But this definition is important in contexts where the primary object of study is some (non-Gaussian) Hilbert space H, and the Gaussian space plays a secondary role.

DEFINITION 1.19. A *Gaussian field on a (real) Hilbert space* H is a linear isometry $h \mapsto \xi_h$ of H into some Gaussian space.

We do not require the isometry to be onto, but evidently its range $\{\xi_h : h \in H\}$ is a Gaussian Hilbert space indexed by H. Conversely, a Gaussian Hilbert space indexed by a Hilbert space H defines a Gaussian field on H.

EXAMPLE 1.20. We may now reinterpret Example 1.10 as an example of a Gaussian Hilbert space indexed by $L^2_{\mathbb{R}}([0,\infty), dt)$, with the defining isometry $f \mapsto \xi_f = \int_0^\infty f(t)\, dB_t$.

Alternatively, this isometry defines a Gaussian field on $L^2_{\mathbb{R}}([0,\infty), dt)$.

Analogously, any Gaussian field on an arbitrary L^2-space $L^2_{\mathbb{R}}(M,\mu)$ may be regarded as a stochastic integral; this will be studied in Section 7.2.

EXAMPLE 1.21. Example 1.16 gives a Gaussian Hilbert space indexed by the Hilbert space $S_{\mathcal{E}}$, and in the particular case of white noise a Gaussian Hilbert space indexed by $L^2_{\mathbb{R}}(\mathbb{R}^d)$. By the remark at the end of the preceding example, the white noise measure thus defines a stochastic integral for $L^2(\mathbb{R}^d)$. We will return to this stochastic integral in Section 7.2.

EXAMPLE 1.22. If H is a real Hilbert space, let $\{e_\alpha\}_{\alpha\in A}$ be an orthonormal basis in H. Let $\{\xi_\alpha\}_{\alpha\in A}$ be a set of independent standard normal variables with the same index set and define a Gaussian Hilbert space G as in Example 1.9. The mapping $\sum a_\alpha e_\alpha \mapsto \sum a_\alpha \xi_\alpha$ is an isometry of H onto G, and thus makes G a Gaussian Hilbert space indexed by H.

Example 1.22 provides one proof of the following fundamental result.

THEOREM 1.23. *If H is a real Hilbert space, then there exists a Gaussian Hilbert space indexed by H, and thus a Gaussian field on H.* □

We continue with more examples. Except the first, they are rather technical, and although they describe important constructions, they may be somewhat misleading here. The central idea in this book is to study Gaussian spaces in general, without caring about how they are constructed. (For our purposes, we can always use the simple construction in Example 1.22.) We may therefore ignore all technical problems concerning for example the existence of Gaussian measures in vector spaces. (Of course, for other purposes it is important to study constructions such as these in detail; see for example the references given below.) In our viewpoint, Gaussian spaces are simple objects by themselves, although they sometimes have complicated constructions and relations to other objects.

EXAMPLE 1.24. If H is a finite-dimensional real Hilbert space, let ξ be an H-valued Gaussian random variable such that $\langle \xi, h \rangle \sim \mathrm{N}(0, \|h\|^2)$ for every $h \in H$. Then $h \mapsto \langle \xi, h \rangle$ defines a Gaussian Hilbert space indexed by H. (To construct such a ξ, choose an orthonormal basis $e_1, \dots, e_n$ in H and let $\xi = \sum_1^n \xi_i e_i$ with $\xi_1, \dots, \xi_n$ independent standard normal. Alternatively, we may identify H with $\mathbb{R}^d$ and let ξ have the standard Gaussian distribution defined in Example 1.14.)

EXAMPLE 1.25. (Rather technical, and may be skipped.) If H is a real Hilbert space of infinite dimension, the construction of Example 1.24 does not work, because no such ξ exists. (Equivalently, in an infinite-dimensional Hilbert space, there is no 'standard Gaussian measure' similar to the one constructed for $\mathbb{R}^d$ in Example 1.14.) Nevertheless, we may find ξ as a random variable in a suitable larger space.

Take for example $H = \ell^2$ and define $\xi = (\xi_1, \xi_2, \dots)$, where the ξ_k are independent standard normal variables. Then a.s. $\xi \notin \ell^2$, but ξ is a well-defined random variable in a larger space, for example the Hilbert space $\{(x_k)_1^\infty : \sum_1^\infty (x_k/k)^2 < \infty\}$ or the product space $\mathbb{R}^\infty$. Moreover, for any $a = (a_k)_1^\infty \in \ell^2$, the inner product $\langle \xi, a \rangle = \sum_1^\infty a_k \xi_k$ is well-defined a.s., and the mapping $a \mapsto \langle \xi, a \rangle$ defines a Gaussian Hilbert space indexed by ℓ^2. (Note that the sum $\sum_1^\infty a_k \xi_k$ converges a.s. for every $a \in \ell^2$, but the exceptional null set depends on a and the union of these null sets is almost the whole probability space because $\xi \notin \ell^2$ a.s.)

The general situation is this. Let $\mathcal{X}$ be a real locally convex topological vector space which contains the Hilbert space H as a dense subspace, with a continuous inclusion mapping $H \to \mathcal{X}$. Then there is a dual inclusion $\mathcal{X}^* \subset H^* \cong H$, so we have a triple $\mathcal{X}^* \subset H \subset \mathcal{X}$. Under suitable topological conditions, there exists a Gaussian random variable ξ in $\mathcal{X}$ such that $\langle \xi, \varphi \rangle \sim \mathrm{N}(0, \|\varphi\|_H^2)$ for every $\varphi \in \mathcal{X}^*$. (Equivalently, there exists a Gaussian measure on $\mathcal{X}$ with the right finite-dimensional marginals.) This holds for example (Gelfand and Vilenkin 1961, Chapter IV) if $\mathcal{X}$ is another Hilbert space such that the inclusion mapping $H \to \mathcal{X}$ is a Hilbert–Schmidt operator, or if $\mathcal{X}$ is the dual of a nuclear space $\mathcal{Y} \subset H$ which is dense in H. (See e.g. Obata (1994), Schaefer (1971) or Treves (1967) for the definition of nuclear spaces.) Examples of such nuclear spaces $\mathcal{Y}$ are the direct sum $\sum_1^\infty \mathbb{R}$, in which case the dual space $\mathcal{X}$ is the the product space $\mathbb{R}^\infty$ mentioned above, and the space $S(\mathbb{R}^d)$ in Example 1.16, in which case $\mathcal{X}$ is the dual space of tempered distributions; in these cases $\mathcal{X}^* = \mathcal{Y}$ because the spaces are reflexive.

Another case where such a Gaussian variable ξ exists is when $\mathcal{X}$ is a Banach space with a norm satisfying a certain tightness (or continuity) condition, see Gross (1967) and Kuo (1975). (Such a space $\mathcal{X}$ is often called an *abstract Wiener space*.)

Suppose that such a variable ξ exists. Then the mapping $\varphi \mapsto \langle \xi, \varphi \rangle_{\mathcal{X}}$ is an isometry of the dense subspace $\mathcal{X}^*$ of H onto the Gaussian linear space

$G = \{\langle\xi,\varphi\rangle_{\mathcal{X}} : \varphi \in \mathcal{X}\}$, and thus φ may be extended by continuity to an isometry of H onto $\overline{G}$, which gives a Gaussian Hilbert space $\overline{G}$ indexed by H. (One interpretation of this extension from $\mathcal{X}^*$ to H is that although only elements of $\mathcal{X}^*$ define functionals defined *everywhere* on $\mathcal{X}$, every element of H defines a functional that is defined *almost everywhere* on $\mathcal{X}$.)

EXAMPLE 1.26. (A special case, but still rather technical.) Let H be the Hilbert space $\{\int_0^t f(s)\,ds : f \in L^2_{\mathbb{R}}([0,1])\}$ consisting of all absolutely continuous functions F on $[0,1]$ with $F(0)=0$ and $F' \in L^2_{\mathbb{R}}([0,1])$; the norm is given by $\|F\|_H = \|F'\|_{L^2}$. Let $\mathcal{X}$ be the Banach space $\{g \in C_{\mathbb{R}}([0,1]) : g(0)=0\}$ of all real continuous functions on $[0,1]$ that vanish at the origin.

This is a case where the construction in Example 1.25 works. As we will verify in Example 8.27, the constructed variable ξ is a Brownian motion and the mapping $h \mapsto \langle h,\xi\rangle$ (defined by extension from $\mathcal{X}^*$) of H onto a Gaussian Hilbert space is given by $F \mapsto \int_0^1 F'(t)\,d\xi(t)$.

EXAMPLE 1.27. A related construction by Itô (1993) gives a canonical Gaussian field for any Hilbert space. Let H be a real Hilbert space and let Ω be its algebraic dual, i.e. the linear space of all (not necessarily continuous) linear functionals on H. Define real functions ξ_h, $h \in H$, on Ω by $\xi_h(\omega) = \omega(h)$ and equip Ω with the σ-field $\mathcal{F}$ generated by the functions ξ_h, $h \in H$. Then there exists a unique probability measure on $(\Omega,\mathcal{F})$ such that $\xi_h \sim \mathrm{N}(0,\|h\|^2)$ for each $h \in H$, and thus $h \mapsto \xi_h$ gives a Gaussian field on H; equivalently, the set $\{\xi_h\}$ yields a Gaussian Hilbert space indexed by H.

This is proved as follows (we give a condensed version): For existence, select a Hamel basis $\{h_\alpha\}_{\alpha\in A}$ in H (i.e. a linear basis of H as a vector space, ignoring the Hilbert structure). The mapping $\omega \mapsto (\omega(h_\alpha))_\alpha$ yields an isomorphism $\Omega \cong \mathbb{R}^A$. By Kolmogorov's theorem there exists a probability measure on $\mathbb{R}^A$ such that the coordinate mappings x_α are centred jointly Gaussian variables with $\operatorname{Cov}(x_\alpha,x_\beta) = \langle h_\alpha,h_\beta\rangle_H$, because this defines consistent finite-dimensional distributions; the isomorphism just described transfers this to a probability measure on $(\Omega,\mathcal{F})$ such that the functions ξ_{h_α} are centred jointly Gaussian variables with $\operatorname{Cov}(\xi_{h_\alpha},\xi_{h_\beta}) = \langle h_\alpha,h_\beta\rangle_H$, and since every $h \in H$ is a finite linear combination of basis elements, $\xi_h \sim \mathrm{N}(0,\|h\|^2)$ follows.

Uniqueness follows by the monotone class theorem, see Example A.3, since all finite-dimensional distributions are determined by the condition.

We end this section with a useful formula for mixed (central) moments of normal variables. (All such moments exist by Hölder's inequality.) In quantum field theory, this formula (like several other related results) is called *Wick's theorem*.

THEOREM 1.28. *Let $\xi_1,\dots,\xi_n$ be centred jointly normal variables. Then*

$$\mathrm{E}(\xi_1\cdots\xi_n) = \sum\prod_k \mathrm{E}(\xi_{i_k}\xi_{j_k}) \tag{1.2}$$

where the sum is over all partitions of $\{1,\dots,n\}$ *into disjoint pairs* $\{i_k, j_k\}$.

REMARK 1.29. If $n=2m$ is even, there are $(2m)!/(2^m m!) = (2m-1)!!$ such partitions; if n is odd there is none and the expectation in (1.2) equals zero (as is obvious by symmetry).

REMARK 1.30. If we take all ξ_i equal in Theorem 1.28 we obtain the well-known formula for moments of a centred normal variable: if $\xi \sim \mathrm{N}(0,\sigma^2)$, then

$$\mathrm{E}\,\xi^n = \begin{cases} (n-1)!!\,\sigma^n, & n \text{ even}, \\ 0, & n \text{ odd}. \end{cases}$$

PROOF. The left hand side of (1.2) is the coefficient of $t_1\cdots t_n$ in the Taylor expansion of

$$\mathrm{E}\Big(\prod_1^n e^{t_i\xi_i}\Big) = \mathrm{E}\big(e^{\sum_1^n t_i\xi_i}\big) = \exp\big(\tfrac12\|\sum_1^n t_i\xi_i\|_2^2\big) = \exp\big(\tfrac12\sum_{i,j} t_it_j\,\mathrm{E}(\xi_i\xi_j)\big).$$

Calculating with power series modulo terms where at least one t_i appears with degree 2 or more (i.e., modulo the ideal generated by $\{t_i^2\}$),

$$\begin{aligned}\exp\big(\tfrac12\sum_{i,j} t_it_j\,\mathrm{E}(\xi_i\xi_j)\big) &\equiv \exp\big(\sum_{i<j} t_it_j\,\mathrm{E}(\xi_i\xi_j)\big)\\ &= \prod_{i<j}\exp\big(t_it_j\,\mathrm{E}(\xi_i\xi_j)\big) \equiv \prod_{i<j}\big(1+t_it_j\,\mathrm{E}(\xi_i\xi_j)\big),\end{aligned}$$

which yields the result since the coefficient of $t_1\cdots t_n$ in the latter term equals the right hand side of (1.2).

Alternatively, we may observe that both sides of (1.2) are symmetric multilinear forms (defined on some Gaussian space, cf. Example 1.8). Hence it suffices by polarization, Theorem D.2, to verify the formula when $\xi_1=\cdots=\xi_n$, which is the well-known case stated in Remark 1.30. □

4. Complex Gaussian spaces

Although our main interest is in real Gaussian variables, it is sometimes convenient to use complex Gaussian variables as well. We collect some basic properties for future reference. (This section can be skipped at the first reading.)

As stated above, a complex random variable ζ is Gaussian or normal if $\operatorname{Re}\zeta$ and $\operatorname{Im}\zeta$ are jointly normal. (This is in accordance with the general definition of vector-valued Gaussian random variables, regarding $\mathbb{C}$ as a vector space over $\mathbb{R}$.) Note that no assumption is made on the covariance between the real and imaginary parts; in particular, every real Gaussian variable is also complex Gaussian by this definition. The distribution of a complex Gaussian variable ζ is thus determined by the five real parameters $\mathrm{E}(\operatorname{Re}\zeta)$,

$\mathrm{E}(\mathrm{Im}\,\zeta)$, $\mathrm{Var}(\mathrm{Re}\,\zeta)$, $\mathrm{Var}(\mathrm{Im}\,\zeta)$, and $\mathrm{Cov}(\mathrm{Re}\,\zeta, \mathrm{Im}\,\zeta)$; or alternatively by the two complex parameters $\mathrm{E}\,\zeta$ and

$$\mathrm{E}(\zeta - \mathrm{E}\,\zeta)^2 = \mathrm{Var}(\mathrm{Re}\,\zeta) - \mathrm{Var}(\mathrm{Im}\,\zeta) + 2i\,\mathrm{Cov}(\mathrm{Re}\,\zeta, \mathrm{Im}\,\zeta) \tag{1.3}$$

and the real

$$\mathrm{E}\,|\zeta - \mathrm{E}\,\zeta|^2 = \mathrm{Var}(\mathrm{Re}\,\zeta) + \mathrm{Var}(\mathrm{Im}\,\zeta).$$

A complex Gaussian variable ζ is *centred* if $\mathrm{E}\,\zeta = 0$. It follows that a centred complex Gaussian distribution is determined by $\mathrm{E}\,\zeta^2$ and $\mathrm{E}\,|\zeta|^2$.

Moreover, we say that a complex Gaussian variable ζ is *symmetric* if $\zeta \overset{\mathrm{d}}{=} \lambda\zeta$ for any complex number λ with $|\lambda| = 1$; some equivalent characterizations are listed in the following proposition. Note that 'centred' and 'symmetric' are *not* equivalent for complex Gaussian variables.

PROPOSITION 1.31. *Let ζ be a complex Gaussian variable. Then the following are equivalent.*

(i) *ζ is symmetric complex Gaussian.*
(ii) $\mathrm{E}\,\zeta = \mathrm{E}\,\zeta^2 = 0$.
(iii) $\mathrm{E}\,\zeta = 0$, $\mathrm{E}(\mathrm{Re}\,\zeta)^2 = \mathrm{E}(\mathrm{Im}\,\zeta)^2$, *and* $\mathrm{E}(\mathrm{Re}\,\zeta\,\mathrm{Im}\,\zeta) = 0$.
(iv) *$\zeta = \xi + i\eta$ where ξ and η are independent centred real Gaussian variables with the same variance.*

PROOF. It is easily seen that (i) $\Leftrightarrow$ (ii) since $\mathrm{E}(\lambda\zeta) = \lambda\,\mathrm{E}\,\zeta$ and $\mathrm{E}(\lambda\zeta)^2 = \lambda^2\,\mathrm{E}\,\zeta^2$ while $\mathrm{E}\,|\lambda\zeta|^2 = \mathrm{E}\,\zeta^2$ when $|\lambda| = 1$. Moreover, (ii) $\Leftrightarrow$ (iii) by (1.3), while (iii) $\Leftrightarrow$ (iv) by standard properties of real Gaussian variables. □

REMARK 1.32. Some authors take one of the equivalent properties in Proposition 1.31 to be the definition of a complex Gaussian variable, thereby giving the term a more restrictive meaning than we do.

We define a *standard complex Gaussian variable* to be a symmetric complex Gaussian variable ζ with $\mathrm{E}\,|\zeta|^2 = 1$. (This is not the only reasonable choice; some authors prefer to define it as a variable ζ with $\mathrm{Re}\,\zeta$ and $\mathrm{Im}\,\zeta$ independent standard normal, which yields $\mathrm{E}\,|\zeta|^2 = 2$ but is in accordance with the definition in Example 1.14 for $\mathbb{R}^2$.) Note that the symmetric complex Gaussian variables are exactly the variables of the form $c\zeta$, where c is a complex number (which may be taken ≥ 0) and ζ is standard complex Gaussian.

A finite family $\{\zeta_i\}_{i=1}^n$ of complex random variables is said to be *jointly normal* if $\sum_i a_i\zeta_i$ is normal for any complex numbers a_i; this is equivalent to the family $\{\mathrm{Re}\,\zeta_i, \mathrm{Im}\,\zeta_i\}_{i=1}^n$ of real random variables being jointly normal. It follows that then also the larger family $\{\zeta_i, \overline{\zeta}_i\}_{i=1}^n$ is jointly normal.

Theorem 1.28 extends by linearity to centred jointly normal complex variables. (Consider the real and imaginary parts of the variables.)

Note that if ζ is a complex symmetric Gaussian variable, then $\mathrm{E}\,\zeta^n = 0$ for any $n > 0$; more generally, $\mathrm{E}\,\zeta^n\bar{\zeta}^m = \delta_{nm} n!(\mathrm{E}\,|\zeta|^2)^n$ for $n, m \geq 0$, for example by Theorem 1.28.

We have defined Gaussian Hilbert spaces as spaces of real Gaussian variables, and these will be the main topic of this book. Let us, nevertheless, for a moment consider the possibilities for complex vector spaces of centred complex Gaussian variables.

One example is the complexification of a real Gaussian space G, i.e. the space

$$G_{\mathbf{C}} = G + iG = \{\xi + i\eta : \xi, \eta \in G\} = \{\zeta : \operatorname{Re}\zeta, \operatorname{Im}\zeta \in G\}.$$

In general, if K is a complex space of centred complex Gaussian variables, then the space $\operatorname{Re} K = \{\operatorname{Re}\zeta : \zeta \in K\}$ is a real Gaussian space and $K \subseteq (\operatorname{Re} K)_{\mathbf{C}}$, but the inclusion may be proper. It is easily seen that if $\overline{K} = \{\bar{\zeta} : \zeta \in K\}$, then always

$$(\operatorname{Re} K)_{\mathbf{C}} = K + \overline{K} = \{\zeta_1 + \bar{\zeta}_2 : \zeta_1, \zeta_2 \in K\}.$$

The spaces that are complexifications of real Gaussian spaces are descibed by the following proposition; we leave the proof as a simple exercise. (Note that no properties of Gaussian variables are needed; this proposition is a general fact about real and complex function spaces.)

PROPOSITION 1.33. *If K is a complex linear space of centred complex Gaussian variables, then the following are equivalent.*

(i) *K equals the complexification of some real Gaussian space.*
(ii) $K = (\operatorname{Re} K)_{\mathbf{C}}$.
(iii) $K = \overline{K}$
(iv) *If $\zeta \in K$, then $\bar{\zeta} \in K$.*
(v) *If $\zeta \in K$, then $\operatorname{Re}\zeta, \operatorname{Im}\zeta \in K$.*

□

Another extreme case is described in the following proposition. Again we leave the proof as an exercise. (This time, Proposition 1.31 is useful.)

PROPOSITION 1.34. *If K is a complex linear space of centred complex Gaussian variables, then the following are equivalent.*

(i) *K is a space of symmetric complex Gaussian variables.*
(ii) *K and $\overline{K}$ are orthogonal.*
(iii) *$(\operatorname{Re} K)_{\mathbf{C}} = K \oplus \overline{K}$, with an orthogonal direct sum.*
(iv) *If $\zeta \in K$, then $\operatorname{Re}\zeta$ and $\operatorname{Im}\zeta$ are independent.*
(v) *The real linear mapping $\zeta \mapsto \sqrt{2}\operatorname{Re}\zeta$ is an isometry of K onto $\operatorname{Re} K$.*

□

Given any complex linear space K of centred complex Gaussian variables, the theory of real Gaussian spaces described in this and the following chapters may be applied to $\operatorname{Re} K$, and the results then transferred to K. For a space closed under complex conjugation as described in Proposition 1.33, there are no complications; this leads to a satisfactory theory for such spaces, but not much is gained that cannot just as well be described using the real Gaussian space $\operatorname{Re} K$, and we will not treat this case further.

For a space of symmetric complex Gaussian variables as described in Proposition 1.34, we also obtain a satisfactory theory by transferring results from $\operatorname{Re} K$. (In this case, there are a few interesting twists; cf. Examples 3.30–3.32 and Remark 4.4.) This is the most interesting case, and it seems reasonable to define a *complex Gaussian Hilbert space* as a Hilbert space of *symmetric* complex Gaussian variables. Note that a complex version of Theorem 1.23 holds, by the same proof: if H is a complex Hilbert space, then there exists an (essentially unique) isometric complex Gaussian Hilbert space.

Finally, for a general space of complex Gaussian variables not satifying the conditions in Proposition 1.33 or 1.34, we do not know any interesting results that do not directly derive from properties of $\operatorname{Re} K$; moreover, we do not know any interesting example of such spaces. Hence they will be ignored in the sequel.

As an example of the transfer of properties from real Gaussian spaces, we observe that except for the first statement in Theorem 1.4, all theorems in this chapter are valid mutatis mutandis for spaces of complex Gaussian variables. The L^p-norms are not proportional in general, however, but they are equivalent, which suffices for the other claims in Theorem 1.4. For symmetric complex Gaussian variables, (1.1) holds with $\kappa(p)$ replaced by $\kappa(p)_{\mathbf{C}} = \Gamma(\frac{p}{2}+1)^{1/p}$, and thus the full Theorem 1.4 extends for them.

5. Feynman diagrams

Sums as in Theorem 1.28 will recur several times in this book, and it will be convenient to use a graphical representation, borrowed from quantum field theory. (Feynman diagrams are used in various versions in theoretical physics. The combinatorial content is essentially the same as here, but the evaluation often involves integration over further variables. We refer to e.g. Nakanishi (1971), Glimm and Jaffe (1981, Chapter 8) and Le Bellac (1991, Chapter 5) for typical examples.)

DEFINITION 1.35. A *Feynman diagram* (of *order* $n \geq 0$ and *rank* $r \geq 0$) is a graph consisting of a set of n vertices and a set of r edges without common endpoints. There are, thus, r disjoint pairs of vertices, each joined by an edge, and $n - 2r$ unpaired vertices. The Feynman diagram is *complete* if $r = n/2$, i.e. if all vertices are paired off, and *incomplete* if $r < n/2$, i.e. if some vertices are unpaired. A *Feynman diagram labelled by n random variables* $\xi_1, \dots, \xi_n$ (defined on the same probability space) is a Feynman diagram of order n with vertices $1, \dots, n$, where we think of ξ_i as attached to vertex i. The *value* of such a labelled Feynman diagram γ with edges (i_k, j_k), $k = 1, \dots, r$, and unpaired vertices $\{i : i \in A\}$ is

$$v(\gamma) = \prod_{k=1}^{r} \mathrm{E}(\xi_{i_k}\xi_{j_k}) \prod_{i\in A} \xi_i. \tag{1.4}$$

The value is thus, in general, a random variable, but the value of a complete Feynman diagram is a (real or complex) number.

We similarly denote the order and rank of a Feynman diagram γ by $n(\gamma)$ and $r(\gamma)$.

We will only consider Feynman diagrams labelled by centred Gaussian variables, or possibly complex centred Gaussian variables.

Note that the variables $\xi_1, \dots, \xi_n$ do not have to be different. We often identify the variables and the vertices in the Feynman diagram, for example talking of an edge as joining two of the variables (or regarding a Feynman diagram as a set of random variables and a family of disjoint pairs of them), but it should be remembered that this should be interpreted as above when some (or all) of the variables coincide; i.e., the variables should be regarded as distinguishable even if they coincide.

There are $n!/2^r r!(n-2r)!$ different Feynman diagrams of rank r labelled by n given variables $\xi_1, \dots, \xi_n$, $0 \le 2r \le n$. In particular, there are $(n-1)!! = n!/2^{n/2}(n/2)!$ complete Feynman diagrams if n is even, and 0 if n is odd.

Theorem 1.28 may now be reformulated.

THEOREM 1.36. *Let $\xi_1, \dots, \xi_n$ be (real or complex) centred jointly normal random variables. Then*

$$\mathrm{E}(\xi_1 \cdots \xi_n) = \sum_{\gamma} v(\gamma)$$

where we sum over all complete Feynman diagrams γ labelled by $\xi_1, \dots, \xi_n$. □

II

Wiener chaos

Let H be a Gaussian Hilbert space defined on a probability space $(\Omega, \mathcal{F}, \mathrm{P})$. Since variables in H belong to L^p for every finite p, Hölder's inequality shows that any finite product of variables in H belongs to L^2 (and, in fact, to every L^p, $p < \infty$).

DEFINITION 2.1. Let, for $n \geq 0$, $\overline{\mathcal{P}}_n(H)$ be the closure in $L^2(\Omega, \mathcal{F}, \mathrm{P})$ of the linear space

$$\mathcal{P}_n(H) = \{p(\xi_1, \dots, \xi_m) : \\ p \text{ is a polynomial of degree} \leq n;\ \xi_1, \dots, \xi_m \in H;\ m < \infty\}$$

and let

$$H^{:n:} = \overline{\mathcal{P}}_n(H) \ominus \overline{\mathcal{P}}_{n-1}(H) = \overline{\mathcal{P}}_n(H) \cap \overline{\mathcal{P}}_{n-1}(H)^{\perp}.$$

For $n = 0$ we let $H^{:0:} = \overline{\mathcal{P}}_0(H)$, the space of constants. (For convenience, we also let $H^{:-1:} = \overline{\mathcal{P}}_{-1}(H) = \{0\}$.)

We may in this definition consider either real or complex spaces. Both cases are used in different applications so we will cover both. We use subscripts $\mathbb{R}$ and $\mathbb{C}$ when necessary to distinguish the real and complex cases, but otherwise all results are valid for both cases. The relation between the real and complex cases is given by the following obvious result.

THEOREM 2.2. *If H is a Gaussian Hilbert space, then $H_{\mathbb{R}}^{:0:} = \mathbb{R}$, $H_{\mathbb{C}}^{:0:} = \mathbb{C}$, $H_{\mathbb{R}}^{:1:} = H$, $H_{\mathbb{C}}^{:1:} = H_{\mathbb{C}} = H + iH$, and more generally $H_{\mathbb{C}}^{:n:} = H_{\mathbb{R}}^{:n:} + iH_{\mathbb{R}}^{:n:}$, the complexification of $H_{\mathbb{R}}^{:n:}$ consisting of all complex random variables whose real and imaginary parts belong to $H_{\mathbb{R}}^{:n:}$.* □

REMARK 2.3. If $\xi_1, \dots, \xi_m \in H$, then there exists an orthonormal sequence $\eta_1, \dots, \eta_l \in H$ such that each ξ_i is a linear combination of the variables η_j. (Choose any orthonormal basis in the linear span of $\{\xi_i\}_1^m$.) Then any polynomial function of $\xi_1, \dots, \xi_m$ can be written as a polynomial function of at most the same degree in $\eta_1, \dots, \eta_l$. Consequently, we may in the definition of $\mathcal{P}_n(H)$ without loss of generality assume that the variables $\xi_1, \dots, \xi_m$ are orthonormal.

REMARK 2.4. If H has finite dimension, then (by the argument of the preceding remark) $\mathcal{P}_n(H)$ equals the space of polynomials of degree at most n in a fixed orthonormal basis $\xi_1, \dots, \xi_m$ of H. Consequently, the space $\mathcal{P}_n(H)$

then also has finite dimension (for any n); in particular it is already closed so $\overline{\mathcal{P}}_n(H) = \mathcal{P}_n(H)$ and it is superfluous to take the closure in the definition of $\overline{\mathcal{P}}_n(H)$.

If H has infinite dimension, on the other hand, taking the closure is essential. If $\{\xi_i\}_1^\infty$ is an orthonormal sequence in H, then $\sum_1^\infty 2^{-i}\xi_i^2$ belongs to $\overline{\mathcal{P}}_2(H)$, but it can be shown that it is not a polynomial in any finite set of variables in H; see Example 6.5.

REMARK 2.5. Let G be a dense subspace of a Gaussian Hilbert space H; for example, let G be any Gaussian linear space and H its completion. Since for any fixed polynomial p, the mapping $(\xi_1, \dots, \xi_m) \mapsto p(\xi_1, \dots, \xi_m)$ is a continuous map $H^m \to L^2$ by Theorem 1.4 and Hölder's inequality, $\mathcal{P}_n(G)$ is a dense subspace of $\mathcal{P}_n(H)$ and thus $\overline{\mathcal{P}}_n(H) = \overline{\mathcal{P}_n(H)} = \overline{\mathcal{P}_n(G)}$. Hence it is sufficient to take polynomials in $\xi_i \in G$ in the definition of $\overline{\mathcal{P}}_n(H)$.

By definition, $\{\overline{\mathcal{P}}_n(H)\}_{n=0}^\infty$ is an increasing sequence of closed subspaces of L^2, while the spaces $H^{:n:}$ are orthogonal. It follows also that

$$\overline{\mathcal{P}}_n(H) = \bigoplus_0^n H^{:k:}$$

and thus

$$\bigoplus_0^\infty H^{:k:} = \overline{\bigcup_0^\infty \overline{\mathcal{P}}_n(H)}. \tag{2.1}$$

The latter space, in fact, consists of all square integrable functions that are measurable with respect to the σ-field generated by H; this is the main assertion of the next theorem.

THEOREM 2.6. *The spaces $H^{:n:}$, $n \geq 0$, are mutually orthogonal, closed subspaces of $L^2 = L^2(\Omega, \mathcal{F}, \mathrm{P})$ and*

$$\bigoplus_0^\infty H^{:n:} = L^2(\Omega, \mathcal{F}(H), \mathrm{P}),$$

where $\mathcal{F}(H)$ is the σ-field generated by the random variables in H.

PROOF. Let $H' = \overline{\bigcup_0^\infty \overline{\mathcal{P}}_n(H)}$. It is evident that $\overline{\mathcal{P}}_n(H) \subseteq L^2(\Omega, \mathcal{F}(H), \mathrm{P})$ and thus $H' \subseteq L^2(\Omega, \mathcal{F}(H), \mathrm{P})$. By (2.1) it remains only to prove that equality holds, or, equivalently, that if $X \in L^2(\Omega, \mathcal{F}(H), \mathrm{P})$ satisfies $X \perp H'$, then $X = 0$.

It is simplest to show this first for the complex case; the real case then follows immediately by Theorem 2.2. Hence we assume in the rest of the proof that we are using complex scalars.

We now observe that if $\xi \in H$, then

$$\left|e^{i\xi} - \sum_0^n \frac{(i\xi)^k}{k!}\right| \le 1 + \sum_0^n \frac{|\xi|^k}{k!} \le 1 + e^{|\xi|} \le 1 + e^{\xi} + e^{-\xi}. \tag{2.2}$$

Since the right hand side of (2.2) belongs to L^2 and the left hand side tends to 0 pointwise as $n \to \infty$, dominated convergence yields that $\sum_0^n \frac{(i\xi)^k}{k!} \to e^{i\xi}$ in L^2. Since $\xi^k \in \mathcal{P}_k(H) \subset H'$, this shows that $e^{i\xi} \in H'$ whenever $\xi \in H$.

Consequently, if $X \perp H'$, then $\mathrm{E}(Xe^{-i\xi}) = \langle X, e^{i\xi}\rangle = 0$ for every $\xi \in H$. The following lemma shows that this implies $X = 0$, which completes the proof. □

LEMMA 2.7. *If $X \in L^1(\Omega, \mathcal{F}(H), \mathrm{P})$ and $\mathrm{E}(Xe^{-i\xi}) = 0$ for every $\xi \in H$, then $X = 0$ a.s.*

PROOF. First assume that H has finite dimension and is spanned by a finite set $\xi_1, \dots, \xi_m$, and let μ be the distribution of $(\xi_1, \dots, \xi_m)$. Then the random variable X in $L^1(\Omega, \mathcal{F}(H), \mathrm{P})$ may be written $\varphi(\xi_1, \dots, \xi_m)$ for some $\varphi \in L^1(\mathbb{R}^m, \mu)$. Hence, for any $(t_1, \dots, t_m) \in \mathbb{R}^m$, we obtain, with $\xi = \sum t_j \xi_j \in H$,

$$\begin{aligned} 0 = \mathrm{E}(Xe^{-i\xi}) &= \mathrm{E}\big(\varphi(\xi_1, \dots, \xi_m) e^{-i\sum t_j \xi_j}\big) \\ &= \int_{\mathbb{R}^m} \varphi(x_1, \dots, x_m) e^{-i\sum t_j x_j}\, d\mu(x) \\ &= \widehat{\varphi\, d\mu}(t_1, \dots, t_m). \end{aligned}$$

Consequently, the complex measure $\varphi\, d\mu$ on $\mathbb{R}^m$ has a Fourier transform that vanishes identically, which implies that $\varphi\, d\mu = 0$ and thus $\varphi = 0$ μ-a.e. and $X = \varphi(\xi_1, \dots, \xi_m) = 0$ a.s., which proves the lemma in the finite-dimensional case.

In general, there exists a countable set of Gaussian variables $\{\xi_k\} \subset H$ such that X is measurable with respect to the σ-field $\mathcal{F}(\{\xi_k\})$ generated by them. Let H_n be the subspace of H spanned by $\xi_1, \dots, \xi_n$ and let $\mathcal{F}_n = \mathcal{F}(\xi_1, \dots, \xi_n) = \mathcal{F}(H_n)$. Then the conditional expectation $\mathrm{E}(X \mid \mathcal{F}_n) \in L^1(\Omega, \mathcal{F}(H_n), \mathrm{P})$ and, for every $\xi \in H_n$,

$$\mathrm{E}\Big(\mathrm{E}(X \mid \mathcal{F}_n) e^{-i\xi}\Big) = \mathrm{E}\big(Xe^{-i\xi}\big) = 0.$$

Consequently, by the case just proved, $\mathrm{E}(X \mid \mathcal{F}_n) = 0$. Moreover, as is well-known, $\mathrm{E}(X \mid \mathcal{F}_n) \to X$ in L^1 (and a.s.) as $n \to \infty$. Hence $X = 0$. □

COROLLARY 2.8. *If the Gaussian Hilbert space H generates the σ-field $\mathcal{F}$, then $L^2 = L^2(\Omega, \mathcal{F}, \mathrm{P})$ has the orthogonal decomposition*

$$L^2 = \bigoplus_0^\infty H^{:n:}.$$

□

This decomposition of $L^2(\Omega, \mathcal{F}, \mathrm{P})$ is called the *Wiener chaos decomposition*, or just *chaos decomposition* (Wiener 1938, Itô 1951, Segal 1956). The same term is also used for the corresponding decomposition

$$X = \sum_0^\infty X_n, \qquad X_n \in H^{:n:}, \tag{2.3}$$

of an element $X \in L^2(\Omega, \mathcal{F}(H), \mathrm{P})$. Similarly, the term *chaos of order* n is used both for the space $\overline{\mathcal{P}}_n(H)$ and for any element of it, and *homogeneous chaos of order* n is used for both the space $H^{:n:}$ and its elements.

EXAMPLE 2.9. Consider as in Example 1.15 the probability space $(\mathbb{R}, \mathcal{B}, \gamma)$ and the Gaussian Hilbert space $H = \{tx : t \in \mathbb{R}\}$ defined on it. In this case, $\overline{\mathcal{P}}_n(H)$ is the space of polynomials of degree at most n. The dimension of $\overline{\mathcal{P}}_n(H)$ is $n+1$, and thus $H^{:n:}$ is one-dimensional. If $H^{:n:}$ is spanned by h_n, then $(h_n)_{n=0}^\infty$ is the sequence of orthogonal polynomials with respect to the standard Gaussian measure $d\gamma = \frac{1}{\sqrt{2\pi}} e^{-x^2/2} dx$, i.e. the Hermite polynomials (up to normalization). Corollary 2.8 says that $\{h_n / \|h_n\|_2\}_0^\infty$ is an orthonormal basis in $L^2(d\gamma)$.

Several normalizations are used in the literature for the Hermite polynomials. We will henceforth let h_n denote the Hermite polynomial with leading coefficient 1, i.e. $h_n(x) = x^n + \dots$.

It is easily verified, see for example the formula (3.11) in Example 3.18 below, that the first Hermite polynomials are given by

$$h_0(x) = 1, \qquad h_1(x) = x, \qquad h_2(x) = x^2 - 1,$$
$$h_3(x) = x^3 - 3x, \qquad h_4(x) = x^4 - 6x^2 + 3, \qquad h_5(x) = x^5 - 10x^3 + 15x.$$

REMARK 2.10. Many authors instead define the Hermite polynomials so that they are orthogonal polynomials with respect to the measure $e^{-x^2} dx$; these polynomials are given by $c_n h_n(\sqrt{2}x)$ in our notation, where the normalization $c_n = 2^{n/2}$ (giving the leading term $2^n x^n$) is usually chosen. In probabilistic settings, this definition seems less natural and convenient than ours.

We define

$$\mathcal{P}(H) = \bigcup_0^\infty \mathcal{P}_n(H),$$

the space of polynomials in elements of H. For simplicity we sometimes call these random variables *polynomial variables*, without explicit mention of H. We also define

$$\overline{\mathcal{P}}_*(H) = \bigcup_0^\infty \overline{\mathcal{P}}_n(H) = \sum_0^\infty H^{:n:},$$

the space of all elements in $L^2(\Omega, \mathcal{F}(H), \mathrm{P})$ having a finite chaos decomposition. (We also use the simplified notations $\mathcal{P}$ and $\overline{\mathcal{P}}_*$, omitting H.) Note that if H has finite dimension, then $\overline{\mathcal{P}}_*(H)$ equals the space $\mathcal{P}(H)$ of polynomial variables, but if H has infinite dimension, then $\overline{\mathcal{P}}_*(H)$ is strictly larger by Remark 2.4 and Example 6.5.

By Theorem 2.6, $\overline{\mathcal{P}}_*(H)$ is a dense subspace of $L^2(\Omega, \mathcal{F}(H), \mathrm{P})$. In Theorem 3.51 we will show that this extends to L^p for every finite p. At this point we show the corresponding result for $\mathcal{P}(H)$.

THEOREM 2.11. *If $0 \le p < \infty$, then the set of polynomial variables $\mathcal{P}(H) = \bigcup_n \mathcal{P}_n(H)$ is a dense subspace of $L^p(\Omega, \mathcal{F}(H), \mathrm{P})$.*

PROOF. The polynomial variables belong to L^p by Hölder's inequality (for $p > 0$; otherwise there is nothing to prove). In order to show denseness for $1 \le p < \infty$, we can use the same proof as for Theorem 2.6, recalling that the dual space of L^p is L^q, where q is the conjugate exponent of p. If p is smaller, including $p = 0$, the result follows from the L^2 case, since $L^2(\Omega, \mathcal{F}(H), \mathrm{P})$ then is a dense subspace of $L^p(\Omega, \mathcal{F}(H), \mathrm{P})$. □

Sometimes it is convenient to use some other simple dense subspaces of L^p that can be constructed from a Gaussian space.

THEOREM 2.12. *If $0 \le p < \infty$, then the set of finite linear combinations of the exponentials e^{ξ} with $\xi \in H$ is a dense subspace of $L^p(\Omega, \mathcal{F}(H), \mathrm{P})$. For complex scalars, it also holds that the set of finite linear combinations of the exponentials $e^{i\xi}$ with $\xi \in H$, and the set of finite linear combinations of e^{ζ} with $\zeta \in H_{\mathbb{C}}$, are dense subspaces of $L^p_{\mathbb{C}}(\Omega, \mathcal{F}(H), \mathrm{P})$.*

PROOF. It is easily checked that all these exponentials belong to L^p.

In order to show density, we begin with the complex case. We may again suppose that $1 \le p < \infty$, and if we let q denote the conjugate exponent, it suffices to prove that if z is one of the numbers 1 and i, and $X \in L^q_{\mathbb{C}}(\Omega, \mathcal{F}(H), \mathrm{P})$ is such that $\mathrm{E}(Xe^{z\xi}) = 0$ for all $\xi \in H$, then $X = 0$. For $z = i$, this follows by Lemma 2.7.

For $z = 1$, we first use the fact that if $\xi \in H$, then $t\xi \in H$ whenever $t \in \mathbb{R}$. Hence $\mathrm{E}(Xe^{t\xi}) = 0$ for every real t. Moreover, it is easily verified that if $\xi \in H$, then $z \mapsto \mathrm{E}(Xe^{z\xi})$ is an entire analytic function. Consequently, if this function vanishes on $\mathbb{R}$, then it vanishes for all complex z, and in particular for $z = i$. Thus $\mathrm{E}(Xe^{i\xi}) = 0$ for every $\xi \in H$, and the result follows from the case just proved.

Finally, for real scalars (and $z = 1$), the result follows immediately from the corresponding complex case. □

REMARK 2.13. An equivalent formulation of Theorem 2.12 is that the given sets of exponentials are total in $L^p(\Omega, \mathcal{F}(H), \mathrm{P})$.

REMARK 2.14. The dense subspaces in Theorems 2.11 and 2.12 are algebras under pointwise multiplication.

REMARK 2.15. For $p = \infty$, we have instead (still by Lemma 2.7) that the linear span of the exponentials $e^{i\xi}$ is weak*-dense in $L^\infty_{\mathbb{C}}(\Omega, \mathcal{F}(H), \mathrm{P})$.

REMARK 2.16. If G is a dense subspace of H, it suffices in Theorems 2.11 and 2.12 to take polynomials and exponentials using elements in G. This follows by the same proofs, since we may choose $\{\xi_k\} \subset G$ in the proof of Lemma 2.7.

We will often use the chaos decomposition, and we will then use the following notation.

DEFINITION 2.17. For $n \geq 0$, π_n denotes the orthogonal projection of L^2 onto $H^{:n:}$, and $\pi_{\leq n}$ the orthogonal projection of L^2 onto $\bigoplus_0^n H^{:k:}$.

In particular, $\pi_0(X) = \mathrm{E}\, X$. For convenience we further define $\pi_n = \pi_{\leq n} = 0$ when $n < 0$.

It follows from the definition that

$$\pi_{\leq n} = \sum_{k=0}^{n} \pi_k. \tag{2.4}$$

The chaos decomposition (2.3) of a random variable $X \in L^2(\Omega, \mathcal{F}(H), \mathrm{P})$ may now be written

$$X = \sum_{n=0}^{\infty} \pi_n(X)$$

with the sum converging in L^2. By (2.4), this implies

$$\pi_{\leq n} X \to X, \qquad \text{as } n \to \infty,$$

in L^2, for every $X \in L^2(\Omega, \mathcal{F}(H), \mathrm{P})$.

We will later (Theorem 5.14) see that the operators π_n and $\pi_{\leq n}$, so far defined on L^2, extend to L^p for every $p > 1$.

III

Wick products

We obtain an important notion by using the Wiener chaos decomposition to orthogonalize the usual product of elements of a Gaussian Hilbert space. Recall that π_n is the orthogonal projection of L^2 onto $H^{:n:}$.

DEFINITION 3.1. If $\xi_1, \dots, \xi_n$ is a finite sequence of elements of a Gaussian Hilbert space H, their *Wick product* $:\xi_1 \cdots \xi_n: \in H^{:n:}$ is given by

$$:\xi_1 \cdots \xi_n: = \pi_n(\xi_1 \cdots \xi_n). \tag{3.1}$$

We include the case $n = 0$:

$$:: = 1 \in H^{:0:}.$$

When dealing with complex random variables, we extend (3.1) to $\xi_1, \dots, \xi_n \in H_{\mathbf{C}}$, the complexification of H.

In fact, this is just a special case of a more general definition.

DEFINITION 3.2. Let H be a Gaussian Hilbert space. Define the *general Wick product* by

$$X \odot Y = \pi_{m+n}(XY), \tag{3.2}$$

if $X \in H^{:m:}$ and $Y \in H^{:n:}$ for some $m, n \geq 0$, and extend $\odot$ by bilinearity to a binary operator on $\overline{\mathcal{P}}_*(H)$.

In order for this to be an honest definition, it must be verified that $XY \in L^2$ for such X and Y. This is clear if X and Y are polynomial variables in $\mathcal{P}(H)$, and thus always when H has finite dimension, but the general case requires an extra argument. We prefer to wait until Section 3 to show this.

We will then also verify that $\xi_1 \odot \cdots \odot \xi_n = :\xi_1 \cdots \xi_n:$ when $\xi_1, \dots, \xi_n \in H_{\mathbf{C}}$, which shows that Definition 3.1 really is a special case of Definition 3.2. We have chosen, however, to use different notations in order to emphasize that the variables $\xi_1, \dots, \xi_n$ are (possibly complex) Gaussian variables belonging to H (or $H_{\mathbf{C}}$) whenever we use the notation of Definition 3.1.

REMARK 3.3. The notation for the Wick product in Definition 3.1 is the standard notation in the literature. (It goes back to Wick (1950) in a different context; the connection between the definition above and Wick's original definition will be explained in Example 13.20.) For the general Wick product in Definition 3.2 various notations are used by different authors, including $\circ$, $\diamond$, $\hat{\otimes}$ and the same notation as in Definition 3.1. Our choice of notation is motivated by Theorem 4.1 below.

The general Wick product may be further extended by continuity to some cases when X or Y does not belong to $\overline{\mathcal{P}}_*(H)$; one case (when $X \in H$) will be studied in Section 13.2 as the 'creation operator' from quantum mechanics. Another, more general, extension will be defined and studied in Section 16.2. However, as we will see in Example 3.48, and more generally in Example 16.30, the general Wick product is not continuous in the L^2-norm, so there is no resonable extension of it to all $X, Y \in L^2$.

We will return to the general Wick product in Section 3, but we will first study the Wick product of Gaussian variables given by (3.1).

1. Wick products of Gaussian variables

Consider a Gaussian Hilbert space H and let ξ and $\xi_1, \xi_2, \dots$ be Gaussian random variables in H, or, more generally, complex Gaussian random variables in $H_{\mathbf{C}}$. (The most important case is for real Gaussian variables, but the results extend immediately to complex Gaussian variables so we will state them in this generality.)

From the definition follow easily

$$:\xi: = \xi, \tag{3.3}$$

$$:\xi_1\xi_2: = \xi_1\xi_2 - \mathrm{E}(\xi_1\xi_2), \tag{3.4}$$

$$:\xi_1\xi_2\xi_3: = \xi_1\xi_2\xi_3 - \mathrm{E}(\xi_2\xi_3)\xi_1 - \mathrm{E}(\xi_1\xi_3)\xi_2 - \mathrm{E}(\xi_1\xi_2)\xi_3. \tag{3.5}$$

These are special cases of a general formula.

THEOREM 3.4. *The Wick product is given by*

$$:\xi_1 \cdots \xi_n: = \sum_\gamma (-1)^{r(\gamma)} v(\gamma), \tag{3.6}$$

where we sum over all Feynman diagrams γ labelled by $\xi_1, \dots, \xi_n$.

PROOF. Denote the right hand side of (3.6) by ψ. Note that $\psi \in \mathcal{P}_n(H)$ and that the term given by the Feynman diagram with no edges equals $\xi_1 \cdots \xi_n$ while all other terms belong to $\mathcal{P}_{n-2}(H)$. Hence $\xi_1 \cdots \xi_n - \psi \in \mathcal{P}_{n-2}(H) \subset \overline{\mathcal{P}}_{n-1}(H)$, and it suffices to show that $\psi \perp \overline{\mathcal{P}}_{n-1}(H)$.

Let $\xi_{n+1}, \dots, \xi_{n+m}$ be $m \geq 0$ random variables in H. It follows from (3.6), (1.4) and Theorem 1.36 that

$$\begin{aligned}\mathrm{E}(\psi \cdot \xi_{n+1} \cdots \xi_{n+m}) &= \sum_\gamma (-1)^{r(\gamma)} \mathrm{E}(v(\gamma)\xi_{n+1} \cdots \xi_{n+m}) \\ &= \sum_\gamma (-1)^{r(\gamma)} \sum_{\gamma'} v(\gamma'),\end{aligned}$$

where we sum over all Feynman diagrams γ labelled by $\xi_1, \dots, \xi_n$ and γ' labelled by $\xi_1, \dots, \xi_{n+m}$ such that γ' is complete and an extension of γ. We interchange the order of summation and note that if γ' has $l \geq 0$ edges that join pairs of variables ξ_i and ξ_j with $i, j \leq n$, then there are 2^l choices of γ and

$$\sum_{\gamma \subseteq \gamma'} (-1)^{r(\gamma)} = (1 + (-1))^l = \begin{cases} 1, & l = 0, \\ 0, & l \geq 1. \end{cases} \tag{3.7}$$

Hence

$$\mathrm{E}(\psi \cdot \xi_{n+1} \cdots \xi_{n+m}) = \sum_{\gamma'} v(\gamma'), \tag{3.8}$$

where we sum over all complete Feynman diagrams γ' labelled by $\xi_1, \dots, \xi_{n+m}$ such that every edge has at most one endpoint in $\{\xi_1, \dots, \xi_n\}$.

If $m \leq n-1$, there are no such diagrams and thus $\mathrm{E}(\psi \cdot \xi_{n+1} \cdots \xi_{n+m}) = 0$. This implies that $\psi \perp \mathcal{P}_{n-1}(H)$, which completes the proof. □

REMARK 3.5. Since $\xi_1 \cdots \xi_n \in \overline{\mathcal{P}}_n(H) = H^{:n:} \oplus \overline{\mathcal{P}}_{n-1}(H)$, $\xi_1 \cdots \xi_n - {:}\xi_1 \cdots \xi_n{:}$ is the orthogonal projection of $\xi_1 \cdots \xi_n$ onto $\overline{\mathcal{P}}_{n-1}(H)$. We now see that this difference actually belongs to $\mathcal{P}_{n-2}(H)$.

An immediate corollary of Theorem 3.4 is the following important fact.

THEOREM 3.6. *The Wick product* ${:}\xi_1 \cdots \xi_n{:}$ *is the same for every Gaussian Hilbert space* H *that contains* $\xi_1, \dots, \xi_n$. *(For complex* ξ_i*, we require instead* $\xi_i \in H_{\mathbf{C}}$.*)* □

Consequently, the Wick product is unambiguously defined for any finite set of random variables with a joint (real or complex) centred normal distribution, without any need for specifying the Gaussian Hilbert space H.

Note also the following simple corollary of Theorem 3.4 on complex conjugation for complex variables. (It also follows easily directly from the definition.)

COROLLARY 3.7. *If* $\zeta_1, \dots, \zeta_n$ *are complex centred jointly normal variables, then* $\overline{{:}\zeta_1 \cdots \zeta_n{:}} = {:}\overline{\zeta}_1 \cdots \overline{\zeta}_n{:}$ □

From the proof of Theorem 3.4, we extract formula (3.8), which gives an extension of Theorem 1.36.

THEOREM 3.8. *If* $\xi_1, \dots, \xi_{n+m}$ *are (real or complex) centred jointly normal variables,* $m, n \geq 0$, *then*

$$\mathrm{E}({:}\xi_1 \cdots \xi_n{:}\, \xi_{n+1} \cdots \xi_{n+m}) = \sum_{\gamma} v(\gamma),$$

where we sum over all complete Feynman diagrams γ *labelled by* $\xi_1, \dots, \xi_{n+m}$ *such that no edge joins any pair* ξ_i *and* ξ_j *with* $i < j \leq n$. □

This leads to the following formula yielding the inner product (or covariance) of two Wick products; $\mathfrak{S}_n$ denotes the symmetric group of the $n!$ permutations of $\{1, \dots, n\}$.

THEOREM 3.9. *If $\xi_1, \dots, \xi_n$ and $\eta_1, \dots, \eta_m$ are (real or complex) centred jointly normal variables, then*

$$\mathrm{E}(:\xi_1 \cdots \xi_n: :\eta_1 \cdots \eta_m:) = \begin{cases} \sum_{\pi \in \mathfrak{S}_n} \prod_{i=1}^n \mathrm{E}(\xi_i \eta_{\pi(i)}), & m = n, \\ 0, & m \neq n. \end{cases}$$

Equivalently,

$$\langle :\xi_1 \cdots \xi_n:, :\eta_1 \cdots \eta_m: \rangle = \begin{cases} \sum_{\pi \in \mathfrak{S}_n} \prod_{i=1}^n \langle \xi_i, \eta_{\pi(i)} \rangle, & m = n, \\ 0, & m \neq n. \end{cases}$$

PROOF. The case $m \neq n$ follows by definition. If $m = n$, we observe that $:\xi_1 \cdots \xi_n: \perp \eta_1 \cdots \eta_n - :\eta_1 \cdots \eta_n:$ (for complex η_i, take the complex conjugate instead). Hence $\mathrm{E}(:\xi_1 \cdots \xi_n: :\eta_1 \cdots \eta_n:) = \mathrm{E}(:\xi_1 \cdots \xi_n: \eta_1 \cdots \eta_n)$. By Theorem 3.8, this can be evaluated by summing the values of all complete Feynman diagrams on $\xi_1, \dots, \xi_n, \eta_1, \dots, \eta_n$ such that every edge joins some ξ_i with some η_j, i.e. the $n!$ diagrams that are perfect matchings of $(\xi_1, \dots, \xi_n)$ with $(\eta_1, \dots, \eta_n)$, which gives the first formula.

The second formula is the same as the first for real variables, and follows in general from the first by Corollary 3.7. □

COROLLARY 3.10. *If $\xi \in \mathrm{N}(0, \sigma^2)$, then*

$$\mathrm{E} :\xi^n:^2 = n!\, \sigma^{2n}.$$

□

COROLLARY 3.11. *If ζ is a complex centred Gaussian variable, then*

$$\mathrm{E}\,|:\zeta^n:|^2 = n!\, (\mathrm{E}\,|\zeta|^2)^n.$$

□

In fact, Theorems 1.36, 3.8 and 3.9 are all special cases of a general formula.

THEOREM 3.12. *Let $Y_i = :\xi_{i1} \cdots \xi_{il_i}:$, where $\{\xi_{ij}\}_{1 \le i \le k, 1 \le j \le l_i}$ are (real or complex) centred jointly normal variables, with $k \ge 0$ and $l_1, \dots, l_k \ge 0$. Then*

$$\mathrm{E}(Y_1 \cdots Y_k) = \sum_\gamma v(\gamma),$$

where we sum over all complete Feynman diagrams γ labelled by $\{\xi_{ij}\}_{ij}$ such that no edge joins two variables $\xi_{i_1 j_1}$ and $\xi_{i_2 j_2}$ with $i_1 = i_2$.

PROOF. Arguing as in the proof of Theorem 3.4 we obtain

$$\mathrm{E}(Y_1 \cdots Y_k) = \sum_{\gamma_1, \dots, \gamma_k} (-1)^{r(\gamma_1) + \cdots + r(\gamma_k)} \sum_{\gamma'} v(\gamma') = \sum_{\gamma'} v(\gamma') \prod_{i=1}^k \sum_{\gamma_i \subseteq \gamma'} (-1)^{r(\gamma_i)}$$

where γ_i ranges over the Feynman diagrams labelled by $\{\xi_{ij}\}_{j=1}^{l_i}$ and γ' over the complete Feynman diagrams labelled by all variables that extend each γ_i in the obvious way. The result follows by a simple extension of (3.7). □

REMARK 3.13. The mixed cumulant (or semi-invariant) $\varkappa(Y_1, \dots, Y_k)$ is obtained similarly by summing in Theorem 3.12 only over the Feynman diagrams that furthermore become connected if the vertices $\{ij\}_{1 \le j \le l_i}$ are merged to a single vertex for each i. We leave the details as an exercise to the interested reader.

REMARK 3.14. Theorem 3.12 suggests an interpretation of the Wick product as a kind of 'renormalization' of the ordinary product with terms corresponding to self-interactions (in some vague sense) eliminated.

The product of two (or several) Wick products of jointly Gaussian variables is a polynomial in those variables. The next theorem gives an explicit decomposition of the product as a linear combination of Wick products. In particular, we obtain the chaos decomposition of the product by collecting the terms according to their degrees.

Recall that the value $v(\gamma)$ of a Feynman diagram is, by (1.4), after relabelling, $c\xi_1 \cdots \xi_m$ for some number c and Gaussian variables $\xi_1, \dots, \xi_m$; we define the *Wick value* to be

$$:v(\gamma): = c :\xi_1 \cdots \xi_m:, \tag{3.9}$$

i.e., we replace the ordinary product $\prod_{i \in A} \xi_i$ in (1.4) by the Wick product $:\prod_{i \in A} \xi_i:$.

THEOREM 3.15. *Let Y_i, $i = 1, \dots, k$, be as in Theorem 3.12. Then*

$$Y_1 \cdots Y_k = \sum_{\gamma} :v(\gamma):, \tag{3.10}$$

where we sum over all Feynman diagrams γ labelled by $\{\xi_{ij}\}_{ij}$ such that no edge joins two variables $\xi_{i_1 j_1}$ and $\xi_{i_2 j_2}$ with $i_1 = i_2$.

PROOF. Let U and V denote the left and right hand sides of (3.10), respectively, and let $W = :\eta_1 \cdots \eta_m:$, where $\eta_1, \dots, \eta_m$, $m \ge 0$, are any variables in the Gaussian Hilbert space H spanned by $\{\xi_{ij}\}$. (In the complex case, let H be the space spanned by $\{\operatorname{Re} \xi_{ij}, \operatorname{Im} \xi_{ij}\}$.)

It follows easily from Theorem 3.12 that both $\operatorname{E} UW$ and $\operatorname{E} VW$ equal the sum of the values of all complete Feynman diagrams labelled by $\{\xi_{ij}\}_{ij} \cup \{\eta_j\}_j$, such that no edge joins any ξ_{ij_1} and ξ_{ij_2} or η_{j_1} and η_{j_2}. Hence $\operatorname{E}(U - V)W = 0$ for all such W, and thus, by linearity, for any polynomial variable W. The result follows by taking $W = \bar{U} - \bar{V}$. □

REMARK 3.16. We have deduced Theorem 3.15 from Theorem 3.12. Conversely, Theorem 3.12 follows immediately from Theorem 3.15 by taking expectations, since the terms for incomplete Feynman diagrams have vanishing expectations.

Taking $Y_i = :\xi_i: = \xi_i$ in Theorem 3.15, we obtain a kind of converse to Theorem 3.4.

COROLLARY 3.17. *The ordinary product of several (real or complex) centred Gaussian variables can be decomposed into Wick products as*

$$\xi_1 \cdots \xi_n = \sum :v(\gamma):,$$

where we sum over all Feynman diagrams γ labelled by $\xi_1, \dots, \xi_n$. □

EXAMPLE 3.18. We make these general results more concrete by looking again at the very simple but fundamental Example 1.15, where $\xi = x$ on the probability space $(\mathbb{R}, \mathcal{B}, (2\pi)^{-1/2} e^{-x^2/2} dx)$. It follows by Definition 3.1 and Example 2.9 that then $:\xi^n: = h_n(\xi)$, where h_n is the nth Hermite polynomial. Theorem 3.4 and simple combinatorics yield

$$h_n(x) = \sum_{r=0}^{n/2} (-1)^r \frac{n!}{2^r r!(n-2r)!} x^{n-2r} \tag{3.11}$$

while Theorem 3.9 (or Corollary 3.10) yields the well-known formula

$$\int_{-\infty}^{\infty} h_m(x) h_n(x) \frac{1}{\sqrt{2\pi}} e^{-x^2/2} dx = n! \delta_{mn}. \tag{3.12}$$

Theorem 3.12 yields a combinatorial formula for the integral of a product of several Hermite polynomials. Theorem 3.15 yields the product formula

$$h_m(x) h_n(x) = \sum_{r=0}^{m \wedge n} \binom{m}{r} \binom{n}{r} r! h_{m+n-2r}(x). \tag{3.13}$$

In particular, $m = 1$ yields $x h_n(x) = h_{n+1}(x) + n h_{n-1}(x)$, or

$$h_{n+1}(x) = x h_n(x) - n h_{n-1}(x); \tag{3.14}$$

the standard recursion formula for Hermite polynomials.

We have shown that Wick powers of a particular standard Gaussian variable are given by Hermite polynomials. Of course, the same must hold for any standard Gaussian variable. (This is easily verified by Theorem 3.4.)

THEOREM 3.19. *If $\xi \sim \mathrm{N}(0, 1)$, then*

$$:\xi^n: = h_n(\xi).$$

□

More generally, it follows that if $\xi \sim \mathrm{N}(0, \sigma^2)$ with $\sigma^2 > 0$, then

$$:\xi^n: = \sigma^n :(\xi/\sigma)^n: = \sigma^n h_n(\xi/\sigma).$$

This gives a complete description of Wick products for a one-dimensional Gaussian space. We can extend the results to general Gaussian spaces by using the following result, which implies that independent variables can be separated when computing Wick products.

THEOREM 3.20. *If $\xi_1, \dots, \xi_n$ and $\eta_1, \dots, \eta_m$ are (real or complex) centred jointly normal variables, such that* $\mathrm{E}\,\xi_i\eta_j = 0$ *for all i and j, then*

$$:\xi_1 \cdots \xi_n \eta_1 \cdots \eta_m: \; = \; :\xi_1 \cdots \xi_n: \, :\eta_1 \cdots \eta_m: .$$

PROOF. By Theorem 3.4,

$$:\xi_1 \cdots \xi_n \eta_1 \cdots \eta_m: = \sum_\gamma v(\gamma),$$

where we sum over all Feynman diagrams γ labelled by $\xi_1, \dots, \xi_n, \eta_1, \dots, \eta_m$. Since it follows from the assumption that $v(\gamma)$ vanishes for every γ that contains an edge joining some ξ_i and η_j, it suffices to sum over the Feynman diagrams that contain no such edges. But these are exactly the Feynman diagrams $\gamma_1 \cup \gamma_2$ that are unions of one Feynman diagram γ_1 labelled by $\xi_1, \dots, \xi_n$ and another γ_2 labelled by $\eta_1, \dots, \eta_m$. Since $v(\gamma_1 \cup \gamma_2) = v(\gamma_1)v(\gamma_2)$ by the definition (1.4), the result follows by Theorem 3.4 again. □

THEOREM 3.21. *Let H be a Gaussian Hilbert space and let $\{\xi_i\}_{i \in \mathcal{I}}$ be an orthonormal basis in H (finite or infinite, possibly even uncountable). If $\alpha = (\alpha_i)_{i \in \mathcal{I}}$ is a multi-index, i.e. a sequence of non-negative integers with only finitely many elements different from 0, then*

$$:\prod_i \xi_i^{\alpha_i}: \; = \prod_i h_{\alpha_i}(\xi_i). \tag{3.15}$$

For each $n \geq 0$, the set $\{(\prod_i \alpha_i!)^{-1/2} :\prod_i \xi_i^{\alpha_i}:\}$, where (α_i) ranges over all multi-indices, is an orthonormal basis in $L^2(\Omega, \mathcal{F}(H), \mathrm{P})$. The subset of all such variables with $|\alpha| = \sum \alpha_i = n$ is an orthonormal basis in $H^{:n:}$, and the subset with $|\alpha| \leq n$ is an orthonormal basis in $\overline{\mathcal{P}}_n(H)$.

We will use the notations $\xi^\alpha = \prod_i \xi_i^{\alpha_i}$ and $\alpha! = \prod_i \alpha_i!$; thus the orthonormal basis given in Theorem 3.21 may be written $\{(\alpha!)^{-1/2} :\xi^\alpha:\}_\alpha$.

PROOF. Theorem 3.20 (and induction) and Theorem 3.19 yield (3.15). Moreover, Theorem 3.9 implies that the Wick products $:\prod \xi_i^{\alpha_i}:$ are orthogonal and $\mathrm{E}(:\prod \xi_i^{\alpha_i}:)^2 = \prod \alpha_i!$. (This follows also from (3.12) and the fact that the ξ_i are independent.) Let V be the linear span of these Wick products and let G be the linear span of $\{\xi_i\}$; then G is a dense subspace of H. It follows, e.g. by Corollary 3.17, that V contains every finite product ξ^α of the elements of the basis. By linearity, V thus contains every product of elements of G, so $\mathcal{P}(G) \subseteq V$. By Theorem 2.11 and Remark 2.16, V is a dense subspace of $L^2(\Omega, \mathcal{F}(H), \mathrm{P})$. (Alternatively, this follows by Theorem 2.6 and Remark 2.5.) This proves that $\{(\alpha!)^{-1/2} :\xi^\alpha:\}_\alpha$ is an orthonormal basis in the latter space.

The final statements follow similarly, or by the fact that π_n preserves ξ^α if $|\alpha| = n$, and annihilates it otherwise. □

REMARK 3.22. Some authors prefer to take this result as the starting point in the definition of the Wiener chaos decomposition and the Wick product. They fix an orthonormal basis in H and show directly that the variables defined in (3.15) constitute a complete orthogonal set in $L^2(\Omega, \mathcal{F}(H), \mathrm{P})$; then $H^{:n:}$ is defined as the space spanned by the set of these variables with $\sum \alpha_i = n$. The Wick product may be defined directly by (3.15) and linearity. One disadvantage of these definitions is that it is not obvious that they do not depend on the chosen basis $\{\xi_i\}$. We therefore prefer the basis-free definitions given above; Theorem 3.21 shows that the definitions are equivalent and allows us to work with a suitable orthonormal basis whenever convenient.

EXAMPLE 3.23. If $\dim H = 1$ and $\xi \in H$ with $\|\xi\|_2^2 = \operatorname{Var} \xi = 1$, then an orthonormal basis in $L^2(\Omega, \mathcal{F}(H), \mathrm{P})$ is given by the random variables

$$(n!)^{-1/2} :\xi^n: = (n!)^{-1/2} h_n(\xi), \qquad n \geq 0.$$

Cf. Examples 2.9 and 3.18.

An easy consequence of Theorem 3.21 is the following.

COROLLARY 3.24. *If the Gaussian Hilbert space H has finite dimension $d \geq 1$, then*

$$\dim H^{:n:} = \binom{n+d-1}{d-1}.$$

□

We use Theorem 3.21 to deduce some technical results (which may also be proved more directly).

COROLLARY 3.25. *If $X \in \mathcal{P}(H)$, then $\pi_n(X) \in \mathcal{P}_n(H)$ for every $n \geq 0$. In particular, the projections π_n preserve the space of polynomial variables.*

PROOF. Let X be a polynomial in the variables $\xi_1, \dots, \xi_m \in H$. We may assume that these variables are orthonormal, and may then extend $\{\xi_i\}_1^m$ to an orthonormal basis $\{\xi_i\}_{i \in \mathcal{I}}$ for some index set $\mathcal{I}$. It is then easy to see that X is orthogonal to all but a finite number of the basis elements $(\alpha!)^{-1/2} :\xi^\alpha:$ given by Theorem 3.21. Hence X is a finite linear combination $\sum_\alpha a_\alpha (\alpha!)^{-1/2} :\xi^\alpha:$ of such basis elements, and $\pi_n(X)$ is the finite linear combination $\sum_{|\alpha|=n} a_\alpha (\alpha!)^{-1/2} :\xi^\alpha:$, which by (3.15) is a polynomial variable of degree $\leq n$. □

COROLLARY 3.26. *The subset of polynomial variables in $H^{:n:}$, i.e. $\mathcal{P}(H) \cap H^{:n:}$, equals $\mathcal{P}_n(H) \cap H^{:n:}$ and is a dense subspace of $H^{:n:}$.*

PROOF. This follows immediately by Corollary 3.25, Theorem 2.11, and the continuity of π_n. □

COROLLARY 3.27. *Let $n \geq 0$ be fixed. The set of Wick products $:\xi_1 \cdots \xi_n:$ with $\xi_1, \dots, \xi_n \in H$ is a total subset of $H^{:n:}$.*

PROOF. Each such Wick product belongs to $H^{:n:}$ by definition, and Theorem 3.21 shows that already the subset with ξ_i in a fixed basis is total. □

By polarization, Theorem D.1, it suffices here to take Wick powers.

COROLLARY 3.28. *Let $n \geq 0$ be fixed. The set of Wick powers $:\xi^n:$ with $\xi \in H$ is a total subset of $H^{:n:}$.* □

The last two results can be compared with the following, where n is unrestricted.

THEOREM 3.29. *The linear span of the set of all Wick powers $:\xi^n:$ with $\xi \in H$ and $n \geq 0$ equals the space $\mathcal{P}(H)$ of polynomial variables. Consequently, the set of all such Wick powers is a total subset in $L^p(\Omega, \mathcal{F}(H), \mathrm{P})$ for any p, $0 \leq p < \infty$.*

The same results hold, a fortiori, for the set of all Wick products $:\xi_1 \cdots \xi_n:$ with $\xi_1, \ldots, \xi_n \in H$ and $n \geq 0$.

PROOF. Immediate by Theorem 3.4, Corollary 3.17, Theorem D.1 and Theorem 2.11. □

We end this section with some comments on Wick products of symmetric complex Gaussian variables (see Section 1.4); the first example will be used later but the others can be skipped at the first reading.

EXAMPLE 3.30. Let ζ be a symmetric complex Gaussian variable, i.e. $\zeta = \xi + i\eta$ where ξ and η are independent centred Gaussian variables with the same variance. Since $\mathrm{E}\,\zeta^2 = 0$, Theorem 3.4 or 3.20 (and induction) yields

$$:\zeta^n: = \zeta^n.$$

This reflects the fact that (unlike the real case) the powers ζ^n are already orthogonal.

EXAMPLE 3.31. More generally, if ζ is a symmetric complex Gaussian variable, Theorem 3.4 yields

$$:\zeta^m\bar{\zeta}^n: = \sum_{r=0}^{m\wedge n} (-1)^r r! \binom{m}{r} \binom{n}{r} \zeta^{m-r}\bar{\zeta}^{n-r} (\mathrm{E}\,|\zeta|^2)^r.$$

In particular, if ζ is a standard complex Gaussian variable, and $m \geq n \geq 0$, then

$$:\zeta^m\bar{\zeta}^n: = \zeta^{m-n} \sum_{r=0}^{n} (-1)^r r! \binom{m}{r} \binom{n}{r} |\zeta|^{2(n-r)}.$$

EXAMPLE 3.32. The proof of Theorem 3.21 shows more generally that if $\{\zeta_i\}_{i\in\mathcal{I}}$ is an orthonormal basis in $H_{\mathbb{C}}$, then $\{(\alpha!)^{-1/2} :\zeta^\alpha:\}_\alpha$ is an orthonormal basis in $L^2_{\mathbb{C}}(\Omega, \mathcal{F}(H), \mathrm{P})$. In particular, if K is a complex Gaussian Hilbert space, and $\{\zeta_i\}_{i\in\mathcal{I}}$ is an orthonormal basis in K, then it follows from Proposition 1.34 that $\{\zeta_i, \bar{\zeta}_i\}_{i\in\mathcal{I}}$ is an orthonormal basis in $K \oplus \overline{K} = H_{\mathbb{C}}$, where $H = \mathrm{Re}(K)$, and thus $\{(\alpha!\beta!)^{-1/2} :\zeta^\alpha \bar{\zeta}^\beta: \}_{\alpha,\beta}$ is an orthonormal basis in $L^2_{\mathbb{C}}(\Omega, \mathcal{F}(H), \mathrm{P}) = L^2_{\mathbb{C}}(\Omega, \mathcal{F}(K), \mathrm{P})$.

2. Wick exponentials

It is sometimes convenient to extend the notation $:\xi^n:$ as follows. If $f(z) = \sum_0^\infty a_n z^n$ is an analytic function and ξ is a (real or complex) centred Gaussian variable, we define

$$:f(\xi): = \sum_0^\infty a_n :\xi^n: \tag{3.16}$$

provided this sum converges in L^2 (i.e., if $\sum |a_n|^2 n! (\mathrm{E}\,|\xi|^2)^n < \infty$). For a trivial example, f can be any polynomial.

The notion (3.16) can be extended without any difficulty to (suitable) entire functions of several variables. (We have already done this in a special case in (3.9).) If $f(z) = \sum_0^\infty f_n(z)$, $z \in \mathbb{C}^k$, is the decomposition with f_n homogeneous of degree n, then

$$:f(\xi_1, \dots, \xi_k): = \sum_0^\infty \pi_n(f_n(\xi_1, \dots, \xi_k)).$$

The most important case of (3.16) is the exponential function, in which case we use the equivalent notations $:\exp(\xi):$ and $:e^\xi:$. By definition, $:e^\xi: = \sum_0^\infty \frac{1}{n!} :\xi^n:$; this evaluates as follows.

THEOREM 3.33. *If ξ is a (real or complex) centred Gaussian variable, then*

$$:e^\xi: = e^{\xi - \mathrm{E}\,\xi^2/2}.$$

PROOF. Theorem 3.4 and simple combinatorics yield, cf. (3.11),

$$:\xi^n: = \sum_{r=0}^{n/2} (-1)^r \frac{n!}{2^r r! (n-2r)!} (\mathrm{E}\,\xi^2)^r \xi^{n-2r}$$

and thus

$$\begin{aligned} :e^\xi: &= \sum_{n=0}^\infty \frac{1}{n!} :\xi^n: = \sum_{n=0}^\infty \sum_{r=0}^{n/2} (-1)^r \frac{1}{2^r r! (n-2r)!} (\mathrm{E}\,\xi^2)^r \xi^{n-2r} \\ &= \sum_{r=0}^\infty \sum_{j=0}^\infty \frac{(-\mathrm{E}\,\xi^2)^r \xi^j}{2^r r! j!} = e^{-\mathrm{E}\,\xi^2/2} e^\xi = e^{\xi - \mathrm{E}\,\xi^2/2}. \end{aligned}$$

(A reader used to combinatorics may recognize this as a standard argument with exponential generating functions.) □

We will give other proofs in Example 4.17 and (for real ξ) Example 7.13.

REMARK 3.34. Note in particular that if ξ is a (real) Gaussian variable, then $:e^{\xi}: > 0$ a.s.

Applying Theorem 3.33 to $t\xi$, where t is a real or complex number, yields the trivial but useful generalization

$$:e^{t\xi}: = e^{t\xi - t^2 \operatorname{E} \xi^2/2}. \tag{3.17}$$

EXAMPLE 3.35. Let ξ be a standard Gaussian variable. Then, by Theorem 3.19, $:\xi^n: = h_n(\xi)$ and thus (3.17) may be written

$$\sum_0^\infty \frac{t^n}{n!} h_n(\xi) = e^{t\xi - t^2/2}. \tag{3.18}$$

The sum actually converges pointwise, and not only in L^2. Indeed, (3.11) yields, by the same argument as in the proof of Theorem 3.33,

$$\sum_0^\infty \frac{t^n}{n!} h_n(x) = e^{tx - t^2/2} \tag{3.19}$$

for any complex t and x. This is the well-known formula for the generating function of the Hermite polynomials.

EXAMPLE 3.36. Let ζ be a symmetric complex Gaussian variable. By Example 3.30, $:\zeta^n: = \zeta^n$. Consequently, if f is entire,

$$:f(\zeta): = f(\zeta).$$

In particular,

$$:e^{\zeta}: = e^{\zeta},$$

a special case of Theorem 3.33.

By Theorem 3.33 or directly from the definition, $\operatorname{E} :e^{\xi}: = 1$ for every Gaussian variable ξ. Higher moments are also easily computed using Theorem 3.33; we obtain for example the following.

COROLLARY 3.37. *If ξ and η have a joint (real or complex) centred Gaussian distribution, then*

$$\operatorname{E}(:e^{\xi}: \, :e^{\eta}:) = e^{\operatorname{E} \xi\eta}$$

or equivalently (taking complex conjugates)

$$\langle :e^{\xi}:, :e^{\eta}: \rangle = e^{\langle \xi, \eta \rangle}.$$

PROOF. By Theorem 3.33,

$$\operatorname{E}(:e^{\xi}: \, :e^{\eta}:) = \operatorname{E} e^{\xi + \eta - \operatorname{E}\xi^2/2 - \operatorname{E}\eta^2/2} = e^{\operatorname{E}(\xi+\eta)^2/2 - \operatorname{E}\xi^2/2 - \operatorname{E}\eta^2/2} = e^{\operatorname{E}\xi\eta}.$$

(This also follows easily from Theorem 3.9.) □

COROLLARY 3.38. *If ξ is a centred Gaussian variable, then*

$$\| :e^{\xi}: \|_p = e^{\frac{p-1}{2}\operatorname{E}\xi^2}, \qquad 0 < p < \infty.$$

More generally, if $\zeta = \xi + i\eta$ is a complex Gaussian variable (with ξ, η real), then

$$\| :e^{\zeta}: \|_p = e^{\frac{p-1}{2}\operatorname{E}\xi^2 + \frac{1}{2}\operatorname{E}\eta^2}, \qquad 0 < p < \infty.$$

PROOF. The real case follows from

$$\operatorname{E}(:e^{\xi}:)^p = \operatorname{E} e^{p\xi - p\operatorname{E}\xi^2/2} = e^{p^2\operatorname{E}\xi^2/2 - p\operatorname{E}\xi^2/2}.$$

The complex case is similar, using

$$| :e^{\zeta}: |^p = e^{p\operatorname{Re}(\zeta - \operatorname{E}\zeta^2/2)} = e^{p\xi - p\operatorname{E}\xi^2/2 + p\operatorname{E}\eta^2/2}.$$

□

We obtain also a product formula.

COROLLARY 3.39. *If ξ and η have a joint (real or complex) centred Gaussian distribution, then*

$$:e^{\xi+\eta}: = e^{-\operatorname{E}\xi\eta} :e^{\xi}: :e^{\eta}:.$$

□

Theorem 3.33 shows that Wick exponentials differ from ordinary exponentials by a constant factor only; hence Theorem 2.12 and Remark 2.16 yield the following.

COROLLARY 3.40. *If G is a dense subspace of a Gaussian Hilbert space H, then the set $\{ :\exp(\xi): \mid \xi \in G\}$ is total in $L^p(\Omega, \mathcal{F}(H), \mathrm{P})$, for any p, $0 \le p < \infty$.* □

Some computations with Wick powers are easily done using the Wick exponential as a generating function.

EXAMPLE 3.41. Let ξ be standard Gaussian and let $a \in \mathbb{C}$ with $\operatorname{Re} a < 1/4$. Then $e^{a\xi^2} \in L^2$, and for every real (or complex) t,

$$\begin{aligned}\sum_{n=0}^{\infty} \frac{t^n}{n!} \langle e^{a\xi^2}, :\xi^n: \rangle &= \operatorname{E}(e^{a\xi^2} e^{t\xi - t^2/2}) = \int_{-\infty}^{\infty} e^{ax^2 + tx - t^2/2} \frac{1}{\sqrt{2\pi}} e^{-x^2/2}\,dx \\ &= \frac{1}{\sqrt{2\pi}} \int_{-\infty}^{\infty} e^{-(\frac{1}{2}-a)x^2 + xt - \frac{1}{2}t^2}\,dx = (1-2a)^{-1/2} e^{\frac{a}{1-2a}t^2} \\ &= \sum_{k=0}^{\infty} (1-2a)^{-1/2} \Big(\frac{a}{1-2a}\Big)^k \frac{t^{2k}}{k!}.\end{aligned}$$

Hence, identifying the coefficients of t^n, $\frac{1}{(2k)!}\langle e^{a\xi^2}, :\xi^{2k}:\rangle = a^k(1-2a)^{-k-1/2}/k!$ for $k \geq 0$, and in any Gaussian space containing ξ, we have the chaos decomposition

$$e^{a\xi^2} = \sum_{k=0}^{\infty} \frac{a^k}{(1-2a)^{k+1/2}k!} :\xi^{2k}:, \qquad \operatorname{Re} a < 1/4.$$

With ξ as in Example 3.18, this yields

$$e^{ax^2} = \sum_{k=0}^{\infty} \frac{a^k}{(1-2a)^{k+1/2}k!} h_{2k}(x), \qquad -\infty < x < \infty,\ \operatorname{Re} a < 1/4; \tag{3.20}$$

it will be shown in Example 4.17 that the sum converges not only in the space $L^2\big((2\pi)^{-1/2}e^{-x^2/2}\,dx\big)$ but also pointwise for every x. In fact, we will see that the sum also converges for complex x and uniformly on compact subsets of the complex plane; hence the sum is an entire analytic function of x, so (3.20) holds for all complex x.

EXAMPLE 3.42. Let ξ and a be as in the preceding example. We obtain similarly, or by the result above and the formula $\xi :\xi^n: = :\xi^{n+1}: + n :\xi^{n-1}:$, cf. Theorem 3.15 and (3.13)–(3.14), the chaos decomposition

$$\xi e^{a\xi^2} = \sum_{k=0}^{\infty} \frac{a^k}{(1-2a)^{k+3/2}k!} :\xi^{2k+1}:, \qquad \operatorname{Re} a < 1/4.$$

It may be shown by the method of Example 4.17 that the sum converges pointwise.

EXAMPLE 3.43. The L^2 convergence of an expansion of the type $\sum a_n :\xi^n: = \sum a_n h_n(\xi)$ (with ξ standard Gaussian) does not automatically entail the pointwise convergence of $\sum a_n h_n(x)$. An explicit example is obtained by taking $a_{2k} = (-1)^k k^{-3/4}(2k)!^{-1/2}$ $(k \geq 1)$, $a_0 = a_{2k-1} = 0$ and $x = 0$. Then

$$\sum_0^{\infty} \|a_n h_n(\xi)\|_2^2 = \sum_1^{\infty} k^{-3/2} < \infty$$

and thus $\sum a_n h_n(\xi)$ converges in L^2.

On the other hand, by (3.11), $h_{2k}(0) = (-1)^k(2k)!/2^k k!$ and thus

$$a_{2k}h_{2k}(0) = k^{-3/4}\frac{(2k)!^{1/2}}{2^k k!} = k^{-3/4}\left(2^{-2k}\binom{2k}{k}\right)^{1/2} \sim \pi^{-1/4}k^{-1},$$

whence $\sum_0^{\infty} a_n h_n(0)$ diverges.

3. General Wick products

Let H be a Gaussian Hilbert space. We begin with an important estimate. Alternative proofs will be given in Chapters 5 and 7, and a generalization to arbitrary powers will follow in Theorem 3.50. In this section we let $c_1(n)$, $c_2(n)$, ... denote constants depending only on n.

LEMMA 3.44. *For each $n \geq 0$, $\overline{\mathcal{P}}_n(H) \subset L^4$, and there exists a constant $c_1(n)$ such that if $X \in \overline{\mathcal{P}}_n(H)$, then $\|X\|_4 \leq c_1(n)\|X\|_2$.*

PROOF. Let $X \in \mathcal{P}_n(H)$ be a polynomial in some variables $\xi_1, \dots, \xi_m \in H$, where by Remark 2.3 we may assume that $\xi_1, \dots, \xi_m$ are orthonormal. By Theorem 3.21, applied to the subspace of H spanned by $\{\xi_i\}_1^m$, there exists a decomposition

$$X = \sum_\alpha a_\alpha :\xi^\alpha:$$

where the sum is a finite sum over multi-indices $\alpha = (\alpha_1, \dots, \alpha_m)$ with $|\alpha| \leq n$.

It follows from Theorem 3.15 that

$$:\xi^\alpha: :\xi^\beta: = \sum_\gamma c_{\alpha\beta\gamma} :\xi^{\alpha-\gamma+\beta-\gamma}:$$

for some combinatorial coefficients $c_{\alpha\beta\gamma}$, where the sum is taken over all multi-indices γ such that $\alpha - \gamma$ and $\beta - \gamma$ are multi-indices (equivalently, $\gamma_i \leq \min(\alpha_i, \beta_i)$ for every i). (Actually, $c_{\alpha\beta\gamma} = \alpha!\beta!/(\alpha-\gamma)!(\beta-\gamma)!\gamma!$, but we do not need an exact formula.) Consequently,

$$\begin{aligned} X^2 &= \sum_{\alpha,\beta} a_\alpha a_\beta :\xi^\alpha: :\xi^\beta: = \sum_{\alpha,\beta,\gamma} a_\alpha a_\beta c_{\alpha\beta\gamma} :\xi^{\alpha-\gamma+\beta-\gamma}: \\ &= \sum_{\alpha,\beta,\gamma} a_{\alpha+\gamma} a_{\beta+\gamma} c_{\alpha+\gamma,\beta+\gamma,\gamma} :\xi^{\alpha+\beta}: \\ &= \sum_\delta b_\delta :\xi^\delta:, \end{aligned}$$

where

$$b_\delta = \sum_{\alpha+\beta=\delta} \sum_\gamma a_{\alpha+\gamma} a_{\beta+\gamma} c_{\alpha+\gamma,\beta+\gamma,\gamma}.$$

Clearly we here only have to consider $\alpha, \beta, \gamma, \delta$ with $|\alpha+\gamma| \leq n$, $|\beta+\gamma| \leq n$ and $|\delta| \leq 2n$. For such multi-indices, $0 \leq c_{\alpha+\gamma,\beta+\gamma,\gamma} \leq c_2(n)$, where $c_2(n)$ is a constant depending only on n.

Moreover, if $|\delta| \leq 2n$, there are at most 2^{2n} pairs (α, β) with $\alpha + \beta = \delta$, and thus, using the Cauchy–Schwarz inequality twice,

$$\begin{aligned} |b_\delta|^2 &\leq 2^{2n} \sum_{\alpha+\beta=\delta} \Big(\sum_\gamma c_2(n) |a_{\alpha+\gamma} a_{\beta+\gamma}|\Big)^2 = c_3(n) \sum_{\alpha+\beta=\delta} \Big(\sum_\gamma |a_{\alpha+\gamma} a_{\beta+\gamma}|\Big)^2 \\ &\leq c_3(n) \sum_{\alpha+\beta=\delta} \sum_\gamma |a_{\alpha+\gamma}|^2 \sum_\gamma |a_{\beta+\gamma}|^2. \end{aligned}$$

Hence,

$$\begin{aligned} \mathrm{E}\,|X|^4 = \|X^2\|_2^2 &= \sum_\delta |b_\delta|^2 \delta! \le (2n)! \sum_\delta |b_\delta|^2 \\ &\le c_4(n) \sum_{\alpha,\beta} \sum_\gamma |a_{\alpha+\gamma}|^2 \sum_\gamma |a_{\beta+\gamma}|^2 = c_4(n) \Big(\sum_{\alpha,\gamma} |a_{\alpha+\gamma}|^2 \Big)^2. \end{aligned}$$

Since each multi-index ε with $|\varepsilon| \le n$ occurs as at most 2^n different sums $\alpha+\gamma$, we find

$$\sum_{\alpha,\gamma} |a_{\alpha+\gamma}|^2 \le 2^n \sum_\varepsilon |a_\varepsilon|^2 \le 2^n \sum_\varepsilon |a_\varepsilon|^2 \varepsilon! = 2^n \|X\|_2^2$$

and thus finally

$$\|X\|_4^4 \le 2^{2n} c_4(n) \|X\|_2^4,$$

which gives the sought inequality

$$\|X\|_4 \le c_1(n) \|X\|_2 \tag{3.21}$$

for $X \in \mathcal{P}_n(H)$.

If $X \in \overline{\mathcal{P}}_n(H)$, then there exists a sequence $X_k \in \mathcal{P}_n(H)$ with $\|X_k - X\|_2 \to 0$. Thus (X_k) is a Cauchy sequence in L^2, and by (3.21) applied to the variables $X_i - X_j$, also a Cauchy sequence in L^4. Since L^4 is complete, the sequence (X_k) converges in L^4, and the limit has to equal X, since X is the limit in L^2. Thus $X \in L^4$, and

$$\|X\|_4 = \lim_{k\to\infty} \|X_k\|_4 \le c_1(n) \lim_{k\to\infty} \|X_k\|_2 = c_1(n)\|X\|_2.$$

□

REMARK 3.45. This proof yields, using the rather crude choice $c_2(n) \le n!^2$, the constant $c_1(n) = 2^n n!(2n)!^{1/4}$. This can be improved, but we leave this to the reader.

The proof in Chapter 5 yields the better value $c_1(n) = 3^{n/2}$, see Theorem 5.10 and Remark 5.11. On the other hand, if $\xi \in H$ with $\xi \ne 0$, then

$$c_1(n) \ge c_5(n) = \| {:}\xi^n{:} \|_4 / \| {:}\xi^n{:} \|_2.$$

which is shown in Remark 5.20 to be of order $n^{-1/4} 3^{n/2}$.

We do not know the best possible constant, but it is shown in Section 7.2 that for $X \in H^{:n:}$, the constant can be taken as $c_5(n)$, which clearly is best possible. We do not know whether the constant $c_5(n)$ suffices for all $X \in \overline{\mathcal{P}}_n(H)$.

THEOREM 3.46. *Let $m, n \ge 0$. If $X \in \overline{\mathcal{P}}_m(H)$ and $Y \in \overline{\mathcal{P}}_n(H)$, then $XY \in \overline{\mathcal{P}}_{m+n}(H)$ and*

$$\|XY\|_2 \le c_6(m,n) \|X\|_2 \|Y\|_2, \tag{3.22}$$

for some constant $c_6(m,n)$ depending on m and n only. Hence the multiplication $(X,Y) \mapsto XY$ is a continuous map $\overline{\mathcal{P}}_m(H) \times \overline{\mathcal{P}}_n(H) \to \overline{\mathcal{P}}_{m+n}(H)$ for every $m,n \geq 0$.

PROOF. By Lemma 3.44, $\overline{\mathcal{P}}_n(H) \subset L^4$. Thus Hölder's inequality shows that the mapping $(X,Y) \mapsto XY$ is a bounded (and thus continuous) bilinear map $\overline{\mathcal{P}}_m(H) \times \overline{\mathcal{P}}_n(H) \to L^2$, and (3.22) holds with $c_6(m,n) = c_1(m)c_1(n)$.

Moreover, if $X \in \mathcal{P}_m(H)$ and $Y \in \mathcal{P}_n(H)$, then $XY \in \mathcal{P}_{m+n}(H)$, and it follows by continuity that if $X \in \overline{\mathcal{P}}_m(H)$ and $Y \in \overline{\mathcal{P}}_n(H)$, then $XY \in \overline{\mathcal{P}}_{m+n}(H)$. □

THEOREM 3.47. *For any $n, m \geq 0$, the general Wick product, as defined in (3.2), is a continuous bilinear operation $H^{:m:} \times H^{:n:} \to H^{:m+n:}$; thus there exists a constant $c_7(m,n)$ such that*

$$\|X \odot Y\|_2 \leq c_7(m,n)\|X\|_2\|Y\|_2, \qquad X \in \overline{\mathcal{P}}_m(H),\ Y \in \overline{\mathcal{P}}_n(H). \tag{3.23}$$

The general Wick product is a commutative, associative operator on $\overline{\mathcal{P}}_*(H)$ with 1 as a unit. Moreover, if $X_i \in H^{:n_i:}$ and $n = \sum_1^k n_i$, then

$$X_1 \odot \cdots \odot X_k = \pi_n(X_1 \cdots X_k). \tag{3.24}$$

In particular,

$$\xi_1 \odot \cdots \odot \xi_n = {:}\xi_1 \cdots \xi_n{:} \tag{3.25}$$

when $\xi_1, \dots, \xi_n \in H_{\mathbb{C}}$.

PROOF. Theorem 3.46 implies that the general Wick product (3.2) is well-defined, and that it is a bounded (and thus continuous) bilinear map $H^{:m:} \times H^{:n:} \to H^{:m+n:}$. Clearly, this defines $\odot$ as a bilinear operator on $\overline{\mathcal{P}}_*(H)$, and $1 \in H^{:0:}$ is a unit.

It is clear that $X \odot Y = Y \odot X$ when $X \in H^{:l:}$ and $Y \in H^{:m:}$; hence, the general Wick product is commutative on $\overline{\mathcal{P}}_*(H)$.

Now suppose $X \in H^{:l:}$, $Y \in H^{:m:}$ and $Z \in H^{:n:}$. Then, by Theorem 3.46, $XY \in \overline{\mathcal{P}}_{l+m}(H)$ and thus $XY - X \odot Y \in \overline{\mathcal{P}}_{l+m-1}(H)$, whence by Theorem 3.46 again, $(XY - X \odot Y)Z \in \overline{\mathcal{P}}_{l+m+n-1}(H)$. Consequently,

$$(X \odot Y) \odot Z = \pi_{l+m+n}((X \odot Y)Z) = \pi_{l+m+n}(XYZ). \tag{3.26}$$

Similarly (or by (3.26) and commutativity),

$$X \odot (Y \odot Z) = \pi_{l+m+n}(XYZ) = (X \odot Y) \odot Z,$$

and thus $\odot$ is associative.

Finally, (3.26) is easily extended by induction to (3.24), and (3.25) follows as a special case. □

It follows from Corollary 3.25 that if $X, Y \in \mathcal{P}(H)$, then $X \odot Y \in \mathcal{P}(H)$. The Wick product thus makes $\mathcal{P}(H)$ and $\overline{\mathcal{P}}_*(H)$ into graded commutative algebras with the constant 1 as unit.

EXAMPLE 3.48. Let ξ be a standard normal variable in H. Then $:\xi^m: \odot :\xi^n: = :\xi^{m+n}:$. Since $\| :\xi^k: \|_2 = \sqrt{k!}$ by Corollary 3.10 this proves that the constant $c_7(m,n)$ in Theorem 3.47 has to be at least $(m+n)!^{1/2}/m!^{1/2}n!^{1/2} = \binom{m+n}{m}^{1/2}$. (We will see in Corollary 4.2 in the next chapter that this value actually suffices for (3.23) when $X \in H^{:m:}$, $Y \in H^{:n:}$. We do not know the best constants in (3.22) and (3.23).)

This example shows that the constant $c_7(m,n)$ in (3.23) cannot be replaced by a constant independent of m and n. Hence the general Wick product $\odot$ does not extend to a continuous bilinear operator on $L^2(\Omega, \mathcal{F}(H), \mathrm{P})$. (For a similar counterexample with L^∞, see Example 16.30.)

REMARK 3.49. There exists a natural locally convex topology on $\overline{\mathcal{P}}_*(H)$, viz. the topology as a locally convex direct sum $\sum_0^\infty H^{:n:}$ or, equivalently, the inductive limit topology as the union $\bigcup_0^\infty \overline{\mathcal{P}}_n(H)$; see for example Schaefer (1971, §II.6) for definitions. A linear map of $\overline{\mathcal{P}}_*(H)$, equipped with this topology, into some locally convex topological vector space is continuous if and only if the restriction to every $H^{:n:}$ is continuous.

It follows from Theorems 3.46 and 3.47 that both the ordinary multiplication and the Wick multiplication are continuous bilinear operations on $\overline{\mathcal{P}}_*(H)$ with this topology; hence $\overline{\mathcal{P}}_*(H)$ is a topological algebra in two different ways.

We will not need this topology in the sequel.

We continue with a couple of other applications of Theorem 3.46. The first is a (partial) extension of Theorem 1.4 to the spaces $H^{:n:}$ and $\overline{\mathcal{P}}_n(H)$. (For other proofs, see Theorem 5.10 and Corollary 7.36.)

THEOREM 3.50. *For each $n \geq 0$, $H^{:n:} \subset \overline{\mathcal{P}}_n(H) \subset \bigcap_{p<\infty} L^p$ and all L^p-norms, for $0 < p < \infty$, are equivalent on $\overline{\mathcal{P}}_n(H)$. More precisely, for every $p, q < \infty$, there exists $c_8(p,q,n) < \infty$ such that if $X \in \overline{\mathcal{P}}_n(H)$, then*

$$\|X\|_q \leq c_8(p,q,n)\|X\|_p. \tag{3.27}$$

Furthermore, $H^{:n:}$ and $\overline{\mathcal{P}}_n(H)$ are closed subspaces of the space $L^p(\Omega, \mathcal{F}, \mathrm{P})$, $0 \leq p < \infty$, and convergence in probability is equivalent to convergence in L^p for elements of $\overline{\mathcal{P}}_n(H)$, for any fixed n and $p < \infty$.

PROOF. The case $p = 2$, $q = 4$ of (3.27) is Lemma 3.44. More generally, by Theorem 3.46 and induction, if $X \in \overline{\mathcal{P}}_n(H)$ and $k \geq 1$, then $X^k \in \overline{\mathcal{P}}_{kn}(H)$ and $\|X^k\|_2 \leq c_9(n,k)\|X\|_2^k$, which yields (3.27) for $p = 2$, $q = 2k$. Since $\|X\|_p$ is an increasing function of p, this yields the result when $2 \leq p \leq q < \infty$.

If $0 < p < 2 \leq q$, choose $r > q$ and define $\theta \in (0,1)$ by $1/q = \theta/p + (1-\theta)/r$. Hölder's inequality and the already proved case yield

$$\|X\|_q \leq \|X\|_p^\theta \|X\|_r^{1-\theta} \leq c_8(q,r,n)^{(1-\theta)} \|X\|_p^\theta \|X\|_q^{1-\theta}.$$

Since $\|X\|_q < \infty$, this implies $\|X\|_q^\theta \le c_8(q,r,n)^{(1-\theta)}\|X\|_p^\theta$, which yields (3.27) with

$$c_8(p,q,n) = c_8(q,r,n)^{(1-\theta)/\theta} = c_8(q,r,n)^{(1/p-1/q)/(1/q-1/r)}.$$

The remaining cases of (3.27) follow trivially; we may take $c_8(p,q,n) = c_8(p,2,n)$ when $q < 2$, and $c_8(p,q,n) = 1$ when $q \le p$.

Now suppose that $\{Y_i\}$ is a sequence in $\overline{\mathcal{P}}_n(H)$ with $Y_i \xrightarrow{\text{p}} Y$ for some random variable Y. Suppose first that $\{Y_i\}$ is not a Cauchy sequence in L^2. Then there exist increasing sequences $j(i)$ and $k(i)$ such that $\|Y_{j(i)} - Y_{k(i)}\|_2 \ge \delta > 0$. Let $Z_i = (Y_{j(i)} - Y_{k(i)})/\|Y_{j(i)} - Y_{k(i)}\|_2$; then $Z_i \xrightarrow{\text{p}} 0$ but $\|Z_i\|_2 = 1$. Since the family $\{Z_i\}$ is bounded in L^2, it is uniformly integrable; hence $Z_i \xrightarrow{\text{p}} 0$ implies $\|Z_i\|_1 \to 0$, which contradicts the first part.

Consequently, $\{Y_i\}$ has to be a Cauchy sequence in L^2. This implies that $\{Y_i\}$ converges in L^2, and since $Y_i \xrightarrow{\text{p}} Y$, the limit has to be Y. Hence $Y \in \overline{\mathcal{P}}_n(H)$, and $Y_i \to Y$ in L^2, and by the first part in every L^p. This shows, for every $p < \infty$, both that $\overline{\mathcal{P}}_n(H)$ is a closed subspace of L^0, and thus of L^p, and that convergence in probability is equivalent to convergence in L^p on $\overline{\mathcal{P}}_n(H)$.

Finally, we recall that $H^{:n:}$ is a closed subspace of $\overline{\mathcal{P}}_n(H)$. □

THEOREM 3.51. *If* $0 < p < \infty$, *then* $\overline{\mathcal{P}}_*(H)$ *is a dense subspace of* $L^p(\Omega, \mathcal{F}(H), \mathrm{P})$.

PROOF. By Theorem 3.50, $\overline{\mathcal{P}}_*(H) \subset L^p$, and by Theorem 2.11, the subspace of polynomial variables is already dense. □

We end this section with a simple but important example of the norm equivalences in Theorem 3.50, where we can give the best constants explicitly.

EXAMPLE 3.52. Let H_1 and H_2 be two orthogonal (and thus independent) subspaces of a Gaussian Hilbert space H, and let $:H_1H_2:$ be the closed linear hull in $L^2_{\mathbb{R}}$ of the set $\{ :\xi\eta: \mid \xi \in H_1, \eta \in H_2\} = \{\xi\eta \mid \xi \in H_1, \eta \in H_2\}$. (It is easily seen that $:H_1H_2:$ may be identified with the tensor product $H_1 \otimes H_2$, cf. Appendix E and the next chapter.) Then $:H_1H_2: \subset H^{:2:}$, and thus all L^p-norms $(0 < p < \infty)$ are equivalent on $:H_1H_2:$ by Theorem 3.50.

In this case we have the explicit estimates, for $X \in :H_1H_2:$,

$$\kappa(p)^2\|X\|_2 \le \|X\|_p \le \kappa(p)\|X\|_2, \qquad 0 < p \le 2, \tag{3.28}$$

$$\kappa(p)\|X\|_2 \le \|X\|_p \le \kappa(p)^2\|X\|_2, \qquad 2 \le p \le \infty, \tag{3.29}$$

where $\kappa(p) = \sqrt{2}\big(\Gamma(\frac{p+1}{2})/\sqrt{\pi}\big)^{1/p}$ as in (1.1).

In order to prove these estimates, we may assume that H_1 and H_2 are defined on different probability spaces $(\Omega_1, \mathcal{F}_1, \mathrm{P}_1)$ and $(\Omega_2, \mathcal{F}_2, \mathrm{P}_2)$, and that $:H_1H_2:$ is defined on the product space $\Omega_1 \times \Omega_2$. Moreover, it suffices to consider the case when X is a finite sum $\sum_1^n \xi_i \eta_i$, with $\xi_i \in H_1$ and $\eta_i \in H_2$. Since then $X(\omega_1, \cdot) = \sum_i \xi_i(\omega_1)\eta_i \in H_2$ for each $\omega_1 \in \Omega_1$, (1.1) yields

$$\|X\|_p = \|X\|_{L^p(\Omega_1;L^p(\Omega_2))} = \kappa(p)\|X\|_{L^p(\Omega_1;L^2(\Omega_2))}.$$

If $p \leq 2$ we thus obtain, by Lyapounov's inequality,

$$\|X\|_p = \kappa(p)\|X\|_{L^p(\Omega_1;L^2(\Omega_2))} \leq \kappa(p)\|X\|_{L^2(\Omega_1;L^2(\Omega_2))} = \kappa(p)\|X\|_2,$$

and by the version of Minkowski's inequality in Proposition C.4,

$$\|X\|_p \geq \kappa(p)\|X\|_{L^2(\Omega_2;L^p(\Omega_1))} = \kappa(p)^2\|X\|_{L^2(\Omega_2;L^2(\Omega_1))} = \kappa(p)^2\|X\|_2.$$

For $p \geq 2$, both these inequalities go in the opposite direction.

This shows (3.28) and (3.29). Moreover, the constants are best possible. In one direction we observe that if $X = \xi\eta$ with $\xi \in H_1$ and $\eta \in H_2$, then $\|X\|_p = \|\xi\|_p\|\eta\|_p = \kappa(p)^2\|\xi\|_2\|\eta\|_2 = \kappa(p)^2\|X\|_2$. On the other hand, if $\dim(H_1) = \dim(H_2) = \infty$, and $\{\xi_i\}_1^\infty$ and $\{\eta_i\}_1^\infty$ are orthonormal families in H_1 and H_2, let $X_n = n^{-1/2}\sum_1^n \xi_i\eta_i$. Then $\|X_n\|_2 = 1$ and, by the central limit theorem, $X_n \xrightarrow{\text{d}} \text{N}(0,1)$ with convergence of all moments as $n \to \infty$. Hence $\|X_n\|_p \to \kappa(p)$ as $n \to \infty$.

IV

Tensor products and Fock space

1. Spaces

We recall some standard facts on tensor products of Hilbert spaces (with real or complex scalars). Further details are given in Appendix E.

If H_1 and H_2 are two Hilbert spaces, their *tensor product* $H_1 \otimes H_2$ is defined to be a Hilbert space together with a bilinear map $H_1 \times H_2 \to H_1 \otimes H_2$, denoted by $(f_1, f_2) \mapsto f_1 \otimes f_2 \in H_1 \otimes H_2$, such that

$$\langle f_1 \otimes f_2, g_1 \otimes g_2 \rangle_{H_1 \otimes H_2} = \langle f_1, g_1 \rangle_{H_1} \langle f_2, g_2 \rangle_{H_2}$$

and, moreover, the closed linear span of the range of this map equals $H_1 \otimes H_2$. The tensor product is defined up to equivalence only, and it may be constructed in several ways. For example, the tensor product of two L^2-spaces $L^2(X_1, \mu_1)$ and $L^2(X_2, \mu_2)$, where μ_1 and μ_2 are σ-finite measures, may be taken as $L^2(X_1 \times X_2, \mu_1 \times \mu_2)$ with $f \otimes g(x_1, x_2) = f(x_1)g(x_2)$.

The tensor product of several Hilbert spaces is defined in the same way. In particular, we may form tensor powers $H^{\otimes n}$ of any Hilbert space.

The *symmetric tensor power* $H^{\odot n}$ of a Hilbert space is similarly defined to be a Hilbert space together with a symmetric multilinear map $H \times \cdots \times H \to H^{\odot n}$, denoted by $(f_1, \dots, f_n) \mapsto f_1 \odot \cdots \odot f_n$, such that

$$\langle f_1 \odot \cdots \odot f_n, g_1 \odot \cdots \odot g_n \rangle = \sum_{\pi \in \mathfrak{S}_n} \prod_{i=1}^{n} \langle f_i, g_{\pi(i)} \rangle$$

(where $\mathfrak{S}_n$ is the symmetric group) and, moreover, the closed linear span of the range of this map equals $H^{\odot n}$. We let $H^{\odot 0}$ be the one-dimensional space of scalars. Note also that $H^{\odot 1} = H$.

As is shown in Proposition E.19, the multiplication

$$(f_1 \odot \cdots \odot f_n) \odot (f_{n+1} \odot \cdots \odot f_{n+m}) = f_1 \odot \cdots \odot f_{n+m},$$

may be extended to a continuous bilinear operation $H^{\odot n} \times H^{\odot m} \to H^{\odot (n+m)}$ for any $n, m \geq 0$. Hence, the direct sum $\Gamma_*(H) = \sum_{n=0}^{\infty} H^{\odot n}$ is a graded commutative algebra which is called the *symmetric tensor algebra* of H. Its completion, the Hilbert space $\Gamma(H) = \bigoplus_{n=0}^{\infty} H^{\odot n}$, is called the (*symmetric*) *Fock space* over H.

Now, let H be a Gaussian Hilbert space. Comparing these definitions with the results in the earlier chapters, in particular Theorem 3.9 and Corollary 3.27, we see immediately that the spaces $H^{:n:}$ are realizations of the symmetric tensor powers of H. More precisely, we obtain the following fundamental result (Segal 1956, Dobrushin and Minlos 1977).

THEOREM 4.1. *If H is a Gaussian Hilbert space, then the map*

$$\xi_1 \odot \cdots \odot \xi_n \mapsto :\xi_1 \cdots \xi_n: \tag{4.1}$$

defines a Hilbert space isometry of $H^{\odot n}$ onto $H_{\mathbb{R}}^{:n:}$. Taken together for all $n \geq 0$, these maps define an algebra isomorphism of the symmetric tensor algebra $\Gamma_*(H)$ onto $\overline{\mathcal{P}}_{*\mathbb{R}}(H)$ with the (general) Wick multiplication given in Definition 3.2; this extends to an isometry of the Fock space $\Gamma(H)$ onto $\bigoplus_{n=0}^{\infty} H_{\mathbb{R}}^{:n:} = L^2_{\mathbb{R}}(\Omega, \mathcal{F}(H), \mathrm{P})$.

Similarly, for complex scalars, (4.1) defines isometries $H_{\mathbb{C}}^{\odot n} \cong H_{\mathbb{C}}^{:n:}$ and $\Gamma(H_{\mathbb{C}}) \cong L^2_{\mathbb{C}}(\Omega, \mathcal{F}(H), \mathrm{P})$, and an algebra isomorphism $\Gamma_*(H_{\mathbb{C}}) \cong \overline{\mathcal{P}}_{*\mathbb{C}}(H)$. □

COROLLARY 4.2. *If H is a Gaussian Hilbert space, then the Wick product satisfies*

$$\|X \odot Y\|_2 \leq \binom{n+m}{m}^{1/2} \|X\|_2 \|Y\|_2, \qquad X \in H^{:m:},\, Y \in H^{:n:}. \tag{4.2}$$

PROOF. This estimate holds for symmetric tensor products by Proposition E.19. □

It was shown in Example 3.48 that equality holds in (4.2) if $X = :\xi^m:$ and $Y = :\xi^n:$. Hence, the constant given in (4.2) is the best possible for every $m, n \geq 0$ (assuming $\dim H \geq 1$).

We have shown that from an abstract point of view, Wick products are the same as tensor products. Indeed, several definitions and results for Gaussian Hilbert spaces in this book may be extended to Fock spaces based on arbitrary Hilbert spaces. (Corollary 4.2 is a typical example.) The theory of Gaussian Hilbert spaces is much richer than the abstract Hilbert space theory, however; the perhaps most important difference is that Wick products are defined as random variables, rather than some abstract 'tensors'. This makes it possible to study (and apply) many kinds of properties with no counterpart for abstract tensor products; for example, we will study (pointwise) positivity and L^p-norms for $p \neq 2$. Hence most of our definitions and results make sense only for Gaussian Hilbert spaces.

In this book, the abstract tensor product theory plays a minor role, but it serves as a source of inspiration and results; all definitions and results for the abstract products may be translated (i.e. specialized) to Wick products. One example is Corollary 4.2, where it seems simplest to formulate the proof in terms of abstract tensor products; as stated above, there are several other examples in this book where a theorem or proof uses only the tensor product structure and thus could have been stated more abstractly, although we prefer to give only a Wick formulation. (Conversely, there are cases in Chapter 13 of such results that could be stated for arbitrary Hilbert spaces and abstract tensor products, where our proof uses the Gaussian Hilbert space structure. We do not state the abstract version, but it follows immediately from the Gaussian case, using an isomorphic Gaussian Hilbert space. Thus the concrete realization of Hilbert space tensor products as Wick products in Gaussian Hilbert spaces may also be useful in proving abstract results.)

We will now study some notions from abstract tensor product theory that yield interesting results when interpreted for Gaussian Hilbert spaces. We begin with a simple example, and continue in the next section with a fundamental construction.

EXAMPLE 4.3. We obtain (for any Hilbert space H) some particularly simple elements of $\Gamma(H)$ by defining, for any $f \in H$,

$$\exp_{\odot}(f) = \sum_{n=0}^{\infty} \frac{f^{\odot n}}{n!}$$

which satisfies

$$\|\exp_{\odot}(f)\|^2_{\Gamma(H)} = \sum_{n=0}^{\infty} \frac{1}{(n!)^2} \|f^{\odot n}\|^2_{H^{\odot n}} = \sum_{n=0}^{\infty} \frac{n!}{(n!)^2} \|f\|^{2n}_H = \exp(\|f\|^2_H).$$

For a Gaussian Hilbert space, the exponentials $\exp_{\odot} \xi$ just defined and $:\exp \xi:$ defined in Section 3.2 coincide under the identification (4.1).

REMARK 4.4. Let us briefly mention the corresponding, but quite different, results for a complex Gaussian Hilbert space K. (This remark should be skipped by the readers that have skipped the optional Section 1.4.)

Let $H = \operatorname{Re} K$. By Proposition 1.34, $H_{\mathbb{C}} = K \oplus \overline{K}$ and thus Theorem 4.1 yields

$$L^2_{\mathbb{C}}(\Omega, \mathcal{F}(K), \mathrm{P}) = L^2_{\mathbb{C}}(\Omega, \mathcal{F}(H), \mathrm{P}) = \Gamma(H_{\mathbb{C}}) = \Gamma(K \oplus \overline{K}).$$

The Fock space $\Gamma(K)$ is thus only a (rather small) subspace of $L^2_{\mathbb{C}}(\Omega, \mathcal{F}(K), \mathrm{P})$.

If $\{\zeta_i\}_{i \in \mathcal{I}}$ is an orthonormal basis in K, then $\{(\alpha!\beta!)^{-1/2} :\zeta^{\alpha}\bar{\zeta}^{\beta}:\}_{\alpha,\beta}$ is an orthonormal basis in $L^2_{\mathbb{C}}(\Omega, \mathcal{F}(K), \mathrm{P})$ by Example 3.32, while the subspace $\Gamma(K)$ has the orthonormal basis $\{(\alpha!)^{-1/2} :\zeta^{\alpha}:\}_{\alpha} = \{(\alpha!)^{-1/2}\zeta^{\alpha}\}_{\alpha}$.

For example, if γ is the standard Gaussian measure $\pi^{-d} e^{-|z|^2}\, d\lambda$ on $\mathbb{C}^d$, where λ denotes the Lebesgue measure, and K is the d-dimensional Gaussian space of linear functions on $\mathbb{C}^d$, then $\Gamma(K \oplus \overline{K}) = L^2_{\mathbb{C}}(\mathbb{C}^d, \gamma)$ while $\Gamma(K)$ is the subspace spanned by $\{z^{\alpha}\}_{\alpha}$. It is not difficult to show that $\Gamma(K) = A(\mathbb{C}^d, \gamma)$, the subspace of $L^2_{\mathbb{C}}(\mathbb{C}^d, \gamma)$ consisting of entire functions.

The space $A(\mathbb{C}^d, \gamma)$ is called the *Segal–Bargmann space*, *Bargmann space* or (conflicting with our terminology) *Fock space*; it is thus a realization of the Fock space $\Gamma(\mathbb{C}^d)$.

2. Operators

In this section we use the important property (see Proposition E.21) that a bounded linear operator A between two Hilbert spaces induces linear operators between their symmetric tensor powers by $A^{\odot n}(f_1 \odot \cdots \odot f_n) = Af_1 \odot \cdots \odot Af_n$, and that $\|A^{\odot n}\| = \|A\|^n$. If A is a contraction, i.e. if $\|A\| \le 1$, these operators furthermore combine to a linear operator $\Gamma(A)$ of norm 1 between the two corresponding Fock spaces. In our context, this gives the following.

THEOREM 4.5. *If $A\colon H_1 \to H_2$ is a bounded linear map between two Gaussian Hilbert spaces, defined on probability spaces $(\Omega_i, \mathcal{F}_i, \mathrm{P}_i)$, $i = 1, 2$, then*

$$:\xi_1 \cdots \xi_n: \ \mapsto \ :A\xi_i \cdots A\xi_n: \tag{4.3}$$

defines a bounded linear operator $A^{:n:}\colon H_1^{:n:} \to H_2^{:n:}$ with $\|A^{:n:}\| = \|A\|^n$. These operators combine to an algebra homomorphism $\overline{\mathcal{P}}_*(H_1) \to \overline{\mathcal{P}}_*(H_2)$, and, provided moreover $\|A\| \le 1$, a linear operator

$$\Gamma(A)\colon L^2(\Omega_1, \mathcal{F}(H_1), \mathrm{P}_1) \to L^2(\Omega_2, \mathcal{F}(H_2), \mathrm{P}_2)$$

with $\|\Gamma(A)\| = 1$.

More generally, for the case of complex spaces, A may here be any bounded complex linear map $H_{1\mathbb{C}} \to H_{2\mathbb{C}}$. □

REMARK 4.6. We also have the functorial property $\Gamma(AB) = \Gamma(A)\Gamma(B)$, if $B\colon H_0 \to H_1$ is another contraction (and H_0 is a third Gaussian Hilbert space); moreover $\Gamma(I) = I$ and $\Gamma(A^*) = \Gamma(A)^*$; see Proposition E.22.

EXAMPLE 4.7. Let H be a Gaussian Hilbert space and let r be a real (or complex) number with $|r| \le 1$. Then, letting I denote the identity operators on the various spaces, $:(rI)^n: \ = r^n I$ on $H^{:n:}$. Hence $\Gamma(rI)$ is the linear operator on $L^2(\Omega, \mathcal{F}(H), \mathrm{P})$ that is given by

$$\Gamma(rI)\Big(\sum_0^\infty X_n\Big) = \sum_0^\infty r^n X_n, \qquad X_n \in H^{:n:},$$

i.e., every $H^{:n:}$ is an eigenspace with eigenvalue r^n. We can express this as $\Gamma(rI) = r^{\mathcal{N}}$, where $\mathcal{N}$ is the unbounded operator in $L^2(\Omega, \mathcal{F}(H), \mathrm{P})$ defined by

$$\mathcal{N}\Big(\sum_0^\infty X_n\Big) = \sum_0^\infty n X_n, \qquad X_n \in H^{:n:},$$

whenever the right hand side converges; note that $\mathcal{N}$ is self-adjoint. $\mathcal{N}$ is known as the *number operator* in quantum field theory and the *Ornstein–Uhlenbeck operator* in stochastic analysis (except that many authors define the Ornstein–Uhlenbeck operator as $-\mathcal{N}$). The operators $\Gamma(e^{-t}I) = e^{-t\mathcal{N}}$, $t \ge 0$, form an operator semigroup known as the *Ornstein–Uhlenbeck semigroup*.

The operator $\Gamma(rI)$ can also be regarded as a generalization of the *Mehler transform* for functions on $\mathbb{R}$, see Example 4.18 and Definition 4.19 below.

EXAMPLE 4.8. If $A\colon H_1 \to H_2$ is a linear map between two Gaussian Hilbert spaces with $\|A\| \le 1$, and $\xi \in H_1$, then

$$\Gamma(A)(:e^{\xi}:) = \Gamma(A)\Big(\sum_0^\infty \frac{1}{n!} :\xi^n:\Big) = \sum_0^\infty \frac{1}{n!} :(A\xi)^n: \; = \; :e^{A\xi}:.$$

In particular, if $|r| \le 1$,

$$\Gamma(rI)(:e^{\xi}:) = :e^{r\xi}:.$$

For ordinary exponentials, we obtain instead, by these formulae and Theorem 3.33,

$$\Gamma(A)(e^{\xi}) = \Gamma(A)(e^{\mathrm{E}\,\xi^2/2} :e^{\xi}:) = e^{A\xi + \mathrm{E}\,\xi^2/2 - \mathrm{E}(A\xi)^2/2} \tag{4.4}$$

and

$$\Gamma(rI)(e^{\xi}) = e^{r\xi + (1-r^2)\,\mathrm{E}\,\xi^2/2}.$$

The same results hold for a complex Gaussian variable $\xi \in H_{1\mathbb{C}}$ and a complex linear map $A\colon H_{1\mathbb{C}} \to H_{2\mathbb{C}}$ or a complex number r.

We proceed to study properties of the operator $\Gamma(A)$ for Gaussian Hilbert spaces, beginning with an important special case.

THEOREM 4.9. *Let H and K be Gaussian Hilbert spaces that are subspaces of a common Gaussian space G, and let P_{HK} be the restriction to H of the orthogonal projection $G \to K$. Then $\Gamma(P_{HK})\colon L^2(\Omega, \mathcal{F}(H), \mathrm{P}) \to L^2(\Omega, \mathcal{F}(K), \mathrm{P})$ equals the conditional expectation $X \mapsto \mathrm{E}(X \mid \mathcal{F}(K))$.*

PROOF. Let $\xi_1, \dots, \xi_n \in H$ and $\eta_1, \dots, \eta_m \in K$. Since we then have $\mathrm{E}(\xi_i\eta_j) = \mathrm{E}\big(P_{HK}(\xi_i)\eta_j\big)$ for all i and j, Theorem 3.9 implies

$$\mathrm{E}(:\xi_1 \cdots \xi_n: \, :\eta_1 \cdots \eta_m:) = \mathrm{E}(:P_{HK}\xi_1 \cdots P_{HK}\xi_n: \, :\eta_1 \cdots \eta_m:).$$

This can be written

$$\mathrm{E}(XY) = \mathrm{E}(\Gamma(P_{HK})X \cdot Y) \tag{4.5}$$

for $X = \,:\xi_1 \cdots \xi_n:$ and $Y = \,:\eta_1 \cdots \eta_m:$. By bilinearity, continuity and Theorem 3.29, (4.5) extends to all $X \in L^2(\Omega, \mathcal{F}(H), \mathrm{P})$ and $Y \in L^2(\Omega, \mathcal{F}(K), \mathrm{P})$, which proves the result. □

This special case is not so special as it might seem; in fact, any operator $\Gamma(A)$ is equivalent to a conditional expectation in the following sense.

THEOREM 4.10. *Let $A\colon H_1 \to H_2$ be a linear operator between two Gaussian Hilbert spaces such that $\|A\| \le 1$. Then there exists a Gaussian Hilbert space G with closed subspaces H_1' and H_2', and isometries U_i of H_i onto H_i', such that $A = U_2^{-1} P_{H_1'H_2'} U_1$ and thus $\Gamma(A) = \Gamma(U_2^{-1})\Gamma(P_{H_1'H_2'})\Gamma(U_1)$, where $\Gamma(P_{H_1'H_2'})$ is a conditional expectation by Theorem 4.9.*

PROOF. Let $H = H_1 \oplus H_2$ and define a symmetric bilinear form on H by

$$\langle(\xi_1,\xi_2),(\eta_1,\eta_2)\rangle_H = \langle\xi_1,\eta_1\rangle_{H_1} - \langle A\xi_1, A\eta_1\rangle_{H_2} + \langle A\xi_1+\xi_2, A\eta_1+\eta_2\rangle_{H_2}. \tag{4.6}$$

Since $\|A\| \le 1$, $\langle(\xi_1,\xi_2),(\xi_1,\xi_2)\rangle_H = \|\xi_1\|^2_{H_1} - \|A\xi_1\|^2_{H_2} + \|A\xi_1+\xi_2\|^2_{H_2} \ge 0$, so this bilinear form is semi-definite.

Let $H_0 = \{h \in H : \langle h,h\rangle_H = 0\}$; the quotient H/H_0 is a pre-Hilbert space and may be completed to a Hilbert space $\widetilde{H}$. Take any Gaussian Hilbert space G of the same dimension as $\widetilde{H}$ and choose an isometry of $\widetilde{H}$ onto G (cf. Example 1.22 and Theorem 1.23). This defines a mapping $T\colon H \to G$ such that $\langle Th, Th'\rangle_G = \langle h,h'\rangle_H$. We define $U_1\xi = T(\xi,0)$ and $U_2\xi = T(0,\xi)$ and note that (4.6) implies that $U_1\colon H_1 \to G$ and $U_2\colon H_2 \to G$ are isometric embeddings. Moreover, if $\xi_i \in H_i$, then

$$\langle U_1\xi_1, U_2\xi_2\rangle_G = \langle(\xi_1,0),(0,\xi_2)\rangle_H = \langle A\xi_1,\xi_2\rangle_{H_2} = \langle U_2A\xi_1, U_2\xi_2\rangle_G$$

and thus $U_2A\xi_1 = P_{U_1(H_1),U_2(H_2)}U_1\xi_1$, which proves the theorem with $H_i' = U_i(H_i)$. ∎

As should be expected, the isometries U_i in this theorem act trivially on the Fock spaces.

THEOREM 4.11. *If U is an isometry of one Gaussian Hilbert space H onto another, H', then the random variables $\Gamma(U)X$ and X have the same distribution, for any $X \in L^2(\Omega,\mathcal{F}(H),\mathrm{P})$; moreover, $|\Gamma(U)X| = \Gamma(U)|X|$ a.s.*

Furthermore, if $\xi_1,\dots,\xi_m \in H$ and φ is any measurable function on $\mathbb{R}^m$ such that $\varphi(\xi_1,\dots,\xi_m) \in L^2$, then

$$\Gamma(U)\big(\varphi(\xi_1,\dots,\xi_m)\big) = \varphi(U\xi_1,\dots,U\xi_m). \tag{4.7}$$

PROOF. We first prove (4.7). Thus, let $\xi_1,\dots,\xi_m \in H$. Since U is an isometry, the random vectors $(\xi_1,\dots,\xi_m)$ and $(U\xi_1,\dots,U\xi_m)$ have the same covariance matrix and thus the same distribution, since both random vectors have centred Gaussian distributions in $\mathbb{R}^m$. Let μ denote this common distribution and let $K = \{\sum_1^m t_i\xi_i : t_i \in \mathbb{R}\}$ be the Gaussian subspace of H spanned by $\xi_1,\dots,\xi_m$. Then $L^2(\Omega,\mathcal{F}(K),\mathrm{P}) = \{\varphi(\xi_1,\dots,\xi_m) : \varphi \in L^2(d\mu)\}$, and Theorem 2.12 implies that linear combinations of exponential functions $\varphi(x_1,\dots,x_m) = e^{\sum_i t_ix_i}$ are dense in $L^2(d\mu)$. However, (4.7) is easily verified for such functions, because U is an isometry and (4.4) thus simplifies to

$$\Gamma(U)(e^{\sum_i t_ix_i}) = e^{U(\sum_i t_ix_i)} = e^{\sum_i t_iUx_i}.$$

Since both sides of (4.7) represent bounded linear operators from $L^2(d\mu)$ into the Fock space $\Gamma(H') = L^2(\Omega',\mathcal{F}(H'),\mathrm{P}')$, (4.7) follows in general by continuity.

It follows immediately that if $X = \varphi(\xi_1, \dots, \xi_m)$, then X and $\Gamma(U)X = \varphi(U\xi_1, \dots, U\xi_m)$ have the same distribution, and, with $\psi = |\varphi|$,

$$\Gamma(U)|X| = \Gamma(U)\bigl(\psi(\xi_1, \dots, \xi_m)\bigr) = \psi(U\xi_1, \dots, U\xi_m) = |\Gamma(U)X|.$$

This proves the first two claims for variables of this type. Since such variables are dense in $L^2(\Omega, \mathcal{F}(H), \mathrm{P})$ (letting m and $\xi_1, \dots, \xi_m$ vary), for example by Theorem 2.11 or 2.12, the general result follows by continuity. (Recall that convergence in L^2 implies convergence in distribution.) □

THEOREM 4.12. *Let $A\colon H_1 \to H_2$ be a linear operator between two Gaussian Hilbert spaces, defined on probability spaces $(\Omega_i, \mathcal{F}_i, \mathrm{P}_i)$, such that $\|A\| \le 1$. Then $\Gamma(A)$ can be (uniquely) extended to a continuous linear operator $L^1(\Omega_1, \mathcal{F}(H_1), \mathrm{P}_1) \to L^1(\Omega_2, \mathcal{F}(H_2), \mathrm{P}_2)$, which we also denote by $\Gamma(A)$. Furthermore,*

(i) *$\|\Gamma(A)X\|_p \le \|X\|_p$ for any $X \in L^p$, $1 \le p \le \infty$;*
(ii) *if $X \ge 0$ a.s., then $\Gamma(A)X \ge 0$ a.s.;*
(iii) *$|\Gamma(A)X| \le \Gamma(A)|X|$ a.s. for any $X \in L^1$.*

PROOF. Follows by the decomposition in Theorem 4.10, since the result is true for the three factors separately by Theorems 4.11 and 4.9, using well-known properties of conditional expectations. □

We give an important sharpening of Theorem 4.12(i) in the next chapter, and a sharpening of (ii) in the next section.

REMARK 4.13. A version of Jensen's inequality holds too: If $A\colon H_1 \to H_2$ is as in Theorem 4.12, $\varphi\colon \mathbb{R} \to \mathbb{R}$ is convex and $X \in L^1_{\mathbb{R}}(\Omega_1, \mathcal{F}(H_1), \mathrm{P}_1)$ with $\varphi(X) \in L^1$, then $\varphi(\Gamma(A)X) \le \Gamma(A)(\varphi(X))$ a.s. This is easily proved by observing that $\varphi(x) = \sup_i l_i(x)$ for some countable family of affine functions $l_i(x) = a_i + b_i x$ (take tangents to φ at all rational x); by Theorem 4.12(ii) $l_i(\Gamma(A)X) = \Gamma(A)(l_i(X)) \le \Gamma(A)(\varphi(X))$ a.s. for every i, and the result follows. The result extends easily to convex functions of several variables and to complex variables.

REMARK 4.14. Not even part (i) of Theorem 4.12 is in general valid for a complex linear $A\colon H_{1\mathbb{C}} \to H_{2\mathbb{C}}$ with $\|A\| \le 1$. For example, assuming that the space is not zero-dimensional, the operator $\Gamma(rI)$ with r purely imaginary is bounded on L^p only if

$$|r|^2 \le \begin{cases} p-1, & 1 \le p \le 2, \\ 1/(p-1), & 2 \le p \le \infty; \end{cases} \tag{4.8}$$

this is easily shown by considering its action on $:e^{t\xi}: = e^{t\xi - t^2/2}$, where ξ is standard Gaussian, and computing L^p-norms for real and imaginary t by Corollary 3.38. (This operator is bounded, and in fact a contraction, when (4.8) is satisfied; see Theorem 5.32.)

REMARK 4.15. If $A\colon H_1 \to H_2$, and H_2 is a closed subspace of a Gaussian Hilbert space H_3, let $\iota\colon H_2 \to H_3$ denote the injection. Then, as a very special case of Theorem 4.9, $\Gamma(\iota)X = X$ for every $X \in L^1(\Omega_2, \mathcal{F}(H_2), \mathrm{P}_2)$. Consequently $\Gamma(\iota A)X = \Gamma(\iota)\Gamma(A)X = \Gamma(A)X$ for every $X \in L^1(\Omega_1, \mathcal{F}(H_1), \mathrm{P}_1)$. In other words, $\Gamma(A)$ does not depend on the choice of the range space H_2, as long as it contains $A(H_1)$.

Similarly, $\Gamma(A)(X)$ is not affected by restricting A to a subspace $H' \subseteq H_1$, as long as $X \in L^1(\Omega_1, \mathcal{F}(H'), \mathrm{P}_1)$.

REMARK 4.16. If U is an isometry as in Theorem 4.11, then (4.7) extends to infinite sequences $\xi_1, \xi_2, \dots$ and measurable $\varphi\colon \mathbb{R}^\infty \to \mathbb{C}$ with $\varphi(\xi_1, \xi_2, \dots)$ in L^1. We omit the details.

EXAMPLE 4.17. Let ξ and η be independent standard Gaussian variables, let H be the two-dimensional space spanned by them and let K be the one-dimensional subspace spanned by ξ. The orthogonal projection $P_{HK}\colon H_{\mathbb{C}} \to K_{\mathbb{C}}$ satisfies $P_{HK}(\xi + i\eta) = \xi$, and thus by Theorems 4.5 and 4.9 and Example 3.30,

$$:\xi^n: = \Gamma(P_{HK})\bigl(:(\xi+i\eta)^n:\bigr) = \mathrm{E}\bigl(:(\xi+i\eta)^n: \mid \xi\bigr) = \mathrm{E}\bigl((\xi+i\eta)^n \mid \xi\bigr).$$

By Theorem 3.19, $:\xi^n: = h_n(\xi)$, so this yields another expression for the Hermite polynomials:

$$h_n(x) = \mathrm{E}(x+i\eta)^n, \tag{4.9}$$

for any $n \geq 0$ and $x \in \mathbb{R}$, with η standard normal. (Since both sides of (4.9) are polynomials, the equation holds for every real (or even complex) x, and not just a.e.)

Similarly, for any real or complex t,

$$:e^{t\xi}: = \Gamma(P_{HK})\bigl(:e^{t(\xi+i\eta)}:\bigr) = \mathrm{E}\bigl(e^{t(\xi+i\eta)} \mid \xi\bigr) = e^{t\xi}\,\mathrm{E}\,e^{it\eta} = e^{t\xi - t^2/2},$$

which gives another proof of (3.18).

We can repeat this argument for a complex centred Gaussian ξ and prove (3.17) and Theorem 3.33 by letting η be an independent copy of ξ, H the Gaussian space spanned by $\operatorname{Re}\xi$, $\operatorname{Im}\xi$, $\operatorname{Re}\eta$, $\operatorname{Im}\eta$, and K the subspace spanned by $\operatorname{Re}\xi$, $\operatorname{Im}\xi$. We omit the details.

Taking absolute values in (4.9), we obtain the estimate

$$|h_n(x)| \leq \mathrm{E}\,|x+i\eta|^n \leq \mathrm{E}(|x|+|\eta|)^n, \tag{4.10}$$

again valid for all complex x and $n \geq 0$. (The right hand side can be estimated explicitly, but we will find the version just given convenient.) For example, for any complex t and x,

$$\sum_0^\infty \Bigl|\frac{t^n h_n(x)}{n!}\Bigr| \leq \sum_0^\infty \frac{|t|^n\,\mathrm{E}(|x|+|\eta|)^n}{n!} = \mathrm{E}\,e^{|t|(|x|+|\eta|)} < \infty,$$

which gives another proof of the fact that the sum in (3.19) converges absolutely for all complex t and x; moreover, the argument shows that the convergence is uniform on compact sets.

Similarly, if $0 \le r < \frac{1}{2}$, then

$$\sum_0^\infty \left|\frac{r^n h_{2n}(x)}{n!}\right| \le \sum_0^\infty \frac{r^n \operatorname{E}(|x|+|\eta|)^{2n}}{n!} = \operatorname{E} e^{r(|x|+|\eta|)^2} < \infty.$$

Taking $r = |a/(1-2a)|$ with $\operatorname{Re} a < 1/4$, which implies $|2a| < |1-2a|$ and thus $r < 1/2$, we see that the series (3.20) in Example 3.41 converges for each x (real or complex). Again, we also obtain uniform convergence on compact sets.

EXAMPLE 4.18. Let ξ be a standard Gaussian variable and let H be the one-dimensional Gaussian space spanned by ξ. Any linear operator on H or $H_{\mathbb{C}}$ equals rI for some real or complex r, and we can compute $\Gamma(rI)$ as follows.

First, let r be real, $-1 < r < 1$. Let η be another standard Gaussian variable, independent of ξ, let G be the Gaussian space spanned by ξ and η, and let K be the subspace spanned by $\xi' = r\xi + \sqrt{1-r^2}\eta$. If $U\colon H \to K$ is the linear isometry defined by $U\xi = \xi'$, and P_{KH} is the orthogonal projection onto H, then $P_{KH}U = rI$ on H, and thus $\Gamma(rI) = \Gamma(P_{KH})\Gamma(U)$. It is easily seen that $L^2 = L^2(\Omega, \mathcal{F}(H), \mathrm{P})$ equals $\{f(\xi) : f \in L^2(d\gamma)\}$, where $d\gamma(x) = (2\pi)^{-1/2}e^{-x^2/2}dx$. Moreover, $\Gamma(U)f(\xi) = f(\xi')$ by Theorem 4.11, while $\Gamma(P_{KH}) = \operatorname{E}(\cdot \mid \xi)$ by Theorem 4.9. Hence

$$\Gamma(rI)f(\xi) = \operatorname{E}\big(f(\xi') \mid \xi\big) = \operatorname{E}\big(f(r\xi + \sqrt{1-r^2}\eta) \mid \xi\big) = \mathcal{M}_r f(\xi), \tag{4.11}$$

where

$$\mathcal{M}_r f(x) = \operatorname{E} f(rx + \sqrt{1-r^2}\eta) = \frac{1}{\sqrt{2\pi}}\int_{-\infty}^{\infty} f(rx + \sqrt{1-r^2}t)e^{-t^2/2}dt; \tag{4.12}$$

a change of variable yields

$$\begin{aligned}\mathcal{M}_r f(x) &= \big(2\pi(1-r^2)\big)^{-1/2}\int_{-\infty}^{\infty} f(y)\exp\Big(-\frac{1}{2}\frac{(y-rx)^2}{1-r^2}\Big)\,dy\\ &= \int_{-\infty}^{\infty} f(y)(1-r^2)^{-1/2}\exp\Big(-\frac{r^2x^2+r^2y^2-2rxy}{2(1-r^2)}\Big)d\gamma(y).\end{aligned} \tag{4.13}$$

This integral operator $\mathcal{M}_r$ is known as the *Mehler transform* (Mehler 1866). Equivalently, the Mehler transform is $\Gamma(rI)$ for the special Gaussian space in Example 1.15. By Theorem 4.12, the Mehler transform is bounded, with norm 1, on every $L^p(d\gamma)$, $1 \le p \le \infty$.

Examples 2.9 and 4.7 imply that the Mehler transform is characterized by $\mathcal{M}_r h_n = r^n h_n$. Since $\{h_n\}$ is an orthogonal basis in $L^2(d\gamma)$ with $\|h_n\|_2^2 = n!$, this means that we also have

$$\mathcal{M}_r f(x) = \int_{-\infty}^{\infty} \sum_{n=0}^{\infty} r^n \frac{h_n(x)h_n(y)}{n!} f(y)\, d\gamma(y), \qquad f \in L^2(\gamma), \tag{4.14}$$

with the sum $\sum (r^n h_n(x) h_n(y)/n!)$ converging in $L^2(d\gamma(x)\, d\gamma(y))$. Clearly, the kernels in (4.13) and (4.14) have to coincide, which shows that the sum equals $(1-r^2)^{-1/2} \exp(-(r^2x^2 + r^2y^2 - 2rxy)/2(1-r^2))$ for a.e. x and y. In order to show that the sum converges pointwise, and that equality holds everywhere, we use (4.10) which implies, since ξ and η are independent standard normal variables and $|r| < 1$,

$$\begin{aligned} \sum_0^\infty \left| \frac{r^n h_n(x) h_n(y)}{n!} \right| &\le \sum_0^\infty \frac{|r|^n \operatorname{E}(|x| + |\xi|)^n \operatorname{E}(|y| + |\eta|)^n}{n!} \\ &= \operatorname{E} e^{|r|(|x|+|\xi|)(|y|+|\eta|)} < \infty. \end{aligned} \tag{4.15}$$

Hence the sum converges absolutely for every x and y; it follows also that the convergence is uniform on compact subsets, and thus the sum is a continuous function of x and y. Consequently, the sum equals the kernel in (4.13) everywhere, which gives the Mehler summation formula (Mehler 1866)

$$\sum_{n=0}^{\infty} \frac{r^n}{n!} h_n(x) h_n(y) = (1-r^2)^{-1/2} \exp\Big(-\frac{r^2x^2 + r^2y^2 - 2rxy}{2(1-r^2)}\Big), \qquad -1 < r < 1. \tag{4.16}$$

Turning to complex r, we note that both sides of (4.16) are analytic functions of r for $|r| < 1$; hence (4.16) extends to all complex r with $|r| < 1$. (It also extends to all complex x and y.) Furthermore, (4.14) remains valid as long as $|r| < 1$, and thus also (4.13) holds for complex r with $|r| < 1$.

When $|r| = 1$ but $r \neq \pm 1$, the integrals in (4.13) do not converge for every $f \in L^2(d\gamma)$. If $f(x)$ does not grow too fast as $|x| \to \infty$, however, for example if f is a polynomial or has compact support, then these integrals are continuous functions of r for $|r| \le 1$, $r \neq \pm 1$, and it follows again that (4.13) holds. (Note that by the definition of $\Gamma(rI)$, $r \mapsto \mathcal{M}_r f$ is a continuous map of the closed unit disc into $L^2(\gamma)$ for every $f \in L^2(\gamma)$.)

Finally, it is easily seen that $\mathcal{M}_1 f = f$ and $\mathcal{M}_{-1} f(x) = f(-x)$; these can of course not be expressed as integral operators.

We remark that $\Gamma(rI)$ can be described by an integral operator as in (4.13) for any finite-dimensional Gaussian space by selecting an orthonormal basis and applying the Mehler kernel to each coordinate, i.e. by taking a tensor power of the one-dimensional Mehler transform. We leave the details to the reader, and refer again to Mehler (1866).

This example suggests the following simplified notation.

DEFINITION 4.19. For any Gaussian Hilbert space and any real (or complex) r, let $M_r = \Gamma(rI)$; we call this operator the *Mehler transform*.

Note that $e^{-t\mathcal{N}} = M_{e^{-t}}$; thus the Ornstein–Uhlenbeck semigroup may be written $\{M_{e^{-t}}\}_{t\geq 0}$ or, with the parameter acting multiplicatively, $\{M_r\}_{0<r\leq 1}$.

THEOREM 4.20. *If $1 \leq p \leq \infty$ and $-1 \leq r \leq 1$, then M_r is a contraction in $L^p(\Omega, \mathcal{F}, \mathrm{P})$ for every Gaussian Hilbert space H. Moreover, if $1 \leq p < \infty$ and $X \in L^p(\Omega, \mathcal{F}(H), \mathrm{P})$, then $M_r X \to X$ in L^p as $r \nearrow 1$.*

PROOF. The first part is a special case of Theorem 4.12. The second part follows easily from the facts that the operators M_r, $0 \leq r < 1$, are uniformly bounded on L^p (by the first part) and that $M_r X \to X$ as $r \to 1$ (trivially) if $X \in \mathcal{P}(H)$, which is dense in $L^p(\Omega, \mathcal{F}(H), \mathrm{P})$ by Theorem 2.11. □

The case of complex r will be studied in Section 5.2.

EXAMPLE 4.21. Again, let ξ and η be independent standard normal variables. For any real or complex x,y and r with $|r| < 1$, (4.9) yields by dominated convergence, cf. (4.15),

$$\sum_0^\infty \frac{r^n}{n!} h_n(x) h_n(y) = \sum_0^\infty \frac{r^n}{n!} \mathrm{E}(x + i\xi)^n (y + i\eta)^n = \mathrm{E}\, e^{r(x+i\xi)(y+i\eta)},$$

which gives another expression for the Mehler kernel.

EXAMPLE 4.22. Although the Mehler transform $\mathcal{M}_r$ defined in Example 4.18 is used mainly for real r, when it has the probabilistic interpretation (4.11), it is also of interest for purely imaginary r, when it is closely related to the Fourier transform. We take $r = i$. (Other imaginary values of r are also interesting in connection with the Fourier transform; see Example 5.33.) By Example 4.18, for suitable f,

$$\mathcal{M}_i f(x) = (4\pi)^{-1/2} e^{x^2/4} \int_{-\infty}^{\infty} e^{ixy/2} e^{-y^2/4} f(y)\, dy.$$

In particular, if $f_n(x) = e^{-x^2/4} h_n(x)$,

$$(4\pi)^{-1/2} \int_{-\infty}^{\infty} e^{ixy/2} f_n(y)\, dy = e^{-x^2/4} \mathcal{M}_i h_n(x) = i^n f_n(x).$$

Changing variables, this gives

$$(2\pi)^{-1/2} \int_{-\infty}^{\infty} e^{ixy} f_n(\sqrt{2}y)\, dy = i^n f_n(\sqrt{2}x),$$

or, if we define the Fourier transform by $\mathcal{F}f(x) = (2\pi)^{-1/2} \int_{-\infty}^{\infty} e^{ixy} f(y)\, dy$ (which is unitary on $L^2_{\mathbb{C}}(\mathbb{R}, dx)$),

$$\mathcal{F}\big(h_n(\sqrt{2}x) e^{-x^2/2}\big) = i^n h_n(\sqrt{2}x) e^{-x^2/2}.$$

We have found the classical sequence of eigenfunctions to the Fourier transform; it follows from Example 3.18 that these eigenfunctions form an orthogonal basis for $L^2(\mathbb{R}, dx)$.

EXAMPLE 4.23. Let $(\xi_t)_{t\in T}$ be a centred Gaussian stochastic process on an arbitrary index set T, and let H be the Gaussian Hilbert space spanned by $(\xi_t)_T$; cf. Example 1.12. We argue as in Example 4.18 to obtain an explicit expression for $M_r = \Gamma(rI)$. Let $-1 \le r \le 1$, let $(\eta_t)_{t\in T}$ be an independent copy of $(\xi_t)_{t\in T}$, and define

$$\xi'_t = r\xi_t + \sqrt{1-r^2}\eta_t.$$

Then $(\xi'_t)_{t\in T}$ is also a Gaussian stochastic process with the same distribution as $(\xi_t)_T$, and thus the mapping $U\colon \xi_t \mapsto \xi'_t$ extends to a linear isometry of H onto the Gaussian Hilbert space K spanned by $(\xi'_t)_T$.

Clearly, H and K are subspaces of the Gaussian Hilbert space spanned by $\{\xi_t\}_T \cup \{\eta_t\}_T$, and if P_{KH} is the orthogonal projection onto H, then by Theorem 4.9, $\Gamma(P_{KH})$ equals the conditional expectation given $(\xi_t)_T$. Moreover, $P_{KH}U(\xi_t) = P_{KH}(\xi'_t) = r\xi_t$ for every $t \in T$, and thus $P_{KH}U = rI$ on H. Consequently, for every $X \in L^2(\Omega, \mathcal{F}(H), \mathrm{P})$,

$$M_r X = \Gamma(rI)X = \Gamma(P_{KH})\Gamma(U)X = \mathrm{E}\big(\Gamma(U)X \mid (\xi_t)_T\big).$$

In particular, if $X = p(\xi_t)$ is a polynomial in a finite number of variables $\xi_{t_1}, \dots, \xi_{t_n}$, then

$$\Gamma(rI)p(\xi_t) = \mathrm{E}(p(\xi'_t) \mid (\xi_t)_T) = \mathrm{E}_\eta\, p(\xi'_t) = \mathrm{E}_\eta\, p(r\xi_t + \sqrt{1-r^2}\eta_t). \tag{4.17}$$

We have so far assumed $-1 \le r \le 1$, but the expressions in (4.17) are all polynomials in r; hence (4.17) holds also for complex r. In particular, $\Gamma(iI)$ is given by (4.17) with

$$\xi'_t = i\xi_t + \sqrt{2}\eta_t.$$

Since the set of such polynomial variables $p(\xi_t)$ is dense in $L^2(\Omega, \mathcal{F}, \mathrm{P})$, (4.17) describes $\Gamma(rI)$, in principle. It is easy to extend (4.17) to suitable analytic functions p, for example exponential functions, but note that the formula does not make sense literally for general functionals p and complex r.

3. Absolute continuity

It can be seen from the results and examples above that operators of the type $\Gamma(A)$ tend to smooth random variables. This is true in at least two different ways; the norm estimates in Theorem 4.12 and (stronger) Chapter 5 below show that large values become less probable, while we in this section show that, except in extreme cases, variables of the form $\Gamma(A)X$ have distributions that are absolutely continuous. (A related but different result on absolute continuity is given in Section 15.4.)

The results in this section are used in quantum theory (Gross 1974, Sloan 1974), but the section can be skipped at the first reading.

THEOREM 4.24. *Let H_1 and H_2 be two Gaussian Hilbert spaces and suppose that $A\colon H_1 \to H_2$ is a linear operator such that $\|A\xi\| < \|\xi\|$ for every $\xi \in H_1$ with $\xi \neq 0$. Let $X \in L^1_{\mathbb{R}}(\Omega_1, \mathcal{F}(H_1), \mathrm{P}_1)$. Then the random variable $\Gamma(A)(X)$ either is a.s. constant or has an absolutely continuous distribution.*

In order to prove this, we will use two lemmas. In this section we let λ denote the Lebesgue measure on $\mathbb{R}^d$ (for any $d \geq 1$), and say that a subset of $\mathbb{R}^d$ is a *null set* if it has Lebesgue measure 0. Recall that a random variable X has an absolutely continuous distribution if and only if $\mathrm{P}(X \in E) = 0$ for every null set $E \subset \mathbb{R}$. (By the Radon–Nikodym theorem, this happens if and only if the distribution has a density function.)

LEMMA 4.25. *Let X be a random variable such that for each null set $E \subset \mathbb{R}$, $\mathrm{P}(X \in E)$ is either 0 or 1. Then X either is a.s. constant or has an absolutely continuous distribution.*

PROOF. If $\mathrm{P}(X \in E) = 0$ for each null set E, then the distribution of X is absolutely continuous by definition. Otherwise, by assumption, there exists a null set E such that $\mathrm{P}(X \in E) = 1$. Let $E_n = E \cap [n, n+1]$, $-\infty < n < \infty$. Then $\mathrm{P}(X \in E_n)$ must be non-zero for at least one n, but since E_n is a null set this implies by assumption $\mathrm{P}(X \in E_n) = 1$. Similarly, if the interval $I_1 = [n, n+1]$ is split into two closed intervals of length $1/2$, we can take I_2 as one of these halves with $\mathrm{P}(X \in E \cap I_2) > 0$ and thus $\mathrm{P}(X \in E \cap I_2) = 1$. Continuing by successive splittings, we find a nested sequence $\{I_k\}$ of closed intervals with lengths $\lambda(I_k) = 2^{1-k}$ and $\mathrm{P}(X \in E \cap I_k) = 1$. The intersection $\bigcap_1^\infty I_k$ consists of a single point x by Cantor's nested set theorem, and $\mathrm{P}(X \neq x) \leq \sum_{k=1}^\infty \mathrm{P}(X \notin I_k) = 0$; hence $X = x$ a.s. □

LEMMA 4.26. *Let $f\colon \mathbb{R}^d \to \mathbb{R}$ be a real analytic function and let $E \subseteq \mathbb{R}$ be a null set. Then the set $f^{-1}(E) = \{x \in \mathbb{R}^d : f(x) \in E\}$ is either $\mathbb{R}^d$ or a null set in $\mathbb{R}^d$.*

PROOF. We employ induction on d.

First, let $d = 1$. The conclusion is clear if f is constant. If not, the set $Z = \{x : f'(x) = 0\}$ is countable. If I is a compact interval in the complement of Z, then $f' \geq c$ or $f' \leq -c$ on I for some $c > 0$. Hence f is injective on I with a Lipschitz continuous inverse, and it follows that $\lambda(\{x \in I : f(x) \in E\}) = 0$. Since the complement of Z can be covered by a countable family of such intervals, and $\lambda(Z) = 0$, $\lambda(f^{-1}(E)) = 0$.

Now assume that $d \geq 2$, and that the lemma holds for $d-1$. If $f^{-1}(E) \neq \mathbb{R}^d$, then there exists $x' \in \mathbb{R}^{d-1}$ and $x'' \in \mathbb{R}$ such that $f(x', x'') \notin E$. The induction hypothesis, applied to the real analytic function $y \mapsto f(y, x'')$ on $\mathbb{R}^{d-1}$, shows that $f(y, x'') \notin E$ for a.e. $y \in \mathbb{R}^{d-1}$. But for each $y \in \mathbb{R}^{d-1}$ with $f(y, x'') \notin E$, the case $d = 1$ similarly shows $f(y, z) \notin E$ for a.e. $z \in \mathbb{R}$. Fubini's theorem yields $f(y, z) \notin E$ for a.e. $(y, z) \in \mathbb{R}^d$, which completes the proof. □

PROOF OF THEOREM 4.24. We will use the spectral theory and functional calculus for self-adjoint operators in Hilbert spaces; see e.g. Conway (1990, Chapter IX). (The results in this reference, as in several others, are stated for complex Hilbert spaces only, but the functional calculus is valid also for a self-adjoint operator A in a real Hilbert space H and real Borel functions f; a simple way to see this is to extend the operator to a complex linear operator on the complexification $H_{\mathbb{C}}$, use the complex theory, and verify that then $f(A)$ maps H into H for real f.)

We begin with some simplifications. (We use Remark 4.15 repeatedly.)

First, the random variable X is measurable with respect to $\mathcal{F}(H_1)$ and thus also with respect to $\mathcal{F}(S)$ for some countable subset S of H_1. Hence, replacing H_1 by the closed linear span of S, we may assume that H_1 is separable.

Secondly, we use the polar decomposition $A = U|A|$, where $|A| = (A^*A)^{1/2}$ is a self-adjoint operator in H_1, and U is an isometry of $\overline{\mathrm{Ran}(|A|)} \subseteq H_1$ onto $\overline{\mathrm{Ran}(A)} \subseteq H_2$ (cf. e.g. Conway 1990, VIII.3.11). ($\mathrm{Ran}(A)$ denotes the range of the operator A.) Since $\Gamma(A)X = \Gamma(U)\Gamma(|A|)X$, and $\Gamma(U)$ does not alter the distribution by Theorem 4.11, it suffices to prove the result for $|A|$. Note that $\| |A|(\xi)\|_{H_1} = \|A(\xi)\|_{H_2}$ for every $\xi \in H_1$; hence the condition in the theorem is satisfied for $|A|$ too.

Consequently, we may assume that A is a positive self-adjoint operator in a separable Gaussian Hilbert space H. The spectrum of A, then, is contained in $[0,1]$, and by assumption, there is no eigenvector with eigenvalue 1.

Define two functions $f, g\colon [0,1] \to [0,1]$ by $f(t) = 1-2^{-n}$ on $[1-2^{-n+1}, 1-2^{-n})$, $n \geq 1$, and $f(1) = 1$, and $g(t) = t/f(t)$. Using the functional calculus for self-adjoint operators in a Hilbert space (Conway 1990, IX.2.3), we define the operators $B = f(A)$ and $C = g(A)$; both have norms ≤ 1 and $BC = fg(A) = A$ and thus $\Gamma(A)X = \Gamma(B)\Gamma(C)X$.

B has a discrete spectrum and there exists an orthonormal basis $\{\xi_k\}_1^\infty$ in H of eigenvectors to B; $B\xi_k = r_k\xi_k$, where $r_k \in \{1-2^{-n} : n \geq 1\}$ and thus $0 < r_k < 1$. (If E is the spectral measure for A, then $\mathrm{Ran}(E([1-2^{-n+1}, 1-2^{-n})))$ is an eigenspace of B with eigenvalue $1-2^{-n}$. These subspaces span H, since $E\{1\} = 0$ by the assumption on A.)

Define, for $n \geq 1$, two bounded operators on H by

$$B_n\xi_k = \begin{cases} r_k\xi_k, & k \leq n, \\ \xi_k, & k > n, \end{cases}$$

$$D_n\xi_k = \begin{cases} \xi_k, & k \leq n, \\ r_k\xi_k, & k > n. \end{cases}$$

Thus $B = B_nD_n$, and $\|B_n\|, \|D_n\| \leq 1$; hence $\Gamma(A)X = \Gamma(B)\Gamma(C)X = \Gamma(B_n)Y_n$, with $Y_n = \Gamma(D_n)\Gamma(C)X \in L^1$. Moreover, Y_n is measurable with respect to $\mathcal{F}(H) = \mathcal{F}(\xi_1, \xi_2, \dots)$, and thus $Y_n = f_n(\xi_1, \dots, \xi_n, \widetilde{\xi})$ for some Borel function f_n on $\mathbb{R}^\infty$, where we for simplicity write $\widetilde{\xi} = (\xi_{n+1}, \xi_{n+2}, \dots)$.

Since the variables ξ_k are orthogonal, they are independent. Letting $\{\eta_k\}_1^n$ be a further set of independent standard normal variables, it follows as in Example 4.23 that

$$\begin{aligned}\Gamma(A)X &= \Gamma(B_n)Y_n \\ &= \mathrm{E}\big(f_n(r_1\xi_1 + \sqrt{1-r_1^2}\eta_1, \dots, r_n\xi_n + \sqrt{1-r_n^2}\eta_n, \widetilde{\xi}) \mid \{\xi_k\}_1^\infty\big) \\ &= g_n(\xi_1, \dots, \xi_n, \widetilde{\xi})\end{aligned}$$

with, by calculations as in (4.12) and (4.13), letting $d\gamma = \frac{1}{\sqrt{2\pi}}e^{-x^2/2}dx$,

$$g_n(x_1, \dots, x_n, \widetilde{\xi}) = \int \cdots \int \prod_{i=1}^n (1-r_i^2)^{-1/2} \exp\Big(-\frac{r_i^2x_i^2 + r_i^2y_i^2 - 2r_ix_iy_i}{2(1-r_i^2)}\Big) \\ \times f_n(y_1, \dots, y_n, \widetilde{\xi})\, d\gamma(y_1) \cdots d\gamma(y_n). \quad (4.18)$$

Let E be a null set in $\mathbb{R}$. For a.e. $\widetilde{\xi}$, the function $f_n(x_1, \dots, x_n, \widetilde{\xi}) \in L^1(\gamma^n)$, and then (4.18) defines g_n as an analytic function of $(x_1, \dots, x_n) \in \mathbb{C}^n$. Hence g_n is a real analytic function of $(x_1, \dots, x_n)$, and Lemma 4.26 shows that, for a.e. $\widetilde{\xi}$, $g_n(x_1, \dots, x_n, \widetilde{\xi}) \in E$ either for all $(x_1, \dots, x_n)$ or for a null set only. It follows that the event $\{\Gamma(A)X \in E\} = \{g_n(\xi_1, \dots, \xi_n, \widetilde{\xi}) \in E\}$ is (stochastically) independent of $\{\xi_1, \dots, \xi_n\}$.

Letting $n \to \infty$, we see by Kolmogorov's 0–1 law that the event $\{\Gamma(A)X \in E\}$ has probability 0 or 1. This holds for any null set $E \subseteq \mathbb{R}$, and the proof is completed by Lemma 4.25. □

REMARK 4.27. The condition $\|A\xi\| < \|\xi\|$, $\xi \neq 0$, in the theorem is also necessary; if $\|A\xi\| = \|\xi\|$ for some $\xi \neq 0$, then $\Gamma(A)(\varphi(\xi)) \stackrel{d}{=} \varphi(\xi)$ for any function φ with $\varphi(\xi) \in L^1$, and this variable may have any distribution.

REMARK 4.28. Theorem 4.24 shows that the random variable $\Gamma(A)X$ has a density function, unless it is degenerate. The density function does not have to be bounded or continuous, however; this is shown by the example $X = {:}\xi^2{:} = \xi^2 - 1$ and $A = rI$, with $\xi \in H$ standard normal and $0 < r < 1$, which yields $\Gamma(A)X = r^2X$. (Recall that ξ^2 has a $\chi^2(1)$-distribution, which has the density function $(2\pi x)^{-1/2}e^{-x/2}\mathbf{1}[x > 0]$; the density functions of X and $\Gamma(A)X$ are simple transformations of this.)

COROLLARY 4.29. *Suppose that A is as in Theorem 4.24. If $X \geq 0$ a.s. and $\mathrm{P}(X \neq 0) > 0$, then $\Gamma(A)X > 0$ a.s.*

PROOF. By Theorem 4.12, $\Gamma(A)X \geq 0$ a.s., while Theorem 4.24 implies that $\mathrm{P}(\Gamma(A)X = 0)$ equals 0 or 1. The latter possibility is ruled out because $\mathrm{E}(\Gamma(A)X) = \mathrm{E}\,X > 0$. □

REMARK 4.30. This property is often expressed by saying that $\Gamma(A)$ is *positivity improving*.

V

Hypercontractivity

Let H be a Gaussian Hilbert space. Theorem 4.12 shows that if $A: H \to H$ is a contraction, then $\Gamma(A)$ is a contraction in each L^p, $1 \le p \le \infty$. If A is a strict contraction, i.e. $\|A\| < 1$, and $1 < p < \infty$, then a substantially stronger property, known as *hypercontractivity*, holds, viz. $\Gamma(A)$ is a contraction of L^p into L^q for some $q > p$.

This theorem has despite its technical appearance many important applications. We give a proof in Section 1, followed by some applications. Further applications are given in Chapters 12, 15 and 16.

In Section 2 we give similar results on hypercontractivity for complex operators. This case is not as important as the real case; nevertheless, these results too will be used in the sequel, in particular in Chapter 15.

Section 3 contains applications of hypercontractivity to the study of a special class of operators on L^p for a Gaussian Hilbert space. The results will be used in Chapter 15.

1. Real hypercontractivity

In this section we consider hypercontractivity for a real operator A. (On the other hand, it does not matter whether the L^p-spaces where $\Gamma(A)$ operates are real or complex.) The precise result is as follows. (Let $0/0 = \infty/\infty = 1$.)

THEOREM 5.1. *Let H_1 and H_2 be Gaussian Hilbert spaces defined on probability spaces $(\Omega_1, \mathcal{F}_1, \mathrm{P}_1)$ and $(\Omega_2, \mathcal{F}_2, \mathrm{P}_2)$. If $1 \le p \le q \le \infty$ and $A: H_1 \to H_2$ is a linear map with norm $\|A\| \le \left(\frac{p-1}{q-1}\right)^{1/2}$, then $\Gamma(A)$ maps $L^p(\Omega_1, \mathcal{F}(H_1), \mathrm{P}_1)$ into $L^q(\Omega_2, \mathcal{F}(H_2), \mathrm{P}_2)$ with norm $\|\Gamma(A)\|_{p,q} = 1$.*

PROOF. This was first proved by Nelson (1973). Several different proofs, using widely different methods, have appeared, see Beckner (1975), Gross (1975), Brascamp and Lieb (1976), Neveu (1976), Simon (1976), Adams and Clarke (1979), Bakry (1994) and the survey Gross (1993). We here present the elementary proof by Beckner (1975) (slightly reformulated in more probabilistic language, cf. also Janson (1983)). Another proof (Neveu 1976), which uses stochastic integration, will be given in Example 7.15.

By the factorization $\Gamma(A) = \Gamma(\|A\|I)\Gamma(A/\|A\|)$ and Theorem 4.12, it suffices to consider the case where $H_1 = H_2$ and $A = rI$ with $0 \le r \le \left(\frac{p-1}{q-1}\right)^{1/2}$.

The main idea is to consider analogous operators for some non-Gaussian variables, and to build up the Gaussian case from simpler cases. We introduce (for this proof only) some definitions. We assume that $1 \le p \le q < \infty$ (the case $q = \infty$ being trivial), and that r is real.

Let $\mathcal{X} = \{X_1, \dots, X_m\}$ be a (finite) set of random variables, defined on some probability space. We assume that all moments are finite. Let $\mathcal{P}(\mathcal{X})$ be the space of all polynomials in these variables, and let $\mathcal{P}_n(\mathcal{X})$ be the subspace of polynomials of degree at most $n \ge 0$. (Let $\mathcal{P}_{-1}(\mathcal{X}) = \{0\}$.) Define the operator $T_{\mathcal{X},r}$ on $\mathcal{P}(\mathcal{X})$ by

$$T_{\mathcal{X},r}\Bigl(\sum_n Y_n\Bigr) = \sum_n r^n Y_n, \qquad Y_n \in \mathcal{P}_n(\mathcal{X}) \cap \mathcal{P}_{n-1}(\mathcal{X})^{\perp}.$$

We say that the set $\{X_1, \dots, X_m\}$ is (p, q, r)-hypercontractive, if

$$\|T_{\mathcal{X},r}(Y)\|_q \le \|Y\|_p, \qquad Y \in \mathcal{P}(\mathcal{X}).$$

For simplicity, we will say that a variable X is (p, q, r)-hypercontractive when the set $\{X\}$ is.

Note that if the variables in $\mathcal{X}$ have a joint Gaussian distribution, then $\mathcal{P}$ and $\mathcal{P}_n$ are as defined in Chapter 2 for the Gaussian Hilbert space spanned by $\mathcal{X}$, and $T_{\mathcal{X},r} = \Gamma(rI)$; hence we want to show that any such set is (p, q, r)-hypercontractive.

We begin with a simpler case, first proved by Bonami (1970).

LEMMA 5.2. *Let X be a random variable with the symmetric two-point distribution* $\mathrm{P}(X = 1) = \mathrm{P}(X = -1) = 1/2$. *If $r^2 \le (p-1)/(q-1)$, then X is (p, q, r)-hypercontractive.*

PROOF. The space $\mathcal{P}(X)$ of polynomial variables is two-dimensional and spanned by 1 and X, with $T_{X,r}\colon a + bX \mapsto a + brX$. Hence the claim can be written

$$\left(\frac{|a + rb|^q + |a - rb|^q}{2}\right)^{1/q} \le \left(\frac{|a + b|^p + |a - b|^p}{2}\right)^{1/p} \tag{5.1}$$

(this is known as the *two-point inequality*). In order to prove (5.1), let us first observe that if $U\colon a + bX \mapsto a - bX$, then $T_{X,r} = \frac{1}{2}\bigl(1 + r + (1-r)U\bigr)$, which implies

$$|T_{X,r}(f(X))| \le \frac{1+r}{2}|f(X)| + \frac{1-r}{2}|U(f(X))| = T_{X,r}(|f(X)|).$$

Hence it suffices to consider the case when $a + bX \ge 0$, i.e. $a \ge 0$ and $-a \le b \le a$. By homogeneity we may further assume $a = 1$ and, by continuity, $|b| < 1$. Using the binomial series expansion, (5.1) is then equivalent to

$$\left(\sum_{k=0}^{\infty} \binom{q}{2k} r^{2k} b^{2k}\right)^{1/q} \le \left(\sum_{k=0}^{\infty} \binom{p}{2k} b^{2k}\right)^{1/p}. \tag{5.2}$$

In the case $1 \leq p \leq q \leq 2$, all binomial coefficients in (5.2) are non-negative and it is easily verified that for every $k \geq 1$

$$\frac{p}{q}\binom{q}{2k}r^{2k} \leq \frac{p}{q}\left(\frac{p-1}{q-1}\right)^k \binom{q}{2k} \leq \binom{p}{2k};$$

moreover, $(1+x)^\lambda \leq 1 + \lambda x$ for $x \geq 0$ and $0 \leq \lambda \leq 1$, and thus

$$\left(1 + \sum_{k=1}^{\infty} \binom{q}{2k} r^{2k} b^{2k}\right)^{p/q} \leq 1 + \frac{p}{q}\sum_{k=1}^{\infty}\binom{q}{2k} r^{2k} b^{2k} \leq \sum_{k=0}^{\infty}\binom{p}{2k} b^{2k},$$

which yields (5.2).

This completes the proof of Lemma 5.2 for $1 \leq p \leq q \leq 2$. By duality we obtain the result for $2 \leq p \leq q \leq \infty$, since $T_{X,r}$ is self-adjoint. The case $p \leq 2 \leq q$ follows from the cases $p \leq q = 2$ and $2 = p \leq q$, using the factorization $T_{X,r_1 r_2} = T_{X,r_1} T_{X,r_2}$. □

We now use another lemma to build up other distributions.

LEMMA 5.3. *Suppose that* $\mathcal{X} = \{X_1, \ldots, X_m\}$ *and* $\mathcal{Y} = \{Y_1, \ldots, Y_n\}$ *are two independent* (p, q, r)*-hypercontractive sets of random variables. Then their union* $\{X_1, \ldots, X_m, Y_1, \ldots, Y_n\}$ *is* (p, q, r)*-hypercontractive.*

PROOF. It is easily seen that every polynomial variable in $\mathcal{P}(\mathcal{X} \cup \mathcal{Y})$ can be written (non-uniquely) as a finite sum $\sum_i f_i(\mathcal{X}) g_i(\mathcal{Y})$, where f_i and g_i are polynomials, and that

$$T_{\mathcal{X}\cup\mathcal{Y},r}\Big(\sum_i f_i(\mathcal{X}) g_i(\mathcal{Y})\Big) = \sum_i T_{\mathcal{X},r} f_i(\mathcal{X}) T_{\mathcal{Y},r} g_i(\mathcal{Y}).$$

Let μ_X and μ_Y denote the distributions of $\mathcal{X}$ and $\mathcal{Y}$. (These are probability measures on $\mathbb{R}^m$ and $\mathbb{R}^n$.) By the assumption that $\mathcal{X}$ and $\mathcal{Y}$ are (p, q, r)-hypercontractive, and the Minkowski type inequality Proposition C.4 (using $p \leq q$),

$$\begin{aligned}
&\|\sum_i T_{\mathcal{X},r} f_i(\mathcal{X}) T_{\mathcal{Y},r} g_i(\mathcal{Y})\|_q = \|\,\|T_{\mathcal{Y},r}\Big(\sum_i T_{\mathcal{X},r} f_i(x) g_i(y)\Big)\|_{L^q(\mu_Y)}\|_{L^q(\mu_X)} \\
&\leq \|\,\|\sum_i T_{\mathcal{X},r} f_i(x) g_i(y)\|_{L^p(\mu_Y)}\|_{L^q(\mu_X)} \leq \|\,\|\sum_i T_{\mathcal{X},r} f_i(x) g_i(y)\|_{L^q(\mu_X)}\|_{L^p(\mu_Y)} \\
&= \|\,\|T_{\mathcal{X},r}\Big(\sum_i g_i(y) f_i(x)\Big)\|_{L^q(\mu_X)}\|_{L^p(\mu_Y)} \leq \|\,\|\sum_i g_i(y) f_i(x)\|_{L^p(\mu_X)}\|_{L^p(\mu_Y)} \\
&= \|\sum_i f_i(\mathcal{X}) g_i(\mathcal{Y})\|_p,
\end{aligned}$$

which completes the proof of the lemma. □

We continue with the proof of Theorem 5.1, fixing p, q and r with $r^2 \leq (p-1)/(q-1)$. Let $X_1, \ldots, X_m$ be independent random variables with the symmetric two-point distribution in Lemma 5.2, and define $S_m = \sum_{i=1}^m X_i$. By Lemmas 5.2 and 5.3, the set $\{X_1, \dots, X_m\}$ is (p,q,r)-hypercontractive.

A polynomial in S_m is a polynomial in $X_1, \dots, X_m$ of the same degree, so $\mathcal{P}_n(S_m) \subset \mathcal{P}_n(X_1, \dots, X_m)$. Moreover, it is easily seen that $\mathcal{P}_n(S_m)$ consists of all symmetric polynomials in $\mathcal{P}_n(X_1, \dots, X_m)$, and thus that if $X \in \mathcal{P}(S_m)$ is orthogonal to $\mathcal{P}_n(S_m)$, then it is also orthogonal to the larger space $\mathcal{P}_n(X_1, \dots, X_m)$. Hence $\mathcal{P}_n(S_m) \cap \mathcal{P}_{n-1}(S_m)^\perp \subseteq \mathcal{P}_n(X_1, \dots, X_m) \cap \mathcal{P}_{n-1}(X_1, \dots, X_m)^\perp$ and $T_{S_m,r}(Y) = T_{\{X_1,\dots,X_m\},r}(Y)$ when $Y \in \mathcal{P}(S_m)$. Consequently the variable S_m is (p,q,r)-hypercontractive.

Neither the spaces $\mathcal{P}_n$ nor the operator T will be affected if we replace the variable S_m by $\tilde{S}_m = S_m/\sqrt{m}$; hence $\tilde{S}_m$ is also (p,q,r)-hypercontractive.

Note that $T_{\tilde{S}_m,r}$ is given by

$$T_{\tilde{S}_m,r}\Big(\sum_{k=0}^{m} a_k p_{m,k}(\tilde{S}_m)\Big) = \sum_{k=0}^{m} a_k r^k p_{m,k}(\tilde{S}_m),$$

where $p_{m,k}$, $k = 0, \dots, m$, are the orthogonal polynomials for the distribution of $\tilde{S}_m$. (Up to a change of variable, these are the orthogonal polynomials for the binomial distribution, i.e. the Krawtchouk polynomials.) Now, let $m \to \infty$. By the central limit theorem, $\tilde{S}_m$ converges in distribution to a standard normal variable ξ, with convergence of all moments. The orthogonal polynomials may be constructed from the sequence $\{1, x, x^2, \dots\}$ by the Gram–Schmidt procedure, and it follows that (for each k, and with suitable normalizations) $p_{m,k}(x) \to h_k(x)$ as $m \to \infty$, where h_k as before denotes the orthogonal polynomials for the standard normal distribution, i.e. the Hermite polynomials. It follows that for any finite sequence $a_0, \dots, a_l$ holds $\sum_k a_k p_{m,k}(\tilde{S}_m) \to \sum_k a_k h_k(\xi)$ in distribution and in every L^p. Consequently, using the (p,q,r)-hypercontractivity of each $\tilde{S}_m$,

$$\begin{aligned}\|T_{\xi,r}\Big(\sum_k a_k h_k(\xi)\Big)\|_q &= \|\sum_k a_k r^k h_k(\xi)\|_q = \lim_{m\to\infty} \|T_{\tilde{S}_m,r}\Big(\sum_k a_k p_{m,k}(\tilde{S}_m)\Big)\|_q \\ &\le \lim_{m\to\infty} \|\sum_k a_k p_{m,k}(\tilde{S}_m)\|_p = \|\sum_k a_k h_k(\xi)\|_p,\end{aligned}$$

which proves that ξ is (p,q,r)-hypercontractive.

Lemma 5.3 now shows that any finite set of independent standard normal variables is (p,q,r)-hypercontractive. If $\{\xi_1, \dots, \xi_m\}$ is any finite set of centred variables with a joint normal distribution, let $\eta_1, \dots, \eta_l$ be an orthogonal basis in the Gaussian space spanned by $\xi_1, \dots, \xi_m$. Then $\eta_1, \dots, \eta_l$ are independent standard normal and they define the same $\mathcal{P}_n$ and T_r as $\xi_1, \dots, \xi_m$. Hence also $\{\xi_1, \dots, \xi_m\}$ is (p,q,r)-hypercontractive.

Finally, if H is a Gaussian Hilbert space and $X \in \mathcal{P}_n(H)$, then $X = f(\xi_1, \dots, \xi_m)$ for some polynomial f and Gaussian variables $\xi_i \in H$, and

$\Gamma(rI)X = T_{\{\xi_1,\dots,\xi_m\},r}X$. Hence,

$$\|\Gamma(rI)X\|_q \le \|X\|_p.$$

Since the set of such variables is dense in L^p by Theorem 2.11, this completes the proof of Theorem 5.1. □

REMARK 5.4. The condition is sharp; if $\|A\| > ((p-1)/(q-1))^{1/2}$, then $\Gamma(A)$ does not even map L^p into L^q.

In fact, if $\Gamma(A)$ maps L^p into L^q, then it is a bounded map by the closed graph theorem. Since $\Gamma(A)(:e^{t\xi}:) = :e^{tA\xi}:$ for any $\xi \in H_1$ and real t by Example 4.8, this yields, computing the norms by Corollary 3.38, for some $C < \infty$,

$$\exp\Bigl(\frac{q-1}{2}t^2\,\mathrm{E}(A\xi)^2\Bigr) \le C\exp\Bigl(\frac{p-1}{2}t^2\,\mathrm{E}\,\xi^2\Bigr).$$

Letting $t \to \infty$ then yields

$$(q-1)\,\mathrm{E}(A\xi)^2 \le (p-1)\,\mathrm{E}\,\xi^2, \qquad \xi \in H_1,$$

and thus $\|A\|^2 \le (p-1)/(q-1)$.

Alternatively, it suffices to consider one-dimensional spaces H_1 and H_2; we may then assume that $H_1 = H_2$ is the space studied in Example 1.15, and the formulae in Example 4.18 imply easily that if $r^2 > (p-1)/(q-1)$, then $\Gamma(rI)(e^{ax^2}) \notin L^q$ for some a such that $e^{ax^2} \in L^p$. We omit the details.

REMARK 5.5. It is essential for Theorem 5.1 that A is a real operator, i.e. that it maps H_1 into H_2. Hypercontractivity holds also for complex linear operators $A\colon H_{1\mathbf{C}} \to H_{2\mathbf{C}}$ under suitable conditions, but the conditions are more complicated, see Section 2 below.

REMARK 5.6. There is a converse inequality for positive variables in the case $0 < q < p < 1$, see Borell (1982) and Borell and Janson (1982).

We give a reformulation of Theorem 5.1 for the one-dimensional case.

COROLLARY 5.7. *Let ξ and η be two jointly normal variables with correlation ρ. Then, for any $p, q \ge 1$ with $(p-1)(q-1) \ge \rho^2$, and any measurable functions f and g,*

$$\mathrm{E}\,|f(\xi)g(\eta)| \le \|f(\xi)\|_p\|g(\eta)\|_q.$$

PROOF. Let H and K be the one-dimensional Gaussian spaces spanned by $\xi - \mathrm{E}\,\xi$ and $\eta - \mathrm{E}\,\eta$, respectively, and let as in Theorem 4.9 P_{HK} be the restriction to H of the orthogonal projection onto K; then $\|P_{HK}\| = |\rho|$. By Theorem 4.9, $\Gamma(P_{HK})$ then equals the conditional expectation given $\mathcal{F}(K) = \mathcal{F}(\eta)$.

If q' is the conjugate exponent of q, then $q' - 1 = (q-1)^{-1}$, and thus Theorem 5.1 shows that $\Gamma(P_{HK})$ is a contraction of L^p into $L^{q'}$. Hence, for all positive $X \in L^2(\mathcal{F}(H))$ and $Y \in L^2(\mathcal{F}(K))$,

$$\mathrm{E}(XY) = \mathrm{E}\bigl(\Gamma(P_{HK})(X)Y\bigr) \le \|\Gamma(P_{HK})(X)\|_{q'}\|Y\|_q \le \|X\|_p\|Y\|_q.$$

The result follows by taking $X = |f(\xi)|$, $Y = |g(\eta)|$, provided these belong to L^2; the general case follows by truncation and monotone convergence. □

Conversely, Theorem 5.1 follows easily from the corollary, applying Lemma 5.3 as at the end of the proof of the theorem above.

The most important case of Theorem 5.1 is $A = rI$, with $r \geq 0$. (This is the only case treated in many references; as shown in the proof above, it easily implies the general case.) We state this case separately; recall that the Mehler transform $M_r = \Gamma(rI)$.

THEOREM 5.8. *If $1 \leq p \leq q \leq \infty$ and $0 \leq r^2 \leq (p-1)/(q-1)$, then M_r maps $L^p(\Omega, \mathcal{F}(H), \mathrm{P})$ into $L^q(\Omega, \mathcal{F}(H), \mathrm{P})$ and has norm 1.* □

REMARK 5.9. Theorem 5.8 shows that $\|M_{(p-1)^{-1/2}}X\|_p$ is a non-increasing function of $p > 1$ (when defined). Taking the derivative at $p = 2$, one obtains after some calculations

$$\mathrm{E}\,|X|^2 \ln |X| \leq \|X\|_2^2 \ln \|X\|_2 + \langle NX, X\rangle \tag{5.3}$$

for suitable X, e.g. $X \in \overline{\mathcal{P}}_*$; it follows easily using Fatou's lemma that (5.3) holds whenever $\langle NX, X\rangle = \sum_1^\infty n\|\pi_n X\|^2 < \infty$. (The set of such X is the Sobolev space $\mathcal{D}^{1,2}$ defined in Chapter 15.) Conversely, Theorem 5.8 (and thus also Theorem 5.1) can be derived from (5.3). For details and other applications of (5.3), see Gross (1975, 1993). (The inequality (5.3) is known as a *logarithmic Sobolev inequality*.)

We give several applications of Theorem 5.1 in this and later chapters. The first is another proof of Theorem 3.50, with a somewhat better constant.

THEOREM 5.10. *For every $p, q < \infty$, there exists $c(p,q) < \infty$ such that if $n \geq 0$ then*

$$\|X\|_q \leq c(p,q)^n \|X\|_p, \qquad X \in \overline{\mathcal{P}}_n(H). \tag{5.4}$$

PROOF. First assume $2 \leq p \leq q < \infty$, and let $X = \sum_{k=0}^n X_k$, with $X_k \in H^{:k:}$. Then $X = \Gamma\big((q-1)^{-1/2}I\big)\sum_0^n (q-1)^{k/2}X_k$ and Theorem 5.1 yields

$$\begin{aligned}\|X\|_q &\leq \Big\|\sum_0^n (q-1)^{k/2}X_k\Big\|_2 = \Big(\sum_0^n (q-1)^k \|X_k\|_2^2\Big)^{1/2} \\ &\leq (q-1)^{n/2}\|X\|_2 \leq (q-1)^{n/2}\|X\|_p,\end{aligned}$$

which proves (5.4) with $c(p,q) = (q-1)^{1/2}$.

If $0 < p < 2 \leq q$, we use the already proved case and Hölder's inequality as in the proof of Theorem 3.50, which yields (5.4) with

$$c(p,q) = c(q,r)^{(1/p-1/q)/(1/q-1/r)} \tag{5.5}$$

for any $r > q$. The remaining cases of (5.4) follow trivially. □

REMARK 5.11. The proof gives the value $c(2,q) = (q-1)^{1/2}$, $q \geq 2$, and it follows from Theorem 5.19 below that this is the best possible value. We do not know the best value of $c(p,q)$ in any other case with $p < q$.

REMARK 5.12. Even in the case $p = 2$, it is not necessarily the case that the best possible constant in (5.4) for a fixed n equals $c(2,q)^n = (q-1)^{n/2}$. If we consider $X \in H^{:n:}$ only, and let q be an even integer, the best constant will be found by a different method in Corollary 7.36; the ratio between the two constants is $\asymp n^{-1/4}$ for fixed $q \geq 4$. (It is possible to give another proof of Theorem 5.10 using Corollary 7.36 instead of Theorem 5.1. We leave the details to the interested reader.)

REMARK 5.13. If $p \leq 2 = q$, we obtain from (5.5),

$$c(p,2) \leq \lim_{r \searrow 2} (r-1)^{(1/p-1/2)/2(1/2-1/r)} = \exp(\tfrac{2}{p} - 1).$$

In particular, we can take $c(1,2) = e$.

We continue with other applications of Theorem 5.1.

THEOREM 5.14. *The projections π_n and $\pi_{\leq n} = \sum_{m=0}^{n} \pi_m$ are (extend to) bounded operators $\pi_n \colon L^p(\Omega, \mathcal{F}, \mathrm{P}) \to H^{:n:}$ and $\pi_{\leq n} \colon L^p(\Omega, \mathcal{F}, \mathrm{P}) \to \overline{\mathcal{P}}_n(H)$ for each $n \geq 0$ and each finite $p > 1$.*

REMARK 5.15. By Theorem 3.50, it does not matter whether $H^{:n:}$ and $\overline{\mathcal{P}}_n(H)$ are given the L^2-norm (as usual) or the L^p-norm here.

PROOF. It suffices to consider the case $1 < p \leq 2$. Let $X \in \overline{\mathcal{P}}_*(H)$ and let $X_k = \pi_k(X)$. Then, using Theorem 5.1 with $q = 2$,

$$\begin{aligned} \|\pi_{\leq n} X\|_2 &= \Big\|\sum_{k=0}^{n} X_k\Big\|_2 = \Big(\sum_{k=0}^{n} \|X_k\|_2^2\Big)^{1/2} \leq (p-1)^{-n/2} \Big(\sum_{k=0}^{\infty} (p-1)^k \|X_k\|_2^2\Big)^{1/2} \\ &= (p-1)^{-n/2} \|\Gamma\big((p-1)^{1/2} I\big)(X)\|_2 \leq (p-1)^{-n/2} \|X\|_p, \end{aligned}$$

which proves that $\pi_{\leq n}$ extends to a bounded linear map $L^p(\Omega, \mathcal{F}(H), \mathrm{P}) \to \overline{\mathcal{P}}_n(H)$. The further extension to $L^p(\Omega, \mathcal{F}, \mathrm{P})$ is obtained by first projecting $L^p(\Omega, \mathcal{F}, \mathrm{P})$ onto $L^p(\Omega, \mathcal{F}(H), \mathrm{P})$, but this projection is just the conditional expectation, which is a contraction on L^p.

The result for π_n follows similarly; alternatively we can use $\pi_n = \pi_{\leq n} - \pi_{\leq n-1}$. □

REMARK 5.16. The argument in the proof yields the estimate $\|\pi_n\|_{p,p} \leq (p-1)^{-n/2}$ for $1 < p \leq 2$; dually (or by a similar argument) $\|\pi_n\|_{p,p} \leq (p-1)^{n/2}$ for $2 \leq p < \infty$. By the proof of Theorem 5.18 below, these estimates are sharp within a power of n (for $n \geq 2$).

REMARK 5.17. Theorem 5.14 fails for $p \leq 1$; π_n and $\pi_{\leq n}$ do not extend to bounded operators on L^1 except in the trivial cases $n = 0$ and $\dim H = 0$. This follows easily from the fact that $\xi^{:n:} \notin L^\infty$ if $0 \neq \xi \in H$, cf. (5.8) and (5.9) below.

The preceding results may give the impression that the chaos decomposition is almost as well-behaved for L^p, $1 < p < \infty$, as it is for L^2. An important difference is seen, however, in the following theorem.

THEOREM 5.18. *Suppose that* $\dim H > 0$. *If* $1 < p < 2$ *or* $2 < p < \infty$, *then there exists* $X \in L^p(\Omega, \mathcal{F}(H), \mathrm{P})$ *such that* $\pi_{\leq n} X \not\to X$ *in* L^p, *and, moreover,* $\sup_n \|\pi_n(X)\|_p = \sup_n \|\pi_{\leq n}(X)\|_p = \infty$. *Hence neither of the families* $\{\pi_n\}_{n \geq 0}$ *and* $\{\pi_{\leq n}\}_{n\geq 0}$ *of bounded projections in* L^p *is uniformly bounded.*

The proof will be based on another application of Theorem 5.1.

THEOREM 5.19. *Let* ξ *be a standard Gaussian variable. For each* $p > 0$, *there exist* $c(p) > 0$ *and* $\gamma(p) \geq 0$ *such that for every* $n \geq 1$,

(i) *if* $0 < p \leq 2$, *then*

$$c(p) n^{-\gamma(p)} \leq \| :\xi^n: \|_p / \sqrt{n!} \leq 1;$$

(ii) *if* $2 \leq p < \infty$, *then*

$$c(p) n^{-\gamma(p)} \leq \| :\xi^n: \|_p / (p-1)^{n/2} \sqrt{n!} \leq 1. \tag{5.6}$$

PROOF. By Corollary 3.10, $\| :\xi^n: \|_2 = \sqrt{n!}$. Hence, taking $c(2) = 1$ and $\gamma(2) = 0$, the result holds for $p = 2$ with equalities everywhere. The upper bound for $p < 2$ now follows from Lyapounov's inequality $\| :\xi^n: \|_p \leq \| :\xi^n: \|_2$, and the upper bound for $p > 2$ follows from Theorem 5.1 which implies $\| :\xi^n: \|_p \leq (p-1)^{n/2} \| :\xi^n: \|_2$.

In order to prove the lower bounds, we first observe that if $t > 0$, then by (3.17), $:e^{it\xi}: = e^{it\xi + t^2/2} \in L^\infty_{\mathbb{C}}$, and thus

$$| \mathrm{E}(:\xi^n: :e^{it\xi}:) | \leq \| :\xi^n: \|_1 \| :e^{it\xi}: \|_\infty = e^{t^2/2} \| :\xi^n: \|_1.$$

Since $\mathrm{E}(:\xi^n: :e^{it\xi}:) = \frac{(it)^n}{n!} \mathrm{E}(:\xi^n:^2) = (it)^n$ by Corollary 3.10, this gives

$$\| :\xi^n: \|_1 \geq t^n e^{-t^2/2}, \qquad t > 0.$$

The right hand side is maximized by taking $t = \sqrt{n}$, which gives, using Stirling's formula,

$$\| :\xi^n: \|_1 \geq n^{n/2} e^{-n/2} = \left(n!/(2\pi n)^{1/2} (1 + O(n^{-1})) \right)^{1/2},$$

which proves the result for $p = 1$ with $\gamma(1) = 1/4$.

The lower bound follows for $1 < p < 2$ by Lyapounov's inequality, and for $0 < p < 1$ by Hölder's inequality which implies

$$\| :\xi^n: \|_1 \leq \| :\xi^n: \|_p^\theta \| :\xi^n: \|_2^{1-\theta}$$

for $\theta = p/(2-p) > 0$.

For $p > 2$, let $m \geq p/2$ an integer. By Theorem 5.1,

$$\| :\xi^n: \|_p \geq (p-1)^{n/2}(2m-1)^{-n/2} \| :\xi^n: \|_{2m},$$

so it suffices to prove the lower bound when $p = 2m$ is an even integer. In this case, we use Theorem 3.12, which shows that $\| :\xi^n: \|_{2m}^{2m} = \mathrm{E}(:\xi^{n}:^{2m})$ equals the number of complete Feynman diagrams labelled by $(\xi_{ik})_{1\leq i\leq 2m, 1\leq k\leq n}$ such that no edge joins two variables with the same first index. Let n_{ij} be the number of edges joining variables with first indices i and j in such a diagram, and let $x_{ij} = n_{ij}/n$. Then $(x_{ij})_{i,j=1}^{2m}$ is a symmetric matrix with all diagonal entries 0 and all row sums 1 such that $0 \leq x_{ij} \leq 1$ and nx_{ij} is an integer for every i and j. Conversely, given every such matrix (x_{ij}), we may construct a corresponding Feynman diagram, and simple combinatorics shows that this may be done in exactly

$$\prod_{i=1}^{2m} \frac{n!}{\prod_{j=1}^{2m} n_{ij}!} \prod_{1\leq i<j\leq 2m} n_{ij}! = \frac{(n!)^{2m}}{\prod_{i<j} n_{ij}!}$$

ways, where $n_{ij} = nx_{ij}$. We choose (x_{ij}) such that $|x_{ij} - x| \leq 1/n$, $i \neq j$, where $x = 1/(2m-1)$, which is always possible, and obtain by Stirling's formula, using $\sum_{i<j} x_{ij} = \frac{1}{2}\sum_{i,j} x_{ij} = \frac{1}{2}\sum_i 1 = m$, for some c_k depending on m and $\beta = 2m(2m-1)/4 - m/2 = m(m-1)$,

$$\begin{aligned}
\| :\xi^n: \|_{2m}^{2m} &\geq (n!)^m \frac{(n^n e^{-n} n^{1/2})^m}{\prod_{i<j} x_{ij}^{x_{ij}n} n^{x_{ij}n} e^{-x_{ij}n}(c_1 n)^{1/2}} \\
&= (n!)^m c_2 n^{-\beta} (n/e)^{nm - n\sum_{i<j} x_{ij}} e^{-n\sum_{i<j} x_{ij}\ln x_{ij}} \\
&= (n!)^m c_2 n^{-\beta} e^{-nm(2m-1)x\ln x + O(1)} \\
&\geq (n!)^m c_3 n^{-\beta} e^{nm\ln(2m-1)}. \qquad (5.7)
\end{aligned}$$

which completes the proof. □

REMARK 5.20. Instead of just choosing one matrix (x_{ij}) in the last part of the proof, we may sum over all (x_{ij}) with $|x_{ij} - x| < n^{-1/2}$. (It is easily seen that the estimate in (5.7) is valid for every such (x_{ij}), using a two-term Taylor expansion of $x_{ij}\ln x_{ij}$.) Counting the degrees of freedom, we see that there are $\asymp (n^{1/2})^{2m^2-3m}$ such (x_{ij}), which gives an additional factor $cn^{m^2-3m/2}$ in (5.7). This leads to (5.6) with $\gamma(2m) = 1/4$, $m \geq 2$. Hence, and by the result above for $p = 1$, we may take $\gamma(p) \leq 1/4$ for every $p \geq 1$.

Moreover, standard estimates show that the sum over all (x_{ij}) is of the same order, which leads to an estimate of $\| :\xi^n: \|_{2m}$ from above of the same order as the lower bound. If $p > 4$ is not an even integer, we combine this estimate for $m = 2$ with Theorem 5.1 again; hence we can for all $p \geq 4$ sharpen (5.6) to

$$c_1(p)n^{-1/4} \leq \| :\xi^n: \|_p/(p-1)^{n/2}\sqrt{n!} \leq c_2(p)n^{-1/4}.$$

We do not know if the best possible values of $\gamma(p)$ are known for any other p.

PROOF OF THEOREM 5.18. It suffices to consider the case when $\dim H = 1$. (Otherwise we consider any one-dimensional subspace; we omit the details.) Let $\xi \in H$ be a standard Gaussian variable. Then $\pi_n\colon \sum_k a_k :\xi^k: \mapsto a_n :\xi^n:$ is given by

$$\pi_n(X) = \mathrm{E}(X :\xi^n:) :\xi^n: / \mathrm{E}(:\xi^n:)^2 \tag{5.8}$$

and thus, if p' is the conjugate exponent defined by $1/p + 1/p' = 1$,

$$\|\pi_n\|_{p,p} = \| :\xi^n: \|_{p'} \| :\xi^n: \|_p / n!. \tag{5.9}$$

If $2 < p < \infty$, then $1 < p' < 2$ and Theorem 5.19 yields

$$\| :\xi^n: \|_p / \sqrt{n!} \geq c(p) n^{-\gamma(p)} (p-1)^{n/2}$$

and

$$\| :\xi^n: \|_{p'} / \sqrt{n!} \geq c(p') n^{-\gamma(p')}$$

and thus, for some constants c and γ depending on p,

$$\|\pi_n\|_{p,p} \geq c n^{-\gamma} (p-1)^{n/2} \to \infty, \qquad \text{as } n \to \infty.$$

Similarly, if $1 < p < 2$, Theorem 5.19 and (5.9) yield

$$\|\pi_n\|_{p,p} \geq c n^{-\gamma} (p'-1)^{n/2} \to \infty, \qquad \text{as } n \to \infty.$$

Hence the family $\{\pi_n\}$ is not uniformly bounded on L^p for $p \neq 2$.

The existence of an $X \in L^p$ with $\sup_n \|\pi_n X\|_p = \infty$ now follows from the uniform boundedness principle. Since $\pi_n = \pi_{\leq n} - \pi_{\leq n-1}$, it follows also that $\sup_n \|\pi_{\leq n} X\|_p = \infty$ and that $\pi_{\leq n} X$ does not converge in L^p. □

REMARK 5.21. It is easy to exhibit an explicit $X \in L^p$ satisfying the conclusions of Theorem 5.18 instead of appealing to the uniform boundedness principle. For example, let $\xi \in H$ be a standard Gaussian variable and let $X = e^{a\xi^2}$ for some a with $\operatorname{Re} a < 1/2p$ (which implies $X \in L^p$). By Example 3.41, if necessary using analytic continuation from $\operatorname{Re} a < 1/4$,

$$\pi_{2k}(X) = \frac{a^k}{(1-2a)^{k+1/2} k!} :\xi^{2k}:$$

and thus, by Theorem 5.19 and Stirling's formula, with $\tilde{p} = \max(2, p)$ and $c_j > 0$,

$$\|\pi_{2k}(X)\|_p \geq c_1 \Big| \frac{a}{1-2a} \Big|^k \frac{\sqrt{(2k)!}}{k!} (\tilde{p}-1)^k k^{-c_2} \geq c_3 \Big| \frac{2a}{1-2a} \Big|^k (\tilde{p}-1)^k k^{-c_4}.$$

If $1 < p < 2$ we may take a real with $1/4 < a < 1/2p$; then $\frac{2a}{1-2a}(\tilde{p}-1) = \frac{2a}{1-2a} > 1$. If $2 < p < \infty$, we have $\tilde{p} = p > 2$ and we may choose a imaginary with $|a|$ so large that $\left|\frac{1-2a}{2a}\right| < \tilde{p} - 1$. In both cases $\|\pi_{2k}(X)\|_p \to \infty$ as $k \to \infty$ as we wanted.

In this example $\pi_n(X) = 0$ when n is odd, but if we modify the example to $X = (1+\xi)e^{a\xi^2}$, it is easily seen, using Example 3.42 also, that $\|\pi_n(X)\|_p \to \infty$ as $n \to \infty$, a slight improvement of Theorem 5.18.

REMARK 5.22. By Example 3.18, the theorems above contain results on L^p-norms with respect to the standard Gaussian measure on $\mathbb{R}$ of Hermite polynomials and Hermite series.

We end this section with another simple application of hypercontractivity, which will be used later as a substitute for the failing convergence of the chaos expansion in Theorem 5.18.

THEOREM 5.23. *Let $1 < p < \infty$. If $X \in L^p(\Omega, \mathcal{F}(H), \mathrm{P})$ and $0 \leq r \leq \min((p-1)^{1/2}, (p-1)^{-1/2})$, then $\pi_{\leq n}(M_r X) \to M_r X$ in L^p as $n \to \infty$. In other words,*

$$M_r X = \sum_{n=0}^{\infty} r^n \pi_n(X), \tag{5.10}$$

with the sum converging in L^p.

PROOF. Let us first consider the case $1 < p \leq 2$. Then, by Theorem 5.8, $M_r X \in L^2$, and thus $\pi_{\leq n}(M_r X) \to M_r X$ in L^2 as $n \to \infty$; but this implies convergence in L^p by Lyapounov's inequality.

For $2 \leq p < \infty$, we argue instead that $X \in L^2$ and thus $\pi_{\leq n} X \to X$ in L^2 as $n \to \infty$. Theorem 5.8 now yields that $\pi_{\leq n}(M_r X) = M_r(\pi_{\leq n} X) \to M_r X$ in L^p.

Finally, note that $\pi_{\leq n}(M_r X) = \sum_{k=0}^{n} r^k \pi_k(X)$. □

REMARK 5.24. It follows from Remark 5.16 that the sum in (5.10) converges absolutely when $0 \leq r < \min((p-1)^{1/2}, (p-1)^{-1/2})$.

2. Complex hypercontractivity

We give here some results on hypercontractivity of $\Gamma(A)$, when A is a complex linear operator $H_{1\mathbb{C}} \to H_{2\mathbb{C}}$. We will use Theorem 5.28 later, but this section is less important than the preceding one. We begin with a simple consequence of the real hypercontractivity theorem.

THEOREM 5.25. *Let H_1 and H_2 be Gaussian Hilbert spaces defined on probability spaces $(\Omega_1, \mathcal{F}_1, \mathrm{P}_1)$ and $(\Omega_2, \mathcal{F}_2, \mathrm{P}_2)$. If $1 \leq p \leq q \leq \infty$ and $A\colon H_{1\mathbb{C}} \to H_{2\mathbb{C}}$ is a complex linear map with norm $\|A\| \leq \big(\min(p-1,1)/\max(q-1,1)\big)^{1/2}$, then $\Gamma(A)$ maps $L^p_{\mathbb{C}}(\Omega_1, \mathcal{F}(H_1), \mathrm{P}_1)$ into $L^q_{\mathbb{C}}(\Omega_2, \mathcal{F}(H_2), \mathrm{P}_2)$ with norm $\|\Gamma(A)\|_{p,q} = 1$.*

PROOF. Let $r_1^2 = \min(p-1, 1)$ and $r_2^2 = \min(1/(q-1), 1)$. Then, by Theorem 5.1 for $1 < p < 2$ and trivially for $p \geq 2$, $\Gamma(r_1 I)$ is a contraction of $L^p_{\mathbb{C}}$ into $L^2_{\mathbb{C}}$, and similarly $\Gamma(r_2 I)$ is a contraction of $L^2_{\mathbb{C}}$ into $L^q_{\mathbb{C}}$; moreover, by Theorem 4.5, $\Gamma(r_1^{-1} r_2^{-1} A)$ is a contraction of $L^2_{\mathbb{C}}(\Omega_1, \mathcal{F}(H_1), \mathrm{P}_1)$ into $L^2_{\mathbb{C}}(\Omega_2, \mathcal{F}(H_2), \mathrm{P}_2)$. The result thus follows from the decomposition $\Gamma(A) = \Gamma(r_2 I)\Gamma(r_1^{-1} r_2^{-1} A)\Gamma(r_1 I)$. □

REMARK 5.26. We have so far defined $\Gamma(A)$ for complex operators A only as an operator on $L^2_{\mathbb{C}}$. Hence, if $p < 2$, the conclusion of the theorem really means that this operator may be (uniquely) extended to a linear operator defined on $L^p_{\mathbb{C}}$ such that the conclusion holds. The same applies to the statements below.

The condition on A in Theorem 5.25 is not optimal. The sharp result was given by Epperson (1989), see also Janson (1997+), and can be stated as follows. If $1 \le p \le q < \infty$ and

$$\|A\zeta\|^2 + (q-2)\|\operatorname{Re}(A\zeta)\|^2 \le \|\zeta\|^2 + (p-2)\|\operatorname{Re}\zeta\|^2, \qquad \text{for all } \zeta \in H_{1\mathbb{C}}, \tag{5.11}$$

then $\|\Gamma(A)\|_{p,q} = 1$; conversely if (5.11) fails, then $\Gamma(A)$ is not even bounded from L^p into L^q. The converse part follows, as in the real case (Remark 5.4), by considering $\Gamma(A)(:e^{t\zeta}:) = :e^{tA\zeta}:$ with $\zeta \in H_{1\mathbb{C}}$ and positive $t \to \infty$, computing the norms by Corollary 3.38; for the other (harder) direction, see the references given above.

The most interesting case is $A = zI$ with $z \in \mathbb{C}$; i.e. the complex Mehler transform M_z, in which case the condition (5.11) for hypercontractivity may be reduced to

$$|zw|^2 + (q-2)|\operatorname{Re}(zw)|^2 \le |w|^2 + (p-2)|\operatorname{Re} w|^2, \qquad \text{for all } w \in \mathbb{C}, \tag{5.12}$$

which is equivalent to the two conditions

$$|z|^2 \le p/q$$
$$(q-1)|z|^4 - (p+q-2)(\operatorname{Re} z)^2 - (pq-p-q+2)(\operatorname{Im} z)^2 + (p-1) \ge 0 \tag{5.13}$$

The Mehler transform with a purely imaginary parameter was first studied by Beckner (1975), who proved hypercontractivity for $z = i\sqrt{p-1}$ and $q = p'$, the conjugate exponent, for $1 \le p \le 2$ as part of his celebrated proof finding the sharp constant in the Hausdorff–Young inequality for the Fourier transform, see Theorem 5.32 and Example 5.33 below. The case of a general complex z was treated by Weissler (1979), who for most values of p and q found a characterization equivalent to (5.13) of the set of z for which M_z is a contraction (or bounded) $L^p_{\mathbb{C}} \to L^q_{\mathbb{C}}$, but the ranges $\frac{3}{2} < p \le q < 2$ and $2 < p \le q < 3$ were left open by his proof.

REMARK 5.27. As a step in the proof, Weissler (1979) proved that, excepting the ranges of p and q just mentioned, the two-point inequality (5.1) holds for a complex parameter z if and only if (5.13) holds. As far as we know, it is still unknown whether the same is true for the exceptional ranges of p and q.

We will not prove the complete characterization given in (5.12) or (5.13) here; for our purposes the following simpler partial result is sufficient. (For the definition of analytic functions with values in a Banach space, see Appendix G.)

THEOREM 5.28. *If $1 < p \le q < \infty$, then there exists a closed set D_{pq} of complex numbers with the following properties, where $r = \sqrt{(p-1)/(q-1)}$.*

(i) *D_{pq} contains the closed interval $[-r, r]$.*
(ii) *The interior D_{pq}° of D_{pq} contains the open interval $(-r, r)$.*
(iii) *The interior D_{pq}° is connected, and its closure equals D_{pq}.*
(iv) *If $z \in D_{pq}$ and H is a Gaussian Hilbert space, then the Mehler transform M_z is a contraction of $L^p_{\mathbb{C}}(\Omega, \mathcal{F}(H), \mathrm{P})$ into $L^q_{\mathbb{C}}(\Omega, \mathcal{F}(H), \mathrm{P})$.*
(v) *If $X \in L^p_{\mathbb{C}}(\Omega, \mathcal{F}(H), \mathrm{P})$, then $z \mapsto M_z X$ defines a function from D_{pq} into $L^q_{\mathbb{C}}(\Omega, \mathcal{F}(H), \mathrm{P})$, which is continuous on D_{pq} and analytic in the interior D_{pq}°.*

PROOF. We use the same method of proof as for Theorem 5.1, and begin by studying the two-point inequality (5.1) for complex values of r. Thus, let, as in Lemma 5.2, X be a random variable with $\mathrm{P}(X = 1) = \mathrm{P}(X = -1) = 1/2$, and define

$$D_{pq} = \{z : \|a + zbX\|_q \le \|a + bX\|_p \quad \text{for all } a, b \in \mathbb{C}\}. \tag{5.14}$$

That D_{pq} is closed is immediate by the continuity of the norms, and (i) was proved in Lemma 5.2. Moreover, D_{pq} is convex by the triangle inequality in L^q (which implies that the function $z \mapsto \|a + zbX\|_q$ is convex for every a and b), and D_{pq} contains by Theorem 5.25 a disc with centre at 0 (and radius $(\min(p-1,1)/\max(q-1,1))^{1/2}$). The properties (ii) and (iii) follow.

The proof of Theorem 5.1 goes through without changes, which shows that (iv) holds.

Finally, if X now denotes any variable in $L^p_{\mathbb{C}}(\Omega, \mathcal{F}(H), \mathrm{P})$, there exists a sequence X_n of polynomial variables in $\mathcal{P}_{\mathbb{C}}(H)$ that converges to X in $L^p_{\mathbb{C}}$. The conclusion of (v) is clear for the functions $M_z X_n$, which are polynomials in z, but by (iv), these functions converge to $M_z X$ uniformly on D_{pq}, and the result follows. □

REMARK 5.29. The significance of the conditions (ii) and (iii) in the theorem is that any function that is continuous on D_{pq} and analytic in the interior, and furthermore vanishes in a small neighbourhood of 0, has to vanish identically on D_{pq}.

REMARK 5.30. It follows easily from (v) that $z \mapsto M_z$ is an analytic function in D_{pq}° with values in the Banach space of bounded linear operators $L^p_{\mathbb{C}} \to L^q_{\mathbb{C}}$.

REMARK 5.31. Note that we defined the set D_{pq} in (5.14) as the set for which the two-point inequality holds, and not directly as the set for which hypercontractivity holds in Gaussian spaces; the reason for this is that it was then immediate that D_{pq} is convex, which implies (iii). In fact, using the result by Epperson (1989), we could instead let D_{pq} be the set defined by (5.12) or (5.13).

The proof above yields an explicit choice of D_{pq}, namely the convex hull of the interval $[-r, r]$ and the disc $\{z : |z| \leq r_1 r_2\}$, with $r^2 = (p-1)/(q-1)$, $r_1^2 = \min(p-1, 1)$ and $r_2^2 = \min(1/(q-1), 1)$. This is not the best possible choice; in fact, for imaginary $z = is$ (s real), this shows hypercontractivity for $|s| \leq r_1 r_2$, while the sharp bound is $\min(r_1, r_2)$, which is larger when $1 < p < 2 < q < \infty$. For completeness we include a proof of this result (Beckner 1975).

THEOREM 5.32. *If $1 \leq p \leq q \leq \infty$, s is real and $H \neq \{0\}$ is a Gaussian Hilbert space, then the Mehler transform M_{is} is a contraction of $L^p_{\mathbb{C}}(\Omega, \mathcal{F}(H), \mathrm{P})$ into $L^q_{\mathbb{C}}(\Omega, \mathcal{F}(H), \mathrm{P})$ (or just bounded) if and only if*

$$s^2 \leq \min\big(p-1, (q-1)^{-1}\big).$$

PROOF. The necessity of this bound is easily shown by applying the operator to $:e^{\xi}:$ or $:e^{i\xi}:$, for $\xi \in H$, and using Corollary 3.38; we leave the details as an exercise.

For sufficiency, it suffices by the monotonicity of the L^p-spaces (Lyapounov's inequality) to consider the case where $1 < p \leq 2 \leq q < \infty$ and $p - 1 = 1/(q-1)$, i.e. p and q are conjugate exponents. This is the case treated by Beckner (1975), and his argument runs as follows.

We will show that the two-point inequality (5.1) holds with $r = is$, or equivalently that is belongs to the set D_{pq} defined in (5.14). The result then follows by the proof of Theorem 5.1, just as in the proof of Theorem 5.28.

Thus, let X be a symmetric ± 1 random variable as in Lemma 5.2 and assume that s is real with $s^2 \leq p - 1 = 1/(q-1)$. We have to prove that $\|a + isbX\|_q \leq \|a + bX\|_p$ for any complex numbers a and b. The case $a = 0$ is clear, so by homogeneity we may assume $a = 1$. Writing $b = u + iv$, we then have

$$|1 + isbX|^2 = |1 + isuX - svX|^2 = (1 - svX)^2 + s^2u^2$$

and thus, using Minkowski's inequality in $L^{q/2}$,

$$\begin{aligned} \|1 + isbX\|_q^2 &= \big\||1 + isbX|^2\big\|_{q/2} = \|(1 - svX)^2 + s^2u^2\|_{q/2} \\ &\leq \|(1 - svX)^2\|_{q/2} + s^2u^2 = \|1 - svX\|_q^2 + s^2u^2. \end{aligned} \tag{5.15}$$

We now apply the real two-point inequality Lemma 5.2 with exponents 2 and q, which gives, because $s^2 \leq (2-1)/(q-1)$,

$$\|1 - svX\|_q \leq \|1 - vX\|_2. \tag{5.16}$$

Combining (5.15) and (5.16), we obtain

$$\|1 + isbX\|_q^2 \leq \|1 - vX\|_2^2 + s^2u^2 = 1 + v^2 + s^2u^2. \tag{5.17}$$

Similarly,

$$\|1+bX\|_p^2 = \||1+bX|^2\|_{p/2} = \|(1+uX)^2+v^2\|_{p/2}.$$

Since $p/2 \le 1$, Minkowski's inequality is valid in the reverse direction (for positive variables), which gives

$$\|1+bX\|_p^2 \ge \|(1+uX)^2\|_{p/2} + v^2 = \|1+uX\|_p^2 + v^2. \tag{5.18}$$

Finally, we again use Lemma 5.2, now with exponents p and 2, and conclude that

$$1+s^2u^2 = \|1+suX\|_2^2 \le \|1+uX\|_p^2. \tag{5.19}$$

The inequalities (5.17), (5.18) and (5.19) yield $\|1+isbX\|_q^2 \le \|1+bX\|_p^2$, which verifies the claim that $is \in D_{pq}$, and thus completes the proof. □

EXAMPLE 5.33. Beckner (1975) used his result in Theorem 5.32 as follows. Let $1 < p \le 2$ and let $q = p/(p-1)$ be the conjugate exponent. Consider the one-dimensional Gaussian Hilbert space $(\mathbb{R}, \mathcal{B}, \gamma)$ from Example 1.15. By Theorem 5.32, the operator $M_{i\sqrt{p-1}}$ is a contraction from $L^p_{\mathbb{C}}(\mathbb{R}, d\gamma)$ into $L^q_{\mathbb{C}}(\mathbb{R}, d\gamma)$; and by Example 4.18, taking $r = i\sqrt{p-1}$ in (4.13), this operator is given by

$$\begin{aligned}\mathcal{M}_{i\sqrt{p-1}}f(x) &= (2\pi p)^{-1/2}\int_{-\infty}^{\infty} f(y)\exp\bigl(-\tfrac{1}{2p}(y-i\sqrt{p-1}x)^2\bigr)\,dy \\ &= (2\pi p)^{-1/2}e^{(p-1)x^2/2p}\int_{-\infty}^{\infty} e^{i\frac{\sqrt{p-1}}{p}xy}e^{-y^2/2p}f(y)\,dy.\end{aligned}$$

Setting $V_pf(x) = (2\pi)^{-1/2p}e^{-x^2/2p}f(x)$, which satisfies $\|V_pf\|_{L^p_{\mathbb{C}}(\mathbb{R},dx)} = \|f\|_{L^p_{\mathbb{C}}(\mathbb{R},d\gamma)}$, this can be written

$$V_q(\mathcal{M}_{i\sqrt{p-1}}f) = T(V_pf)$$

where the operator T is defined by

$$Tg(x) = (2\pi)^{-1/2-1/2q+1/2p}p^{-1/2}\int_{-\infty}^{\infty} e^{i\frac{\sqrt{p-1}}{p}xy}g(y)\,dy. \tag{5.20}$$

(Cf. Example 4.22, where the case $p = q = 2$ was treated.) Thus this operator T is a contraction from $L^p_{\mathbb{C}}(\mathbb{R}, dx)$ into $L^q_{\mathbb{C}}(\mathbb{R}, dx)$.

The operator T is essentially the Fourier transform. In order to be more precise, let us this time define the Fourier transform by $\hat{g}(x) = \int e^{2\pi ixy}g(y)\,dy$, which yields a unitary operator on $L^2_{\mathbb{C}}(\mathbb{R}, dx)$ and satisfies the Hausdorff–Young inequality in the standard form

$$\|\hat{g}\|_{L^q_{\mathbb{C}}(\mathbb{R},dx)} \le \|g\|_{L^p_{\mathbb{C}}(\mathbb{R},dx)}. \tag{5.21}$$

Then (5.20) yields, using $1/p + 1/q = 1$ and $p/\sqrt{p-1} = \sqrt{pq}$,

$$\hat{g}(x) = (2\pi)^{1/q}p^{1/2}Tg(2\pi\sqrt{pq}x).$$

Consequently, we obtain the sharper Hausdorff–Young inequality

$$\|\hat{g}\|_{L^q_{\mathbb{C}}(\mathbb{R},dx)} = p^{1/2}(pq)^{-1/2q}\|Tg\|_{L^q_{\mathbb{C}}(\mathbb{R},dx)} \leq p^{1/2p}q^{-1/2q}\|g\|_{L^p_{\mathbb{C}}(\mathbb{R},dx)}, \tag{5.22}$$

which improves (5.21) for $1 < p < 2$. Moreover, (5.22) is sharp, since we obtain equality everywhere above by choosing $f = 1$ above, which gives a Gaussian function g. In fact, by rescaling, it follows that (5.22) is an equality for any Gaussian function g. (This is, perhaps, the most memorable way to state the sharp Hausdorff–Young inequality.)

Some similar investigations of $\Gamma(A)$ for an arbitrary complex linear operator on $H_{\mathbb{C}}$, where H is finite-dimensional and the operators are expressible as integral operators on $\mathbb{R}^n$, are given by Peetre (1980).

REMARK 5.34. If K is a complex Gaussian Hilbert space (see Section 1.4) and $A\colon K \to K$ is complex linear, the results above may be applied to $H = \operatorname{Re} K$ and the real linear operator $A_{\mathbb{R}}\colon \xi + \overline{\eta} \mapsto A\xi + \overline{A\eta}$ in $H_{\mathbb{C}} = K \oplus \overline{K}$. For variables in the Fock space $\Gamma(K) \subset \Gamma(H_{\mathbb{C}})$, see Remark 4.4, a stronger result holds, viz. $\|\Gamma(A)X\|_q \leq \|X\|_p$ when $X \in \Gamma(K)$, $0 < p \leq q < \infty$ and $\|A\|^2 \leq p/q$; in particular $\|M_z X\|_q \leq \|X\|_p$ when $|z|^2 \leq p/q$. As for the real Gaussian Hilbert spaces studied above, this follows easily from the one-dimensional case, which may be written

$$\|f(z\zeta)\|_q \leq \|f(\zeta)\|_p, \qquad |z| \leq \sqrt{p/q},$$

if ζ is a standard complex Gaussian variable, f is entire and $0 < p \leq q < \infty$. For proofs, see Janson (1983) and Zhou (1991).

Unfortunately, we do not know any applications of this result.

3. Multipliers

If $f\colon \mathbb{Z}_+ = \{0, 1, 2, \dots\} \to \mathbb{R}$ (or $\mathbb{C}$) is any real (or complex) function, then $f(\mathcal{N})$ is the operator formally defined by $f(\mathcal{N})X = \sum_{n=0}^{\infty} f(n)\pi_n(X)$, cf. Example 4.7. This formula always defines a linear operator in $\mathcal{P}$ and $\overline{\mathcal{P}}_*$. In this section we study, following Meyer (1984) and Watanabe (1984), boundedness on L^p. The results of this section will be used in Chapter 15, but this section may be skipped at the first reading.

REMARK 5.35. Suppose $1 \leq p < \infty$. Then $\mathcal{P}$ is dense in L^p by Theorem 2.11, and thus $f(\mathcal{N})$ is bounded on L^p in the sense that $\|f(\mathcal{N})X\|_p \leq C\|X\|_p$ for some $C < \infty$ and all $X \in \mathcal{P}$ (or, equivalently, all $X \in \overline{\mathcal{P}}_*$) if and only if $f(\mathcal{N})$ extends to a bounded linear operator in $L^p(\Omega, \mathcal{F}(H), \mathrm{P})$. Moreover, $f(\mathcal{N})$ may then be extended to a bounded linear operator $L^p(\Omega, \mathcal{F}, \mathrm{P}) \to L^p(\Omega, \mathcal{F}(H), \mathrm{P}) \subseteq L^p(\Omega, \mathcal{F}, \mathrm{P})$ by composing it (on the right) with the conditional expectation $\mathrm{E}(\cdot \mid \mathcal{F}(H))$ (recall that the projections π_n act in this way on $L^p(\Omega, \mathcal{F}, \mathrm{P})$). Hence we may without ambiguity say that $f(\mathcal{N})$ is bounded on L^p, with all these interpretations at once.

REMARK 5.36. If $1 < p < \infty$ and $f(\mathcal{N})$ is bounded on L^p, then, as an easy consequence of the continuity of π_n, $\pi_n(f(\mathcal{N})X) = f(n)\pi_n(X)$ for every $X \in L^p$ and $n \geq 0$. This relation gives a characterization of $f(\mathcal{N})$ on L^p.

The case $p = 2$ is simple. Since $L^2(\Omega, \mathcal{F}(H), \mathrm{P})$ is the orthogonal sum of $H^{:n:}$ (Theorem 2.6), $f(\mathcal{N})$ is bounded on L^2 whenever f is bounded. (Conversely, except in the trivial case $H = \{0\}$, f has to be bounded if $f(\mathcal{N})$ maps $L^2(\Omega, \mathcal{F}(H), \mathrm{P})$ into itself.)

This is not always true for $p \neq 2$, as follows easily from Theorem 5.18, but we will give some important examples where $f(\mathcal{N})$ is bounded on L^p.

Note that $\mathrm{E}\big(f(\mathcal{N})(X)Y\big) = \mathrm{E}\big(Xf(\mathcal{N})(Y)\big)$, for $X, Y \in \overline{\mathcal{P}}_*$, say, and that the adjoint of $f(\mathcal{N})$ is $\bar{f}(\mathcal{N})$. Hence, by duality, $f(\mathcal{N})$ is bounded on L^p if and only if is it bounded on $L^{p'}$, where p' is the conjugate exponent. (This also shows that $f(\mathcal{N})$ defines a bounded operator on L^∞ for f such that $f(\mathcal{N})$ is bounded on L^1.)

EXAMPLE 5.37. If $f(n) = r^n$, with $-1 \leq r \leq 1$, then $f(\mathcal{N}) = r^{\mathcal{N}} = \Gamma(rI)$, which is bounded, with norm 1, on L^p for $1 \leq p \leq \infty$ by Theorem 4.12. (This is the Mehler transform M_r again.) By Theorem 5.28, if $1 < p < \infty$, the same holds for complex r in a neighbourhood D_{pp} of $(-1, 1)$.

EXAMPLE 5.38. If $f(n) = \mathbf{1}[n = m]$, for some $m \geq 0$, then $f(\mathcal{N}) = \pi_m$, which is bounded on L^p for $1 < p < \infty$ by Theorem 5.14, but in general not for $p = 1$ (or $p = \infty$), see Remark 5.17.

THEOREM 5.39. *If $\alpha \geq 0$ and $1 \leq p \leq \infty$, then $(\mathcal{N} + 1)^{-\alpha}$ is bounded on L^p, with norm* 1.

PROOF. The result is trivial if $\alpha = 0$. Otherwise, define, for $X \in \overline{\mathcal{P}}_*$,

$$TX = \tfrac{1}{\Gamma(\alpha)} \int_0^\infty t^{\alpha-1} e^{-t} e^{-t\mathcal{N}} X \, dt,$$

where $\Gamma(\alpha)$ is the Gamma function. Since $e^{-t\mathcal{N}} = M_{e^{-t}}$ has norm 1 on L^p by Example 5.37,

$$\|TX\|_p \leq \tfrac{1}{\Gamma(\alpha)} \int_0^\infty t^{\alpha-1} e^{-t} \|X\|_p \, dt = \|X\|_p.$$

On the other hand, if $X \in H^{:n:}$, then

$$TX = \tfrac{1}{\Gamma(\alpha)} \int_0^\infty t^{\alpha-1} e^{-t} e^{-nt} X \, dt = (n+1)^{-\alpha} X;$$

thus $T = (\mathcal{N} + 1)^{-\alpha}$.

Finally the norm of $(\mathcal{N} + 1)^{-\alpha}$ is exactly 1, since constants are preserved. □

REMARK 5.40. The same argument shows that $(\mathcal{N}+1)^{-\alpha}$ is bounded also for complex α with $\operatorname{Re} \alpha > 0$.

THEOREM 5.41. *Suppose that* $-1 < r < 1$, $k \geq 0$, $-\infty < \alpha < \infty$ *and* $1 < p < \infty$. *Then* $\mathcal{N}^k r^{\mathcal{N}}$ *and* $(\mathcal{N}+1)^\alpha r^{\mathcal{N}}$ *are bounded on* L^p.

PROOF. We use complex scalars. Let $X \in \overline{\mathcal{P}}_*$. By Theorem 5.28, $z \mapsto \Phi(z) = z^{\mathcal{N}} X$ is an analytic function of D^0_{pp} into L^p that is bounded by $\|X\|_p$. Hence, by Cauchy's estimate, letting $\delta(z)$ denote the distance from z to the boundary of D_{pp},

$$\|\frac{d}{dz}\Phi(z)X\|_p \leq \delta(z)^{-1}\|X\|_p, \qquad z \in D_{pp}.$$

It is easily seen that $z\frac{d}{dz}\Phi(z)X = \mathcal{N}z^{\mathcal{N}}X$. Hence the operator $\mathcal{N}z^{\mathcal{N}}$ is bounded on L^p for $z \in D_{pp}$, in particular for $z \in (-1,1)$.

Taking higher derivatives, we similarly see that $\mathcal{N}^k z^{\mathcal{N}}$ is bounded on L^p for $z \in D_{pp}$ for any integer $k \geq 0$.

If now α is real, choose $k \geq \max(\alpha, 0)$. Then, for $z \in D_{pp}$, $(\mathcal{N}+1)^k z^{\mathcal{N}}$ is bounded on L^p by the result just proved, while $(\mathcal{N}+1)^{\alpha-k}$ is bounded on L^p by Theorem 5.39. Hence the composite $(\mathcal{N}+1)^\alpha z^{\mathcal{N}}$ is bounded too. □

LEMMA 5.42. *If* $1 < p < \infty$ *and* $N \geq 1$, *there exists a constant* $K < \infty$ *such that if* $R_N = \sum_{n=N}^\infty \pi_n$, *then*

$$\|M_r R_N X\|_p \leq K r^N \|X\|_p$$

for all $X \in \overline{\mathcal{P}}_*$ *and* $r \in [0,1]$.

PROOF. Assume first that $1 < p \leq 2$, and let $r_0 = (p-1)^{1/2}$. Then, M_{r_0} maps L^p into L^2 by Theorem 5.1, and thus, for $0 \leq r \leq r_0$,

$$\|M_r R_N X\|_p \leq \|M_r R_N X\|_2 = \Big(\sum_{n=N}^\infty r^{2n}\|\pi_n(X)\|_2\Big)^{1/2}$$

$$\leq \Big(\frac{r}{r_0}\Big)^N \Big(\sum_{n=0}^\infty r_0^{2n}\|\pi_n(X)\|_2\Big)^{1/2} = r^N r_0^{-N}\|M_{r_0}X\|_2 \leq r^N r_0^{-N}\|X\|_p.$$

Moreover, $R_N = I - \pi_{\leq N-1}$ is bounded on $L^p(\Omega, \mathcal{F}(H), \mathrm{P})$ by Theorem 5.14, and thus for $r_0 < r \leq 1$, with some $K_1 < \infty$,

$$\|M_r R_N X\|_p \leq \|R_N X\|_p \leq K_1\|X\|_p \leq K_1 r^N r_0^{-N}\|X\|_p.$$

This yields the result with $K = \max(K_1, 1) r_0^{-N}$.

The case $p > 2$ follows by duality (or by a similar argument). □

THEOREM 5.43. *Suppose that there exists an analytic function* $\varphi(z)$ *in a neighbourhood of* 0 *such that* $f(n) = \varphi(1/n)$ *for all large* n. *Then* $f(\mathcal{N})$ *is bounded on* L^p *for* $1 < p < \infty$.

PROOF. Let $\rho > 0$ be such that φ is analytic in $\{z : |z| < \rho\}$, let n_0 be such that $f(n) = \varphi(1/n)$ for $n \geq n_0$, and choose an integer $N > \max(n_0, 1/\rho)$. Then

$$f(n) = \varphi(1/n) = \sum_{k=0}^{\infty} a_k n^{-k}, \qquad n \geq N,$$

for some a_k with $\sum_0^\infty |a_k| N^{-k} < \infty$.

Let R_N and K be as in Lemma 5.42. Then, for $k \geq 1$ and $X \in \overline{\mathcal{P}}_*$,

$$\tfrac{1}{k!} \int_0^\infty t^{k-1} M_{e^{-t}} R_N X \, dt = \tfrac{1}{k!} \int_0^\infty \sum_{n=N}^{\infty} t^{k-1} e^{-nt} \pi_n(X) \, dt = \sum_{n=N}^{\infty} n^{-k} \pi_n(X),$$

and thus, by Lemma 5.42,

$$\begin{aligned} \|\sum_{n=N}^{\infty} n^{-k} \pi_n(X)\|_p &\leq \tfrac{1}{k!} \int_0^\infty t^{k-1} \|M_{e^{-t}} R_N X\|_p \, dt \leq \tfrac{1}{k!} \int_0^\infty t^{k-1} K e^{-Nt} \|X\|_p \, dt \\ &= K N^{-k} \|X\|_p. \end{aligned}$$

This estimate holds also for $k = 0$, by taking $r = 1$ in Lemma 5.42. Consequently,

$$\begin{aligned} \|\sum_{n=N}^{\infty} f(n) \pi_n(X)\|_p &= \|\sum_{k=0}^{\infty} a_k \sum_{n=N}^{\infty} n^{-k} \pi_n(X)\|_p \\ &\leq \sum_{k=0}^{\infty} |a_k| K N^{-k} \|X\|_p = C_1 \|X\|_p, \end{aligned}$$

with $C_1 < \infty$. Moreover, by Theorem 5.14 and Minkowski's inequality, we also have

$$\|\sum_{n=0}^{N-1} f(n) \pi_n(X)\|_p \leq C_2 \|X\|_p,$$

for some C_2, and consequently $\|f(\mathcal{N})X\|_p \leq (C_1 + C_2)\|X\|_p$. □

EXAMPLE 5.44. The operator $\mathcal{N}^{-1/2}$ is not even defined for all $X \in \mathcal{P}$, since it corresponds to $f(\mathcal{N})$ with $f(0) = \infty$. However, the operator is well-defined on $\{X \in \overline{\mathcal{P}}_* : \mathrm{E} X = 0\}$, and for such X it coincides with $f(\mathcal{N})$ for

$$f(n) = \begin{cases} n^{-1/2}, & n \geq 1, \\ 0, & n = 0. \end{cases}$$

We can write $f(n) = (n+1)^{-1/2} f_1(n)$, where for $n \geq 1$, $f_1(n) = (1 + 1/n)^{1/2}$. Since the function $(1+z)^{1/2}$ is analytic for $|z| < 1$, Theorem 5.43 shows that $f_1(\mathcal{N})$ is bounded on L^p for $1 < p < \infty$. By Theorem 5.39, so is $(\mathcal{N}+1)^{-1/2}$, and thus also $f(\mathcal{N}) = (\mathcal{N}+1)^{-1/2} f_1(\mathcal{N})$.

Consequently, for $1 < p < \infty$, $\mathcal{N}^{-1/2}$ is bounded on $\{X \in L^p(\Omega, \mathcal{F}(H), \mathrm{P}) : \mathrm{E}\, X = 0\}$.

We end this section with some remarks on the case when $f(\mathcal{N})$ is not bounded. (Important examples are the number or Ornstein–Uhlenbeck operator $\mathcal{N}$ itself and its powers.) In that case, $f(\mathcal{N})$ cannot be extended to all of $L^p(\Omega, \mathcal{F}(H), \mathrm{P})$, but the following theorem, which extends Remark 5.36, shows that there exists a natural extension to a subspace of L^p larger than $\overline{\mathcal{P}}_*$.

THEOREM 5.45. *Let $1 < p < \infty$, and let f be any real or complex function. Then the following are equivalent, for $X, Y \in L^p(\Omega, \mathcal{F}(H), \mathrm{P})$.*

(i) *There exists a sequence $(X_k)_1^\infty$ of polynomial variables in $\mathcal{P}$ such that $X_k \to X$ and $f(\mathcal{N})X_k \to Y$ in L^p as $k \to \infty$.*
(ii) *$\pi_n(Y) = f(n)\pi_n(X)$ for every $n \geq 0$.*

PROOF. If (i) holds, then

$$\pi_n(Y) = \lim_{k\to\infty} \pi_n(f(\mathcal{N})X_k) = \lim_{k\to\infty} f(n)\pi_n(X_k) = f(n)\pi_n(X),$$

because π_n is continuous on L^p by Theorem 5.14.

Conversely, suppose that (ii) holds. Let $V_0 = \{(Z, f(\mathcal{N})Z) : Z \in \mathcal{P}\} \subset L^p \times L^p$. Then $(Z, f(\mathcal{N})Z) \in \overline{V_0}$ for every $Z \in \overline{\mathcal{P}}_*$; in particular $(\pi_n(X), \pi_n(Y)) \in \overline{V_0}$.

Consider the function $z \to \Phi(z) = (M_z X, M_z Y) \in L^p \times L^p$. By Theorem 5.28, Φ is an analytic function of D^0_{pp} into $L^p \times L^p$. Moreover, if $r \geq 0$ is small enough, then by Theorem 5.23,

$$(M_r X, M_r Y) = \sum_{k=0}^{\infty} r^k(\pi_k(X), \pi_k(Y)) \in \overline{V_0}.$$

Theorem G.1 shows that $\Phi(z) \in \overline{V_0}$ for every $z \in D_{pp}$; in particular $(X, Y) = \Phi(1) \in \overline{V_0}$. □

DEFINITION 5.46. Let $1 < p < \infty$. We say that $f(\mathcal{N})X$ *exists in* L^p, and $f(\mathcal{N})X = Y$, if the conditions of Theorem 5.45 are satisfied. This defines $f(\mathcal{N})$ as a closed densely defined operator in $L^p(\Omega, \mathcal{F}(H), \mathrm{P})$.

Again, the L^2 case is simple: $f(\mathcal{N})X$ exists in L^2 if and only if $\sum_n (1 + |f(n)|^2)\|\pi_n(X)\|_2^2 < \infty$.

THEOREM 5.47. *Let $1 < p < \infty$ and suppose that f and g are two functions on $\mathbb{Z}_+$ such that there exists an analytic function φ in a neighbourhood of 0 with $g(n)/f(n) = \varphi(1/n)$ for all sufficiently large n.*

If $X \in L^p$ and $f(\mathcal{N})X$ exists in L^p, then $g(\mathcal{N})X$ exists in L^p, and

$$\|g(\mathcal{N})X\|_p \leq C(\|f(\mathcal{N})X\|_p + \|X\|_p),$$

for some constant C (independent of X). Moreover, the following hold, where c and C are constants that depend on f, g and p only.

(i) *If* $g(n) = 0$ *for every* n *such that* $f(n) = 0$, *then*

$$\|g(\mathcal{N})X\|_p \leq C\|f(\mathcal{N})X\|_p.$$

(ii) *If* $\lim_{n\to\infty}(g(n)/f(n)) \neq 0$, *then* $f(\mathcal{N})X$ *exists in* L^p *if and only if* $g(\mathcal{N})X$ *exists in* L^p.

(iii) *If* $\{n : f(n) = 0\} = \{n : g(n) = 0\}$ *and* $\lim_{n\to\infty}(g(n)/f(n)) \neq 0$, *then*

$$c\|f(\mathcal{N})X\|_p \leq \|g(\mathcal{N})X\|_p \leq C\|f(\mathcal{N})X\|_p.$$

PROOF. Let $h(n) = g(n)/f(n)$, defined as 0 when $f(0) = 0$, and let $g_1(n) = h(n)f(n)$. Then $g_1(n) = g(n)$ except when n belongs to the finite set $S = \{n : f(n) = 0 \neq g(n)\}$.

By Theorem 5.43, $h(\mathcal{N})$ is bounded on L^p. Hence, if $X, f(\mathcal{N})X \in L^p$, then $g_1(\mathcal{N})X = h(\mathcal{N})f(\mathcal{N})X \in L^p$ and, using Theorem 5.14 also, $g(\mathcal{N})X = g_1(\mathcal{N})X + \sum_{n\in S} g(n)\pi_n(X) \in L^p$, with

$$\|g(\mathcal{N})X\|_p \leq C\|f(\mathcal{N})X\|_p + \sum_{n\in S} |g(n)|\|\pi_n(X)\|_p \leq C\|f(\mathcal{N})X\|_p + C\|X\|_p.$$

This proves the first claim.

Furthermore, if the condition in (i) holds, then $S = \varnothing$, $g = g_1$ and

$$\|g(\mathcal{N})X\|_p = \|h(\mathcal{N})f(\mathcal{N})X\|_p \leq C\|f(\mathcal{N})X\|_p.$$

Finally, note that $\varphi(0) = \lim_{n\to\infty}\varphi(1/n) = \lim_{n\to\infty}(g(n)/f(n))$. If this limit is non-zero, then $1/\varphi$ is analytic in a neighbourhood of 0 with $f(n)/g(n) = 1/\varphi(1/n)$ for all large n; hence we may interchange f and g above, which yields (ii) and (iii). □

VI

Variables with finite chaos decompositions

In this chapter, we assume that H is a Gaussian Hilbert space and study the distributions of random variables belonging to the spaces $\overline{\mathcal{P}}_n(H)$ and $H^{:n:}$. Such variables are very special, but they appear naturally, for example in limit theorems for U-statistics, see Chapter 11 and in particular Theorem 11.3.

Some results of this chapter will be used in Chapter 11.

For $n = 0$ or 1, the situation is trivial. By the definitions, an element of $P_0(H) = H^{:0:}$ is constant, i.e. has a degenerate distribution, an element of $H^{:1:}$ has a centred normal distribution, and an element of $P_1(H) = H^{:0:} \oplus H^{:1:}$ has a normal distribution. (For the complex case we obtain complex normal distributions.)

For $n = 2$, the situation is more complicated, but (in the real case) there is a canonical description of the possible distributions given by (possibly infinite) linear combinations of independent χ^2-distributed variables. We will use the spectral theorem for compact symmetric (i.e. self-adjoint) operators in a Hilbert space. We recall that there is a 1–1 correspondence between bounded operators and bounded bilinear forms on a real Hilbert space, where the operator T corresponds to the form $\langle Tx, y\rangle$, $x, y \in H$. The operator is symmetric if and only if the bilinear form is, and we freely use this correspondence to talk about eigenvalues, compactness and so on of the bilinear form as well as of the operator.

THEOREM 6.1. *If $X \in H_{\mathbb{R}}^{:2:}$, then there exists a finite or infinite sequence $(\alpha_j)_{j=1}^N$, $0 \le N \le \infty$, of non-zero real numbers, unique up to rearrangement, such that $\sum \alpha_j^2 < \infty$ and*

$$X \overset{\mathrm{d}}{=} \sum_{j=1}^{N} \alpha_j(\xi_j^2 - 1), \tag{6.1}$$

where ξ_j are independent standard normal variables. Moreover, $\sum \alpha_j^2 = \frac{1}{2} \operatorname{E} X^2$. The characteristic function of X is given by

$$\varphi(t) = \operatorname{E} e^{itX} = \prod_{j=1}^{N} (1 - 2i\alpha_j t)^{-1/2} e^{-i\alpha_j t} \tag{6.2}$$

and is analytic in the strip $|\operatorname{Im} t| < \frac{1}{2}\big(\max_j |\alpha_j|\big)^{-1}$. The square $\varphi(t)^2$ is meromorphic in the complex plane with poles at $\{-i/2\alpha_j\}_{j\ge1}$. The numbers

α_j are the non-zero eigenvalues (counted with multiplicities) of the compact symmetric bilinear form

$$B(\xi,\eta) = \tfrac{1}{2}\,\mathrm{E}(X\xi\eta), \qquad \xi,\eta \in H, \tag{6.3}$$

or, equivalently, of the corresponding compact symmetric operator $\widetilde{B}\colon \xi \mapsto \frac{1}{2}\pi_1(X\xi)$ on H. If $\dim H < \infty$, then there are thus at most $\dim H$ numbers α_j.

PROOF. This reflects the general fact that the symmetric tensor square $H^{\odot 2}$ of a (real) Hilbert space may be identified with the space of symmetric Hilbert–Schmidt bilinear forms (or, equivalently, symmetric Hilbert–Schmidt operators) on the Hilbert space, cf. Example E.13. In our case we can just as easily argue directly.

First we check that B and $\widetilde{B}$ indeed correspond by

$$\langle \widetilde{B}\xi,\eta\rangle = \tfrac{1}{2}\langle X\xi,\eta\rangle = B(\xi,\eta), \qquad \xi,\eta \in H.$$

Next we rewrite the definition of B as

$$B(\xi,\eta) = \tfrac{1}{2}\,\mathrm{E}(X\xi\eta) = \tfrac{1}{2}\langle X,\xi\eta\rangle = \tfrac{1}{2}\langle X, :\!\xi\eta\!:\rangle.$$

Let $\{\xi_j\}$ be any orthonormal basis in H. By Theorem 3.21, $\{\frac{1}{\sqrt{2}} :\!\xi_j^2\!: \}_j \cup \{ :\!\xi_j\xi_k\!: \}_{j<k}$ is an orthonormal basis in $H^{:2:}$. Hence

$$\begin{aligned} X &= \sum_j \tfrac{1}{2}\langle X, :\!\xi_j^2\!:\rangle :\!\xi_j^2\!: + \sum_{j<k}\langle X, :\!\xi_j\xi_k\!:\rangle :\!\xi_j\xi_k\!: \\ &= \sum_j B(\xi_j,\xi_j) :\!\xi_j^2\!: + \sum_{j<k} 2B(\xi_j,\xi_k) :\!\xi_j\xi_k\!: \\ &= \sum_{j,k} B(\xi_j,\xi_k) :\!\xi_j\xi_k\!: \end{aligned} \tag{6.4}$$

and, similarly,

$$\begin{aligned} \sum_{j,k} |B(\xi_j,\xi_k)|^2 &= \sum_j \tfrac{1}{2}\langle X, \tfrac{1}{\sqrt{2}} :\!\xi_j^2\!:\rangle^2 + \sum_{j<k} \tfrac{1}{2}\langle X, :\!\xi_j\xi_k\!:\rangle^2 \\ &= \tfrac{1}{2}\|X\|_2^2 < \infty. \end{aligned} \tag{6.5}$$

The latter equality shows that B and $\widetilde{B}$ are Hilbert–Schmidt and thus compact, see Appendix H. By the spectral theorem (Conway 1990, Theorem II.5.1), we can choose the basis $\{\xi_j\}$ such that it consists of eigenvectors; $\widetilde{B}\xi_j = \alpha_j\xi_j$ for some real α_j, and thus

$$B(\xi_j,\xi_k) = \langle \widetilde{B}\xi_j,\xi_k\rangle = \alpha_j\delta_{jk}.$$

For this basis we thus have, by (6.4),

$$X = \sum_j \alpha_j :\!\xi_j^2\!: = \sum_j \alpha_j(\xi_j^2 - 1),$$

where we only have to keep the (countably many) terms with $\alpha_j \neq 0$. This proves (6.1), since the standard normal variables ξ_j are orthogonal and thus independent. By (6.5), for this basis, $\sum \alpha_j^2 = \frac{1}{2}\|X\|_2^2 < \infty$.

The expression (6.2) for the characteristic function follows from (6.1), using the well-known characteristic function for the $\chi^2(1)$-distributed variables ξ_j^2. The infinite product in (6.2) converges whenever $\mathrm{Re}(2i\alpha_j t) < 1$ for all j, and $\varphi(t)^2 = \prod_j (1-2i\alpha_j t)^{-1} e^{-2i\alpha_j t}$, which converges everywhere except at the poles $-i/2\alpha_j$. Since the multiplicity of the pole equals the number of repetitions of α_j, this shows also that the sequence (α_j) is determined up to rearrangement by $\varphi(t)$, and thus by the distribution of X. □

As a very special case we obtain a (centred) χ^2-distribution when all eigenvalues are 0 or 1, i.e. when $\widetilde{B}$ is a projection.

We can easily extend the theorem to $\overline{\mathcal{P}}_2$.

THEOREM 6.2. *If $X \in \overline{\mathcal{P}}_{2\mathbb{R}}$, then there exist sequences $(\alpha_j)_1^N$ and $(\beta_j)_1^N$, $N \leq \infty$, of real numbers such that $\sum \alpha_j^2 < \infty$, $\sum \beta_j^2 < \infty$ and*

$$X \overset{\mathrm{d}}{=} \mathrm{E}\,X + \sum_{j=1}^{N} (\alpha_j \xi_j^2 + \beta_j \xi_j - \alpha_j) \tag{6.6}$$

where ξ_j are independent standard normal variables.

The characteristic function of X is given by

$$\varphi(t) = \mathrm{E}\,e^{itX} = e^{i\,\mathrm{E}Xt} \prod_{j=1}^{N} (1-2i\alpha_j t)^{-1/2} \exp\Big(-i\alpha_j t - \frac{\beta_j^2}{2(1-2i\alpha_j t)} t^2\Big) \tag{6.7}$$

and is analytic in the strip $|\mathrm{Im}\, t| < \frac{1}{2}(\max_j |\alpha_j|)^{-1}$. The square $\varphi^2(t)$ is analytic in the complex plane except at the points $-i/2\alpha_j$, where it has singularities. The numbers α_j (except possibly 0) are the eigenvalues (counted with multiplicities) of the compact symmetric bilinear form

$$B(\xi, \eta) = \frac{1}{2}\mathrm{E}((X - \mathrm{E}\,X)\xi\eta), \qquad \xi, \eta \in H.$$

We may further assume that each $\beta_j \geq 0$, with $\beta_j > 0$ when $\alpha_j = 0$, and that whenever $\alpha_j = \alpha_k$ for two distinct indices j and k, at most one of β_j and β_k is non-zero. Then the sequences (α_j) and (β_j) are uniquely determined, up to simultaneous rearrangement, by the distribution of X.

PROOF. We decompose $X = X_0 + X_1 + X_2$, with $X_k \in H^{:k:}$. Then $X_0 = \mathrm{E}\,X$, and $X_2 = \sum_j \alpha_j(\xi_j^2 - 1)$ as in Theorem 6.1, for a suitable orthonormal basis (ξ_j) in H (allowing uncountably many ξ_j as well as some $\alpha_j = 0$). Since $X_1 \in H$, we then also have $X_1 = \sum_j \beta_j \xi_j$ with $\sum \beta_j^2 = \|X_1\|_2^2$, which yields (6.6). (There are only countably many ξ_j with $\alpha_j \neq 0$ or $\beta_j \neq 0$.)

The expression (6.7) for the characteristic function follows by elementary calculations. Moreover, the infinite product in (6.7) converges whenever

$\operatorname{Re}(2i\alpha_j t) < 1$ for all j, and the corresponding product for $\varphi(t)^2$ converges everywhere except at $t = -i/2\alpha_j$. If $\alpha \in \{\alpha_j\} \setminus \{0\}$, let $J = \{j : \alpha_j = \alpha\}$ and $z_0 = -i/2\alpha$. It follows from (6.7) that as $\varepsilon \to 0$ with ε real, for some $C > 0$,

$$|\varphi^2(z_0 + \varepsilon)| \sim C\varepsilon^{-|J|},$$

$$|\varphi^2(z_0 + i\varepsilon)| \sim C\varepsilon^{-|J|} \exp\left(\frac{\sum_{j\in J}\beta_j^2}{8\alpha^3\varepsilon}\right).$$

Hence φ determines both the multiplicity $|J|$ and the sum $\sum_J \beta_j^2$. Since $\operatorname{Var} X = \|X_2\|_2^2 + \|X_1\|_2^2 = 2\sum_j \alpha_j^2 + \sum_j \beta_j^2$, φ determines also $\sum_{\alpha_j=0}\beta_j^2$.

We finally observe that we may normalize β_j as stated in the theorem by choosing the basis (ξ_j) of eigenvectors of B carefully; in each eigenspace such that the projection η of X_1 onto the eigenspace is non-zero, we choose a basis containing $\eta/\|\eta\|$. □

REMARK 6.3. The summand $\alpha_j\xi_j^2 + \beta_j\xi_j - \alpha_j$ in (6.6) may be rewritten as $\alpha_j(\xi_j + \beta_j/2\alpha_j)^2 - \alpha_j \operatorname{E}(\xi_j + \beta_j/2\alpha_j)^2$ when $\alpha_j \neq 0$. Hence X may be given as a (possibly infinite) linear combination of (centred) squares of independent normal variables, plus a normal term corresponding to $\alpha_j = 0$. A special case is a non-central χ^2-distribution.

REMARK 6.4. The proofs above give representations where (6.1) and (6.6) are equalities of random variables (and not only of their distributions), with (ξ_j) being (a subset of) an orthonormal basis in H consisting of eigenvectors of B.

EXAMPLE 6.5. We may now easily verify the claim made in Remark 2.4 that if $\dim H = \infty$, then $\overline{\mathcal{P}}_2(H) \neq \mathcal{P}_2(H)$. Let $(\xi_j)_1^\infty$ be an orthonormal sequence in H, and let $X = \sum_{j=1}^\infty 2^{-j}(\xi_j^2 - 1) \in H^{:2:}$. Then the bilinear form $B(\xi, \eta) = \frac{1}{2}\operatorname{E}(X\xi\eta)$ has eigenvectors $(\xi_j)_{j\geq 1}$ with eigenvalues $(2^{-j})_{j\geq 1}$. Suppose now that $X \in \mathcal{P}_2$, or, more generally, that $X \in \mathcal{P}$; then X is a polynomial in some finite set $\{\eta_i\}$ of variables in H. Let $H_1 \subset H$ be the subspace spanned by $\{\eta_i\}$. It follows from Theorem 3.8 that if $\xi \in H_1^\perp$, then

$$B(\xi, \eta) = \tfrac{1}{2}\operatorname{E}(:\xi\eta:X) = 0.$$

Hence B has finite rank, and thus only a finite number of non-zero eigenvalues, a contradiction.

We know no similar canonical description of the distributions of elements of $\overline{\mathcal{P}}_n$ or $H^{:n:}$ for $n \geq 3$ (or even for $H_{\mathbb{C}}^{:2:}$). We can, however, give qualitative results.

THEOREM 6.6. *If $X \in \overline{\mathcal{P}}_{n\mathbb{R}}$ for some $n \geq 0$, then either X is a.s. constant or X has an absolutely continuous distribution.*

PROOF. This follows by Theorem 4.24, since $X = \Gamma(\frac{1}{2}I)Y$ with $Y = \Gamma(2I)X \in \overline{\mathcal{P}}_{n\mathbb{R}} \subset L^1_{\mathbb{R}}$. □

The example $X = :\xi^2:$, with ξ standard normal, shows that the density function of the distribution of X does not have to be continuous or bounded, see Remark 4.28.

Further qualitative results are based on the hypercontractive inequalities of Chapter 5.

THEOREM 6.7. *For each $n \geq 1$ there exists a universal constant $c_n > 0$ such that for every $X \in \overline{\mathcal{P}}_n$ and $t \geq 2$, $\mathrm{P}(|X| > t\|X\|_2) \leq \exp(-c_n t^{2/n})$.*

PROOF. We may assume that $\|X\|_2 = 1$. By Theorem 5.10 (and Remark 5.11) we then have, for all $q \geq 2$,

$$\|X\|_q \leq (q-1)^{n/2}$$

and thus

$$\mathrm{P}(|X| > t) \leq t^{-q}\,\mathrm{E}\,|X|^q \leq t^{-q}(q-1)^{nq/2} = \exp\Bigl(-q\ln t + \frac{nq}{2}\ln(q-1)\Bigr).$$

We choose $q = 1 + t^{2/n}e^{-1}$ so $\frac{n}{2}\ln(q-1) = -\frac{n}{2} + \ln t$, and obtain for $t \geq e^{n/2}$,

$$\mathrm{P}(|X| > t) \leq \exp(-\frac{nq}{2}) \leq \exp(-\frac{n}{2e}t^{2/n}).$$

For the range $2 \leq t < e^{n/2}$ we use Chebyshev's inequality and obtain, provided c_n is small enough,

$$\mathrm{P}(|X| > t) \leq t^{-2} \leq \exp(-c_n t^{2/n}).$$

□

REMARK 6.8. We may replace $\|X\|_2$ in Theorem 6.7 by $\mathrm{E}\,|X|$ or any other $\|X\|_p$, $p > 0$, at the expense of changing c_n. This follows from Theorem 5.10.

We turn to estimates from below of the tails of the distribution of an element of $\overline{\mathcal{P}}_*$.

THEOREM 6.9. *For each $n \geq 1$ there exists a universal constant $\delta_n > 0$ such that if $X \in \overline{\mathcal{P}}_n$, then*

$$\mathrm{P}(|X| \geq \tfrac{1}{2}\,\mathrm{E}\,|X|) \geq \delta_n. \tag{6.8}$$

PROOF. Let $A = \{|X| \geq \frac{1}{2}\,\mathrm{E}\,|X|\}$. Then $\mathrm{E}(|X|(1-\mathbf{1}_A)) = \mathrm{E}\bigl(|X|\mathbf{1}(|X| < \frac{1}{2}\,\mathrm{E}\,|X|)\bigr) \leq \frac{1}{2}\,\mathrm{E}\,|X|$ and thus $\mathrm{E}(|X|\mathbf{1}_A) \geq \frac{1}{2}\,\mathrm{E}\,|X|$. The Cauchy–Schwarz inequality and Theorem 5.10 yield

$$\tfrac{1}{2}\,\mathrm{E}\,|X| \leq \mathrm{E}(|X|\mathbf{1}_A) \leq \|X\|_2\,\mathrm{P}(A)^{1/2} \leq c(1,2)^n\,\mathrm{E}\,|X|\,\mathrm{P}(A)^{1/2},$$

which yields (6.8) with $\delta_n = \frac{1}{4}c(1,2)^{-2n}$; by Remark 5.13, we may thus take $\delta_n = \frac{1}{4}e^{-2n}$. □

REMARK 6.10. We may replace $\mathrm{E}\,|X|$ by any $\|X\|_p$ in (6.8), at the expense of changing δ_n. This follows similarly, now using Theorem 5.10 with p and $2p$.

COROLLARY 6.11. *A family of random variables in $\overline{\mathcal{P}}_n$ is tight if and only if it is bounded.* $\square$

We can now prove that the estimate in Theorem 6.7 is sharp; this shows that random variables with finite chaos decomposition have rather large tails, except in the simplest cases.

THEOREM 6.12. *If $n \geq 1$ and $X \in \overline{\mathcal{P}}_n$ but $X \notin \overline{\mathcal{P}}_{n-1}$, then there exist $a, b, t_0 > 0$ such that*

$$\exp(-at^{2/n}) \leq \mathrm{P}(|X| > t) \leq \exp(-bt^{2/n}), \qquad t \geq t_0. \tag{6.9}$$

PROOF. The upper bound is given by Theorem 6.7. For the lower bound, let us first assume that X is a polynomial variable, $X = q(\xi_1, \dots, \xi_m)$, where q is a polynomial of degree n in m variables. We may assume that $\xi_1, \dots, \xi_m$ are independent standard normal variables; we may also assume that $m \geq 2$, since we otherwise may add an extra variable.

Write $\xi = (\xi_1, \dots, \xi_m)$ and $\xi' = \xi/|\xi|$. There exist $\xi'_0 \in \mathbb{R}^m$ with $|\xi'_0| = 1$, $\varepsilon > 0$ and $\delta > 0$, such that if $|\xi' - \xi'_0| < \varepsilon$ and $|\xi|$ is large enough, then $X = |q(\xi)| \geq \delta |\xi|^n$. Consequently, for large t,

$$\begin{aligned}\mathrm{P}(|X| > t) &\geq \mathrm{P}(|\xi| \geq (t/\delta)^{1/n},\ |\xi' - \xi'_0| < \varepsilon) \\ &= c\int_{(t/\delta)^{1/n}}^{\infty} r^{m-1} e^{-r^2/2} dr \\ &\geq c\int_{(t/\delta)^{1/n}}^{\infty} r e^{-r^2/2} dr = ce^{-c_1 t^{2/n}}\end{aligned} \tag{6.10}$$

with $c, c_1 > 0$, which gives the result for any $a > c_1$.

Now consider a general $X \in \overline{\mathcal{P}}_n(H)$. If $\dim H < \infty$, then $\overline{\mathcal{P}}_n(H) = \mathcal{P}_n(H)$ so X is a polynomial variable, the case just treated. If $\dim H = \infty$, we use the easily seen fact that $X \in \overline{\mathcal{P}}_n(H_1)$ for some subspace $H_1 \subseteq H$ of countable dimension. We may assume that $H = H_1$ and select an orthonormal basis $\{\xi_i\}_{i=1}^{\infty}$ in H.

By Theorem 3.21, the variables $:\xi^\alpha: \; = \; :\prod_i \xi_i^{\alpha_i}:$, where $\alpha = (\alpha_i)_1^\infty$ ranges over all multi-indices with $|\alpha| \leq n$, constitute an orthogonal basis in $\overline{\mathcal{P}}_n(H)$, and thus

$$X = \sum_{|\alpha| \leq n} a_\alpha :\xi^\alpha:. \tag{6.11}$$

We truncate the series (6.11) and define, for each $N \geq 0$,

$$X_N = \sum_{\substack{|\alpha| \leq n \\ \alpha_i = 0 \text{ when } i > N}} a_\alpha :\xi^\alpha:. \tag{6.12}$$

Thus each X_N is a polynomial variable in $\mathcal{P}_n(H)$, and $X_N \to X$ in L^2 as $N \to \infty$. Fix N so large that $X_N \notin \overline{\mathcal{P}}_{n-1}(H)$. Let $M \geq N$. Then X_M is a polynomial of degree $\leq n$ in the independent standard normal variables

$\xi_1, \dots, \xi_M$. If we condition on some fixed values of $\xi_1, \dots, \xi_N$, we have a polynomial variable in $\overline{\mathcal{P}}_n(\operatorname{span}\{\xi_{N+1}, \dots, \xi_M\})$, whose expectation is

$$\mathrm{E}(X_M \mid \xi_1, \dots, \xi_N) = X_N,$$

as is easily seen from (6.12). We apply Theorem 6.9 to this conditioned variable, and obtain

$$\begin{aligned}\mathrm{P}\big(|X_M| \geq \tfrac{1}{2}|X_N| \mid \xi_1, \dots, \xi_N\big) &\geq \mathrm{P}\big(|X_M| \geq \tfrac{1}{2}\,\mathrm{E}(|X_M| \mid \xi_1, \dots \xi_N) \mid \xi_1, \dots, \xi_N\big) \\ &\geq \delta_n.\end{aligned}$$

Consequently, for every $t > 0$,

$$\mathrm{P}(|X_M| \geq t) \geq \delta_n \, \mathrm{P}(|X_N| \geq 2t). \tag{6.13}$$

Let $M \to \infty$. Then $X_M \to X$ in L^2 and thus in distribution, and (6.13) yields

$$\mathrm{P}(|X| \geq t) \geq \delta_n \, \mathrm{P}(|X_N| \geq 2t).$$

The result follows by the estimate (6.10) for X_N. □

COROLLARY 6.13.

(i) *If* $X \in \overline{\mathcal{P}}_1$, *then* $\mathrm{E}\, e^{t|X|} < \infty$ *for every* $t < \infty$.
(ii) *If* $X \in \overline{\mathcal{P}}_2 \setminus \overline{\mathcal{P}}_1$, *then* $\mathrm{E}\, e^{t|X|} < \infty$ *for small* $t > 0$ *but not for large* t.
(iii) *If* $X \in \overline{\mathcal{P}}_* \setminus \overline{\mathcal{P}}_2$, *then* $\mathrm{E}\, e^{t|X|} = \infty$ *for every* $t > 0$.

□

REMARK 6.14. The upper bound in Theorem 6.7 holds uniformly in X for a fixed n. Hence the constant b in Theorem 6.12 may be taken independent of X, provided we normalize by assuming for example $\|X\|_2 = 1$. This is not true for the lower bound, except in the trivial case $n = 1$, even if we assume $X \in H^{:n:}$. For example, suppose that $\dim H = \infty$ and let $\{\xi_j\}_1^\infty$ be an orthonormal sequence in H, fix $n \geq 2$ and define $X_m = m^{-1/2} \sum_{j=1}^m (n!)^{-1/2} :\xi_j^n:$. Then $X \in H^{:n:}$ with $\|X_m\|_2 = 1$, and $X_m \xrightarrow{\mathrm{d}} \mathrm{N}(0,1)$ as $m \to \infty$ by the central limit theorem (recall that the variables ξ_j are independent). Hence there exist no a and t_0 such that (6.9) holds for all X_m. If $\dim H < \infty$, however, $\dim \overline{\mathcal{P}}_n(H) < \infty$ by Theorem 3.21, and a compactness argument shows that we may choose a, b and t_0 uniformly for all $X \in H^{:n:}$ with $\|X\|_2 = 1$.

REMARK 6.15. If $X \in \overline{\mathcal{P}}_{2\mathbf{R}}$ we thus see that the Laplace transform $\mathrm{E}\, e^{tX}$ exists in some interval $(-a, b)$ containing 0, but not on the entire line unless $X \in \overline{\mathcal{P}}_1$. This follows also from Theorem 6.2, which implies that $\mathrm{E}\, e^{tX} < \infty \iff t \in (-a, b)$ with $a = (2\sup\{|\alpha_j| : \alpha_j < 0\})^{-1}$ and $b = (2\sup\{\alpha_j : \alpha_j > 0\})^{-1}$, where $\{\alpha_j\}$ is the set of eigenvalues of the bilinear form in (6.3). In particular, the Laplace transform exists on a half-line if and only if all non-zero eigenvalues have the same sign.

REMARK 6.16. For real variables, there are one-sided estimates similar to Theorem 6.9. For example, by arguments very similar to the proof of Theorem 6.9, it is easily shown that if $X \in \overline{\mathcal{P}}_{n\mathbb{R}}$ with $\operatorname{E} X = 0$, then $\operatorname{P}(X \geq 0) \geq \delta_n$, and moreover $\operatorname{P}(X \geq \frac{1}{3}\operatorname{E}|X|) \geq \delta_n$, for some universal constants δ_n. There is, however, no such one-sided version of Theorem 6.12 because for example $-:\xi^2: \in H^{:2:}$ but is bounded above, for any $\xi \in H$.

If E is a set of random variables, let $\mathcal{L}(E) = \{\mathcal{L}(X) : X \in E\}$ denote the set of distributions of the variables in E. Thus $\mathcal{L}(E)$ is a subset of the set of probability measures on $\mathbb{R}$ (or, if we are studying complex variables, $\mathbb{C}$). Theorem 6.12 implies that the sets $\mathcal{L}(\overline{\mathcal{P}}_n \setminus \overline{\mathcal{P}}_{n-1})$, $n \geq 0$, are disjoint; in particular, the sets $\mathcal{L}(H^{:n:})$, $n \geq 0$, are disjoint.

Note, in contrast, that elements of $L^2 \setminus \overline{\mathcal{P}}_*$ may have any distribution with finite variance; for example, if $0 \neq \xi \in H$, then the variable

$$X = \begin{cases} \xi, & \text{when } |\xi| \leq 1, \\ -\xi, & \text{when } |\xi| > 1, \end{cases}$$

has a normal distribution but $X + \xi$ is not normal so $X \notin \overline{\mathcal{P}}_1$ and by Theorem 6.12, $X \notin \overline{\mathcal{P}}_*$.

We also observe that $\mathcal{L}(\overline{\mathcal{P}}_0) = \mathcal{L}(H^{:0:})$, $\mathcal{L}(\overline{\mathcal{P}}_1)$ and $\mathcal{L}(H^{:1:})$ are closed, since the limit of a sequence of normal distributions has to be normal. Moreover, if (X_k) is a sequence in $\overline{\mathcal{P}}_n$ that converges in distribution to some distribution μ, then Theorem 6.9 implies that (X_k) is bounded; $\sup_k \|X_k\|_2 < \infty$. If $\dim H < \infty$, it follows by compactness that some subsequence $(X_{k(i)})$ converges in $\overline{\mathcal{P}}_n$. Hence $\mu \in \mathcal{L}(\overline{\mathcal{P}}_n)$, and thus $\mathcal{L}(\overline{\mathcal{P}}_n)$ and similarly $\mathcal{L}(H^{:n:})$ are closed when $\dim H < \infty$. On the other hand, if $\dim H = \infty$ and $n \geq 2$, the example in Remark 6.14 shows that $\mathcal{L}(H^{:n:})$ is not closed. We do not know whether $\mathcal{L}(\overline{\mathcal{P}}_n)$ is always closed or not.

VII

Stochastic integration

The theory of Gaussian Hilbert spaces developed in this book has strong connections to stochastic integration, in particular to Itô integrals with respect to Brownian motion. We treat these Itô integrals in the first section, and some extensions and related results in the following ones: stochastic integrals over general measure spaces in Section 2, the Skorohod integral in Section 3, and complex stochastic integrals and measures in Section 4.

Our treatment is self-contained, and we do not require that the reader has any prior knowledge of stochastic integration. On the other hand, such a knowledge would certainly be useful; we treat only those parts of stochastic integration theory that are directly relevant to the subject of this book, and many important topics are not included. For example, we consider only stochastic integrals with respect to Gaussian processes. Moreover, even for Brownian motion we do not include the fundamental Itô formula.

Hence, this chapter will perhaps be best understood in connection and comparison with other, more direct and complete, treatments of stochastic integration; see for example McKean (1969) and Protter (1990).

1. Brownian motion and Itô integrals

In this section, we assume that B_t, $0 \leq t < \infty$, is a standard Brownian motion and consider, as in Example 1.10, the Gaussian Hilbert space $H = H(B)$ spanned by $\{B_t\}_{t \geq 0}$. Note that the σ-field $\mathcal{F}(H)$ generated by H equals the σ-field $\mathcal{F}(B)$ generated by (B_t). (All results below remain valid if we consider the interval $0 \leq t \leq 1$ only, cf. Theorem 7.10. We may also, without essential differences, consider a Brownian motion on the entire real axis $(-\infty, \infty)$, defined by letting $(B_t)_{t\geq 0}$ and $(B_{-t})_{t \geq 0}$ be independent Brownian motions on $[0, \infty)$, but we leave this case to the reader.)

Recall that the Brownian motion has the property that the variables B_t have a joint normal distribution with $\operatorname{E} B_t = 0$ and

$$\operatorname{Cov}(B_t, B_u) = t \wedge u, \qquad t, u \geq 0. \tag{7.1}$$

Equivalently, $B_0 = 0$ and

$$B_t - B_u \sim \mathrm{N}(0, |t-u|), \qquad t, u \geq 0. \tag{7.2}$$

Note that (7.1) may be written

$$\mathrm{E}\, B_t B_u = \int_0^\infty \mathbf{1}_{[0,t]}(s)\mathbf{1}_{[0,u]}(s)\, ds.$$

Hence the linear map $I\colon \sum a_i \mathbf{1}_{[0,t_i]} \mapsto \sum a_i B_{t_i}$ is an isometry of the subspace of $L^2_{\mathbb{R}}([0,\infty))$ consisting of step functions onto the (incomplete) Gaussian space spanned by $\{B_t\}$. Since the step functions are dense in $L^2_{\mathbb{R}}([0,\infty))$, I extends to an isometry $I\colon L^2_{\mathbb{R}}([0,\infty)) \to H$.

This isometry I is known as the stochastic (Itô) integral and is usually written

$$f \mapsto \int_0^\infty f(t)\, dB_t.$$

(The notation makes sense, because the defining property of I then becomes $\int_0^\infty \mathbf{1}_{[0,t]}(s)\, dB_s = B_t = B_t - B_0$.) We also use the notation

$$\int_0^t f(s)\, dB_s = \int_0^\infty f(s)\mathbf{1}_{[0,t]}(s)\, dB_s.$$

THEOREM 7.1. *With notations as above,*

$$H = \Big\{ \int_0^\infty f(t)\, dB_t : f \in L^2_{\mathbb{R}}([0,\infty)) \Big\}$$

and the mapping $f \mapsto \int_0^\infty f(t)\, dB_t$ is an isometry of $L^2_{\mathbb{R}}([0,\infty))$ onto H. □

REMARK 7.2. Conversely, any family $(B_t)_{t\geq 0}$ of centred jointly normal variables satisfying (7.1), or equivalently (7.2) and $B_0 = 0$, is a Brownian motion. Hence we may construct a Brownian motion by reversing the procedure above: we start with an isometry I of $L^2_{\mathbb{R}}([0,\infty))$ into a Gaussian Hilbert space and define $B_t = I(\mathbf{1}_{[0,t)})$.

For example, we may (as in Example 1.22) take an orthonormal basis $(\varphi_j)_1^\infty$ in $L^2_{\mathbb{R}}([0,\infty))$ and a sequence $(\xi_j)_1^\infty$ of independent standard Gaussian variables; we then define I by

$$I(f) = \sum_j \int_0^\infty \langle f, \varphi_j \rangle \xi_j = \sum_j \int_0^\infty f\varphi_j\, dt \cdot \xi_j,$$

and thus $B_t = \sum_j \Phi_j(t)\xi_j$ with $\Phi_j(t) = \int_0^t \varphi_j(s)\, ds$.

The same construction may be used on a finite interval $[0,T]$. If we take $\{\varphi_j\}$ to be the trigonometrical basis on $[0,T]$, it may further be shown that the sum a.s. converges uniformly for $0 \leq t \leq T$, which gives one proof of the fact that (a version of) the Brownian motion has continuous sample paths (Kahane 1985, Section 16.3).

We can now apply the theory of (symmetric) tensor products, cf. Chapter 4 and Appendix E. Since I is an isometry of $L^2_{\mathbb{R}}([0,\infty))$ onto H, the tensor power $I^{\odot n}$ is an isometry of $L^2_{\mathbb{R}}([0,\infty))^{\odot n}$ onto $H^{\odot n}$, which may be identified with $H^{:n:}_{\mathbb{R}}$ by Theorem 4.1. Similarly, for complex scalars, I extends to an isometry of $L^2_{\mathbb{C}}([0,\infty))$ onto $H_{\mathbb{C}}$, and the tensor power gives an isomorphism of $L^2_{\mathbb{C}}([0,\infty))^{\odot n}$ onto $H_{\mathbb{C}}^{:n:}$.

Moreover, using Proposition E.16, $L^2([0,\infty))^{\odot n}$ may be identified with the space of symmetric functions in $L^2([0,\infty)^n, \frac{1}{n!}dx)$. If we define

$$D_n = \{(t_1,\dots,t_n) : 0 < t_1 < t_2 < \cdots < t_n < \infty\}, \tag{7.3}$$

equipped with the usual Lebesgue measure, then this space may be identified with $L^2(D_n)$. (We let D_0 be a set with one element, of measure 1; hence $L^2(D_0)$ is the space of scalars.) Combining these identifications, we arrive at the following result.

THEOREM 7.3. *For each $n \geq 0$, there exists an isometry*

$$I_n\colon L^2(D_n) \to H^{:n:}$$

such that

$$I_n(f_1 \odot \cdots \odot f_n) = :\int f_1\,dB \cdots \int f_n\,dB:,$$

where $f_1 \odot \cdots \odot f_n$ is given by

$$f_1 \odot \cdots \odot f_n(x_1,\dots,x_n) = \sum_{\pi\in\mathfrak{S}_n} \prod_{i=1}^{n} f_i(x_{\pi(i)}). \tag{7.4}$$

□

(Here I_0 is the mapping $I_0(c) = c$ for every constant c.)

We can regard I_n as a multiple Itô integral. Our next aim is to show that I_n equals an iterated Itô integral. We have so far only considered stochastic integrals of deterministic functions, but now we will use integrals of random functions X_t (cf. Appendix B). For the reader's convenience, we recall the basic definitions. See for example Protter (1990) for a detailed and more general treatment.

Let $\mathcal{F}_t$ denote the σ-field generated by $\{B_s : s \leq t\}$. (It is customary in the general theory of stochastic processes and integrals to complete $\mathcal{F}_t$ with all null sets in $\mathcal{F}$. For our purposes this makes no difference.) A random function X_t, $t \geq 0$, is an *elementary predictable process* if it can be written as a finite sum

$$X_t = \sum_{i=1}^{N} Y_i \mathbf{1}_{(t_i,u_i]}(t), \tag{7.5}$$

where for each i, Y_i is a random variable that is measurable with respect to $\mathcal{F}_{t_i}$, and $0 \leq t_i < u_i$. The stochastic integral $\int_0^\infty X_t\,dB_t$ of the elementary predictable process X_t in (7.5) is defined to be the random variable

$$\sum_i Y_i(B_{u_i} - B_{t_i}); \tag{7.6}$$

it is easily seen that this defines $\int_0^\infty X_t\,dB_t$ uniquely for every elementary predictable process X_t, even though the representation (7.5) is not unique. Moreover, $B_{u_i} - B_{t_i}$ is independent of every B_s with $s \le t_i$, because then $\operatorname{Cov}(B_{u_i} - B_{t_i}, B_s) = s - s = 0$ by (7.1), and thus $B_{u_i} - B_{t_i}$ is independent of every $\mathcal{F}_{t_i}$-measurable variable. By choosing a representation (7.5) with the intervals $(t_i, u_i]$ disjoint, it is now easy to see that

$$\mathrm{E}\Big|\int_0^\infty X_t\,dB_t\Big|^2 = \int_0^\infty \mathrm{E}\,|X_t|^2\,dt \tag{7.7}$$

for every elementary predictable process X_t such that the right hand side of (7.7) is finite. This right hand side equals (by Fubini's theorem) the square of the norm of the function $(t,\omega) \mapsto X_t(\omega)$ in $L^2(\mathbb{R}_+ \times \Omega, dt\,d\mathrm{P})$.

We define the space Π^2 of *square integrable predictable processes* to be the closure of the set of elementary predictable processes in $L^2(\mathbb{R}_+ \times \Omega, dt\,d\mathrm{P})$. (We use the L^2-norm, so Π^2 becomes a Hilbert space.) We now extend the definition of the stochastic integral to Π^2 by continuity; hence $\int_0^\infty X_t\,dB_t$ is defined (as a random variable in L^2) for every square integrable predictable process X_t and (7.7) holds. We also remark that if X_t is an elementary predictable process as in (7.5), then $\mathrm{E}\,Y_i(B_{u_i} - B_{t_i}) = \mathrm{E}\,Y_i\,\mathrm{E}(B_{u_i} - B_{t_i}) = 0$. Hence

$$\mathrm{E}\int_0^\infty X_t\,dB_t = 0 \tag{7.8}$$

for every elementary predictable process X_t, and thus by continuity for every square integrable predictable process.

REMARK 7.4. More generally, a process X_t is said to be *predictable* if it is measurable with respect to the σ-field on $\mathbb{R}_+ \times \Omega$ generated by the set of elementary predictable processes; it is easily seen (using e.g. the monotone class theorem in Appendix A) that X_t is a square integrable predictable process if and only if it is predictable and $\int \mathrm{E}\,|X_t|^2 < \infty$. The stochastic integral can be further extended beyond the square integrable case, but we will not need such extensions.

Now suppose that $F \in L^2(D_n)$, with $n \ge 1$. Define

$$F_t(t_1, \dots, t_{n-1}) = \begin{cases} F(t_1, \dots, t_{n-1}, t), & 0 < t_1 < t_2 < \dots < t, \\ 0, & \text{otherwise.} \end{cases}$$

Then $F_t \in L^2(D_{n-1})$ for a.e. t, and $\int \|F_t\|^2_{L^2(D_{n-1})} = \|F\|^2_{L^2(D_n)}$. We can thus define the stochastic process $t \mapsto I_{n-1}(F_t)$. (If $n = 1$, then $I_{n-1}(F_t) = F_t = F(t)$, so this process is just the deterministic function F.)

THEOREM 7.5. *If $n \ge 1$ and $F \in L^2(D_n)$, then $t \mapsto I_{n-1}(F_t)$ is a square integrable predictable process, and*

$$I_n(F) = \int_0^\infty I_{n-1}(F_t)\,dB_t.$$

PROOF. Consider first the case when F is the indicator function of a box $\prod_1^n (a_i, b_i] \subset D_n$; thus $0 \le a_1 < b_1 \le a_2 < \cdots < b_{n-1} \le a_n < b_n$. By (7.4), for $(t_1, \dots, t_n) \in D_n$,

$$\begin{aligned}\mathbf{1}_{(a_1,b_1]} \odot \cdots \odot \mathbf{1}_{(a_n,b_n]}(t_1, \dots, t_n) &= \sum_{\pi \in \mathfrak{S}_n} \prod_{i=1}^n \mathbf{1}_{(a_i,b_i]}(t_{\pi(i)}) \\ &= \prod_1^n \mathbf{1}_{(a_i,b_i]}(t_i) = F(t_1, \dots, t_n).\end{aligned}$$

Moreover, $\int \mathbf{1}_{(a_i,b_i]}\, dB = B_{b_i} - B_{a_i}$ and these variables are orthogonal because the functions $\mathbf{1}_{(a_i,b_i]}$ are. Consequently, by Theorems 7.3 and 3.4,

$$\begin{aligned}I_n(F) &= :\int \mathbf{1}_{(a_1,b_1]}\, dB \cdots \int \mathbf{1}_{(a_n,b_n]}\, dB: \\ &= :\prod_1^n (B_{b_i} - B_{a_i}): = \prod_1^n (B_{b_i} - B_{a_i}).\end{aligned} \tag{7.9}$$

Furthermore, $F_t = 0$ unless $t \in (a_n, b_n]$, in which case $F_t = \prod_1^{n-1} \mathbf{1}_{(a_i,b_i]}$. Hence (7.9) implies

$$I_{n-1}(F_t) = \prod_1^{n-1} (B_{b_i} - B_{a_i}) \mathbf{1}_{(a_n,b_n]}(t).$$

Since B_{a_i} and B_{b_i}, $i \le n-1$, are $\mathcal{F}_{a_n}$-measurable, this is an elementary predictable process, and (7.6) yields

$$\int_0^\infty I_{n-1}(F_t)\, dB_t = \prod_1^n (B_{b_i} - B_{a_i}) = I_n(F).$$

Hence the conclusions hold for any F that is a linear combination of indicator functions of boxes. Such functions are dense in $L^2(D_n)$, and the general result follows by continuity because I_n and I_{n-1} are isometries and thus

$$\int_0^\infty \mathrm{E}\, |I_{n-1}(F_t)|^2\, dt = \int_0^\infty \|F_t\|^2_{L^2(D_{n-1})}\, dt = \|F\|^2_{L^2(D_n)}.$$

□

In particular, for $n = 2$ Theorem 7.5 yields

$$I_2(F(s,t)) = \int_0^\infty I_1(F(\cdot, t))\, dB_t = \int_0^\infty \int_0^t F(s,t)\, dB_s\, dB_t,$$

where the inner integral exists for a.e. t and defines a square integrable predictable process. For general n, we obtain similarly the following result.

THEOREM 7.6. *Let $n \ge 1$. If $F \in L^2(D_n)$, then*

$$I_n(F) = \int_0^\infty \int_0^{t_n} \cdots \int_0^{t_2} F(t_1, \dots, t_n)\, dB_{t_1} \cdots dB_{t_n}. \tag{7.10}$$

Consequently, the n-fold iterated stochastic integral is an isometry of $L^2(D_n)$ onto $H^{:n:}$, and the combined mapping

$$(F_n)_{n=0}^{\infty} \mapsto F_0 + \sum_{n=1}^{\infty} \int_0^{\infty} \int_0^{t_n} \cdots \int_0^{t_2} F_n(t_1, \dots, t_n)\, dB_{t_1} \cdots dB_{t_n} \tag{7.11}$$

is an isometry of $\bigoplus_{n=0}^{\infty} L^2(D_n)$ onto $L^2(\Omega, \mathcal{F}(B), \mathrm{P})$.

PROOF. Theorem 7.5 and induction yield (7.10). The final statements follow by Theorems 7.3 and 4.1. □

REMARK 7.7. Each integral $\int_0^{t_k} \dots dB_{t_{k-1}}$ in (7.10) is a square integrable predictable process, at least for a.e. $(t_{k+1}, \dots, t_n)$, which makes the iterated integral well-defined.

Another consequence of Theorem 7.5 is that essentially every random variable can be represented as a stochastic integral.

THEOREM 7.8. *The stochastic integral $X_t \mapsto I(X_t) = \int_0^\infty X_t\, dB_t$ is an isometry mapping the space Π^2 of square integrable predictable processes onto $L_0^2(\Omega, \mathcal{F}(B), \mathrm{P}) = \{X \in L^2(\Omega, \mathcal{F}(B), \mathrm{P}) : \mathrm{E}\, X = 0\}$. Consequently, if $X \in L^2(\Omega, \mathcal{F}(B), \mathrm{P})$, then*

$$X = \mathrm{E}\, X + \int_0^\infty Y_t\, dB_t \tag{7.12}$$

for some (unique) square integrable predictable process Y_t.

PROOF. By (7.7) and the discussion after it, the mapping I is an isometry of the Hilbert space Π^2 into $L^2(\Omega, \mathcal{F}(B), \mathrm{P})$, and by (7.8) the range is a subset of $L_0^2(\Omega, \mathcal{F}(B), \mathrm{P})$. Since the range of an isometry is closed, it suffices to show that it is dense.

Now suppose that $X \in H^{:n:}$, $n \geq 1$. By Theorem 7.3, $X = I_n(F)$ for some $F \in L^2(D_n)$, and by Theorem 7.5

$$X = I(I_{n-1}(F_t)),$$

with $I_{n-1}(F_t) \in \Pi^2$. Hence the range of I contains $H^{:n:}$, $n \geq 1$, and the result follows because $L_0^2 = \bigoplus_1^\infty H^{:n:}$ by Theorem 4.1. □

REMARK 7.9. Somewhat more concretely, if X is given by the sum of integrals in (7.11), then (7.12) holds with

$$Y_t = F_1(t) + \sum_{n=1}^{\infty} \int_0^t \int_0^{t_n} \cdots \int_0^{t_2} F_{n+1}(t_1, \dots, t_n, t)\, dB_{t_1} \cdots dB_{t_n}.$$

Nothing is changed if we consider the Brownian motion on a finite interval only. We can relate the results as follows.

THEOREM 7.10. *Let H_t be the closed subspace of $H(B)$ spanned by $\{B_s : s \leq t\}$ and let*

$$D_{n,t} = \{(t_1, \dots, t_n) : 0 < t_1 < \dots < t_n < t\}.$$

Then I_n restricts to an isometry of $L^2(D_{n,t})$ onto $H_t^{:n:}$, and $\bigoplus_0^\infty I_n$ restricts to an isometry of $\bigoplus_0^\infty L^2(D_{n,t})$ onto $L^2(\Omega, \mathcal{F}_t, \mathrm{P})$. If $X = \sum_0^\infty I_n(F_n) \in L^2(\Omega, \mathcal{F}(H), \mathrm{P})$, with $F_n \in L^2(D_n)$, then

$$\mathrm{E}(X \mid \mathcal{F}_t) = \sum_0^\infty I_n(F_n \mathbf{1}_{D_{n,t}}). \tag{7.13}$$

PROOF. It is easily seen that I_1 maps $L^2_{\mathbb{R}}(D_{1,t}) = L^2_{\mathbb{R}}([0,t))$ onto H_t. It follows by Theorems 7.3 and 4.1 (applied to H_t), that I_n maps $L^2(D_{n,t}) = L^2(D_{1,t})^{\odot n}$ onto $H_t^{:n:}$, and that $\bigoplus_0^\infty I_n$ maps $\bigoplus_0^\infty L^2(D_{n,t})$ onto $\bigoplus H_t^{:n:} = L^2(\Omega, \mathcal{F}_t, \mathrm{P})$.

If $X = \sum_0^\infty I_n(F_n)$, then the right hand side of (7.13) converges to an element Y of $L^2(\mathcal{F}_t)$, and if $Z \in L^2(\mathcal{F}_t)$, then $Z = \sum_0^\infty I_n(G_n)$ with $G_n \in L^2(D_{n,t})$, and thus

$$\mathrm{E}(YZ) = \sum_0^\infty \int_{D_n} F_n \mathbf{1}_{D_{n,t}} G_n = \sum_0^\infty \int_{D_n} F_n G_n = \mathrm{E}(XZ).$$

Since Z is arbitrary, $Y = \mathrm{E}(X \mid \mathcal{F}_t)$. $\square$

COROLLARY 7.11. *If $F \in L^2(D_n)$, then*

$$X_t = \int_0^t \int_0^{t_n} \cdots \int_0^{t_2} F(t_1, \dots, t_n)\, dB_{t_1} \cdots dB_{t_n}$$

is the martingale $\mathrm{E}(I_n(F) \mid \mathcal{F}_t)$. $\square$

EXAMPLE 7.12. Let $t > 0$ be fixed, and consider the indicator functions

$$f_n(t_1, \dots, t_n) = \mathbf{1}[0 < t_1 < \dots < t_n < t].$$

In particular, $f_1 = \mathbf{1}_{(0,t)}$, and (7.4) yields

$$f_1^{\odot n} = n! f_n.$$

Since $\int_0^\infty f_1\, dB = \int_0^t dB = B_t$, we obtain from Theorems 7.6 and 7.3

$$\int_0^t \int_0^{t_n} \cdots \int_0^{t_2} dB_{t_1} \cdots dB_{t_n} = I_n(f_n) = \frac{1}{n!} I_n(f_1^{\odot n}) = \frac{1}{n!} :B_t^n:.$$

By this equation for n and $n+1$ (or by Theorem 7.5), we obtain

$$\int_0^t :B_s^n:\, dB_s = \frac{1}{n+1} :B_t^{n+1}:, \qquad n \geq 0. \tag{7.14}$$

In particular, we have proved the well-known formula

$$\int_0^t B_s\, dB_s = \frac{1}{2} :B_t^2: = \frac{1}{2}(B_t^2 - t). \tag{7.15}$$

Note that B_t is not a square integrable predictable process, but the restriction to any finite interval is. Hence the integral in (7.15) is defined, although the infinite integral $\int_0^\infty B_s\,dB_s$ diverges. The same remark applies e.g. to (7.14).

EXAMPLE 7.13. Consider the Wick exponential

$$:\exp B_t: = \sum_0^\infty \frac{1}{n!}:B_t^n:.$$

It is easily seen, using Corollary 3.10, that this sum converges in $L^2(\Omega)$ for every fixed t, and in $L^2([0,u]\times\Omega, dt\,d\mathrm{P})$ for any fixed u. Hence $:\exp B_t:$ is a square integrable predictable process on every finite interval, and by (7.14),

$$\int_0^t :\exp B_s:\,dB_s = \sum_0^\infty \frac{1}{(n+1)!}:B_t^{n+1}: = :\exp B_t: - 1.$$

In other words, the stochastic integral equation $1 + \int_0^t X_s\,dB_s = X_t$ is solved by

$$:\exp B_t: = \exp(B_t - \mathrm{E}\,B_t^2/2) = \exp(B_t - t/2),$$

where we evaluate the Wick exponential using Theorem 3.33. This is a well-known fact in the theory of stochastic integration, usually proved by Itô's formula.

Conversely, if we assume this fact, and that the solution moreover is unique, we obtain another proof of Theorem 3.33 for real ξ. (Obviously, it suffices to prove Theorem 3.33 with $\xi = B_t$.)

The last examples show that Itô integrals of Wick powers and Wick exponentials behave as ordinary integrals of ordinary powers and exponentials. Consequently, the 'correction terms' in the Wick and Itô calculi (i.e. the terms not appearing in ordinary calculus), must be closely related. For example, it should be possible to deduce Itô's formula from the results above and the formulae for in Chapter 3. We will not attempt to do this in general, but give a simple case as an illustration.

EXAMPLE 7.14. Suppose that $\xi = \int_0^\infty f(t)\,dB_t$ and $\eta = \int_0^\infty g(t)\,dB_t$ are two elements of $H(B)$. Consider the stochastic processes

$$\xi_t = \int_0^t f(s)\,dB_s \quad \text{and} \quad \eta_t = \int_0^t g(s)\,dB_s.$$

By Theorems 7.3 and Theorem 7.6,

$$\begin{aligned} :\xi\eta: &= I_2(f\odot g) = I_2\big(f(s)g(t) + f(t)g(s)\big) \\ &= \int_0^\infty \int_0^t \big(f(s)g(t) + g(s)f(t)\big)\,dB_s dB_t \\ &= \int_0^\infty \xi_t g(t)\,dB_t + \int_0^\infty \eta_t f(t)\,dB_t. \end{aligned}$$

Introducing the standard notations $d\xi_t = f(t)dB_t$ and $d\eta_t = g(t)dB_t$, we can write this as $:\xi\eta: = \int_0^\infty \xi_t \, d\eta_t + \int_0^\infty \eta_t \, d\xi_t$, and thus

$$\xi\eta = :\xi\eta: + \mathrm{E}\,\xi\eta = \int_0^\infty \xi_t \, d\eta_t + \int_0^\infty \eta_t \, d\xi_t + \int_0^\infty f(t)g(t)\, dt.$$

More generally, by substituting $f\mathbf{1}_{[0,t]}$ and $g\mathbf{1}_{[0,t]}$ above,

$$\xi_t\eta_t = \int_0^t \xi_s \, d\eta_s + \int_0^t \eta_s \, d\xi_s + \int_0^t f(s)g(s)\, ds,$$

a simple instance of Itô's formula.

EXAMPLE 7.15. As a final example of the connections with the Itô calculus, we give here the proof of Theorem 5.1 by Neveu (1976). (This example requires knowledge of Itô's formula and other features not treated above.)

We will actually prove Corollary 5.7. The full Theorem 5.1 follows easily as remarked in Chapter 5.

We may assume that $1 \le p, q < \infty$, and that $f, g \ge 0$. By standard approximation arguments, we may further assume that f and g are bounded and $f, g \ge \delta$ for some $\delta > 0$. Let B_t and B'_t be two independent Brownian motions, and define $B''_t = \rho B_t + \sqrt{1-\rho^2} B'_t$. Then B''_t is another Brownian motion, and $\mathrm{Cov}(B_t, B''_t) = \rho t$. We will prove the result with $\xi = B_1$ and $\eta = B''_1$.

The random variable $X = f(B_1)^p$ is bounded and $\mathcal{F}_1$-measurable, and thus by Theorems 7.8 and 7.10 there exists a square integrable predictable process Z_t such that

$$X = \mathrm{E}\,X + \int_0^1 Z_t \, dB_t.$$

We define

$$X_t = \mathrm{E}\,X + \int_0^t Z_s \, dB_s.$$

Thus $X_1 = X$ and $X_0 = \mathrm{E}X$. By Theorem 7.10, $X_t = \mathrm{E}(X \mid \mathcal{F}_t)$ and thus $0 < \delta^p \le X_t \le C^p < \infty$ a.s., for some C and every t.

We similarly define $Y = g(B''_1)^q$ and have $Y = Y_1$ with

$$Y_t = \mathrm{E}\,Y + \int_0^t W_s \, dB''_s.$$

Now define $U_t = X_t^{1/p} Y_t^{1/q}$. Then $U_1 = f(\xi)g(\eta)$ and

$$U_0 = (\mathrm{E}\,X)^{1/p}(\mathrm{E}\,Y)^{1/q} = \|f(\xi)\|_p \|g(\eta)\|_q,$$

so we want to prove $\operatorname{E} U_1 \le U_0$. We apply Itô's formula and obtain

$$\begin{aligned} U_1 = U_0 &+ \int_0^1 \tfrac{1}{p} X_t^{-1} U_t \, dX_t + \int_0^1 \tfrac{1}{q} Y_t^{-1} U_t \, dY_t \\ &+ \tfrac{1}{2} \int_0^1 \tfrac{1}{p}(\tfrac{1}{p} - 1) X_t^{-2} U_t \, d[X,X]_t + \int_0^1 \tfrac{1}{p}\tfrac{1}{q} X_t^{-1} Y_t^{-1} U_t \, d[X,Y]_t \\ &+ \tfrac{1}{2} \int_0^1 \tfrac{1}{q}(\tfrac{1}{q} - 1) Y_t^{-2} U_t \, d[Y,Y]_t, \end{aligned} \tag{7.16}$$

where $dX_t = Z_t \, dB_t$, $dY_t = W_t \, dB''_t$, $d[X,X]_t = Z_t^2 \, d[B,B]_t = Z_t^2 \, dt$, $d[X,Y]_t = Z_t W_t \, d[B,B'']_t = Z_t W_t \rho \, dt$ and $d[Y,Y]_t = W_t^2 \, dt$. The first two integrals in (7.16) have mean zero, while the remaining three may be combined into

$$\tfrac{1}{2} \int_0^1 \left(-(p-1) \left(\frac{Z_t}{pX_t} \right)^2 + 2\rho \frac{Z_t}{pX_t} \frac{W_t}{qY_t} - (q-1) \left(\frac{W_t}{qY_t} \right)^2 \right) U_t \, dt \le 0,$$

because the quadratic form $(p-1)x^2 - 2\rho xy + (q-1)y^2$ is positive semi-definite when $\rho^2 \le (p-1)(q-1)$. Hence $\operatorname{E} U_1 \le \operatorname{E} U_0 = U_0$, as we wanted to show.

2. Stochastic integrals on general measure spaces

We saw in the preceding section that the stochastic integral $f \mapsto \int_0^\infty f \, dB$ is an isometry of $L^2_{\mathbb{R}}([0,\infty), dt)$ onto the Gaussian Hilbert space $H(B)$. We can use this property as the definition of (one kind of) stochastic integrals on general measure spaces.

DEFINITION 7.16. A *Gaussian stochastic integral* on a measure space $(M, \mathcal{M}, \mu)$ is a linear isometry I of $L^2_{\mathbb{R}}(M, \mathcal{M}, \mu)$ into a Gaussian Hilbert space H.

In other words, a Gaussian stochastic integral on $(M, \mathcal{M}, \mu)$ is the same as a Gaussian field on $L^2_{\mathbb{R}}(M, \mathcal{M}, \mu)$. Since the range of I is a closed subspace of H, and thus another Gaussian Hilbert space, we may without essential loss assume that I maps onto H; then H is a Gaussian Hilbert space indexed by $L^2_{\mathbb{R}}(M, \mathcal{M}, \mu)$.

We will shortly see that the stochastic integral defined above can also be defined using a certain stochastic measure. We let $\mathcal{M}_\mu = \{A \in \mathcal{M} : \mu(A) < \infty\}$ denote the family of measurable sets with finite measures.

DEFINITION 7.17. A *Gaussian stochastic measure* on a measure space $(M, \mathcal{M}, \mu)$ is a family $Z(A)$, $A \in \mathcal{M}_\mu$, of random variables (defined on a common probability space) such that

(i) if $A \in \mathcal{M}_\mu$, then

$$Z(A) \sim \mathrm{N}(0, \mu(A)); \tag{7.17}$$

(ii) if $A_1, \dots, A_n$ is a finite family of disjoint sets in $\mathcal{M}_\mu$, then the variables $Z(A_i)$ are independent and

$$Z\Big(\bigcup_1^n A_i\Big) = \sum_1^n Z(A_i) \qquad \text{a.s.} \tag{7.18}$$

REMARK 7.18. We assume only finite additivity in (ii), but this turns out to be equivalent to σ-additivity. In fact, if $A = \bigcup_1^\infty A_i$ where $(A_i)_1^\infty$ is a sequence of disjoint measurable sets such that $A \in \mathcal{M}_\mu$, then

$$\mathrm{E}\,\Big|Z(A) - \sum_1^n Z(A_i)\Big|^2 = \mathrm{E}\,\Big|Z(A \setminus \bigcup_1^n A_i)\Big|^2 = \mu(A \setminus \bigcup_1^n A_i) \to 0,$$

as $n \to \infty$. Hence $\sum_1^\infty Z(A_i)$ converges to $Z(A)$ in L^2, which implies convergence a.s. because the summands are independent.

REMARK 7.19. It is easily seen, e.g. using Theorem 7.20 below, that a family $Z(A)$, $A \in \mathcal{M}_\mu$, of random variables is a Gaussian stochastic measure if and only if the variables are jointly centred Gaussian and

$$\operatorname{Cov}(Z(A), Z(B)) = \mu(A \cap B), \qquad A, B \in \mathcal{M}_\mu.$$

THEOREM 7.20. *If $I\colon L^2_{\mathbb{R}}(M, \mathcal{M}, \mu) \to H$ is a Gaussian stochastic integral, then $Z(A) = I(\mathbf{1}_A)$, $A \in \mathcal{M}_\mu$, defines a Gaussian stochastic measure on $(M, \mathcal{M}, \mu)$.*

Conversely, every Gaussian stochastic measure corresponds in this way to a unique Gaussian stochastic integral.

PROOF. If $Z(A) = I(\mathbf{1}_A)$, where I is a Gaussian stochastic integral, then (7.17) and (7.18) follow immediately from the assumption that I is a linear isometry into H. Moreover, if $A_i \cap A_j = \varnothing$, then $\operatorname{Cov}(Z(A_i), Z(A_j)) = \int \mathbf{1}_{A_i}\mathbf{1}_{A_j} = 0$, which implies that $Z(A_i)$ and $Z(A_j)$ are independent.

Conversely, suppose that $A \mapsto Z(A)$ is a Gaussian stochastic measure. If

$$f = \sum_1^n c_i \mathbf{1}_{A_i}, \qquad A_i \in \mathcal{M}_\mu, \tag{7.19}$$

is a real integrable simple function, define

$$I(f) = \sum_1^n c_i Z(A_i).$$

It is easily seen from (7.18) that $I(f)$ is well-defined and independent of the representation (7.19). Moreover, if we choose a representation with the sets A_i disjoint, then $I(f)$ is expressed as a sum of independent centred Gaussian variables so $I(f)$ is centred Gaussian and

$$\mathrm{E}\, I(f)^2 = \sum c_i^2\, \mathrm{E}(Z(A_i)^2) = \sum c_i^2 \mu(A_i) = \int f^2 d\mu.$$

Hence I is a linear isometry of the subspace of simple functions in $L^2_{\mathbb{R}}(M, \mathcal{M}, \mu)$ onto the Gaussian linear space $G = \{I(f) : f \text{ simple}\}$. By Theorem 1.3, the closed hull $\overline{G}$ is a Gaussian Hilbert space, and since the simple functions are dense in L^2, we may extend I to a Gaussian stochastic integral $L^2_{\mathbb{R}}(M, \mathcal{M}, \mu) \to \overline{G}$. By definition, $I(\mathbf{1}_A) = Z(A)$, and it is clear that this determines I uniquely. □

The stochastic integral corresponding to a (Gaussian) stochastic measure Z is denoted by $f \mapsto \int f\,dZ$.

REMARK 7.21. By Theorem 1.23, there exists a Gaussian stochastic integral on any measure space. Hence there also exists a Gaussian stochastic measure on any measure space. Obviously, all Gaussian stochastic measures on the same measure space have the same distribution.

EXAMPLE 7.22. A Brownian motion B_t on $\mathbb{R}_+$ defines by Section 1 a Gaussian stochastic integral $\int f\,dB$ on $L^2(\mathbb{R}_+, dt)$. Hence we obtain a Gaussian stochastic measure, also denoted by B, on $(\mathbb{R}_+, dt)$. Clearly,

$$B((s,t]) = B_t - B_s, \qquad 0 \leq s < t < \infty.$$

EXAMPLE 7.23. Let M consist of d copies of the positive half-line $\mathbb{R}_+$ with Lebesgue measure; formally $M = \{1, \dots, d\} \times [0, \infty)$. Then $L^2(M)$ can be identified with $L^2(\mathbb{R}_+, dt)^d$. It is easily seen, generalizing Theorem 7.1, that if $B^1_t, \dots, B^d_t$ are d independent Brownian motions, then $(f_j)_1^d \mapsto \sum_1^d \int_0^\infty f_j\,dB^j_t$ defines a Gaussian stochastic integral on M.

EXAMPLE 7.24. The white noise measure on $S'(\mathbb{R}^d)$ in Example 1.16 defines a Gaussian stochastic integral on $\mathbb{R}^d$ (equipped with Lebesgue measure), which we may write as

$$f \mapsto \int_{\mathbb{R}^d} f\,dB.$$

The corresponding Gaussian stochastic measure is

$$B(A) = \int_{\mathbb{R}^d} \mathbf{1}_A\,dB$$

and the Gaussian process

$$B(t_1, \dots, t_d) = B\Big(\prod_1^d [0, t_i]\Big) = \int_0^{t_1} \cdots \int_0^{t_d} dB, \qquad t_i \geq 0,$$

is a kind of d-dimensional generalization of the Brownian motion, sometimes called a *Brownian sheet*. (Warning: The process usually called d-parameter Brownian motion is different, see Example 8.10.)

Note that the rather complicated construction in Example 1.16 defines the random variables given by this stochastic integral and the corresponding stochastic measure or stochastic process as functions on the space $S'(\mathbb{R}^d)$ of distributions. If we do not care about this, and allow the variables to be any random variables (defined on an unspecified probability space), we can use the much simpler construction of Example 1.22 instead, cf. Remark 7.21.

We may define multiple stochastic integrals using tensor products as in Section 1. Multiple stochastic integrals of this type were introduced (using an equivalent definition) by Itô (1951).

Observe first that I extends to an isometry $L^2_{\mathbb{C}}(M,\mathcal{M},\mu) \to H_{\mathbb{C}}$; hence it does not matter whether we use real or complex scalars. Proposition E.16 and Theorem 4.1 then yield the following analogue of Theorem 7.3, where $L^2(M^n,\mathcal{M}^{\odot n},\mu^{\odot n})$ is the subspace of symmetric functions in $L^2(M^n,\mu^{\odot n}) = L^2(M^n,\frac{1}{n!}\mu^n)$.

THEOREM 7.25. *Suppose that* $I\colon L^2_{\mathbb{R}}(M,\mathcal{M},\mu) \to H$ *is a Gaussian stochastic integral on a* σ*-finite measure space. Then there exist isometries* $\hat{I}_n\colon L^2(M^n,\mathcal{M}^{\odot n},\mu^{\odot n}) \to H^{:n:}$, $n \geq 0$, *and* $\bigoplus_0^\infty \hat{I}_n\colon \bigoplus_0^\infty L^2(M^n,\mathcal{M}^{\odot n},\mu^{\odot n}) \to L^2(\Omega,\mathcal{F}(H),\mathrm{P})$ *such that*

$$\hat{I}_n(f_1\odot\cdots\odot f_n) = {:}I(f_1)\cdots I(f_n){:}. \tag{7.20}$$

These isometries are onto if I *is.* □

We may extend $\hat{I}_n$ to $L^2(M^n,\mathcal{M}^n,\mu^n)$ by symmetrization. Define, for a function f on M^n and a permutation $\pi \in \mathfrak{S}_n$,

$$f\circ\pi(x_1,\dots,x_n) = f(x_{\pi(1)},\dots,x_{\pi(n)}).$$

Define further

$$\operatorname{Sym} f = \frac{1}{n!}\sum_{\pi\in\mathfrak{S}_n} f\circ\pi;$$

for $f \in L^2(M^n)$; this is the orthogonal projection onto the subspace of symmetric functions. We can thus define $\hat{I}_n$ on $L^2(M^n)$ by

$$\hat{I}_n(f) = \hat{I}_n(\operatorname{Sym} f)$$

and obtain

$$\mathrm{E}\,|\hat{I}_n(f)|^2 = \frac{1}{n!}\int_{M^n} |\operatorname{Sym} f|^2\,d\mu^n \leq \frac{1}{n!}\int_{M^n} |f|^2\,d\mu^n.$$

In the special case treated in the preceding section, where $M = [0,\infty)$, we interpreted $\hat{I}_n$ as a multiple stochastic integral over the domain $D_n \subset [0,\infty)^n$, which can be regarded as an $(n!)$th of the whole product space $[0,\infty)^n$. In the present generality, we cannot select any corresponding subset and it is more natural to regard $\hat{I}_n$ as $1/n!$ times a multiple stochastic integral over M^n. We thus define

$$I_n(f) = n!\,\hat{I}_n(f), \qquad f \in L^2(M^n).$$

If Z is the Gaussian stochastic measure corresponding to I by Theorem 7.20, we also write $I_n(f)$ as $\int_M\cdots\int_M f(x_1,\dots,x_n)\,dZ(x_1)\cdots dZ(x_n)$ or, shorter,

$\int_{M^n} f\, dZ^n$; we further write $\int_A f\, dZ^n = \int_{M^n} \mathbf{1}_A f\, dZ^n$ when $A \subseteq M^n$ is measurable.

We can rewrite the definition as

$$\int_{M^n} f\, dZ^n = I_n(f) = n!\, \hat{I}_n(\operatorname{Sym} f) = \sum_{\pi \in \mathfrak{S}_n} \hat{I}_n(f \circ \pi). \tag{7.21}$$

In particular, by (E.7) and (7.20), for $f_1, \dots, f_n \in L^2(M)$,

$$\int_M \cdots \int_M f_1(x_1) \cdots f_n(x_n)\, dZ(x_1) \cdots dZ(x_n) = \hat{I}_n\Big(\sum_{\pi \in \mathfrak{S}_n} f_1(x_{\pi(1)}) \cdots f_n(x_{\pi(n)})\Big)$$
$$= \hat{I}_n(f_1 \odot \cdots \odot f_n) = :\int_M f_1\, dZ \cdots \int_M f_n\, dZ:.$$

We collect from the above the basic properties of the multiple stochastic integral.

THEOREM 7.26. *Let Z be a Gaussian stochastic measure on a σ-finite measure space $(M, \mathcal{M}, \mu)$. Let H be the Gaussian space spanned by $\{Z(A)\}$. Then the mapping*

$$f \mapsto I_n(f) = \int_{M^n} f\, dZ^n = \int_M \cdots \int_M f\, dZ(x_1) \cdots dZ(x_n) \tag{7.22}$$

is a bounded linear operator $L^2(M^n, \mu^n) \to H^{:n:} \subset L^2(\Omega, \mathcal{F}, \mathrm{P})$ such that for any $f_1, \dots, f_n \in L^2(M, \mu)$,

$$\int_M \cdots \int_M f_1(x_1) \cdots f_n(x_n)\, dZ(x_1) \cdots dZ(x_n) = :\int_M f_1\, dZ \cdots \int_M f_n\, dZ:. \tag{7.23}$$

Moreover,

$$\mathrm{E}\Big|\int_M \cdots \int_M f\, dZ(x_1) \cdots dZ(x_n)\Big|^2 = n! \int_{M^n} |\operatorname{Sym} f|^2\, d\mu^n$$
$$= \sum_{\pi \in \mathfrak{S}_n} \int_{M^n} \overline{f} f \circ \pi\, d\mu^n$$
$$\leq n! \int_{M^n} |f|^2\, d\mu^n, \tag{7.24}$$

and the multiple integral (7.22) maps the subspace of symmetric functions in $L^2(M^n)$ isomorphically onto $H^{:n:}$ (multiplying the norms by $\sqrt{n!}$). Hence every random variable $X \in L^2(\Omega, \mathcal{F}(H), \mathrm{P})$ has a (unique) expansion

$$X = \sum_0^\infty \int_{M^n} f_n\, dZ^n,$$

where each f_n is a symmetric function on M^n and

$$\sum_0^\infty n! \int_{M^n} |f_n|^2 \, d\mu^n = \mathrm{E}\,|X|^2 < \infty.$$

□

Note that we have a Wick product and not an ordinary product on the right hand side of (7.23). For example, with $n = 2$, using (3.4),

$$\begin{aligned}\iint f(x)g(y)\,dZ(x)dZ(y) &= \int f\,dZ \int g\,dZ - \mathrm{E}\Big(\int f\,dZ \int g\,dZ\Big) \\ &= \int f\,dZ \int g\,dZ - \int fg\,d\mu.\end{aligned}$$

REMARK 7.27. If the functions f_i have supports in disjoint subsets of M, then the integrals $\int f_i\,dZ$ are independent, and thus the Wick product in (7.23) coincides with the ordinary product. Note that linear combinations of products $f_1(x_1)\cdots f_n(x_n)$ with disjointly supported $f_1, \dots, f_n$ are dense in $L^2(M^n \setminus \Delta, \mu^n)$, where $\Delta = \{(x_i)_1^n : x_i = x_j \text{ for some } i < j\}$ is the union of the 'diagonals'. If, further, μ is non-atomic, then $\mu^n(\Delta) = 0$ and thus $L^2(M \setminus \Delta) = L^2(M)$. This suggests that the 'renormalization' involved in using the Wick product in (7.23) can be loosely interpreted as integrating outside the diagonals only. (But observe that if μ has atoms, then the diagonals cannot be eliminated completely.)

REMARK 7.28. We could have defined the multiple stochastic integral by (7.23). In fact, the right hand side of (7.23) defines a multilinear operator on $L^2(M)^n$, and it is easy to show that this operator defines a unique bounded linear operator on the tensor product $L^2(M)^{\otimes n} = L^2(M^n)$.

Some authors, following Itô (1951), begin with (7.23) for indicator functions $f_i = \mathbf{1}_{A_i}$ with disjoint supports, in which case the right hand side is just the product $\prod Z(A_i)$. If μ is non-atomic, the multiple integral may then be extended by linearity and continuity to $L^2(M^n)$, but if μ has atoms, then an extra step is needed in the definition. Note that our definition, using Wick products, avoids the need for a special treatment when μ has atoms.

For Brownian motion, we have now defined two multiple stochastic integrals, viz. the integrals over D_n in Section 1 and the integrals over $\mathbb{R}_+^n$ defined here. The difference is mainly notational, and the exact relation is as follows.

THEOREM 7.29. *Let B_t, $t \geq 0$, be a Brownian motion. If $f \in L^2(\mathbb{R}_+^n)$, then*

$$\int_{\mathbb{R}_+^n} f\,dB^n = \int_{D_n} \tilde{f}\,dB_{t_1} \cdots dB_{t_n}, \tag{7.25}$$

where

$$\tilde{f} = n!\,\mathrm{Sym}\,f = \sum_\pi f \circ \pi.$$

In particular, if f vanishes outside D_n, then

$$\int_{D_n} f\,dB_{t_1}\cdots dB_{t_n} = \int_{\mathbb{R}_+^n} f\,dB^n = \int_{D_n} f\,dB^n. \tag{7.26}$$

PROOF. This is an exercise in our notation. $\widetilde{f}$ is a symmetric function in $L^2(\mathbb{R}_+^n)$ so $\widetilde{f} \in L^2(\mathbb{R}_+^n, \mathcal{B}^{\odot n}, dt^{\odot n})$; moreover, $\widetilde{f}$ and its restriction to D_n correspond to the same element of $L^2(\mathbb{R}_+, dt)^{\odot n}$ under the two identifications used in this section and the preceding one. Consequently,

$$\int_{\mathbb{R}_+}\cdots\int_{\mathbb{R}_+} f\,dB(t_1)\cdots dB(t_n) = \hat{I}_n(\widetilde{f}) = \int_0^\infty\cdots\int_0^{t_2} \widetilde{f}\,dB_{t_1}\cdots dB_{t_2},$$

which proves (7.25).

An alternative, more concrete argument is that if $f(x) = f_1(x_1)\cdots f_n(x_n)$, with $f_i \in L^2(\mathbb{R}_+)$, then both sides of (7.25) equal $:\int f_1\,dB\cdots\int f_n\,dB$: by Theorems 7.26 and 7.3, and such functions are total in $L^2(\mathbb{R}_+^n)$.

Finally, if f vanishes outside D_n, then $\widetilde{f}$ equals f on D_n, and (7.26) follows. □

REMARK 7.30. In other words, if we identify $L^2(D_n)$ with the subspace of $L^2(\mathbb{R}_+^n)$ consisting of the functions that vanish outside D_n, then the operator $I_n\colon L^2(D_n) \to H^{:n:}$ defined in Section 1 is the restriction of the operator $I_n\colon L^2(\mathbb{R}_+^n) \to H^{:n:}$ defined above. (This is why we allow ourselves to use the same notation.)

We will next study the algebra of multiple stochastic integrals. We first need some definitions.

If f is a function on M^2, define its *contraction* to be

$$Cf = \int_M f(x,x)\,d\mu$$

(provided this integral exists). More generally, if f is a function on M^n, and $1 \le i < j \le n$, the (i,j) contraction of f is the function

$$C_{ij}f = \int f(x_1,\dots,x_i,\dots,x_{j-1},x_i,x_{j+1},\dots,x_n)\,d\mu(x_i)$$

on M^{n-2}. Now, let f be a function on M^n. We will use Feynman diagrams as in Section 1.5, but labelled by the coordinates $x_1,\dots,x_n$ in M^n instead of random variables. Given such a Feynman diagram γ, with edges (i_k, j_k), $k = 1,\dots,r$, we define the corresponding contraction by

$$C_\gamma f = C_{i_1j_1}\cdots C_{i_rj_r}f. \tag{7.27}$$

Note that $C_\gamma f$ is a function on M^{n-2r}; in particular, if γ is a complete Feynman diagram, then $C_\gamma f$ is a real or complex number. We will only consider the case when $C_\gamma|f| < \infty$ a.e. on M^{n-2r}; then $C_\gamma f$ is defined a.e. and, by Fubini's theorem, the order of the contractions in (7.27) does not matter.

LEMMA 7.31. *Suppose that $f_i \in L^2(M^{n_i})$, $i = 1, \dots, k$, and that $f = f_1 \otimes \cdots \otimes f_k$ is the function on M^n, where $n = \sum_1^k n_i$, given by*

$$f(x_{11}, \dots, x_{1n_1}, \dots, x_{k1}, \dots, x_{kn_k}) = \prod_i f_i(x_{i1}, \dots, x_{in_i}). \tag{7.28}$$

Then, for any Feynman diagram γ labelled by $\{x_{ij}\}_{ij}$ such that no edge connects two variables $x_{i_1j_1}$ and $x_{i_2j_2}$ with $i_1 = i_2$, we have $C_\gamma|f| < \infty$ a.e., $C_\gamma f \in L^2(M^{n-2r(\gamma)})$, and

$$\|C_\gamma f\|_{L^2} \le \prod_{i=1}^k \|f_i\|_{L^2}. \tag{7.29}$$

PROOF. It suffices to prove (7.29), since this applied to $|f|$, for which all integrals are well-defined, shows that $C_\gamma|f| < \infty$ a.e., and thus $C_\gamma f$ is well-defined.

We use induction on k. The case $k = 1$ is trivial since then γ has no edges and thus $C_\gamma(f) = f$.

Now suppose $k = 2$. By permuting the coordinates, we may assume that the $r = r(\gamma)$ edges in γ join $(1, j)$ and $(2, j)$, $j = 1, \dots, r$. Then, with $x' \in M^{n_1-r}$, $x'' \in M^{n_2-r}$ and $y \in M^r$,

$$C_\gamma f(x', x'') = \int_{M^r} f_1(y, x') f_2(y, x'') \, d\mu^r(y),$$

and thus, by the Cauchy–Schwarz inequality,

$$|C_\gamma f(x', x'')|^2 \le \int_{M^r} |f_1(y, x')|^2 \, d\mu^r(y) \int_{M^r} |f_2(y, x'')|^2 \, d\mu^r(y),$$

which gives $\|C_\gamma f\|_{L^2}^2 \le \|f_1\|_{L^2}^2 \|f_2\|_{L^2}^2$ by integration. (The cases $n_1 - r = 0$, $n_2 - r = 0$ or $r = 0$ are covered by the same argument with suitable interpretations.)

For $k > 2$, let $f' = f_2 \otimes \cdots \otimes f_k$. Further, let γ_1 be the subdiagram of γ containing only the edges with at least one endpoint in $\{(1, j)\}_{j=1}^{n_1}$, and let γ_2 be the restriction of γ to $\{(i, j) : i \ge 2\}$. Then

$$C_\gamma f = C_{\gamma_1}(f_1 \otimes C_{\gamma_2} f')$$

and thus, by the case $k = 2$ and induction,

$$\|C_\gamma f\|_{L^2} \le \|f_1\|_{L^2} \|C_{\gamma_2} f'\|_{L^2} \le \|f_1\|_{L^2} \prod_{i=2}^k \|f_i\|_{L^2}.$$

□

REMARK 7.32. It is important that we consider only f as in (7.28) (i.e. tensor products) in Lemma 7.31. The simple case $k = 2$, $n_1 = n_2 = 1$, $C_\gamma f = \int f(x, x) \, d\mu(x)$ shows that C_γ is in general not a bounded operator on L^2 (and indeed not even well-defined for general $f \in L^2$).

We can now state a multiplication formula.

THEOREM 7.33. *Suppose that Z is a Gaussian stochastic measure on a σ-finite measure space $(M, \mathcal{M}, \mu)$. Let $f_i \in L^2(M^{n_i})$, $i = 1, \dots, k$, and let $Y_i = \int f_i \, dZ^{n_i}$. Then*

$$Y_1 \cdots Y_k = \sum_\gamma \int C_\gamma(f) \, dZ^{n-2r(\gamma)}, \tag{7.30}$$

where $n = \sum_1^k n_i$, $f = f_1 \otimes \cdots \otimes f_k$ is the function on M^n given by (7.28), and the sum is taken over all Feynman diagrams γ labelled by $\{x_{ij}\}_{ij}$ such that no edge connects two variables $x_{i_1 j_1}$ and $x_{i_2 j_2}$ with $i_1 = i_2$.

In particular,

$$\mathrm{E}(Y_1 \cdots Y_k) = \sum_\gamma C_\gamma(f), \tag{7.31}$$

where we sum over all complete such Feynman diagrams.

PROOF. Lemma 7.31 shows that all contractions and stochastic integrals are defined. It follows also, by Theorem 7.26, that the right hand side of (7.30) is a continuous multilinear operator $\prod_i L^2(M^{n_i}) \to L^2(\Omega, \mathcal{F}, \mathrm{P})$; and so also is the left hand side by Theorems 7.26 and 3.46.

Hence it suffices to prove (7.30) in the special case $f_i = f_{i1} \otimes \cdots \otimes f_{in_i}$, with $f_{ij} \in L^2(M)$. In this case, let $\xi_{ij} = \int f_{ij} \, dZ$. Then (7.23) yields $Y_i = {:}\xi_{i1} \cdots \xi_{in_i}{:}$. Similarly, it is easily seen that if we label the Feynman diagram γ by $\{\xi_{ij}\}$ and compute its Wick value, then

$${:}v(\gamma){:} = \int C_\gamma(f) \, dZ^{n-2r(\gamma)}.$$

Hence (7.30) is, in this special case, an instance of Theorem 3.15. This completes the proof of (7.30), and (7.31) follows by taking expectations, since all non-trivial stochastic integrals on the right hand side have vanishing expectations. □

REMARK 7.34. For the general Wick product we similarly obtain, using the same notations, simply

$$Y_1 \odot \cdots \odot Y_k = \int f \, dZ^n.$$

We can obtain similar formulae for products of the iterated Itô integrals defined in Section 1. By Theorem 7.29, those integrals coincide with the multiple integrals treated here, so we may compute the product by (7.30), and then express each multiple stochastic integral as an iterated Itô integral by (7.25). The need for symmetrization makes the result a little more complicated than (7.30). Therefore we give only an example, and leave the general formulation to the reader.

EXAMPLE 7.35. For $k = 2$, $n_1 = 2$, $n_2 = 1$ we obtain from Theorem 7.33

$$\begin{aligned}\int f(s,t)\,dZ^2 \int g(u)\,dZ &= \int f(s,t)g(u)\,dZ^3 + \iint f(s,t)g(s)\,d\mu(s)\,dZ(t) \\ &\qquad + \iint f(t,s)g(s)\,d\mu(s)\,dZ(t) \\ &= \int f(s,t)g(u)\,dZ^3 + \iint \big(f(s,t)+f(t,s)\big)g(s)\,d\mu(s)\,dZ(t).\end{aligned}$$

For iterated Itô integrals with respect to Brownian motion, this yields for $f \in L^2(D_2)$ and $g \in L^2(\mathbb{R}_+)$, extending f by 0 outside D_2 and using (7.26) and (7.25),

$$\begin{aligned}&\int_0^\infty \int_0^t f(s,t)\,dB_s\,dB_t \int_0^\infty g(u)\,dB_u \\ &\quad = \int_0^\infty \int_0^u \int_0^t \big(f(s,t)g(u) + f(s,u)g(t) + f(t,u)g(s)\big)\,dB_s\,dB_t\,dB_u \\ &\qquad + \int_0^\infty \int_0^t f(s,t)g(s)\,ds\,dB_t + \int_0^\infty \int_t^\infty f(t,u)g(u)\,du\,dB_t.\end{aligned}$$

COROLLARY 7.36. *If H is a Gaussian Hilbert space and $Y \in H^{:n:}$, then*

$$\|Y\|_{2m} \le c(2m,n)^{1/2m}(n!)^{-1/2}\|Y\|_2, \tag{7.32}$$

where $c(2m,n)$ is the number of complete Feynman diagrams on the vertices $\{(i,j)\}$, $1 \le i \le 2m$, $1 \le j \le n$, such that no edge connects two vertices (i,j_1) and (i,j_2) with the same first coordinate.

Equality holds in (7.32) *when $Y = {:}\xi^n{:}$ for some $\xi \in H$.*

PROOF. We may without loss of generality assume that H is the range of a Gaussian stochastic integral $f \mapsto \int f\,dZ$ on some measure space $(M, \mathcal{M}, \mu)$. Then $Y = \int f\,dZ^n$ for some symmetric $f \in L^2(M^n)$. Evaluate $\mathrm{E}\,|Y|^{2m} = \mathrm{E}\,Y^m\overline{Y}^m$ by (7.31). There are $c(2m,n)$ terms and each is by (7.29) bounded by $\|f\|_2^{2m} = n!^{-m}\|Y\|_2^{2m}$.

That equality holds when $Y = {:}\xi^n{:}$ follows by the same argument, or by Theorem 3.12. ∎

The sharp bound given by the corollary is better than the bound $(2m-1)^{n/2}$ provided by Theorem 5.10 and Remark 5.11. For fixed $m \ge 2$, we see by Theorem 5.19 and Remark 5.20 that the ratio between the two bounds grows as $n^{1/4}$.

REMARK 7.37. The corollary shows that if $n \ge 1$ and p is an even integer ≥ 2, then $\|X\|_p/\|X\|_2$ is maximized for $X \in H^{:n:}$ by taking X to be a Wick power ${:}\xi^n{:}$. We do not know whether this holds for other values of p.

REMARK 7.38. Let I be a Gaussian stochastic integral on $(M, \mathcal{M}, \mu)$ and define, for suitable functions f on M^n, $n \geq 0$,

$$\mathring{I}_n(f) = \sum_{\gamma} I_{n-2r(\gamma)}(C_\gamma(f)),$$

summing over all Feynman diagrams γ on $\{1, \dots, n\}$; then $\mathring{I}_n$ may be regarded as a multiple *Stratonovich integral* (Hu and Meyer 1988, Budhiraja and Kallianpur 1995, Betounes and Redfern 1996).

It follows from Theorem 7.33 that if $f_1, \dots, f_n \in L^2(M)$, then $\mathring{I}_n(f_1(x_1) \cdots f_n(x_n)) = I(f_1) \cdots I(f_n)$ (with ordinary multiplication; cf. the similar (7.23) with Wick multiplication for the Itô type integral). This formula (possibly with $f_1, \dots, f_n$ restricted to indicator functions) can be used to define the multiple Stratonovich integral, see the references just given.

Note also the easily proved inversion formula (generalizing Theorem 3.4)

$$I_n(f) = \sum_{\gamma} (-1)^{r(\gamma)} \mathring{I}_{n-2r(\gamma)}(C_\gamma(f)).$$

3. Skorohod integrals

Just as for the Itô integrals in Section 1, it is possible to extend the stochastic integrals considered in Section 2 to random functions in such a way that the multiple integrals equal iterated integrals. The relevant stochastic integral is known as the *Skorohod integral*, and was introduced by Skorohod (1975). We refer to Nualart (1995), Nualart and Pardoux (1988), and the references in them for many results that are not covered here.

The Skorohod integral may be defined as follows. (Two other, more general, definitions will be given in Chapters 15 and 16.)

Suppose that $I \colon L^2_{\mathbb{R}}(M, \mathcal{M}, \mu) \to H$ is a Gaussian stochastic integral on a σ-finite measure space, and let Z denote the Gaussian stochastic measure given by Theorem 7.20. For simplicity we assume that I maps onto H.

Let (cf. Appendices B and C) $t \mapsto X_t$ be a random function belonging to $L^2(M, \mathcal{M}, \mu; L^2(\Omega, \mathcal{F}(H), \mathrm{P})) = L^2(M \times \Omega, \mathcal{M} \times \mathcal{F}(H), \mu \times \mathrm{P})$, i.e. a measurable function $M \to L^2(\Omega, \mathcal{F}(H), \mathrm{P})$ with $\int_M \mathrm{E}\,|X_t|^2 d\mu(t) < \infty$.

Suppose first that $X_t \in H^{:n:}$ for some n and all $t \in M$. By Theorem 7.25, $X_t = \hat{I}_n(f_t)$ for some $f_t \in L^2(M^n, \mathcal{M}^{\odot n}, \mu^{\odot n})$. Define $F_t = (n!)^{-1} f_t$, so that

$$X_t = I_n(F_t) = \int_{M^n} F_t \, dZ^n, \tag{7.33}$$

and define the function F on M^{n+1} by

$$F(t_1, \dots, t_{n+1}) = F_{t_{n+1}}(t_1, \dots, t_n). \tag{7.34}$$

Since I_n^{-1} is continuous, the map $t \mapsto F_t = I_n^{-1}(X_t)$ is a measurable function $M \to L^2(M^n, \mu^n)$, and thus (a suitable version of) F is a measurable function on M^{n+1}, see Appendix C. Moreover,

$$\int_{M^{n+1}} |F|^2 d\mu^{n+1} = \int_M \|F_t\|^2_{L^2(M^n,\mu^n)}\, d\mu(t) = \int_M \frac{1}{n!}\, \mathrm{E}\,|X_t|^2 d\mu(t), \qquad (7.35)$$

and thus $F \in L^2(M^{n+1}, \mu^{n+1})$. We define the Skorohod integral of X_t by

$$\int_M X_t\, dZ(t) = \int F\, dZ^{n+1} \in H^{:n+1:} \qquad (7.36)$$

and have by (7.24) and (7.35),

$$\mathrm{E}\left|\int_M X_t\, dZ(t)\right|^2 \leq (n+1)! \int_{M^{n+1}} |F|^2 d\mu^{n+1} = (n+1) \int_M \mathrm{E}\,|X_t|^2 d\mu. \quad (7.37)$$

For a general $X_t \in L^2(M, \mathcal{M}, \mu; L^2(\Omega, \mathcal{F}(H), \mathrm{P}))$ we use the orthogonal decomposition $X_t = \sum_0^\infty \pi_n(X_t)$ and define the Skorohod integral by

$$\int_M X_t\, dZ(t) = \sum_{n=0}^{\infty} \int_M \pi_n(X_t)\, dZ(t), \qquad (7.38)$$

provided the sum converges in L^2. (Otherwise the Skorohod integral of X_t is not defined.) Note that the terms in the sum in (7.38) are orthogonal by (7.36).

THEOREM 7.39. *The Skorohod integral $\int X_t\, dZ(t)$ is defined as a random variable belonging to $L^2(\Omega, \mathcal{F}(H), \mathrm{P})$ for random functions X_t in a dense subspace of $L^2(M, \mathcal{M}, \mu; L^2(\Omega, \mathcal{F}(H), \mathrm{P}))$.*

(i) *The integral is linear:*

$$\int (\alpha X_t + \beta Y_t)\, dZ(t) = \alpha \int X_t\, dZ(t) + \beta \int Y_t\, dZ(t)$$

when the integrals on the right hand side are defined.

(ii) *The domain contains every function X_t such that*

$$|||X|||^2 = \sum_0^\infty (n+1) \int_M \mathrm{E}\,|\pi_n(X_t)|^2 d\mu(t) = \int_M \big(\langle \mathcal{N} X_t, X_t\rangle + \|X_t\|^2\big)\, d\mu(t) < \infty, \qquad (7.39)$$

where $\mathcal{N} = \sum_0^\infty n\pi_n$ is the number operator, and then

$$\mathrm{E}\left|\int_M X_t\, dZ(t)\right|^2 \leq |||X|||^2. \qquad (7.40)$$

(iii) *If $f \in L^2(M, \mathcal{M}, \mu)$ is a deterministic function, then the Skorohod integral of f equals the Gaussian stochastic integral $\int f\, dZ = I(f)$.*

(iv) *If $n \geq 1$ and $F \in L^2(M^n, \mu^n)$, then*

$$\int_{M^n} F\, dZ^n = \int_M \cdots \int_M F(t_1, \dots, t_n)\, dZ(t_1) \cdots dZ(t_n), \tag{7.41}$$

where the left hand side is the multiple integral defined in Section 2 and the right hand side is an iterated Skorohod integral. Hence also, if $F_t(t_1, \dots, t_{n-1}) = F(t_1, \dots, t_{n-1}, t)$,

$$\int_M \int_{M^{n-1}} F_t\, dZ^{n-1} dZ(t) = \int_{M^n} F\, dZ^n. \tag{7.42}$$

PROOF. The linearity is clear. The existence of $\int X_t\, dZ(t)$ when $|||X||| < \infty$ and the estimate (7.40) follow by (7.38), (7.37) and (7.36). Since $|||X||| < \infty$ if $X \in L^2(M; \overline{\mathcal{P}}_n(H))$ for some $n < \infty$, and the set of all such X is dense in $L^2(M; L^2(\Omega, \mathcal{F}(H), \mathrm{P}))$, the domain of the Skorohod integral is dense.

If $f \mapsto f_t$ is a deterministic function in $L^2(M; \mathcal{M}, \mu)$, then $f_t \in H^{:0:}$ for each t so $f_t = I_0(f_t)$ and an inspection of the definition above for $n = 0$ yields $F(t) = F_t = f_t$ and the Skorohod integral equals by (7.36) the stochastic integral $\int f\, dZ(t) = I(f)$.

Similarly, (7.42) follows directly from the definition if F is symmetric in the first $n - 1$ coordinates, and the general case follows (for $n > 2$) by symmetrization in these coordinates.

Finally, (7.41) follows by (7.42) and induction. □

The following simple special case shows again the connection between the Skorohod integral and the Wick product. An extension will be given in Theorem 16.51.

THEOREM 7.40. *Suppose that $X \in \overline{\mathcal{P}}_*(H)$ and $f \in L^2(M, \mathcal{M}, \mu)$. Then*

$$\int_M f(t) X\, dZ_t = X \odot \int_M f(t)\, dZ_t. \tag{7.43}$$

PROOF. Suppose first that $X = {:}\xi_1 \cdots \xi_n{:}$, with $\xi_i \in H$. Then $\xi_i = \int f_i\, dZ$ for some $f_i \in L^2(M, \mathcal{M}, \mu)$, and by Theorem 7.26,

$$X = {:}\xi_1 \cdots \xi_n{:} = \int_{M^n} f_1(t_1) \cdots f_n(t_n)\, dZ(t_1) \cdots dZ(t_n).$$

Hence,

$$f(t)X = \int_{M^n} f_1(t_1) \cdots f_n(t_n) f(t)\, dZ(t_1) \cdots dZ(t_n)$$

and by Theorem 7.39, with $\xi = \int f\, dZ$,

$$\int f(t) X\, dZ_t = \int_{M^{n+1}} f_1(t_1) \cdots f_n(t_n) f(t_{n+1})\, dZ^{n+1} = {:}\xi_1 \cdots \xi_n \xi{:} = X \odot \xi,$$

using Theorem 7.26 again and Theorem 3.47. For fixed f, both sides of (7.43) are linear functions of X and, by Theorem 7.39 and Theorem 3.47, continuous in $X \in H^{:n:}$. Hence (7.43) holds for all $X \in H^{:n:}$, and by linearity again for all $X \in \overline{\mathcal{P}}_*(H)$. □

In the remainder of this section we restrict ourselves to the case of a Brownian motion B_t on $[0, \infty)$. (Similar results hold on $[0,1]$.) We defined in Section 1 the Itô integral $\int X_t \, dB_t$ for a square integrable predictable process X_t. The next theorem shows that the Skorohod integral is an extension of the Itô integral. (This is why we allow ourselves to use the same notation.)

THEOREM 7.41. *Let B_t, $t \geq 0$, be a Brownian motion and let X_t be a square integrable predictable process. Then the Skorohod integral $\int X_t \, dB_t$ exists and equals the Itô integral defined in Section* 1.

PROOF. We will in this proof use the notations $\int_{(I)}$ for the Itô integral and $\int_{(S)}$ for the Skorohod integral, both taken over $[0, \infty)$.

If $n \geq 1$ and $F_n \in L^2(D_n)$, let $F_{n,t}(t_1, \dots, t_{n-1}) = F_n(t_1, \dots, t_{n-1}, t)$, interpreted as 0 when $t_{n-1} \geq t$, and $X_{n,t} = I_{n-1}(F_{n,t})$. Then, by Theorem 7.5, $t \mapsto X_{n,t}$ is a square integrable predictable process and the Itô integral $\int_{(I)} X_{n,t} \, dB_t = I_n(F_n)$. Moreover, by (7.42) and Theorems 7.6 and 7.29, this is also true for the Skorohod integral. Hence the Itô and Skorohod integrals of $X_{n,t}$ coincide.

Now let X_t be any square integrable predictable process. Let Y be the Itô integral $\int_{(I)} X_t \, dB_t$ and let $Y_n = \pi_n(Y)$, $n \geq 0$. Note that $Y_0 = \mathrm{E}\, Y = 0$ by Theorem 7.8, while if $n \geq 1$, then $Y_n = I_n(F_n)$ for some $F_n \in L^2(D_n)$ by Theorem 7.3.

If $X_{n,t} = I_{n-1}(F_{n,t})$ as above, then by Theorem 7.3,

$$\sum_{n=1}^{\infty} \int_0^{\infty} \mathrm{E}\, |X_{n,t}|^2 \, dt = \sum_{n=1}^{\infty} \int_0^{\infty} \|F_{n,t}\|_{L^2(D_{n-1})} \, dt$$
$$= \sum_{n=1}^{\infty} \|F_n\|_{L^2(D_n)} = \sum_{n=1}^{\infty} \mathrm{E}\, |Y_n|^2 = \mathrm{E}\, |Y|^2 < \infty.$$

Thus the sum $U_t = \sum_{n=1}^{\infty} X_{n,t}$ converges in L^2 for a.e. t. The sum converges also in $L^2(\mathbb{R}_+ \times \Omega, dt\, d\mathrm{P})$, which implies that U_t is a square integrable predictable process, and that the Itô integral

$$\int_{(I)} U_t \, dB_t = \sum_{n=1}^{\infty} \int_{(I)} X_{n,t} \, dB_t = \sum_{n=1}^{\infty} Y_n = Y = \int_{(I)} X_t \, dB_t.$$

By the uniqueness part of Theorem 7.8, we must have $X_t = U_t = \sum_{n=1}^{\infty} X_{n,t}$ a.e. For the Skorohod integral we thus have, by the definition (7.38),

$$\int_{(S)} X_t \, dB_t = \sum_{n=1}^{\infty} \int_{(S)} X_{n,t} \, dB_t = \sum_{n=1}^{\infty} \int_{(I)} X_{n,t} \, dB_t = \int_{(I)} X_t \, dB_t,$$

since we have already shown that the Skorohod integral $\int_{(S)} X_{n,t} dB_t$ coincides with the Itô integral. □

The Skorohod integral, in contrast to the Itô integral, is also defined for many unpredictable processes. (Such processes are usually called *anticipating.*) The reader should be warned, however, that many of the nice properties of the Itô integral do not extend to the Skorohod integral, at least not without changes. See Nualart (1995) and Nualart and Pardoux (1988) where for example an extension of Itô's formula is given.

One important difference between the Skorohod integral and the Itô integral (and most other integrals) is that the integral of a constant function over an interval, i.e. $\int_a^b X\,dB_t = \int_0^\infty \mathbf{1}_{[a,b]}(t) X\,dB_t$, in general differs from $X \int_a^b dB_t = X(B_b - B_a)$. As was seen more generally in Theorem 7.40, equality holds, however, if we replace the product by the Wick product. In the Brownian case, this result may be written as follows.

THEOREM 7.42. *Suppose that $X \in \overline{\mathcal{P}}_*(H)$ and $f \in L^2(\mathbb{R}_+, dt)$. Then*

$$\int_0^\infty f(t) X\,dB_t = X \odot \int_0^\infty f(t)\,dB_t.$$

□

REMARK 7.43. This suggests that a more suggestive notation for the Skorohod integral would be

$$\int X_t \odot dB_t.$$

See e.g. Lindstrøm, Øksendal and Ubøe (1992) and Holden, Øksendal, Ubøe and Zhang (1996) for further results showing how the Wick product is implicit in Itô and Skorohod integrals.

COROLLARY 7.44. *If $X \in L^2(\Omega, \mathcal{F}(H), \mathrm{P})$ and $f \in L^2(\mathbb{R}_+, dt)$, then*

$$\int_0^\infty f(t) X\,dB(t) = \sum_{n=0}^\infty \Big(\int_0^\infty f\,dB \Big) \odot \pi_n(X)$$

in the sense that the Skorohod integral exists if and only if the series on the right hand side converges in L^2, i.e. if and only if

$$\sum_{n=0}^\infty \Big\| \Big(\int_0^\infty f\,dB \Big) \odot \pi_n(X) \Big\|_2^2 < \infty.$$

□

EXAMPLE 7.45. Let $(a_n)_0^\infty$ be a sequence of real numbers with $\sum a_n^2 < \infty$, and define $X = \sum_0^\infty a_n (n!)^{-1/2} :B_1^n: \in L^2$. (Recall that $(n!)^{-1/2} :B_1^n:$ are orthonormal elements of L^2 by Theorem 3.9.) Then the Skorohod integral

$$\int_0^1 X\,dB_t = \sum_0^\infty a_n(n!)^{-1/2}\,{:}B_1^n{:} \odot B_1 = \sum_0^\infty a_n(n!)^{-1/2}\,{:}B_1^{n+1}{:}$$

exists if and only if

$$\sum_0^\infty a_n^2(n!)^{-1}\,\mathrm{E}({:}B_1^{n+1}{:})^2 = \sum_0^\infty (n+1)a_n^2 = \langle \mathcal{N}X, X\rangle + \|X\|_2^2 < \infty.$$

This shows that the Skorohod integral is not defined for every stochastic process in $L^2(\mathbb{R}_+\times\Omega, dt\,d\mathrm{P})$; it shows also that the factors $n+1$ in (7.39) are best possible. Note, however, that with the same X,

$$\int_1^2 X\,dB_t = \sum_0^\infty a_n(n!)^{-1/2}\,{:}B_1^n{:}\,(B_2-B_1)$$

exists for every such sequence $(a_n)_0^\infty$. (This follows also by Theorem 7.41.) Hence (7.39) is not necessary for the existence of the Skorohod integral.

For Itô integrals we have the norm equality

$$\Big\|\int X_t\,dB_t\Big\|_2^2 = \int_0^\infty \|X_t\|_2^2\,dt.$$

The preceding example shows that for Skorohod integrals, the left hand side can be much larger than the right hand side. It can also be much smaller.

EXAMPLE 7.46. Let

$$X(t) = \begin{cases} B_2 - B_1, & \text{if } 0 \le t \le 1,\\ B_0 - B_1, & \text{if } 1 < t \le 2,\\ 0, & \text{if } t > 2.\end{cases}$$

Then

$$\int_0^\infty X(t)\,dB_t = (B_2-B_1)\odot(B_1-B_0) + (B_0-B_1)\odot(B_2-B_1) = 0.$$

4. Complex Gaussian stochastic measures

In the sections above we assumed that we had an isometry of $L^2_{\mathbb{R}}(M,\mathcal{M},\mu)$ into H, for some measure space $(M,\mathcal{M},\mu)$ and a Gaussian Hilbert space H. Such an isometry extends uniquely to an isometry of the complex space $L^2_{\mathbb{C}}(M,\mathcal{M},\mu)$ into $H_{\mathbb{C}}$, but there are also other such complex isometries which are not induced by real isometries. As long as we use complex scalars, we can repeat the development in the preceding sections for any complex isometry.

DEFINITION 7.47. A *complex Gaussian stochastic integral* on a measure space $(M,\mathcal{M},\mu)$ is a linear isometry I of $L^2_{\mathbb{C}}(M,\mathcal{M},\mu)$ into the complexification $H_{\mathbb{C}}$ of a Gaussian Hilbert space H.

The corresponding definition of stochastic measures has to be reformulated.

DEFINITION 7.48. A *complex Gaussian stochastic measure* on a measure space $(M, \mathcal{M}, \mu)$ is a family $Z(A)$, $A \in \mathcal{M}_\mu = \{A \in \mathcal{M} : \mu(A) < \infty\}$, of random variables with a joint centred complex Gaussian distribution such that for any sets $A, B \in \mathcal{M}_\mu$,

$$\mathrm{E}\, Z(A)\overline{Z(B)} = \mu(A \cap B). \tag{7.44}$$

REMARK 7.49. If $Z(A)$ is a real random variable for every $A \in \mathcal{M}_\mu$, then the definition above is equivalent to the one in Section 2, see Remark 7.19.

REMARK 7.50. If Z is a complex Gaussian stochastic measure, then for any finite family of disjoint sets $A_1, \dots, A_n$ in $\mathcal{M}_\mu$, just as in the real case (Definition 7.17),

$$Z\Big(\bigcup_1^n A_i\Big) = \sum_1^n Z(A_i) \qquad \text{a.s.,}$$

because (7.44) easily implies $\mathrm{E}\,|Z(\bigcup_1^n A_i) - \sum_1^n Z(A_i)|^2 = 0$.

On the other hand, in contrast to the real case, it is not necessarily true that the variables $Z(A_1), \dots, Z(A_n)$ are independent. For example, if Z is as in Example 7.58 below, then $Z((-\pi, 0)) = \overline{Z((0, \pi))}$, so these (non-constant) variables are not independent.

REMARK 7.51. In contrast to the real case, the distribution of a complex Gaussian stochastic measure is *not* determined by the measure space. For a trivial example, if $Z(A)$ is a real Gaussian stochastic measure, then $Z(A)$ and $iZ(A)$ are two complex Gaussian stochastic measures. Moreover, if $Z_1(A)$ and $Z_2(A)$ are two independent real Gaussian stochastic measures, then $2^{-1/2}(Z_1(A) + iZ_2(A))$ is a symmetric complex Gaussian stochastic measure. Another example is given in Remark 7.60 below.

We can now extend everything in Sections 2 and 3 to this, more general, case.

THEOREM 7.52. *Let $(M, \mathcal{M}, \mu)$ be a measure space. There is a one-to-one correspondence between complex Gaussian stochastic integrals and complex Gaussian stochastic measures on $(M, \mathcal{M}, \mu)$, given by $Z(A) = I(\mathbf{1}_A)$, $A \in \mathcal{M}_\mu$. Given a complex Gaussian stochastic integral or measure, we may define isometries $\hat{I}_n \colon L^2_{\mathbb{C}}(M^n, \mathcal{M}^{\odot n}, \mu^{\odot n}) \to H^{:n:}_{\mathbb{C}}$ and multiple integrals $I_n \colon L^2_{\mathbb{C}}(M^n, \mu^n) \to H^{:n:}_{\mathbb{C}}$ such that the results in Sections 2 and 3, in particular Theorems 7.25, 7.26, 7.33 and 7.39, hold provided we consider complex spaces only.* □

An example is provided by the theory of stationary stochastic processes. Let $(X_n)_{n=-\infty}^{\infty}$ be a (real) stationary Gaussian process, and assume for convenience that $\mathrm{E}\, X_n = 0$. Recall that then $\mathrm{Cov}(X_n, X_m) = r(n - m)$ for some function r, and that r is semi-definite since $\sum_{n,m} a_n \bar{a}_m r(n-m) = \operatorname{Var}(\sum_n a_n X_n) \geq 0$ for any finite sequence (a_n), which by Bochner's theorem implies the existence of a unique finite positive measure μ on $\mathbb{T} = (-\pi, \pi]$, the *spectral measure* of (X_n), such that

$$r(n) = \widehat{\mu}(n) = \int_{\mathbb{T}} e^{int} d\mu(t), \qquad n \in \mathbb{Z}.$$

Thus, for every $m, n \in \mathbb{Z}$,

$$\langle e^{int}, e^{imt} \rangle_{L^2(\mu)} = \int_{\mathbb{T}} e^{int-imt}\, d\mu(t) = r(n-m) = \mathrm{E}\, X_n X_m = \mathrm{E}\, X_n \overline{X_m}.$$

This proves that the mapping

$$I \colon \sum a_n e^{int} \mapsto \sum a_n X_n$$

is an isometry of the space of trigonometrical polynomials into $H_{\mathbb{C}}$, where H is the Gaussian Hilbert space spanned by $\{X_n\}$.

If $f \in L^2_{\mathbb{C}}(\mu)$ is orthogonal to the trigonometric polynomials, then

$$\int_{\mathbb{T}} e^{-int} f(t)\, d\mu(t) = \langle f, e^{int} \rangle_{L^2(\mu)} = 0, \qquad n \in \mathbb{Z},$$

so the Fourier transform of the finite complex measure $f\, d\mu$ vanishes; thus $f = 0$ μ-a.e. Hence the trigonometric polynomials are dense in $L^2(\mu)$ and I may be extended to an isometry of $L^2_{\mathbb{C}}(\mu)$ onto $H_{\mathbb{C}}$, i.e. a complex Gaussian stochastic integral. We denote the corresponding complex stochastic measure on $(\mathbb{T}, \mu)$ by Z, and note that it is characterized by

$$X_n = \int_{\mathbb{T}} e^{int}\, dZ(t).$$

This measure is called the *stochastic spectral measure* corresponding to (X_n) (Major 1981).

Since r is symmetric, the spectral measure μ is symmetric so $f \to \check{f}(x) = f(-x)$ is an isometry of $L^2(\mu)$ onto itself. Moreover, complex conjugation is an anti-linear isometry in both $L^2(\mu)$ and $H_{\mathbb{C}}$. Hence the mapping $f \mapsto I'(f) = \overline{I(\overline{f(-x)})}$ is a linear isometry of $L^2(\mu)$ into H. However, since X_n is real,

$$I'(e^{inx}) = \overline{I(\overline{e^{-inx}})} = \overline{X_n} = X_n = I(e^{inx})$$

and thus I' and I coincide for every trigonometric polynomial, and thus for every function in $L^2(\mu)$. In other words,

$$\overline{I(f)} = I(\overline{f(-x)}).$$

The argument above extends verbatim to the case of a stationary Gaussian process on $\mathbb{Z}^d$, $d > 1$; the spectral measure and the stochastic spectral measure are now defined on $\mathbb{T}^d$. Moreover, the argument is valid also for stationary Gaussian process on $\mathbb{R}^d$, $d \geq 1$, provided only that $t \mapsto X_t$ is continuous in probability (and thus, by Theorem 1.4, in L^2); now the spectral measure is defined on $\mathbb{R}^d$.

REMARK 7.53. More generally, the results hold for continuous Gaussian processes on any locally compact Abelian group, with the spectral measure defined on the dual group and e^{inx} replaced by the pairing $\langle x, \chi \rangle \to \chi(x)$ of the group and its dual.

We collect the results above, letting $-A = \{-x : x \in A\}$ for a subset A of an Abelian group. Let, throughout this section, Γ be the group $\mathbb{Z}^d$ or $\mathbb{R}^d$, $d \geq 1$, and let $\widehat{\Gamma}$ be the dual group $\mathbb{T}^d$ or $\mathbb{R}^d$, respectively.

THEOREM 7.54. *If $(X_t)_{t \in \Gamma}$ is a (real) stationary Gaussian stochastic process on Γ that is continuous in probability, then there exist a unique symmetric, finite, positive Borel measure μ on $\widehat{\Gamma}$ (the spectral measure) and a unique complex Gaussian stochastic measure Z on $(\widehat{\Gamma}, \mathcal{B}, \mu)$ such that*

$$X_t = \int_{\widehat{\Gamma}} e^{it \cdot x} \, dZ(x), \qquad t \in \Gamma. \tag{7.45}$$

Moreover, for every $f \in L^2(\widehat{\Gamma}, \mu)$,

$$\overline{\int f(x) \, dZ(x)} = \int \overline{f(-x)} \, dZ(x); \tag{7.46}$$

in particular

$$\overline{Z(A)} = Z(-A), \qquad A \in \mathcal{B}. \tag{7.47}$$

Consequently,

$$\mathrm{E}\, Z(A)\overline{Z(B)} = \mu(A \cap B), \qquad A, B \in \mathcal{B}, \tag{7.48}$$

and

$$\mathrm{E}\, Z(A) Z(B) = \mu(A \cap (-B)), \qquad A, B \in \mathcal{B}. \tag{7.49}$$

PROOF. The only remaining claims are (7.47), which follows by taking $f = \mathbf{1}_A$ in (7.46), and (7.49), which follows by (7.47) and (7.48). □

REMARK 7.55. If $f \in L^1(\Gamma) \cap L^2(\Gamma)$, then

$$\int_{\widehat{\Gamma}} \widehat{f}(x) \, dZ(x) = I\Big(\int_{\Gamma} f(t) e^{it \cdot x} \, dt \Big) = \int_{\Gamma} f(t) I(e^{it \cdot x}) \, dt = \int_{\Gamma} f(t) X_t \, dt.$$

Hence, if $f \to \tilde{f}$ denotes the inverse Fourier transform (by the Fourier inversion formula, $\tilde{f}(x) = c\widehat{f}(-x)$, with $c = 1$ for $\mathbb{Z}^d$ and $c = (2\pi)^{-d}$ for $\mathbb{R}^d$), we obtain, for suitable f,

$$\int_{\widehat{\Gamma}} f \, dZ = \int_{\Gamma} \tilde{f}(t) X_t \, dt.$$

REMARK 7.56. It is easily seen, by taking $B = -A$ in (7.49), that any complex Gaussian stochastic measure satisfying (7.48) and (7.49) must also satisfy (7.47); the converse was observed in the proof above. Moreover, (7.48) and (7.49) determine the distribution of Z (i.e. the joint distribution of the $Z(A)$). Equivalently, the distribution of a complex Gaussian stochastic measure on $L^2(\mu)$ that satisfies (7.47) is uniquely determined by μ.

EXAMPLE 7.57. Let Z and μ be as above. For any measurable sets A and B,

$$\mathrm{E}\big(\operatorname{Re} Z(A) \operatorname{Im} Z(B)\big) = \frac{1}{4i}\mathrm{E}\big(Z(A) + \overline{Z(A)}\big)\big(Z(B) - \overline{Z(B)}\big)$$
$$= \frac{1}{4i}\big(\mu(A \cap (-B)) + \mu(A \cap B) - \mu(A \cap B) - \mu(A \cap (-B))\big) = 0.$$

Hence the families $\{\operatorname{Re} Z(A)\}_{A \in \mathcal{B}}$ and $\{\operatorname{Im} Z(A)\}_{A \in \mathcal{B}}$ are independent. Similarly,

$$\mathrm{E}\big(\operatorname{Re} Z(A)\big)^2 = \tfrac{1}{2}\big(\mu(A) + \mu(A \cap (-A))\big)$$

and

$$\mathrm{E}\big(\operatorname{Im} Z(A)\big)^2 = \tfrac{1}{2}\big(\mu(A) - \mu(A \cap (-A))\big), \tag{7.50}$$

so for every $A \in \mathcal{B}$, $\operatorname{Re} Z(A)$ and $\operatorname{Im} Z(A)$ are independent centred Gaussian random variables with these variances.

In particular, by Proposition 1.31, $Z(A)$ is a symmetric complex Gaussian variable if and only if $\mu(A \cap (-A)) = 0$.

EXAMPLE 7.58. If (X_n) is a sequence of independent standard normal variables, then $r(n) = 0$ when $n \neq 0$, and μ is the normalized Lebesgue measure on $\mathbb{T}$. By the preceding example, the restriction of the corresponding stochastic spectral measure to subsets of $(0, \pi)$ is a symmetric complex Gaussian stochastic measure; moreover, this restriction determines $Z(A)$ for every A by (7.47) and the additivity.

We consider briefly converses to the results above.

THEOREM 7.59. *Let μ be a symmetric finite positive measure on $\widehat{\Gamma}$. Then there exists a stationary Gaussian stochastic process with spectral measure μ. There exists also a complex Gaussian stochastic measure on $(\widehat{\Gamma}, \mu)$ that satisfies* (7.47).

PROOF. Consider the closed hull V in $L^2(\mu)$ of the set of trigonometric polynomials $\sum a_t e^{itx}$ with real coefficients a_t. This is a real Hilbert space, and by Theorem 1.23 there exists a Gaussian Hilbert space H with an isometry $I\colon V \to H$. If we define $X_t = I(e^{itx})$, then

$$\operatorname{Cov}(X_t, X_s) = \langle e^{itx}, e^{isx} \rangle_{L^2(\mu)} = \int e^{itx} e^{-isx}\, d\mu = \hat{\mu}(t - s),$$

which shows that (X_t) is a stationary Gaussian stochastic process with spectral measure μ. Note that X_t is real by definition, and that $t \mapsto X_t$ is continuous because $t \mapsto e^{itx}$ is a continuous map into $L^2(\mu)$ by dominated convergence. The second claim follows by Theorem 7.54. $\square$

REMARK 7.60. As noted above, the distribution of this complex stochastic measure is determined by μ. It is easily seen, using (7.50), that this measure is not real except in the trivial case when μ is supported on the set $\{x : x = -x\}$. (For $\mathbb{R}^d$, this means that μ is a point mass at the origin, $r(t) = \mu(\{0\})$ is independent of t, and $X_t = X_0$ a.s., for every $t \in \mathbb{R}^d$; for $\mathbb{Z}^d$, μ has to be a sum of point masses in $\{0, \pi\}^d \subset \mathbb{T}^d$, which implies that $r(2n) = r(0)$ and $X_{2n} = X_0$ a.s. for every $n \in \mathbb{Z}^d$.)

On the other hand, there also exists a real Gaussian stochastic measure on $(\widehat{\Gamma}, \mu)$ by Remark 7.21. This gives another example of the non-uniqueness of the distribution of complex Gaussian stochastic measures.

THEOREM 7.61. *Let μ be a symmetric finite measure, and Z a complex Gaussian stochastic measure on $\widehat{\Gamma}$ such that* (7.48) *and* (7.49) *hold, or, equivalently,* (7.47) *and* (7.48). *Then* (7.45) *defines a real stationary Gaussian process which is continuous in L^2.*

PROOF. It is easily seen that (7.47) implies (7.46) by continuity, and thus

$$\overline{X_t} = \int_{\widehat{\Gamma}} \overline{e^{-itx}}\, dZ(x) = \int_{\widehat{\Gamma}} e^{itx}\, dZ(x) = X_t.$$

Hence (X_t) is a real Gaussian process. The same argument as in the proof of Theorem 7.59 shows that (X_t) is stationary and continuous. $\square$

Finally, we consider multiple integrals.

THEOREM 7.62. *Let $(X_t)_{t\in\Gamma}$, μ and Z be as in Theorem 7.54, and let H be the Gaussian Hilbert space spanned by $\{X_t\}_{t\in\Gamma}$. Then the multiple integral*

$$f \mapsto \int_{\widehat{\Gamma}^n} f\, dZ^n$$

maps $L^2_{\mathbb{C}}(\widehat{\Gamma}^n, \mu^n)$ onto $H^{:n:}_{\mathbb{C}}$, and

$$\mathrm{E}\left|\int f\, dZ^n\right|^2 \leq n! \int |f|^2\, d\mu^n,$$

with equality if f is symmetric. Furthermore,

$$\overline{\int_{\widehat{\Gamma}^n} f(x)\, dZ^n(x)} = \int_{\widehat{\Gamma}^n} \overline{f(-x)}\, dZ^n(x). \tag{7.51}$$

In particular, the stochastic integral $\int f(x)\, dZ^n(x)$ is a real random variable whenever $f(-x) = \overline{f(x)}$ on $\widehat{\Gamma}^n$.

Every $X \in L^2_{\mathbb{C}}(\Omega, \mathcal{F}((X_t)_{t\in\Gamma}), \mathrm{P})$ *can be uniquely represented as*

$$X = \sum_0^\infty \int_{\widehat{\Gamma}^n} f_n \, dZ^n,$$

where f_n *is a symmetric (complex) function on* $\widehat{\Gamma}^n$ *and*

$$\sum_0^\infty n! \int_{\widehat{\Gamma}^n} |f_n|^2 \, d\mu^n = \mathrm{E}\,|X|^2 < \infty;$$

X *is real if and only if* $f_n(-x) = \overline{f_n(x)}$ *a.e. for every* $n \geq 0$.

PROOF. We use Theorem 7.54 and the complex version of Theorem 7.26. In order to verify (7.51), it suffices (by linearity and continuity) to consider the case $f(x_1, \dots, x_n) = f_1(x_1) \cdots f_n(x_n)$, but this case follows by (7.23), (7.46) and Corollary 3.7. The rest follows directly. □

VIII

Gaussian stochastic processes

There is a vast literature on stochastic processes in general, and on Gaussian stochastic processes in particular. We will here treat only a few topics; for other aspects and further results we refer to for example Gihman and Skorohod (1971), Hida and Hitsuda (1976), Ibragimov and Rozanov (1970), Kahane (1985), Neveu (1968).

In this chapter, we assume that $\{\xi_t\}_{t\in T}$ is a Gaussian stochastic process, with some (completely arbitrary) index set T. We consider only real-valued processes, although there are similar results for complex Gaussian stochastic processes. For convenience we assume also that the process is centred, i.e. $\mathrm{E}\,\xi_t = 0$ for each t. In other words, $\{\xi_t\}_{t\in T}$ is a collection of centred jointly Gaussian variables.

As is described in detail in Appendix B, the process may also, with some provisos, be regarded as a function on $T \times \Omega$, or as a random real-valued function on T (i.e. a random element of the product space $\mathbb{R}^T$). In particular, the distribution of the process is the induced probability measure on $\mathbb{R}^T$.

1. The covariance function

DEFINITION 8.1. The covariance function of the stochastic process ξ_t is the function on $T \times T$ given by

$$\rho(s,t) = \operatorname{Cov}(\xi_s, \xi_t) = \langle \xi_s, \xi_t \rangle.$$

THEOREM 8.2. *Let T be any set. The covariance function of a Gaussian stochastic process is symmetric and semi-definite. In other words,*

$$\rho(s,t) = \rho(t,s), \qquad s,t \in T,$$

and, for every finite sequence $\{t_i\}_1^n$ in T and real numbers λ_i,

$$\sum_{i,j} \lambda_i \lambda_j \rho(t_i, t_j) \geq 0. \tag{8.1}$$

Conversely, every symmetric semi-definite function on $T \times T$ is the covariance function of some centred Gaussian stochastic process on T. The covariance function and the distribution of the process determine each other uniquely.

PROOF. The symmetry of ρ is obvious from the definition, and (8.1) follows because

$$\sum_{i,j} \lambda_i \lambda_j \rho(t_i, t_j) = \operatorname{Var}(\sum_i \lambda_i \xi_i) \geq 0.$$

Clearly, the covariance function is determined by the distribution of the process. Conversely, it is well-known that (since the variables are centred Gaussian) the covariance function determines the distribution of every finite set $\{\xi_{t_i}\}$ with $t_i \in T$, and these finite-dimensional marginal distributions determine in turn the full distribution of the process (see Example A.3). □

It remains to show that any semi-definite function ρ is a covariance function for some centred Gaussian stochastic process $\{\xi_t\}$. In order to do this, we first find a real Hilbert space H and vectors $v_t \in H$ with $(v_s, v_t) = \rho(s,t)$ for $s, t \in T$; this is always possible by Lemma F.6. We then take a Gaussian Hilbert space G of the same dimension as H and an isometry $\varphi\colon H \to G$ and define $\xi_t = \varphi(v_t)$ (cf. Theorem 1.23).

EXAMPLE 8.3. Brownian motion on $[0, \infty)$ is a centred Gaussian stochastic process with covariance function

$$\rho(s,t) = \min(s,t), \qquad s, t \geq 0.$$

It is possible to verify directly that this function is positive definite, which yields another way to construct a Brownian motion. The construction in Remark 7.2 is simpler, however.

EXAMPLE 8.4. The Ornstein–Uhlenbeck process is a centred Gaussian stochastic process on $(-\infty, \infty)$ with covariance function

$$\rho(s,t) = e^{-|s-t|}.$$

It is easy to see that such a process exists; one construction is to let $B(t)$, $t \geq 0$, be a Brownian motion and let $\xi_t = e^{-t} B(e^{2t})$.

EXAMPLE 8.5. A Gaussian field on a Hilbert space H (see Definition 1.19) is a centred Gaussian stochastic process with the Hilbert space as index set and covariance function equal to the inner product:

$$\rho(h,k) = \langle \xi_h, \xi_k \rangle = \langle h, k \rangle_H, \qquad h, k \in H.$$

EXAMPLE 8.6. Any Gaussian linear space G may be regarded as a centred Gaussian stochastic process indexed by itself; we define $X_\eta = \eta$, $\eta \in G$. The covariance function is given by

$$\rho(\xi, \eta) = \langle X_\xi, X_\eta \rangle = \langle \xi, \eta \rangle.$$

(When G is complete, this is a special case of Example 8.5.)

EXAMPLE 8.7. By Remark 7.19, a Gaussian stochastic measure on a measure space $(M, \mathcal{M}, \mu)$ is a centred Gaussian stochastic process defined on the set $\mathcal{M}_\mu$ with covariance function $\rho(A, B) = \mu(A \cap B)$, $A, B \in \mathcal{M}_\mu$.

2. The pseudometric

The Gaussian stochastic process defines a pseudometric on T by

$$d(s,t) = \|\xi_s - \xi_t\|_2, \qquad s,t \in T.$$

(Note that $d(s,t) = 0$ is possible also when $s \neq t$, since we allow $\xi_s = \xi_t$ for $s \neq t$. If this is excluded, d is a metric on T.)

Obviously, d is determined by the covariance function ρ by the formula

$$d^2(s,t) = \rho(s,s) + \rho(t,t) - 2\rho(s,t). \tag{8.2}$$

Conversely, if there exists a t_0 with $\xi_{t_0} = 0$ a.s. (which is equivalent to $\rho(t_0,t_0) = 0$), then

$$\rho(s,t) = \tfrac{1}{2}(d^2(s,t_0) + d^2(t,t_0) - d^2(s,t)),$$

and ρ is determined by d. Hence, given $\xi_{t_0} = 0$, the distribution of the process is determined by d.

Consequently, ρ and d are closely related, but they have somewhat different applications. Often ρ is natural; for example, ρ has nice linearity properties. In other cases it is convenient to consider the (pseudo)metric space (T,d).

REMARK 8.8. We may always regard d as a true metric on the quotient space $\tilde{T}$ obtained by identifying any elements s and t in T such that $\xi_s = \xi_t$ a.s. Note that the resulting metric space $(\tilde{T},d)$ is isometric to the set $\{\xi_t\}_{t\in T}$, regarded as a subset of the metric space L^2, and it is mainly a matter of taste whether one uses $(\tilde{T},d)$ or $\{\xi_t\}_{t\in T}$.

EXAMPLE 8.9. Brownian motion on $[0,\infty)$ (or on $[0,1]$) has the pseudometric

$$d(s,t) = |s-t|^{1/2}.$$

EXAMPLE 8.10. Lévy (1948) similarly defined a *multiparameter Brownian motion* to be a Gaussian stochastic process (ξ_t), $t \in \mathbb{R}^n$, with pseudometric $|s-t|^{1/2}$, for example normalized by $\xi_0 = 0$. (By the comments above, this determines the distribution of the process.) One way to see that such a process exists is to define the functions $f_s(x) = |s-x|^{1/2-n/2}$ on $\mathbb{R}^n$, $n \geq 2$. Then $f_s \notin L^2(\mathbb{R}^n, dx)$, but $g_s = f_s - f_0 \in L^2(\mathbb{R}^n, dx)$, and by translation invariance and homogeneity,

$$\|g_s - g_t\|_{L^2(\mathbb{R}^n,dx)} = \|f_{s-t} - f_0\|_{L^2(\mathbb{R}^n,dx)} = c|s-t|^{1/2},$$

for some constant $c > 0$. Now let I be a Gaussian stochastic integral on $L^2(\mathbb{R}^n, dx)$, and define $\xi_t = I(c^{-1}g_t)$.

EXAMPLE 8.11. *Fractal Brownian motion* is similarly a Gaussian stochastic process on $\mathbb{R}^n$, $n \geq 1$, with pseudometric $|s-t|^\beta$, where $0 < \beta < 1$. Such a process can be constructed as for the case $\beta = 1/2$ treated in the preceding example, now taking $f_s(x) = |s-x|^{\beta-n/2}$ on $\mathbb{R}^n$ (for $\beta \neq 1/2$, this works also for $n = 1$). See further Kahane (1985, Chapter 18) and Yor (1988).

3. Continuity

If the index set T is a topological space, it makes sense to talk about continuity. There are several different notions of continuity for stochastic processes; the three most important are continuity in probability, in L^2, and almost surely. (The latter means, more precisely, that the process has a continuous version, cf. Appendix B.)

For Gaussian stochastic processes, the first two notions are equivalent; moreover they can easily be characterized using the covariance function or the pseudometric.

THEOREM 8.12. *Let $\{\xi_t\}$ be a centred Gaussian stochastic process defined on a topological space T. Then the following are equivalent.*

(i) *$t \mapsto \xi_t$ is continuous in probability.*
(ii) *$t \mapsto \xi_t$ is continuous in L^2.*
(iii) *The covariance function $\rho(s,t)$ is a continuous function on $T \times T$.*
(iv) *The pseudometric $d(s,t)$ is a continuous function on $T \times T$.*
(v) *The covariance function ρ is separately continuous on $T \times T$ and $\rho(t,t)$ (or equivalently $\|\xi_t\|_2$) is continuous on T.*
(vi) *The pseudometric d is separately continuous on $T \times T$.*

PROOF. (i) $\Leftrightarrow$ (ii) by Theorem 1.4.

(ii) $\Rightarrow$ (iii) by the definition of ρ, since the inner product is continuous on $L^2 \times L^2$.

(iii) $\Rightarrow$ (iv) and (v) $\Rightarrow$ (vi) follow by (8.2).

(iii) $\Rightarrow$ (v) and (iv) $\Rightarrow$ (vi) are trivial.

(vi) $\Rightarrow$ (ii) because (vi) implies that for every $t \in T$ and $\varepsilon > 0$, there exists a neighbourhood U of t such that $d(s,t) = |d(s,t) - d(t,t)| < \varepsilon$ when $s \in U$. But this means $\|\xi_s - \xi_t\|_2 < \varepsilon$ for $s \in U$. □

REMARK 8.13. It is *not* sufficient to assume that ρ is separately continuous. For example, let T be the compact set $\{1/n\}_{n=1}^{\infty} \cup \{0\}$ and let $\{\xi_{1/n}\}$ be independent standard normal variables and $\xi_0 = 0$; then $\rho(s,t) = 1$ when $s = t \neq 0$ and $\rho(s,t) = 0$ otherwise, which is separately continuous but not continuous. (The reader who prefers an example with $T = [0,1]$ may fill in the gaps by linear interpolation.)

The question of a.s. continuity is much more subtle, and although conditions on ρ and d are known, they are more complicated. We refer to for example Kahane (1985, Chapter 15) and Ledoux and Talagrand (1991).

4. The Cameron–Martin space

Every Gaussian stochastic process defines a space of functions on the index set, which we call the *Cameron–Martin space.* This space is important in several contexts. It was initially studied in connection with translations of Brownian motion (Cameron and Martin 1944, Maruyama 1950); we will return to that problem in Section 14.2. We let H be the Gaussian Hilbert space spanned by $\{\xi_t\}_{t\in T}$.

DEFINITION 8.14. Given a stochastic process $\{\xi_t\}_{t\in T}$, let, for each $\xi \in H$, $R(\xi)$ be the function on T given by

$$R(\xi)(t) = \langle \xi, \xi_t \rangle_H = \mathrm{E}(\xi\xi_t).$$

The Cameron–Martin space is the linear space $R(H) = \{R(\xi) : \xi \in H\}$ of real functions on T.

The next theorem shows that this space of functions is a Hilbert space, and moreover a reproducing Hilbert space; see Appendix F for definitions.

THEOREM 8.15. *The Cameron–Martin space $R(H)$ is a Hilbert space of real functions on T with the inner product given by*

$$\langle f, g \rangle_{R(H)} = \langle R^{-1}f, R^{-1}g \rangle_H, \qquad f, g \in R(H).$$

(This is well-defined because R is injective.) Moreover, $R(H)$ is the reproducing Hilbert space with reproducing kernel $\rho(s,t)$.

The mapping R is a linear isometry of H onto $R(H)$.

PROOF. By the definition of H, the set $\{\xi_t\}_{t\in T}$ is total in H. Hence Theorem F.5 applies, and the result follows. □

COROLLARY 8.16. *The functions $\rho_s(t) = \rho(s,t)$, where ρ is the covariance function, form a total subset of the Cameron–Martin space $R(H)$. The inner product in $R(H)$ is determined by*

$$\langle \rho_s, \rho_t \rangle_{R(H)} = \rho(s,t).$$

PROOF. This follows by the general Theorem F.3, but it is also a direct consequence of Example 8.17 below. □

We continue with some examples.

EXAMPLE 8.17. For any centred Gaussian stochastic process we have

$$R(\xi_t) = \rho_t, \qquad t \in T.$$

In fact, by the definitions,

$$R(\xi_t)(s) = \langle \xi_t, \xi_s \rangle_H = \rho(t,s) = \rho_t(s).$$

EXAMPLE 8.18. Consider a Brownian motion B_t on $[0,\infty)$. By Theorem 7.1, $H = \{I(f) : f \in L^2_{\mathbb{R}}([0,\infty))\}$, where $I(f) = \int_0^\infty f(t)\,dB_t$ is the stochastic integral. Moreover, $B_t = I(\mathbf{1}_{[0,t]})$, and thus for every $f \in L^2_{\mathbb{R}}([0,\infty))$,

$$\begin{aligned} R(I(f))(t) = \langle I(f), B_t \rangle_H &= \langle \int_0^\infty f(s)\,dB_s, \int_0^t dB_s \rangle_H \\ &= \langle f, \mathbf{1}_{[0,t]} \rangle_{L^2([0,\infty))} = \int_0^t f(s)\,ds. \end{aligned}$$

Consequently,

$$R(H) = \{\int_0^\cdot f(s)\,ds : f \in L^2_{\mathbb{R}}([0,\infty))\},$$

the Sobolev space of all real (locally) absolutely continuous functions F on $[0,\infty)$ with $F(0)=0$ and $F' \in L^2([0,\infty))$, and

$$R^{-1}(F) = I(F') = \int_0^\infty F'(t)\,dB_t.$$

Note that

$$\langle F, G\rangle_{R(H)} = \int_0^\infty F'(t)G'(t)\,dt.$$

EXAMPLE 8.19. Cameron and Martin (1944) considered Brownian motion on $[0,1]$. In this case we similarly have, using Theorem 7.10, $H = \{I(f) : f \in L^2_{\mathbb{R}}([0,1])\}$,

$$R(H) = \{\int_0^\cdot f(s)\,ds : f \in L^2_{\mathbb{R}}([0,1])\},$$

the space of all real absolutely continuous functions F on $[0,1]$ with $F(0)=0$ and $F' \in L^2([0,1])$; furthermore

$$R^{-1}(F) = \int_0^1 F'(t)\,dB_t.$$

EXAMPLE 8.20. Consider a Gaussian stochastic measure Z on a measure space $(M, \mathcal{M}, \mu)$, cf. Definition 7.17. Let H denote the Gaussian Hilbert space spanned by Z, and let $I\colon L^2_{\mathbb{R}}(M,\mu) \to H$ be the Gaussian stochastic integral given by Theorem 7.20.

We assume for simplicity that μ is a finite measure (leaving the general case to the reader) and regard Z as a Gaussian stochastic process on $\mathcal{M}$ as in Example 8.7. Then, for any $f \in L^2_{\mathbb{R}}(M,\mu)$ and $A \in \mathcal{M}$,

$$R(I(f))(A) = \langle I(f), Z(A)\rangle_H = \langle I(f), I(\mathbf{1}_A)\rangle_H = \langle f, \mathbf{1}_A\rangle_{L^2(M,\mu)} = \int_A f\,d\mu.$$

Consequently, if $\xi \in H$, then the set function $R(\xi)$ on $\mathcal{M}$ is a finite signed measure, and $R(H)$ is the space of all signed measures on $(M,\mathcal{M})$ that are absolutely continuous with respect to μ and have square integrable Radon–Nikodym derivatives. (Cf. e.g. Cohn (1980, Chapter 4) for the measure theory used here.)

When T is a topological space, there is a connection between continuity of the process and continuity of the functions in the Cameron–Martin space.

THEOREM 8.21. *Suppose that T is a topological space and that the Gaussian stochastic process $\{\xi_t\}_T$ is continuous in probability. Then the Cameron–Martin space is a space of continuous functions on T.*

Conversely, if all functions in the Cameron–Martin space are continuous, and furthermore $t \mapsto \|\xi_t\|_2$ *is continuous, then the process is continuous in probability and in* L^2.

PROOF. If the process is continuous in probability, then $t \to \xi_t$ is continuous in L^2 by Theorem 8.12, and thus $t \mapsto R(\xi)(t) = \langle \xi, \xi_t \rangle$ is continuous for every $\xi \in H$.

Conversely, if every function in $R(H)$ is continuous, then in particular $t \mapsto \rho(s,t) = R(\xi_s)(t)$ is continuous for every $s \in T$. Thus ρ is separately continuous, and the proof is completed by another application of Theorem 8.12 □

It is easily verified that for the example in Remark 8.13, all functions in $R(H)$ are continuous on T; hence the extra condition in the converse part of the theorem is really needed.

The Cameron–Martin space leads also to the following representation theorem.

THEOREM 8.22. *Suppose that* $\{e_i\}_1^\infty$ *is a countable orthonormal basis in the Cameron–Martin space* $R(H)$. *Then there exist independent standard normal variables* $\{\eta_i\}_1^\infty$ *such that for each* $t \in T$,

$$\xi_t = \sum_1^\infty e_i(t)\eta_i \tag{8.3}$$

with the sum converging in L^2 *and a.s.*

Conversely, for any sequence $\{\eta_i\}$ *of independent standard normal variables, the sum in* (8.3) *converges for every* t *and defines a centred Gaussian stochastic process with the same distribution as* $\{\xi_t\}$.

REMARK 8.23. The result remains true (with finite sums) if H has a finite basis; it also remains true if H is non-separable and thus has an uncountable basis, with the sums interpreted as limits of nets converging in L^2.

PROOF. The expansion of $\rho_t \in R(H)$ in the orthonormal basis $\{e_i\}$ is given by

$$\rho_t = \sum_1^\infty \langle \rho_t, e_i \rangle e_i = \sum_1^\infty e_i(t) e_i,$$

with the sum converging in $R(H)$ and

$$\sum_1^\infty e_i(t)^2 = \|\rho_t\|_{R(H)}^2 = \|\xi_t\|_H^2 < \infty.$$

Applying R^{-1}, we find

$$\xi_t = R^{-1}(\rho_t) = \sum_1^\infty e_i(t) R^{-1}(e_i),$$

converging in $H \subseteq L^2$, which is (8.3) with $\eta_i = R^{-1}(e_i) \in H$. Since $\{e_i\}$ is an orthonormal basis in $R(H)$, these η_i form an orthonormal basis in H, which implies that they are independent and standard normal.

It follows also, for example by the martingale convergence theorem, that the sum in (8.3) converges a.s. (this is the only place where we need the basis to be countable).

The converse is clear, since replacing $\{\eta_i\}$ by another set with the same distribution does not affect the distribution of the sum in (8.3), nor the joint distribution for a set of $t \in T$. □

REMARK 8.24. The theorem asserts only that the series (8.3) converges a.s. for every fixed t; it does not claim anything about convergence in some function space. For example, suppose T is a compact metric topological space and that ξ_t is continuous in probability. It follows from Theorem 8.21 that $R(H) \subseteq C(T)$, the Banach space of continuous functions on T. In particular, $e_i(t) \in C(T)$ and thus each partial sum of (8.3) is a random element of $C(T)$. It follows easily from the Itô–Nisio theorem (Ledoux and Talagrand 1991, Theorem 2.4) that the sum converges a.s. in $C(T)$ (i.e., uniformly) if and only if the process has a continuous version.

REMARK 8.25. The isometry $R^{-1}\colon R(H) \to H$ exhibits H in a canonical way as a Gaussian Hilbert space indexed by the Hilbert space $R(H)$ of functions on T. Examples 8.18 and 8.19 show that although this representation is canonical, it is not always the most convenient; usually it is simpler to regard the Gaussian Hilbert spaces in Examples 8.18 and 8.19 as indexed by $L^2([0,\infty))$ and $L^2([0,1])$, respectively, using the stochastic integral I as the defining isomorphism.

EXAMPLE 8.26. Let, as in Example 1.13, μ be a symmetric Gaussian measure on some real locally convex topological vector space $\mathcal{X}$. Then the dual space $\mathcal{X}^*$ is a Gaussian linear space, which we regard as a centred Gaussian stochastic process indexed by itself as in Example 8.6. The Gaussian Hilbert space H is the completion of this Gaussian space. (It may happen that μ is concentrated on some closed subspace of $\mathcal{X}$, in which case $x^* = 0$ a.s. for some non-trivial $x^* \in \mathcal{X}^*$ and the Gaussian space is a quotient space of $\mathcal{X}^*$, but we still regard it as a Gaussian stochastic process indexed by $\mathcal{X}^*$.) Thus

$$\rho(f,g) = \langle f,g\rangle_{L^2(\mu)}, \qquad f,g \in \mathcal{X}^*,$$

and for every $\xi \in H$,

$$R(\xi)(f) = \langle \xi, f\rangle_{L^2(\mu)}, \qquad f \in \mathcal{X}^*. \tag{8.4}$$

By definition, $R(\xi)$ is a function on $\mathcal{X}^*$, and (8.4) shows that it is a linear function. Hence $R(H)$ is a subspace of the algebraic dual of the dual of $\mathcal{X}$. In many important cases, each function $R(\xi)$ is actually continuous in the weak*-topology on $\mathcal{X}^*$, and is thus given by an element of $\mathcal{X}$. In this case we may thus regard R as a mapping of H into $\mathcal{X}$, and $R(H)$ as a subspace of

$\mathcal{X}$. (This happens for example if $\mathcal{X}$ is a separable Banach space, but it is not true for all spaces and measures, see Bogachev (1996) and Talagrand (1983).)

As a concrete example, let us consider the Wiener measure on $\mathcal{X} = C_{\mathbb{R}}([0,1])$. We denote, as usual, the point evaluations by B_t, $0 \leq t \leq 1$; this is the standard Brownian motion.

The dual space of $C_{\mathbb{R}}([0,1])$ is isomorphic to the space $M_{\mathbb{R}}([0,1])$ of finite signed measures; a measure ν acts by integration and corresponds thus to the random variable $\int_0^1 B_t \, d\nu(t)$. The Gaussian Hilbert space H spanned by these variables is easily seen to coincide with the Gaussian Hilbert space spanned by $\{B_t\}_{0 \leq t \leq 1}$, which by Theorems 7.1 and 7.10 equals $\{I(f) : f \in L^2_{\mathbb{R}}([0,1])\}$, where $I(f) = \int_0^1 f(t) \, dB_t$.

Let $\xi \in H$. Then $R(\xi)$ is the function on $M_{\mathbb{R}}([0,1])$ given by

$$R(\xi)(\nu) = \langle \xi, \int_0^1 B_t \, d\nu(t) \rangle_H = \mathrm{E}\left(\xi \int_0^1 B_t \, d\nu(t)\right) = \int_0^1 \mathrm{E}(\xi B_t) \, d\nu(t).$$

This is the linear functional on $M_{\mathbb{R}}([0,1]) = C_{\mathbb{R}}([0,1])^*$ induced by the function $\mathrm{E}(\xi B_t) = \langle \xi, B_t \rangle_H \in C_{\mathbb{R}}([0,1])$. Consequently, this is a case where $R(H) \subseteq \mathcal{X}$; moreover, as a mapping into $C_{\mathbb{R}}([0,1])$, R is given by $R(\xi)(t) = \langle \xi, B_t \rangle_H$.

Note that although both Example 8.19 and this example treat Wiener measure and Brownian motion, they have somewhat different points of view. In Example 8.19, Brownian motion was studied as a Gaussian stochastic process on $[0,1]$, but here it is used to define a Gaussian stochastic process on $M_{\mathbb{R}}([0,1])$. Nevertheless (using the standard identification of $C_{\mathbb{R}}([0,1])$ as a subset of its bidual), we have found the same Gaussian Hilbert space H and the same operator R, and thus the same Cameron–Martin space $R(H)$ for the two versions.

EXAMPLE 8.27. (Rather technical and may be skipped.) We may now verify the assertions in Example 1.26. Let us somewhat more generally suppose that T is a compact metric space, and that $\{\xi_t\}$ is a centred Gaussian stochastic process on T that has a continuous version. This continuous version defines a random element ξ of $C_{\mathbb{R}}(T)$ such that $\xi(t) = \xi_t$ a.s. (The Borel σ-field on $C_{\mathbb{R}}(T)$ is generated by the point evaluations, which can be seen directly or as a consequence of Blackwell's theorem (Cohn 1980, Section 8.6).)

If $t_n \to t$ in T, then $\xi_{t_n} \to \xi_t$ a.s. and thus in probability. Hence the process is continuous in probability, and by Theorem 8.21, $R(H) \subseteq C_{\mathbb{R}}(T)$. If $\mu \in M_{\mathbb{R}}(T) \cong C_{\mathbb{R}}(T)^*$, then the real random variable $\langle \xi, \mu \rangle_{C(T)} = \int_T \xi(t) \, d\mu(t)$ belongs to H, for example because it equals the H-valued Bochner integral $\int_T \xi_t \, d\mu(t)$, cf. Appendix C. In particular, each $\langle \xi, \mu \rangle$ is Gaussian, so ξ is a

Gaussian random variable in $C_{\mathbb{R}}(T)$. Moreover, for $\mu \in M_{\mathbb{R}}(T)$ and $t \in T$,

$$\begin{aligned} R(\langle \xi, \mu \rangle)(t) &= \Big\langle \int_T \xi(s)\, d\mu(s), \xi_t \Big\rangle = \mathrm{E}\Big(\int_T \xi(s)\, d\mu(s) \cdot \xi(t) \Big) \\ &= \int_T \mathrm{E}\big(\xi(s)\xi(t)\big)\, d\mu(s) = \int_T \rho(s,t)\, d\mu(s). \end{aligned} \tag{8.5}$$

Next, let i denote the embedding $R(H) \to C_{\mathbb{R}}(T)$ and let $i^*\colon M_{\mathbb{R}}(T) \to R(H)^* \cong R(H)$ be the adjoint map. Then, using (8.5), for $t \in T$,

$$\begin{aligned} i^*(\mu)(t) &= \langle i^*(\mu), \rho_t \rangle_H = \langle \mu, i(\rho_t) \rangle_{C(T)} \\ &= \langle \mu, \rho_t \rangle_{C(T)} = \int_T \rho(t,s)\, d\mu(s) = R(\langle \xi, \mu \rangle)(t). \end{aligned}$$

Thus

$$i^*(\mu) = R(\langle \xi, \mu \rangle). \tag{8.6}$$

Consequently,

$$||\langle \xi, \mu \rangle||_H = ||R(\langle \xi, \mu \rangle)||_{R(H)} = ||i^*(\mu)||_{R(H)}$$

and hence

$$\langle \xi, \mu \rangle \sim \mathrm{N}(0, ||i^*(\mu)||^2_{R(H)}).$$

If $R(H)$ is dense in $C_{\mathbb{R}}(T)$, this is an example of the situation in Example 1.25, with $\mathcal{X} = C_{\mathbb{R}}(T)$ (and $R(H)$ instead of H).

In general, let $\mathcal{X}$ be the closed hull of $R(H)$ in $C_{\mathbb{R}}(T)$. Since $C_{\mathbb{R}}(T)$ is separable, the quotient space $C_{\mathbb{R}}(T)/\mathcal{X}$ is separable too, and thus there exists a countable subset A of $(C_{\mathbb{R}}(T)/\mathcal{X})^* = \mathcal{X}^\perp \subseteq M_{\mathbb{R}}(T)$ such that $F \in C_{\mathbb{R}}(T)/\mathcal{X}$ and $\langle F, \mu \rangle = 0$ for every $\mu \in A$ implies $F = 0$; in other words, $A \subset M_{\mathbb{R}}(T)$ and

$$\mathcal{X} = \{f \in C_{\mathbb{R}}(T) : \langle f, \mu \rangle = 0,\ \mu \in A\}.$$

If $\mu \in A$, then (8.5) shows that $R(\langle \xi, \mu \rangle)(t) = 0$ for every $t \in T$, because $\rho_t \in R(H) \subseteq \mathcal{X}$; thus $R(\langle \xi, \mu \rangle) = 0$ and hence $\langle \xi, \mu \rangle = 0$ a.s. Since A is countable, this shows that a.s. $\langle \xi, \mu \rangle = 0$ for every $\mu \in A$, and thus $\xi \in \mathcal{X}$ a.s.

We thus always have an example of the situation in Example 1.25, with $R(H)$ instead of H. (Note that by the Hahn–Banach theorem, each continuous linear functional on $\mathcal{X}$ is given by a (non-unique) measure μ as above.) The canonical map $R(H) \to H$ defined in Example 1.25 is determined by $i^*(\mu) \mapsto \langle \xi, \mu \rangle$ for $\mu \in M_{\mathbb{R}}(T)$, and is thus by (8.6) given by R^{-1}.

For the Wiener process as in Example 1.26, with $T = [0,1]$, it is easily seen that $\mathcal{X} = \{f \in C_{\mathbb{R}}([0,1]) : f(0) = 0\}$ and the assertions of Example 1.26 follow by the results above and Example 8.19.

IX

Conditioning

1. Introduction

Conditioning in Gaussian spaces is very simple because orthogonal centred Gaussian variables are independent; hence conditioning reduces to orthogonal projections.

For example, suppose that ξ and η are two variables in a Gaussian space, with η non-degenerate. We can write

$$\xi = \xi' + \xi'', \tag{9.1}$$

where ξ' is the orthogonal projection of ξ onto the one-dimensional subspace $\mathbb{R}\eta$ spanned by η, and $\xi'' = \xi - \xi'$; thus $\xi' = a\eta$ for some real constant a and $\xi'' \perp \eta$, where a is given explicitly by

$$a = \langle \xi, \eta \rangle / \|\eta\|^2 = \operatorname{Cov}(\xi, \eta) / \operatorname{Var}(\eta). \tag{9.2}$$

Consequently, ξ' is determined by η, while ξ'' is a centred Gaussian variable independent of η; thus conditioning becomes almost trivial.

We consider two different problems.

First, let us consider the conditional expectation of ξ given η; we obtain

$$\operatorname{E}(\xi \mid \eta) = \operatorname{E}(\xi' \mid \eta) + \operatorname{E}(\xi'' \mid \eta) = \xi' + 0 = \xi'. \tag{9.3}$$

Since $\xi' = a\eta$, we can also write this result as

$$\operatorname{E}(\xi \mid \eta = y) = ay, \tag{9.4}$$

where y is any real number. There is a minor technical complication here: in (9.4) we condition on an event of probability 0, which needs justification to be well-defined. In our case, (9.3) is really equivalent to (9.4) for *almost every* real y. (This is quite general; for any two random variables X and Y, the conditional expectation $\operatorname{E}(X \mid Y)$, when it exists, defines $\operatorname{E}(X \mid Y = y)$ as a function of y defined μ-a.e., where μ is the distribution of Y. However, this function is defined only up to equality μ-a.e.; in particular, when Y has a continuous distribution, the value for any specific y is not defined without further specification. Note in this connection that there is an 'almost every' hidden in (9.3) by our convention to use the equality sign for a.s. equal random variables.) Nevertheless, we may regard (9.4) as valid for all y, either by regarding the equation as a definition of the left hand side for each individual y, or by defining the left hand side as the limit of $\operatorname{E}\big(\xi \mid |\eta - y| < \varepsilon\big)$ as $\varepsilon \to 0$.

Secondly, let us again condition on $\eta = y$, but this time we ask for the conditional distribution $\mathcal{L}(\xi \mid \eta = y)$ and not only the conditional expectation. The decomposition (9.1) may be written $\xi = a\eta + \xi''$, with ξ'' independent of η, and thus conditioning on $\eta = y$ means that we replace η by y but leave ξ'' unaffected. Hence,

$$\mathcal{L}(\xi \mid \eta = y) = \mathcal{L}(ay + \xi'') = \mathrm{N}(ay, \|\xi''\|^2). \tag{9.5}$$

Again the general theory of conditioning (in this case equivalent to disintegration of the joint distribution of ξ and η) defines the left hand side of (9.5) only for a.e. y, but we may regard the equation as valid for each y by a limit procedure (or by fiat).

In the following two sections we repeat these considerations in greater detail and generality.

2. Conditional expectations

In this section and the following, we suppose that $\{\eta_i\}_{i\in\mathcal{I}}$ is a finite or infinite set of Gaussian random variables in a Gaussian Hilbert space H; we want to study conditioning of one or several other random variables with respect to the values of the variables η_i.

Let H' be the closed subspace of H spanned by $\{\eta_i\}$ and let $H'' = H \ominus H'$ be its orthogonal complement. Further, let P' denote the orthogonal projection onto H' and $P'' = I - P'$ the orthogonal projection onto H''.

If now ξ is any variable in H, we have the orthogonal decomposition

$$\xi = P'\xi + P''\xi. \tag{9.6}$$

THEOREM 9.1. *If $\xi \in H$, then its conditional expectation given $\{\eta_i\}$ is*

$$\mathrm{E}(\xi \mid \{\eta_i\}) = P'\xi.$$

PROOF. Since $P'\xi \in H'$, $P'\xi$ is measurable with respect to the σ-field $\mathcal{F}(H') = \mathcal{F}(\{\eta_i\})$, and thus

$$\mathrm{E}(P'\xi \mid \{\eta_i\}) = P'\xi.$$

Moreover, $P''\xi \in H''$ is orthogonal to $\{\eta_i\}$ and is thus independent of $\{\eta_i\}$. Hence,

$$\mathrm{E}(P''\xi \mid \{\eta_i\}) = \mathrm{E}\, P''\xi = 0.$$

The result follows by (9.6) and additivity. □

REMARK 9.2. If $\{\eta_i\}$ is finite, then $P'\xi \in H'$ is a linear combination of the variables η_i. In general, $P'\xi$ is at least a limit (converging in L^2 and a.s.) of finite linear combinations of variables in $\{\eta_i\}$.

This theorem is easily extended to conditional expectations of general, not necessarily Gaussian, random variables. (We still condition on Gaussian variables only.) In fact, we have already done this in Theorem 4.9.

THEOREM 9.3. *If $X \in L^1(\Omega, \mathcal{F}(H), \mathrm{P})$, then its conditional expectation given $\{\eta_i\}$ is*

$$\mathrm{E}(X \mid \{\eta_i\}) = \Gamma(P')X. \tag{9.7}$$

PROOF. For $X \in L^2$, this is just Theorem 4.9 with $K = H'$ and $G = H$. The extension to $L^1(\Omega, \mathcal{F}(H), \mathrm{P})$ is valid because both sides of (9.7) define continuous linear operators on $L^1(\Omega, \mathcal{F}(H), \mathrm{P})$. (Actually, (9.7) was used implicitly in Theorem 4.12 and its proof as the definition of the extension of $\Gamma(P')$ to $L^1(\Omega, \mathcal{F}(H), \mathrm{P})$.) □

This theorem may be made more explicit if we use a suitable basis in H.

COROLLARY 9.4. *Let $\{\xi_j\}_j$ be an orthonormal basis in H such that each $\xi_j \in H' \cup H''$. (Such a basis may be constructed by taking the union of any two orthonormal bases in H' and H''.) If $X \in L^2(\Omega, \mathcal{F}(H), \mathrm{P})$, then $X = \sum_\alpha a_\alpha :\xi^\alpha:$ for some coefficients a_α, with summation over all multi-indices α; in this case*

$$\mathrm{E}(X \mid \{\eta_i\}) = \sum_\alpha{}' a_\alpha :\xi^\alpha:, \tag{9.8}$$

where $\sum'$ denotes the sum over all terms involving only $\xi_j \in H'$. The sums converge in L^2.

PROOF. The existence of such a decomposition of X follows by Theorem 3.21; note that the terms are orthogonal which implies that the sum in (9.8) of a subset of the terms also converges in L^2. The formula (9.8) now follows from Theorem 9.3 and the observation (following directly from the definition (4.3))

$$\Gamma(P') :\xi_1 \cdots \xi_n: = \begin{cases} :\xi_1 \cdots \xi_n:, & \text{if } \xi_1, \dots, \xi_n \in H', \\ 0, & \text{if some } \xi_k \in H'' \text{ and thus } P'\xi_k = 0. \end{cases}$$

□

EXAMPLE 9.5. Let $\{\xi_t\}_{t \in T}$ be a centred Gaussian stochastic process, indexed by a subset $T \subseteq \mathbb{R}$, with the parameter interpreted as time; the present being represented by time 0, say. Then the best prediction of a future observation ξ_t given the observed values $\{\xi_s\}_{s \le 0}$ up to the present is given by the conditional expectation $\mathrm{E}(\xi_t \mid \{\xi_s\}_{s \le 0}) = P_0^- \xi_t$, where P_0^- denotes the orthogonal projection onto the closed subspace H_0^- of H spanned by $\{\xi_s\}_{s \le 0}$. This simple observation is the basis of much work in prediction theory.

Assume now, for simplicity, that $\mathrm{E}\,\xi_t^2 = 1$ and let $P_0^- \xi_t = r\eta$, where $-1 \le r \le 1$ and $\eta \in H_0^-$ is normalized by $\mathrm{E}\,\eta^2 = 1$. Then Theorem 9.3 yields, for every $n \ge 0$,

$$\mathrm{E}(:\xi_t^n: \mid \{\xi_s\}_{s \le 0}) = r^n :\eta^n:.$$

Furthermore, it follows as in Example 4.18 that

$$\mathrm{E}(f(\xi_t) \mid \{\xi_s\}_{s\leq 0}) = \mathcal{M}_r f(\eta),$$

where $\mathcal{M}_r$ is the Mehler transform defined in (4.12), i.e. the Mehler transform for the standard one-dimensional Gaussian Hilbert space on $(\mathbb{R}, \gamma)$.

EXAMPLE 9.6. More generally, let $\{\xi_t\}_{t\in T}$ be as in the preceding example and let, for each $t \in T$, H_t^-, H_t^+ and H_t be the closed subspaces of H spanned by $\{\xi_s\}_{s\leq t}$, $\{\xi_s\}_{s\geq t}$ and ξ_t (thus $\dim H_t \leq 1$); further, let P_t^-, P_t^+ and P_t denote the corresponding projections in H.

Now suppose that for every $t, u \in T$ with $t \leq u$,

$$P_t^- \xi_u \in H_t. \tag{9.9}$$

Then $P_t^- \xi_u = P_t\xi_u$, $u \geq t$, and thus $P_t^- = P_t$ on H_t^+. Consequently, for every $X \in L^1(\mathcal{F}(H_t^+))$,

$$\mathrm{E}(X \mid \{\xi_s\}_{s\leq t}) = \Gamma(P_t^-)X = \Gamma(P_t)X = \mathrm{E}(X \mid \xi_t);$$

in other words, the process is Markov.

Conversely, (9.9) is evidently also necessary for the process to be Markov.

EXAMPLE 9.7. Let $\{\xi_t\}$ be the Ornstein–Uhlenbeck process defined in Example 8.4. If $t \geq 0$, then it follows from the formula for the covariance function that $\xi_t - e^{-t}\xi_0 \perp \xi_s$ for every $s \leq 0$; thus $P_0^- \xi_t = e^{-t}\xi_0$, $t \geq 0$. Since the process is stationary, $P_t^- \xi_u = e^{t-u}\xi_t$ whenever $t \leq u$, and by the preceding example, the process is Markov.

Furthermore, by Example 9.5,

$$\mathrm{E}(f(\xi_t) \mid \{\xi_s\}_{s\leq 0}) = \mathcal{M}_{e^{-t}} f(\xi_0),$$

which shows the relation between the Ornstein–Uhlenbeck process and the Ornstein–Uhlenbeck semigroup $\{M_{e^{-t}}\}$.

Let us next try to interpret the results above as conditioning on the event $\{\eta_i = y_i \text{ for every } i \in \mathcal{I}\}$, where y_i are given real numbers.

Recall that $P'X$ above (and as a special case $P'\xi$) is measurable with respect to the σ-field $\mathcal{F}(H') = \mathcal{F}(\{\eta_i\})$, and thus can be written as some Borel function f of $\{\eta_i\}$. Plugging in the values y_i in f, we obtain an interpretation of $\mathrm{E}(X \mid \eta_i = y_i,\, i \in \mathcal{I})$. This is, however, in accordance with the general theory referred to in Section 1, well-defined only a.e. with respect to the distribution of $\{\eta_i\}$, since f may be arbitrarily changed on a null set; in particular, it is not defined for any individual $\{y_i\}$ (except in the trivial case when all η_i and y_i vanish). In general, we cannot do better, but in some cases it is possible to assign well-defined values for all $\{y_i\}$.

The simplest case is when $\{\eta_i\}$ is a finite set of linearly independent variables, and we condition a Gaussian variable $\xi \in H$. (Recall that a finite set of centred jointly Gaussian variables is linearly independent if and only if the joint distribution is non-singular.) In this case,

$$P'\xi = \sum a_i \eta_i \tag{9.10}$$

for some coefficients a_i, and we may uniquely define, for any y_i,

$$\mathrm{E}(\xi \mid \eta_i = y_i,\, i \in \mathcal{I}) = \sum_i a_i y_i. \tag{9.11}$$

(This includes (9.4) as a special case.)

The next simplest case is when $\{\eta_i\}$ is finite, but linearly dependent. In this case, (9.10) still holds, but the coefficients a_i are not unique; hence (9.11) is not a valid definition for all y_i. We avoid this problem by restricting the set of allowed $\{y_i\}$. We say that the numbers y_i are *compatible* if

$$\sum_i b_i y_i = 0 \text{ for any numbers } b_i \text{ such that } \sum_i b_i \eta_i = 0 \text{ a.s.}, \tag{9.12}$$

and we may then use (9.11) for any compatible family $\{y_i\}$; although the coefficients a_i in (9.10) are non-unique, (9.12) implies that $\sum a_i y_i$ is the same for every choice. (The reader should try to convince him/herself that, in any case, it is reasonable to condition on compatible y_i only.)

We summarize these cases.

THEOREM 9.8. *If $\{\eta_i\}_{i\in\mathcal{I}}$ is a finite set, and $\xi \in H$, then* $\mathrm{E}(\xi \mid \eta_i = y_i,$ $i \in \mathcal{I})$ *may be defined by* (9.11) *for any numbers y_i that are compatible in the sense of* (9.12) (*the latter condition is vacuous if $\{\eta_i\}$ is linearly independent*), *where a_i are any numbers satisfying* (9.10). *In particular, the conditional expectation is then a linear function of $\{y_i\}$.* □

If $\{\eta_i\}$ is infinite, the situation is more complicated. Even if the variables η_i are independent, in which case $P'\xi$ can always be expressed uniquely as an infinite sum $\sum a_i \eta_i$, convergent in L^2, the corresponding sum $\sum a_i y_i$ does not in general converge for all real numbers y_i, so any extension of Theorem 9.8 to infinite sets has to restrict $\{y_i\}$ further.

The situation for general random variables is similar. If X is a random variable such that $X \in L^2(\Omega, \mathcal{F}(H), \mathrm{P})$, we may try to define $\mathrm{E}(X \mid \eta_i = y_i,\, i \in \mathcal{I})$ as follows: select an orthonormal basis $\{\xi_j\}$ in H' consisting of linear combinations of the η_i (such a basis may be constructed by the Gram–Schmidt procedure), and then use (9.8), expressing each Wick power $:\xi^\alpha:$ as a polynomial in $\{\eta_i\}$ and substituting y_i for η_i. However, there is no guarantee that the resulting numerical sum converges (except when there are only finitely many non-zero terms in the sum in (9.8)), and in general we have to settle for the conditional expectation being defined for almost every $\{y_i\}$. (For an explicit example of divergence, let X be the sum in Example 3.43 and condition on ξ; then $P'X = X \in L^2$, but the procedure above does not define $\mathrm{E}(X \mid \xi = 0)$ as a convergent sum.)

3. Conditional distributions

Again let $\xi \in H$. The decomposition (9.6) may be written, using Theorem 9.1,

$$\xi = \mathrm{E}(\xi \mid \{\eta_i\}) + P''\xi,$$

where $P''\xi$ is a centred Gaussian variable independent of $\{\eta_i\}$. Consequently, we obtain the following result by conditioning on $\eta_i = y_i$, $i \in \mathcal{I}$, which does not affect $P''\xi$.

THEOREM 9.9. *For all sequences* $\{y_i\}$ *such that* $m(\{y_i\}) = \mathrm{E}(\xi \mid \eta_i = y_i, i \in \mathcal{I})$ *is defined, the conditional distribution of* ξ *given* $\eta_i = y_i$, $i \in \mathcal{I}$, *is*

$$\mathcal{L}(\xi \mid \eta_i = y_i, i \in \mathcal{I}) = \mathrm{N}\big(m(\{y_i\}), \sigma^2\big),$$

where $\sigma^2 = \mathrm{Var}(P''\xi) = \mathrm{Var}(\xi) - \mathrm{Var}\big(\mathrm{E}(\xi \mid \{\eta_i\})\big)$. *In particular,* σ^2 *is independent of* $\{y_i\}$. □

In general, this applies to almost every $\{y_i\}$ (with respect to the distribution of $\{\eta_i\}$), and indeed, the conditional distribution of ξ given $\eta_i = y_i$, $i \in \mathcal{I}$, is defined only up to equivalence (arbitrary modifications on null sets) and not for any particular $\{y_i\}$. Nevertheless, the discussion in the preceding section shows that sometimes there is a reasonable definition of $\mathcal{L}(\xi \mid \eta_i = y_i, i \in \mathcal{I})$ for given numbers y_i also. One such case is, by Theorem 9.8, when $\{\eta_i\}$ is a finite set and y_i are compatible; then $m(\{y_i\}) = \sum_i a_i y_i$ is a linear function. Another important case is when every $y_i = 0$; we may then unambiguously set $\mathrm{E}(\xi \mid \eta_i = 0, i \in \mathcal{I}) = 0$ (for example by symmetry), and thus

$$\mathcal{L}(\xi \mid \eta_i = 0, i \in \mathcal{I}) = \mathrm{N}(0, \sigma^2),$$

with σ^2 as above.

REMARK 9.10. In general, a conditioned random variable is not defined on the same probability space $(\Omega, \mathcal{F}, \mathrm{P})$ as the original variable, but for Gaussian spaces it follows from the discussion above that we can interpret the random variable $P''\xi + m(\{y_i\}) \in H \oplus \mathbb{R}$ (when $m(\{y_i\})$ is defined) as ξ conditioned on $\eta_i = y_i$, $i \in \mathcal{I}$. Hence we can regard the conditioned random variable as defined on the original probability space.

In particular, ξ conditioned on $\eta_i = 0$, $i \in \mathcal{I}$, may be identified with $P''\xi$, which belongs to the original Gaussian space H.

This special feature of Gaussian conditioning is often convenient.

For a more general random variable $X \in L^2(\Omega, \mathcal{F}(H), \mathrm{P})$, we may select a basis $\{\xi_j\}$ as in Corollary 9.4, express X as a sum $\sum a_\alpha :\xi^\alpha:$, expand each $:\xi^\alpha:$ as a polynomial in $\{\xi_j\}$, substitute the value specified by $\{y_i\}$ for each $\xi_j \in H'$, and finally try to sum the result (for example in L^2) to obtain a function of $\{\xi_j : \xi_j \in H''\}$. This succeeds at least for almost every $\{y_i\}$, and then defines X conditioned on $\eta_i = y_i$, $i \in \mathcal{I}$, as a random variable in $L^2(\Omega, \mathcal{F}(H''), \mathrm{P}) \subset L^2(\Omega, \mathcal{F}, \mathrm{P})$. Note that unlike the Gaussian case just treated, the conditional distribution of X given $\eta_i = y_i$, $i \in \mathcal{I}$, in general depends on $\{y_i\}$ even if we subtract its mean. (This is due to terms in (9.8) like $:\xi'\xi'':$, involving basis vectors in both $\{\xi_j\} \cap H'$ and $\{\xi_j\} \cap H''$.)

REMARK 9.11. Since all our considerations are independent of the underlying probability space $(\Omega, \mathcal{F}, \mathrm{P})$, we may assume that actually $(\Omega, \mathcal{F}, \mathrm{P})$ is the product of two probability spaces $(\Omega', \mathcal{F}', \mathrm{P}')$ and $(\Omega'', \mathcal{F}'', \mathrm{P}'')$, with the variables in H' being functions of $\omega' \in \Omega'$ and the variables in H'' functions of $\omega'' \in \Omega''$. We may further assume that $\mathcal{F}'$ is generated by $\{\eta_i\}$. In that case, a random variable $X \in L^1(\Omega, \mathcal{F}(H), \mathrm{P})$ is a function of (ω', ω''), the conditional expectation $\mathrm{E}(X \mid \{\eta_i\})$ is a function of ω', obtained by averaging over ω'', and the random variable obtained by conditioning X on $\eta_i = y_i$, $i \in \mathcal{I}$, is a function of ω'', obtained by fixing ω'.

EXAMPLE 9.12. Let B_t, $0 \le t \le 1$, be a Brownian motion, and condition on $B_1 = 0$. In this case, H' is the one-dimensional space spanned by B_1, and $\|B_1\|_H = 1$; thus $P'B_t = \langle B_t, B_1 \rangle B_1 = tB_1$ and $P''B_t = B_t - tB_1$. Consequently, by Remark 9.10, the Brownian motion conditioned on $B_1 = 0$ is given by the process

$$B_t - tB_1, \qquad 0 \le t \le 1.$$

This stochastic process is known as the *Brownian bridge*; it is a Gaussian stochastic process with covariance function $\rho(s,t) = s(1-t)$, $0 \le s \le t \le 1$.

4. Example: Random potential in a network

As an example of conditioning, we study in detail a model that is sometimes called a *random Gaussian potential* in a network $\mathcal{G}$. This model can be described as follows. (See Remark 9.14 below for a different, but equivalent, definition.)

Let $\mathcal{G}$ be a connected graph, and let, for each edge $e \in \mathcal{G}$, ξ_e be a standard normal random variable such that the variables for different edges are independent. We want to interpret ξ_e as the difference between the endpoints of e (taken in some order) of a potential defined on the vertices of $\mathcal{G}$. However, this is not possible (unless the graph is a tree), since the sum of the 'potential differences' ξ_e around a closed circuit in general does not vanish. This obstacle is removed by conditioning on the event that all such sums around closed curves do vanish; this produces a new set of random variables $\{\widetilde{\xi}_e\}$ (no longer independent) that can be interpreted as the differences along the edges of a random potential on $\mathcal{G}$. We want to study this random potential; a typical problem is to find the distribution of the potential difference between two given vertices.

REMARK 9.13. $\mathcal{G}$ may have loops and multiple edges. Loops, however, do not affect the result at all and may as well be ignored. (Actually, some statements below require trivial modifications for loops.)

In order to attack this problem, we first repeat the construction just given in greater detail, at the same time introducing more notations.

Let $\mathcal{G}$ be a connected graph, with vertex set $V(\mathcal{G})$ and edge set $E(\mathcal{G})$. For simplicity, we will assume that the graph is finite, although the arguments below extend to many infinite graphs with minor modifications. (See e.g. Soardi (1994) for the required theory of electric currents on infinite networks.)

We assign an orientation to each edge; thus an edge e has a specific starting vertex $\alpha(e)$ and an end vertex $\omega(e)$, and we say that the edge goes from $\alpha(e)$ to $\omega(e)$. Moreover, we assign independent standard normal variables ξ_e to the edges of $\mathcal{G}$.

A *path* is a sequence $v_0e_1v_1e_2\dots e_nv_n$ ($n \geq 1$) of vertices v_i and edges e_i such that, for each $i = 1, \dots, n$, e_i goes either from v_{i-1} to v_i or from v_i to v_{i-1}. (Note that this definition is independent of the orientations of the edges, we are just talking about paths in the undirected graph $\mathcal{G}$.) We say the path *goes from* v_0 *to* v_n. The path is *closed* if $v_n = v_0$.

For each path $\gamma = v_0e_1\dots e_nv_n$, we define

$$\xi_\gamma = \sum_{i=1}^{n} \varepsilon_i \xi_{e_i}, \tag{9.13}$$

where $\varepsilon_i = 1$ if e_i goes from v_{i-1} to v_i and $\varepsilon_i = -1$ if e_i goes in the opposite direction. In other words, ξ_γ is the sum of the variables ξ_e along the path, taking the orientations into account.

We now condition on the event $\{\xi_\gamma = 0 \text{ for every closed path } \gamma\}$; we denote the conditioned variables corresponding to ξ_e and ξ_γ by $\widetilde{\xi}_e$ and $\widetilde{\xi}_\gamma$, respectively. By Section 3 above, these conditioned random variables are well-defined (although we condition on an event of probability 0) and they belong to the Gaussian space spanned by $\{\xi_e\}_{e\in E(\mathcal{G})}$. Note that by (9.13), $\widetilde{\xi}_\gamma = \sum_{i=1}^n \varepsilon_i \widetilde{\xi}_{e_i}$, with γ, e_i and ε_i as in (9.13).

If v and w are any two vertices in $\mathcal{G}$, and γ_1 and γ_2 are two paths from v to w, then we may form a closed path λ by following first γ_1 from v to w and then γ_2 backwards back to v. It is easily seen that $\xi_\lambda = \xi_{\gamma_1} - \xi_{\gamma_2}$ and $\widetilde{\xi}_\lambda = \widetilde{\xi}_{\gamma_1} - \widetilde{\xi}_{\gamma_2}$, but $\widetilde{\xi}_\lambda = 0$ by definition since λ is closed; hence $\widetilde{\xi}_{\gamma_1} = \widetilde{\xi}_{\gamma_2}$. In other words, the variable $\widetilde{\xi}_\gamma$ depends only on the endpoints of γ, and we may define $\widetilde{\xi}_{vw} = \widetilde{\xi}_\gamma$, for any path γ from v to w. (Such paths always exist, since $\mathcal{G}$ is assumed to be connected.)

This defines random variables $\widetilde{\xi}_{vw}$ for any vertices $v, w \in V(\mathcal{G})$. We claim that, for any $v, w, u \in V(\mathcal{G})$,

$$\widetilde{\xi}_{vv} = 0, \tag{9.14}$$

$$\widetilde{\xi}_{wv} = -\widetilde{\xi}_{vw}, \tag{9.15}$$

$$\widetilde{\xi}_{vw} + \widetilde{\xi}_{wu} = \widetilde{\xi}_{vu}. \tag{9.16}$$

These relations are almost obvious. Indeed, (9.14) follows because $\widetilde{\xi}_\gamma = 0$ for any closed path γ; (9.15) follows by reversing any path γ from v to w (or by (9.14) and (9.16)); (9.16) follows by taking two paths γ_1 from v to w and γ_2 from w to u and combining them into a path $\gamma_1\gamma_2$ from v to u such that $\xi_{\gamma_1\gamma_2} = \xi_{\gamma_1} + \xi_{\gamma_2}$ and thus $\widetilde{\xi}_{\gamma_1\gamma_2} = \widetilde{\xi}_{\gamma_1} + \widetilde{\xi}_{\gamma_2}$.

The final step in the construction is to show that there exist random variables $U(v)$, $v \in V(\mathcal{G})$, such that

$$\widetilde{\xi}_{vw} = U(v) - U(w), \qquad v, w \in V(\mathcal{G}); \tag{9.17}$$

they constitute our random potential U.

Note that (9.17) determines $U(\cdot)$ only up to an additive (random) constant; thus a further normalization is needed in order to define U uniquely. As long as we only study potential differences, the choice of normalization is irrelevant, so we may choose any possibility. Perhaps the simplest normalization is to fix a 'reference vertex' v_0 and require $U(v_0) = 0$. (This means that the potential is measured relative to the reference vertex.) It follows from (9.17) that we then have to have

$$U(v) = \widetilde{\xi}_{vv_0}, \qquad v \in V(\mathcal{G});$$

and, indeed, it follows immediately from (9.16) that this choice satisfies (9.17).

REMARK 9.14. The probability density of the random vector $(\xi_e)_{e\in E(\mathcal{G})}$ in $\mathbb{R}^{E(\mathcal{G})}$ is $c\exp\bigl(-\frac{1}{2}\sum_e x_e^2\bigr)$ for some constant c. We condition on the event that this random vector belongs to a certain subspace M of $\mathbb{R}^{E(\mathcal{G})}$, and it is easily seen that this implies that the resulting conditioned distribution also has a density $c_1 \exp\bigl(-\frac{1}{2}\sum_e x_e^2\bigr)$ (for another constant c_1) with respect to Lebesgue measure in the subspace M. If we, as above, normalize the potential U by $U(v_0) = 0$, then (9.17) defines a linear bijection of $\mathbb{R}^{V(\mathcal{G})\setminus\{v_0\}}$ onto M, and thus U has a density in $\mathbb{R}^{V(\mathcal{G})\setminus\{v_0\}}$ which is given by

$$c_2 \exp\Bigl(-\tfrac{1}{2}\sum_e \bigl(U(\alpha(e)) - U(\omega(e))\bigr)^2\Bigr)$$

for a normalization constant c_2. This is a distribution of the standard type encountered in statistical physics, corresponding to an energy function on the space of potentials given by

$$\tfrac{1}{2}\sum_e \bigl(U(\alpha(e)) - U(\omega(e))\bigr)^2 = \tfrac{1}{2}\|\operatorname{Grad} U\|^2, \tag{9.18}$$

using notation introduced below.

Conversely, this may be taken directly as the definition of the distribution of U. In this approach, no conditioning is involved in the definition. Instead, conditioning may be used as a technical tool, enabling us to use the first definition above and the analysis below to obtain results on this distribution.

In order to study the random potential, we begin by letting H be the Gaussian Hilbert space spanned by $\{\xi_e\}_{e\in E(\mathcal{G})}$. Obviously, H is the set of all sums $\sum_e a_e\xi_e$ with real coefficients a_e. Furthermore, by our assumption that ξ_e be independent standard normal, $\|\sum_e a_e\xi_e\|_H^2 = \sum_e a_e^2$.

We also introduce the Euclidean spaces $F = \mathbb{R}^{E(\mathcal{G})}$ and $G = \mathbb{R}^{V(\mathcal{G})}$ of all real functions on $E(\mathcal{G})$ and $V(\mathcal{G})$, respectively. There is a natural isometry $\Phi\colon F \to H$ given by $\Phi(\varphi) = \sum_e \varphi(e)\xi_e$; hence we can regard H as a Gaussian Hilbert space indexed by F.

We interpret elements of F as flows on the graph $\mathcal{G}$. The *energy* of a flow φ is defined to be $\|\varphi\|_F^2 = \sum_e \varphi(e)^2$. Moreover, we define the *divergence* of a flow φ to be the function on $V(\mathcal{G})$ given by

$$\operatorname{Div}(\varphi)(v) = \sum_{e:\alpha(e)=v} \varphi(e) - \sum_{e:\omega(e)=v} \varphi(e). \tag{9.19}$$

Thus $\operatorname{Div}(\varphi) \in G$ and Div is a linear operator $F \to G$.

We also define a dual operator $\operatorname{Grad}\colon G \to F$ by

$$\operatorname{Grad}(\psi)(e) = \psi(\alpha(e)) - \psi(\omega(e)), \qquad \psi \in G; \tag{9.20}$$

it follows easily from (9.19) and (9.20) that

$$\langle \operatorname{Div}\varphi, \psi\rangle_G = \langle \varphi, \operatorname{Grad}\psi\rangle_F, \qquad \varphi\in F,\ \psi\in G. \tag{9.21}$$

For two distinct vertices v_1 and v_2 in $\mathcal{G}$, we say that a flow φ is a *unit flow* from v_1 to v_2 if $\operatorname{Div}(\varphi)(v_1) = 1$, $\operatorname{Div}(\varphi)(v_2) = -1$ and $\operatorname{Div}(\varphi)(v) = 0$ for $v \neq v_1, v_2$.

We define some special flows as follows. If e is an edge, then φ_e, the *unit flow along* e, is given by

$$\varphi_e(e') = \begin{cases} 1, & e' = e, \\ 0, & e' \neq e, \end{cases}$$

and if $\gamma = v_0e_1v_1\dots e_nv_n$ is a path, then φ_γ, the *unit flow along* γ, is $\sum_{i=1}^n \varepsilon_i\varphi_{e_i}$, where ε_i is as in (9.13); these are the elements of F corresponding to ξ_e and ξ_γ in H. Note that φ_γ is a unit flow from v_0 to v_n, unless the path is closed in which case $\operatorname{Div}(\varphi_\gamma) = 0$ everywhere.

After these lengthy preliminaries, we return to the conditioning. To condition on the event $\{\xi_\gamma = 0 \text{ when } \gamma \text{ is a closed path}\}$ is by Section 3 the same as to project onto the orthogonal complement $H_0^\perp$ of the subspace H_0 spanned by $\{\xi_\gamma : \gamma \text{ is a closed path}\}$.

The subspace H_0 corresponds to the subspace F_0 of F spanned by $\{\varphi_\gamma : \gamma \text{ is a closed path}\}$, and thus $H_0^\perp$ corresponds to $F_0^\perp$.

In order to find F_0 and $F_0^\perp$, we begin with the latter. Clearly, $\varphi \in F_0^\perp$ if and only if $\langle\varphi, \varphi_\gamma\rangle_F = 0$ for every closed path γ. For any path γ, $\langle\varphi,\varphi_\gamma\rangle_F$ equals the sum of $\varphi(e)$ along the path γ (taking the orientations into account), and the same argument as for $\widetilde{\xi}_\gamma$ above shows that if $\varphi \in F_0^\perp$, then $\langle\varphi,\varphi_\gamma\rangle_F$ depends on the endpoints of γ only, and, moreover, if these endpoints are v

and w, it equals $\psi(v) - \psi(w)$ for some function ψ on $V(\mathcal{G})$. In particular, taking the path γ to be a single edge e,

$$\varphi(e) = \langle \varphi, \varphi_e \rangle_F = \psi(\alpha(e)) - \psi(\omega(e))$$

or

$$\varphi = \operatorname{Grad} \psi$$

for some $\psi \in G$.

Conversely, every gradient $\operatorname{Grad}(\psi)$ belongs to $F_0^\perp$, for example by

$$\langle \operatorname{Grad}(\psi), \varphi_\gamma \rangle_F = \langle \psi, \operatorname{Div}(\varphi_\gamma) \rangle_G = \langle \psi, 0 \rangle_G = 0$$

whenever γ is a closed path.

We have shown that

$$F_0^\perp = \operatorname{Ran}(\operatorname{Grad}) = \{\operatorname{Grad}(\psi) : \psi \in G\}.$$

Hence $\varphi \in F_0$ if and only if $\langle \varphi, \operatorname{Grad}(\psi) \rangle_F = 0$ for any $\psi \in G$. But by (9.21), this holds if and only if $\langle \operatorname{Div} \varphi, \psi \rangle_G = 0$ for every $\psi \in G$, i.e. if and only if $\operatorname{Div} \varphi = 0$. Consequently,

$$F_0 = \{\varphi : \operatorname{Div}(\varphi) = 0\}, \tag{9.22}$$

the set of divergence-free flows.

Now, let v and w be two distinct vertices and let γ be a path from v to w. Further, let $\widetilde{\varphi}_\gamma$ be the projection of φ_γ onto $F_0^\perp$; this is the element of F corresponding to $\widetilde{\xi}_\gamma \in H$. Then $\widetilde{\varphi}_\gamma$ can be characterized as the element of $\varphi_\gamma + F_0 = \{\varphi_\gamma + \varphi_0 : \varphi_0 \in F_0\}$ which has minimal norm. Since φ_γ is a unit flow from v to w, it follows from (9.22) that $\varphi_\gamma + F_0$ is the set of all unit flows from v to w. This yields the following characterization:

CLAIM. *$\widetilde{\varphi}_\gamma$ is the unit flow from v to w which has minimal energy.*

In particular, $\widetilde{\varphi}_\gamma$ depends only on the endpoints v and w of γ; this is also a consequence of the corresponding fact established earlier for $\widetilde{\xi}_\gamma = \widetilde{\xi}_{vw}$.

Now it is a standard fact from the theory of electrical networks that the unit flow of minimal energy between two vertices is the flow exhibited by the electric current in the network when the two vertices are connected to an external source of suitable voltage, assuming that each edge has unit resistance. (It is also easy to show directly that $\widetilde{\varphi}_\gamma$ obeys Kirchhoff's laws. One law follows from $\widetilde{\varphi}_\gamma \in \varphi_\gamma + F_0$ and the other from $\widetilde{\varphi}_\gamma \in F_0^\perp$.) Moreover, the energy of the flow as defined above, which in physical terms is really the energy dissipation of the electric current, equals the effective resistance between the two vertices. Since this energy is $\|\widetilde{\varphi}_\gamma\|_F^2 = \|\widetilde{\xi}_{vw}\|_H^2 = \operatorname{Var}(\widetilde{\xi}_{vw})$, we obtain the following result.

THEOREM 9.15. *In the model described above, the potential difference $U(v) - U(w) = \widetilde{\xi}_{vw}$ between two vertices in $\mathcal{G}$ has a centred normal distribution $\mathrm{N}(0, \sigma_{vw}^2)$, where the variance σ_{vw}^2 equals the effective resistance between*

v and w if $\mathcal{G}$ is regarded as an electrical network with unit resistances along the edges.

Moreover, $\widetilde{\xi}_{vw}$ may be expressed as a linear combination $\varphi(e)\xi_e$, where $\varphi(e)$ is the component along e of a unit electric current flowing from v to w through the network. □

Furthermore, the different potential differences $U(v) - U(w) = \widetilde{\xi}_{vw}$, $v, w \in V(\mathcal{G})$, are jointly normal. There are a lot of obvious linear relations between these variables, cf. (9.14)–(9.16), but there are no other relations between them.

COROLLARY 9.16. *For any fixed vertex $v_0 \in V(\mathcal{G})$, the random variables $U(v) - U(v_0)$, $v \neq v_0$, are jointly normal and linearly independent. In other words, their joint distribution is a non-singular normal distribution in $\mathbb{R}^{V(\mathcal{G})\setminus\{v_0\}}$.*

PROOF. Joint normality is clear, since all variables belong to H. Now suppose that there exists a linear relation $\sum_{v\neq v_0} a_v \widetilde{\xi}_{vv_0} = 0$. Then also $\sum a_v \widetilde{\varphi}_{vv_0} = 0$, and thus for every $w \neq v_0$,

$$0 = \operatorname{Div}\Big(\sum_{v\neq v_0} a_v \widetilde{\varphi}_{vv_0}\Big)(w) = \sum_{v\neq v_0} a_v \operatorname{Div}(\widetilde{\varphi}_{vv_0})(w) = a_w,$$

since $\operatorname{Div}(\widetilde{\varphi}_{vv_0})(w)$ vanishes when $v \neq w$ and equals 1 when $v = w$. Hence $a_v = 0$ for every $v \neq v_0$. □

The connection with an electrical network can be carried further.

THEOREM 9.17. *Let $v_1, w_1, v_2, w_2 \in V(\mathcal{G})$. The covariance*

$$\operatorname{Cov}\big(U(v_1) - U(w_1), U(v_2) - U(w_2)\big) = \langle \widetilde{\xi}_{v_1w_1}, \widetilde{\xi}_{v_2w_2} \rangle_H$$

equals the potential difference between v_2 and w_2 for a unit electric current flowing from v_1 to w_1.

REMARK 9.18. By symmetry, this covariance also equals the potential difference between v_1 and w_1 for a unit electric current flowing from v_2 to w_2.

PROOF. Let γ_i be a path from v_i to w_i, $i = 1, 2$. Then, by the argument above,

$$\langle \widetilde{\xi}_{v_1w_1}, \widetilde{\xi}_{v_2w_2} \rangle_H = \langle \widetilde{\xi}_{\gamma_1}, \widetilde{\xi}_{\gamma_2} \rangle_H = \langle \widetilde{\varphi}_{\gamma_1}, \widetilde{\varphi}_{\gamma_2} \rangle_F.$$

However, $\widetilde{\varphi}_{\gamma_1}$ is the unit electric flow from v_1 to w_1, and if ψ is the corresponding potential we have $\widetilde{\varphi}_{\gamma_1} = \operatorname{Grad}\psi$ and thus, by (9.21),

$$\langle \widetilde{\varphi}_{\gamma_1}, \widetilde{\varphi}_{\gamma_2} \rangle_F = \langle \operatorname{Grad}\psi, \widetilde{\varphi}_{\gamma_2} \rangle_F = \langle \psi, \operatorname{Div}\widetilde{\varphi}_{\gamma_2} \rangle_G = \psi(v_2) - \psi(w_2).$$

□

We can continue and condition further in our model. The random potential above was restricted to be 0 at a certain reference vertex. What happens if we fix the values at several vertices?

Thus, let W be a non-empty subset of $V(\mathcal{G})$ and let $u\colon W \to \mathbb{R}$ be a given function on W. Choose a reference vertex $v_0 \in W$ and define the random potential U as above but with $U(v_0) = u(v_0)$ (which just means that the constant $u(v_0)$ is added to all values of U); then condition on $U(w) = u(w)$ for all $w \in W$. Note that this conditioning is well-defined by Corollary 9.16 and the results of Sections 2 and 3. It is easily seen that the result does not depend on the choice of $v_0 \in W$, for example since we may as well instead condition on $U(w_1) - U(w_2) = u(w_1) - u(w_2)$ for all pairs $w_1, w_2 \in W$; we thus may forget v_0 and call the resulting random function U the *random potential conditioned by* $U(w) = u(w)$, $w \in W$.

REMARK 9.19. The random potential conditioned by $U(w) = u(w)$, $w \in W$, may also be obtained by choosing a random normalization $U(v_0) = \eta$, for some $v_0 \in V(\mathcal{G})$ and a non-degenerate normal variable η independent of all ξ_e, and then conditioning on $U(w) = u(w)$, $w \in W$.

Furthermore, the same distribution may be obtained without conditioning at all by the definition in Remark 9.14, fixing the values $U(w) = u(w)$, $w \in W$.

The results above are easily extended to this case as follows. By a unit electric current flowing from a vertex v to the set W we mean the current obtained by connecting v to one pole of a power source and all $w \in W$ to the other (thus short-circuiting W), adjusting the potential of the power source such that a unit current flows; the required potential difference is the effective resistance between v and W.

THEOREM 9.20. *Let W be a non-empty subset of $V(\mathcal{G})$ and let $u\colon W \to \mathbb{R}$ be any real function on W. If U denotes the random potential conditioned by $U(w) = u(w)$, $w \in W$, then the variables $U(v)$, $v \in V(\mathcal{G})$, have a joint normal distribution. Moreover, if $\mathcal{G}$ is regarded as an electrical network with unit resistances,*

$$U(v) \sim \mathrm{N}\left(u(v), \sigma^2(v)\right),$$

where $u(v)$ equals the potential at v in $\mathcal{G}$ obtained when the potential is kept at $u(w)$ for $w \in W$, and $\sigma^2(v)$ equals the effective resistance between v and W.

Finally, the covariance $\operatorname{Cov}(U(v_1), U(v_2))$ *between the conditioned random potentials at two vertices $v_1, v_2 \in V(\mathcal{G})$ equals the potential at v_2 of a unit electric current flowing from v_1 to W (keeping W earthed).*

PROOF. Let us first consider the case when $u = 0$ identically on W; we thus require that the potential vanishes on W. If W is a single vertex, this is the situation studied above. In the general case, it is easily seen that we obtain the same conditioned random potential U as if we merge the vertices in W into a single vertex v^* and then study the random potential on the resulting reduced network, with v^* as reference vertex. Theorem 9.15 now gives $U(v) \sim \mathrm{N}(0, \sigma^2(v))$, where $\sigma^2(v)$ is the resistance between v and v^* in the reduced network, which equals the resistance between v and W in $\mathcal{G}$.

Similarly, when $u = 0$, the assertion about covariances follows by applying Theorem 9.17 to the reduced network.

We next observe that by Theorem 9.9, the variance of $U(v)$ does not depend on u; the same holds for the covariance $\mathrm{Cov}(U(v_1), U(v_2))$, for example by applying Theorem 9.9 to $U(v_1) + U(v_2)$. Consequently, the claims about variance and covariance hold for any u, and it remains only to verify that $\mathrm{E}\, U(v) = u(v)$ for any $v \in V(\mathcal{G})$.

In order to verify this claim, we first observe that it follows from Theorem 9.8 that $\mathrm{E}\, U(v)$ is a linear function of $\{u(w) : w \in W\}$. Since the same is true for the potential u, it suffices to verify the claim when $u(w_1) = 1$ for some vertex $w_1 \in W$ and $u(w) = 0$ for $w \in W' = W \setminus \{w_1\}$. Moreover, since the case when W consists of only a single vertex follows by Theorem 9.15, we may assume that W' is non-empty. It is now, however, easily seen that we may merge W' into a single vertex without affecting $U(v)$ or $u(v)$; hence it suffices to verify $\mathrm{E}\, U(v) = u(v)$ when W consists of two vertices w_0 and w_1, and $u(w_0) = 0$, $u(w_1) = 1$.

Let us therefore study this case, and let U' denote the random Gaussian potential on W, normalized by $U'(w_0) = 0$ but without further conditioning. Thus U equals U' conditioned on $\{U'(w_1) = 1\}$. By Theorem 9.17, $U'(v)$ and $U'(w_1)$ are two centred jointly Gaussian random variables, and $\mathrm{Cov}(U'(v), U'(w_1))$ equals the potential at v for a unit current flowing from w_1 to w_0, with w_0 earthed. Since the potential at w_1 of this current equals $\mathrm{Cov}(U'(w_1), U'(w_1)) = \mathrm{Var}(U'(w_1))$, by taking $v = w_1$ in the last result or by Theorem 9.15, this potential equals $\mathrm{Var}(U'(w_1))u$. Thus,

$$\mathrm{Cov}(U'(v), U'(w_1)) = \mathrm{Var}(U'(w_1))u(v),$$

and (9.2), (9.4) yield

$$\mathrm{E}\, U(v) = \mathrm{E}\big(U'(v) \mid U'(w_1) = 1\big) = \frac{\mathrm{Cov}(U'(v), U'(w_1))}{\mathrm{Var}(U'(w_1))} \cdot 1 = u(v).$$

□

REMARK 9.21. It is well-known, and easy to see, that the function u in Theorem 9.20 is harmonic on $V(\mathcal{G}) \setminus W$, in the sense that its value at a vertex $v \notin W$ equals the average of its values at the neighbours of v (or equivalently, using the notations introduced above, $\mathrm{Div}\,\mathrm{Grad}\, u = 0$ on $V(\mathcal{G}) \setminus W$); moreover, u is the unique function with the given values on W that is harmonic on $V(\mathcal{G}) \setminus W$.

REMARK 9.22. The results above are easily generalized by allowing the variables ξ_e to have different variances, $\xi_e \sim \mathrm{N}(0, \sigma_e^2)$ (still assuming them to be independent). The only difference in the results is that $\mathcal{G}$ should be regarded as a network with resistance $R_e = \sigma_e^2$ on edge e. In the proofs, the main difference is that F should now be normed by $\|\varphi\|_F^2 = \sum_e R_e \varphi(e)^2$ and that Grad has to be replaced by the new adjoint of Div, which is $\psi \mapsto R^{-1}\operatorname{Grad}(\psi)$, defined by

$$R^{-1}\operatorname{Grad}(\psi)(e) = R_e^{-1} \cdot \operatorname{Grad}(\psi)(e) = R_e^{-1}\big(\psi(\alpha(e)) - \psi(\omega(e))\big).$$

The alternative definition in Remark 9.14 remains valid, with the energy function (9.18) modified to

$$\tfrac{1}{2}\sum_e R_e^{-1}\big(U(\alpha(e)) - U(\omega(e))\big)^2 = \tfrac{1}{2}\|R^{-1}\operatorname{Grad}(U)\|_F^2.$$

X

Pairs of Gaussian subspaces

In this chapter we consider two subspaces H and K of a Gaussian Hilbert space G. Our main objective is to calculate, or at least estimate, some measures of dependence between the σ-fields $\mathcal{F}(H)$ and $\mathcal{F}(K)$ generated by the subspaces.

In the sequel we will assume that H and K are closed subspaces of G, and thus themselves Hilbert spaces; this is no real loss of generality since $\mathcal{F}(H)$ and $\mathcal{F}(K)$ remain the same if H and K are replaced by their closed hulls.

1. Projections and singular numbers

Let P_H and P_K denote the orthogonal projections of G onto H and K, and let P_{HK} be the restriction of P_K to H.

It is easily seen that all probabilistic properties of the pair of Gaussian spaces (H, K) are embodied in the operator P_{HK}. In fact, if (H', K') is another such pair and $P_{H',K'}$ is unitarily equivalent to P_{HK}, then there exist isometries $U_H\colon H \to H'$ and $U_K\colon K \to K'$ such that $P_{H'K'} = U_K P_{HK} U_H^{-1}$, and it is easily seen that U_H and U_K combine to an isometry of $H + K$ onto $H' + K'$; this isometry preserves the joint distribution of the variables in H and K.

REMARK 10.1. If we interchange H and K, we instead obtain the operator $P_{KH}\colon K \to H$, which equals the adjoint P_{HK}^*.

A useful way to extract and express properties of P_{HK} is to use the singular numbers defined in Appendix H; we thus define

$$\rho_k = \rho_k(H, K) = s_k(P_{HK}), \qquad k = 1, 2, \dots.$$

Appendix H and Remark 10.1 yield immediately the following properties.

(i) $\rho_1 \geq \rho_2 \geq \dots \geq 0$.
(ii) $\rho_1 = \|P_{HK}\|$, the operator norm.
(iii) $\sum_1^\infty \rho_k^2 = \|P_{HK}\|_{HS}^2$, the square of the Hilbert–Schmidt norm.
(iv) $\lim_{k\to\infty} \rho_k = 0$ if and only if P_{HK} is a compact operator.
(v) $\rho_k(H, K) = \rho_k(K, H)$.

REMARK 10.2. $\{\rho_k\}_1^\infty$ equals the sequence of singular numbers of the operator $P_K P_H$ or its adjoint $P_H P_K$ in G. If P_{HK} or $P_H P_K$ is compact, then $\{\rho_k\}_1^\infty$ also equals the sequence of square roots of the eigenvalues of $(P_K P_H)^* P_K P_H = P_H P_K P_H$.

If P_{HK} is compact, we can diagonalize it by Theorem H.2, which now takes the following form.

THEOREM 10.3. *If P_{HK} is compact, and has N non-zero singular numbers $\{\rho_k\}_1^N$, with $0 \le N \le \infty$, then there exist orthonormal bases $\{\xi_k\}_{k=1}^N \cup \{\xi_i'\}_{i\in\mathcal{I}}$ in H and $\{\eta_k\}_{k=1}^N \cup \{\eta_j'\}_{j\in\mathcal{J}}$ in K such that $\langle \xi_k, \eta_k\rangle = \mathrm{E}\,\xi_k\eta_k = \rho_k$, $1 \le k \le N$, while all other scalar products of two different elements of these bases vanish. In particular, the three families $\{\xi_k, \eta_k\}_{k=1}^N$, $\{\xi_i'\}_{i\in\mathcal{I}}$ and $\{\eta_j'\}_{j\in\mathcal{J}}$ of standard normal variables are mutually independent, and the two-dimensional normal variables $\{(\xi_k, \eta_k)\}_{k=1}^N$ are independent.*

PROOF. If $\xi \in H$ and $\eta \in K$, then $\langle \xi, \eta\rangle = \langle P_{HK}\xi, \eta\rangle$. The result now follows from Theorem H.2 and the fact that orthogonal families of jointly Gaussian variables are independent. □

EXAMPLE 10.4. Given any sequence $\{\rho_k\}_1^\infty$ such that $1 \ge \rho_1 \ge \rho_2 \ge \dots \ge 0$, we can construct two Gaussian subspaces H and K of some Gaussian space G such that $\rho_k(H, K) = \rho_k$ for every $k \ge 1$. One possibility is to take a sequence of independent two-dimensional Gaussian random variables (ξ_k, η_k), $k = 1, 2, \dots$, such that ξ_k and η_k are standard normal with covariance ρ_k, and let H and K be the closed linear hulls of $\{\xi_k\}_1^\infty$ and $\{\eta_k\}_1^\infty$, respectively.

Theorem 10.3 shows that if $\rho_k \to 0$, this example yields the essentially only possibility (up to addition of subspaces independent of the remaining variables). Consequently, in the compact case, the probabilistic structure of H and K is essentially determined by $\{\rho_k\}_1^\infty$.

2. Measures of dependence

A *measure of dependence* assigns a number to each pair of sub-σ-fields $\mathcal{F}_1$, $\mathcal{F}_2$ of the σ-field $\mathcal{F}$ in a probability space $(\Omega, \mathcal{F}, \mathrm{P})$, such that this number somehow reflects the degree of dependence between the σ-fields.

Each measure of dependence δ defines a corresponding mixing condition. Consider, for simplicity, a stationary sequence $\{X_k\}_{-\infty}^\infty$ of random variables. The sequence is said to be *δ-mixing* if $\delta\big(\mathcal{F}(\{X_k\}_{-\infty}^0), \mathcal{F}(\{X_k\}_n^\infty)\big) \to 0$ as $n \to \infty$.

Several different measures of dependence have been defined and studied. We will here consider three important ones.

DEFINITION 10.5. The *strong mixing coefficient* α is defined by

$$\alpha(\mathcal{F}, \mathcal{G}) = \sup_{A\in\mathcal{F}, B\in\mathcal{G}} |\mathrm{P}(A \cap B) - \mathrm{P}(A)\,\mathrm{P}(B)|.$$

The corresponding mixing condition is known as *strong mixing*.

DEFINITION 10.6. The *maximal correlation coefficient* ρ is defined by

$$\rho(\mathcal{F}, \mathcal{G}) = \sup_{X\in L^2(\mathcal{F}), Y\in L^2(\mathcal{G})} \frac{|\operatorname{Cov}(X,Y)|}{\|X\|_2\|Y\|_2} = \sup_{X\in L^2(\mathcal{F}), Y\in L^2(\mathcal{G})} |\operatorname{Corr}(X,Y)|,$$

where $\operatorname{Corr}(X,Y) = \operatorname{Cov}(X,Y)/(\operatorname{Var}(X)\operatorname{Var}(Y))^{1/2}$ is the correlation between X and Y. The corresponding mixing condition is known as *ρ-mixing* or *complete regularity*.

DEFINITION 10.7. The *absolute regularity coefficient* β is defined by

$$\beta(\mathcal{F}_1,\mathcal{F}_2) = \tfrac{1}{2}\sup\sum_{i,j} |\operatorname{P}(A_i \cap B_j) - \operatorname{P}(A_i)\operatorname{P}(B_j)|,$$

with the supremum taken over all pairs of finite measurable partitions $\{A_i\} \subseteq \mathcal{F}_1$ and $\{B_j\} \subseteq \mathcal{F}_2$. The corresponding mixing condition is known as *absolute regularity*.

If δ is one of these measures of dependence and $\mathcal{U}$ and $\mathcal{V}$ are two sets of random variables (on the same probability space), we define $\delta(\mathcal{U},\mathcal{V}) = \delta(\mathcal{F}(\mathcal{U}),\mathcal{F}(\mathcal{V}))$. It is easily seen that this quantity depends only on the joint distribution of the variables in $\mathcal{U} \cup \mathcal{V}$, and not on the underlying probability space.

We observe that $0 \le \alpha \le 1$ and $0 \le \beta \le 1$, while $0 \le \rho \le \infty$; all three coefficients vanish if and only if the two σ-fields are independent.

Since $\operatorname{P}(A \cap B) - \operatorname{P}(A)\operatorname{P}(B) = \operatorname{Cov}(\mathbf{1}_A, \mathbf{1}_B)$, it follows that

$$\alpha(\mathcal{F},\mathcal{G}) \le \rho(\mathcal{F},\mathcal{G}). \tag{10.1}$$

Similarly, using the partitions $\{A, \Omega \setminus A\}$ and $\{B, \Omega \setminus B\}$ in the definition of β, we have

$$\alpha(\mathcal{F},\mathcal{G}) \le \tfrac{1}{2}\beta(\mathcal{F},\mathcal{G}).$$

There is no similar general inequality relating ρ and β.

The absolute regularity coefficient has a useful alternative description. We define the *total variation distance* between two probability measures P_1 and P_2 on the same measurable space $(\Omega,\mathcal{F})$ by

$$d_{\mathrm{TV}}(\operatorname{P}_1,\operatorname{P}_2) = \sup_{A\in\mathcal{F}} |\operatorname{P}_1(A) - \operatorname{P}_2(A)|.$$

It is easily seen that $d_{\mathrm{TV}}(\operatorname{P}_1,\operatorname{P}_2) = \frac{1}{2}\|\operatorname{P}_1 - \operatorname{P}_2\|$, where $\|\cdot\|$ is the total variation (i.e., the standard norm on signed measures), and that if μ is any σ-finite measure on $(\Omega,\mathcal{F})$ such that P_1 and P_2 are absolutely continuous with respect to μ, then

$$d_{\mathrm{TV}}(\operatorname{P}_1,\operatorname{P}_2) = \tfrac{1}{2}\int_\Omega \left|\frac{d\operatorname{P}_1}{d\mu} - \frac{d\operatorname{P}_2}{d\mu}\right| d\mu. \tag{10.2}$$

PROPOSITION 10.8. *Let the random variables $\{X_i\}_\mathcal{I}$ and $\{Y_j\}_\mathcal{J}$ have the joint distribution* P *(on $\mathbb{R}^\mathcal{I} \times \mathbb{R}^\mathcal{J}$); let further the distribution of $\{X_i\}_\mathcal{I}$ be* P_1 *(on $\mathbb{R}^\mathcal{I}$) and let the distribution of $\{Y_j\}_\mathcal{J}$ be* P_2 *(on $\mathbb{R}^\mathcal{J}$). Then*

$$\beta(\{X_i\}_\mathcal{I},\{Y_j\}_\mathcal{J}) = d_{\mathrm{TV}}(\operatorname{P},\operatorname{P}_1\times\operatorname{P}_2).$$

PROOF. Let $\mu = \mathrm{P} + \mathrm{P}_1 \times \mathrm{P}_2$ and let F be the set of all measurable functions $f\colon \mathbb{R}^{\mathcal{I}} \times \mathbb{R}^{\mathcal{J}} \to \mathbb{R}$ with $|f| \le 1$. By (10.2), $d_{\mathrm{TV}}(\mathrm{P}, \mathrm{P}_1 \times \mathrm{P}_2)$ equals

$$\tfrac{1}{2} \sup_{f \in F} \int_\Omega \Bigl(\frac{d\mathrm{P}}{d\mu} - \frac{d(\mathrm{P}_1 \times \mathrm{P}_2)}{d\mu}\Bigr) f \, d\mu,$$

while it is easily seen that we obtain $\beta(\{X_i\}_{\mathcal{I}}, \{Y_j\}_{\mathcal{J}})$ if we restrict the supremum to the subset $F \cap S$ of F, where S is the set of finite sums $\sum a_i \mathbf{1}_{A_i \times B_i}$ with $A_i \in \mathcal{B}(\mathbb{R}^{\mathcal{I}})$ and $B_i \in \mathcal{B}(\mathbb{R}^{\mathcal{J}})$. The set S is dense in $L^1(\mu)$ by a standard application of the monotone class theorem. (Use for example Corollary A.2, with $\mathcal{A}$ equal to the closure in $L^1(\mu)$ of S, and $\mathcal{C} = \{\mathbf{1}_{A \times B}\}$.) It follows easily that $F \cap S$ is dense in F, regarded as a subset of $L^1(\mu)$, and thus the two suprema coincide. □

COROLLARY 10.9. *If the family $\{X'_k\}_{\mathcal{K}}$ is independent of $\{X_i\}_{\mathcal{I}} \cup \{Y_j\}_{\mathcal{J}}$, then $\beta(\{X_i\}_{\mathcal{I}} \cup \{X'_k\}_{\mathcal{K}}, \{Y_j\}_{\mathcal{J}}) = \beta(\{X_i\}_{\mathcal{I}}, \{Y_j\}_{\mathcal{J}})$.* □

REMARK 10.10. A similar approximation argument shows that if $\{X_i\}_{i \in \mathcal{I}}$ and $\{Y_j\}_{j \in \mathcal{J}}$ are sets of random variables, then $\beta(\{X_i\}_{\mathcal{I}}, \{Y_j\}_{\mathcal{J}})$ equals the supremum of $\beta(\{X_i\}_{\mathcal{I}'}, \{Y_j\}_{\mathcal{J}'})$ over finite subsets $\mathcal{I}' \subset \mathcal{I}$ and $\mathcal{J}' \subset \mathcal{J}$.

Let us now consider the Gaussian case. We assume as above that H and K are closed subspaces of some Gaussian Hilbert space.

THEOREM 10.11. *The maximal correlation coefficient is given by*

$$\rho(H, K) = \rho_1(H, K) = \|P_{HK}\|.$$

PROOF. The maximal correlation coefficient equals the norm of the operator $X \mapsto \mathrm{E}(X \mid \mathcal{F}(K)) - \mathrm{E}\, X$ of $L^2(\mathcal{F}(H))$ into $L^2(\mathcal{F}(K))$. By the Wiener chaos decomposition and Theorem 4.9, this operator decomposes as the orthogonal sum of the operators $P_{HK}^{:n:}\colon H^{:n:} \to K^{:n:}$, $n \ge 1$, and $P_{HK}^{:0:} - I = 0\colon H^{:0:} \to K^{:0:}$ (recall that $\mathrm{E}\, X$ vanishes on $H^{:n:}$, $n \ge 1$). Hence the norm equals $\max_{n \ge 1} \|P_{HK}^{:n:}\| = \max_{n \ge 1} \|P_{HK}\|^n = \|P_{HK}\| = \rho_1$. □

LEMMA 10.12. *Let ξ and η be two non-degenerate centered jointly normal variables and let $\rho = \mathrm{E}\, \xi\eta / \|\xi\|_2 \|\eta\|_2$ be their correlation. Then*

$$\mathrm{P}(\xi > 0, \eta > 0) = \tfrac{1}{2\pi} \arcsin \rho + \tfrac{1}{4},$$

and thus

$$\mathrm{Cov}(\mathbf{1}[\xi > 0], \mathbf{1}[\eta > 0]) = \tfrac{1}{2\pi} \arcsin \rho.$$

PROOF. We may assume that ξ and η have norm 1, and thus are standard normal. The results are trivially true for $\rho = \pm 1$; we may thus suppose that $|\rho| < 1$. Then ξ and $\zeta = (1 - \rho^2)^{-1/2}(\eta - \rho\xi)$ are independent standard normal, so the complex Gaussian variable $\xi + i\zeta$ is symmetric and

$$\begin{aligned} \mathrm{P}(\xi > 0, \eta > 0) &= \mathrm{P}(\xi > 0, \zeta > -\rho(1 - \rho^2)^{-1/2}\xi) \\ &= \mathrm{P}\bigl(-\arcsin \rho < \arg(\xi + i\zeta) < \tfrac{\pi}{2}\bigr) = \tfrac{1}{2\pi}\bigl(\tfrac{\pi}{2} + \arcsin \rho\bigr). \end{aligned}$$

□

THEOREM 10.13. *The strong mixing coefficient* $\alpha(H,K) \asymp \rho_1(H,K) = \|P_{HK}\|$. *More precisely,*

$$\tfrac{1}{2\pi}\rho_1(H,K) \le \alpha(H,K) \le \rho_1(H,K).$$

PROOF. The upper bound follows from Theorem 10.11 and the inequality (10.1).

For the lower bound, let $\xi \in H$ and $\eta \in K$ be two standard normal variables and let $\rho = \mathrm{E}\,\xi\eta$ be their covariance. Take $A = \{\xi > 0\}$ and $B = \{\eta > 0\}$. Then, by Lemma 10.12,

$$\mathrm{P}(A \cap B) - \mathrm{P}(A)\,\mathrm{P}(B) = \mathrm{Cov}(\mathbf{1}[\xi > 0], \mathbf{1}[\eta > 0]) = \tfrac{1}{2\pi}\arcsin\rho.$$

Since $A \in \mathcal{F}(H)$ and $B \in \mathcal{F}(K)$, this implies (when $\rho \ge 0$)

$$\alpha(H,K) \ge \tfrac{1}{2\pi}\arcsin\rho \ge \tfrac{1}{2\pi}\rho = \tfrac{1}{2\pi}\,\mathrm{E}\,\xi\eta,$$

and the result follows by taking the supremum over all such ξ and η. ☐

For the absolute regularity coefficient, we know only more complicated estimates. We follow essentially Ibragimov and Rozanov (1970). (It follows from the proof and Example 3.52 that the constants $\frac{1}{\pi}$ and $\frac{1}{\sqrt{2\pi}}$ in (10.3) are the best possible.)

THEOREM 10.14. *The absolute regularity coefficient is estimated by*

$$\frac{1}{\pi}\|P_{HK}\|_{HS} - \frac{\|P_{HK}\|_{HS}^2 + \frac{1}{4}\|P_{HK}\|_{HS}^4}{(1-\|P_{HK}\|)^2} \le \beta(H,K)$$
$$\le \frac{1}{\sqrt{2\pi}}\|P_{HK}\|_{HS} + \frac{\|P_{HK}\|_{HS}^2 + \frac{1}{16}\|P_{HK}\|_{HS}^4}{(1-\|P_{HK}\|)^2} \tag{10.3}$$

and thus

$$\tfrac{1}{\pi}\|P_{HK}\|_{HS} + O(\|P_{HK}\|_{HS}^2) \le \beta(H,K) \le \tfrac{1}{\sqrt{2\pi}}\|P_{HK}\|_{HS} + O(\|P_{HK}\|_{HS}^2). \tag{10.4}$$

Furthermore,

$$\|P_{HK}\|_{HS}^2 \le 8|\ln(1-\beta(H,K))|. \tag{10.5}$$

PROOF. We first observe that if P_{HK} is not Hilbert–Schmidt, then (10.3) and (10.4) are trivial; moreover we can find finite-dimensional subspaces H' of H with arbitrarily large $\|P_{H'K}\|_{HS}$. Since $\beta(H,K) \ge \beta(H',K)$, the inequality (10.5) in this case, which means $\beta(H,K) = 1$, follows from (10.5) applied to H' and K.

Consequently, it suffices to consider the case when P_{HK} is Hilbert–Schmidt. Since then P_{HK} is compact, we can apply Theorem 10.3. Using the orthonormal bases given by that theorem, and applying Corollary 10.9 twice, we have

$$\beta(H,K) = \beta(\{\xi_k\}_1^N \cup \{\xi_i'\}_{\mathcal{I}}, \{\eta_k\}_1^N \cup \{\eta_j'\}_{\mathcal{J}}) = \beta(\{\xi_k\}_1^N, \{\eta_k\}_1^N).$$

In other words, we may assume that H and K have orthonormal bases $\{\xi_k\}_1^N$ and $\{\eta_k\}_1^N$ such that $\langle \xi_k, \eta_l \rangle = \delta_{kl}\rho_k$.

Moreover, if $N = \infty$, we use Remark 10.10, which implies that

$$\beta(\{\xi_k\}_1^\infty, \{\eta_k\}_1^\infty) = \lim_{n\to\infty} \beta(\{\xi_k\}_1^n, \{\eta_k\}_1^n);$$

thus it suffices to prove the theorem for the case $N < \infty$, when H and K have finite bases $\{\xi_k\}_1^N$ and $\{\eta_k\}_1^N$. We may further assume $1 \le N < \infty$, since the case $N = 0$ is trivial.

If $\|P_{HK}\| = \rho_1 = \langle \xi_1, \eta_1 \rangle = 1$, then (10.3) and (10.4) are trivial; moreover $\xi_1 = \eta_1$ and it is easily seen, using $\beta(H, K) \ge \beta(\xi_1, \eta_1)$ and Proposition 10.8, that $\beta(H, K) = 1$; hence also (10.5) holds.

We may thus assume $\rho_1 < 1$, and hence $0 < \rho_k < 1$ for each k. The two-dimensional variable (ξ_k, η_k) then has the density function

$$p(x, y; \rho_k) = (2\pi)^{-1}(1-\rho_k^2)^{-1/2} \exp\Bigl(-\frac{x^2 + y^2 - 2\rho_k xy}{2(1-\rho_k^2)}\Bigr).$$

Recall that these two-dimensional variables are independent; thus the $2N$ variables $\xi_1, \dots, \xi_N, \eta_1, \dots, \eta_N$ have a joint distribution in $\mathbb{R}^{2N}$ with the density function $p(\boldsymbol{x}, \boldsymbol{y}) = \prod_k p(x_k, y_k; \rho_k)$, where we use the notations $\boldsymbol{x} = (x_k)_1^N$ and $\boldsymbol{y} = (y_k)_1^N$.

We denote this Gaussian probability distribution on $\mathbb{R}^{2N}$ by P; the two marginal distributions P_1 and P_2 on $\mathbb{R}^N$ are both the distribution of N independent standard normal variables, so their product, which we denote by P_0, is the standard Gaussian probability measure on $\mathbb{R}^{2N}$ with density function $p_0(\boldsymbol{x}, \boldsymbol{y}) = \prod_k p(x_k, y_k; 0)$.

By Proposition 10.8,

$$\beta(H, K) = \beta(\{\xi_k\}_1^N, \{\eta_k\}_1^N) = d_{\mathrm{TV}}(\mathrm{P}, \mathrm{P}_0) = \tfrac{1}{2} \int_{\mathbb{R}^{2N}} \bigl|p(\boldsymbol{x}, \boldsymbol{y}) - p_0(\boldsymbol{x}, \boldsymbol{y})\bigr| \, d\boldsymbol{x}\, d\boldsymbol{y}. \tag{10.6}$$

We introduce the random variable

$$Z = \ln \frac{p(\{\xi_k\}_1^N, \{\eta_k\}_1^N)}{p_0(\{\xi_k\}_1^N, \{\eta_k\}_1^N)} = \sum_{k=1}^N \Bigl(-\frac{\rho_k^2\xi_k^2 + \rho_k^2\eta_k^2 - 2\rho_k\xi_k\eta_k}{2(1-\rho_k^2)} - \tfrac{1}{2}\ln(1-\rho_k^2)\Bigr) \tag{10.7}$$

and begin by writing (10.6) as

$$\beta(H, K) = \tfrac{1}{2} \int_{\mathbb{R}^{2N}} \Bigl|\frac{p(\boldsymbol{x}, \boldsymbol{y})}{p_0(\boldsymbol{x}, \boldsymbol{y})} - 1\Bigr| p_0(\boldsymbol{x}, \boldsymbol{y}) \, d\boldsymbol{x}\, d\boldsymbol{y} = \tfrac{1}{2} \mathrm{E}_0 \, |e^Z - 1|,$$

where E_0 denotes expectation with respect to the probability measure P_0, for which all variables ξ_k and η_l are independent standard normal.

It is easily seen, considering the cases $Z > 0$ and $Z < 0$ separately, that

$$|Z| - \tfrac{1}{2}Z^2 \le |e^Z - 1| \le |Z| + \tfrac{1}{2}Z^2 e^Z;$$

thus

$$\tfrac{1}{2}\operatorname{E}_0|Z| - \tfrac{1}{4}\operatorname{E}_0 Z^2 \le \beta(H,K) \le \tfrac{1}{2}\operatorname{E}_0|Z| + \tfrac{1}{4}\operatorname{E}_0 Z^2 e^Z = \tfrac{1}{2}\operatorname{E}_0|Z| + \tfrac{1}{4}\operatorname{E} Z^2.$$

We next estimate the terms $\operatorname{E}_0 Z^2 = \operatorname{Var}_0 Z + (\operatorname{E}_0 Z)^2$ and $\operatorname{E} Z^2 = \operatorname{Var} Z + (\operatorname{E} Z)^2$, where Var_0 denotes variance for the distribution P_0.

The definition (10.7) yields, since $\|P_{HK}\|_{HS}^2 = \sum_1^N \rho_k^2$,

$$\begin{aligned}|\operatorname{E}_0(Z)| &= \Big|\sum_k -\frac{\rho_k^2+\rho_k^2}{2(1-\rho_k^2)} - \tfrac{1}{2}\ln(1-\rho_k^2)\Big| \\ &\le \sum_k \frac{\rho_k^2}{1-\rho_k^2} \le \frac{\|P_{HK}\|_{HS}^2}{1-\rho_1^2} \le \frac{\|P_{HK}\|_{HS}^2}{1-\rho_1}.\end{aligned}$$

Similarly,

$$\begin{aligned}|\operatorname{E}(Z)| &= \Big|\sum_k\Big(-\frac{\rho_k^2+\rho_k^2-2\rho_k^2}{2(1-\rho_k^2)} - \tfrac{1}{2}\ln(1-\rho_k^2)\Big)\Big| \\ &= \sum_k \tfrac{1}{2}|\ln(1-\rho_k^2)| \le \sum_k \tfrac{1}{2}\frac{\rho_k^2}{1-\rho_k^2} \le \tfrac{1}{2}\frac{\|P_{HK}\|_{HS}^2}{1-\rho_1^2} \le \tfrac{1}{2}\frac{\|P_{HK}\|_{HS}^2}{1-\rho_1}.\end{aligned}$$

For the variance calculations we observe that if ξ',ξ'',η',η'' are any four centered jointly Gaussian variables, then by Theorem 3.9

$$\operatorname{Cov}(\xi'\xi'',\eta'\eta'') = \operatorname{Cov}(:\xi'\xi'':, :\eta'\eta'':) = \operatorname{E}(\xi'\eta')\operatorname{E}(\xi''\eta'') + \operatorname{E}(\xi'\eta'')\operatorname{E}(\xi'\eta'').$$

It follows that

$$\operatorname{Var}_0(Z) = \sum_{k=1}^N \frac{2\rho_k^4+2\rho_k^4+4\rho_k^2}{4(1-\rho_k^2)^2} = \sum_{k=1}^N \frac{\rho_k^2(1+\rho_k^2)}{(1-\rho_k)^2(1+\rho_k)^2} \le \frac{\|P_{HK}\|_{HS}^2}{(1-\rho_1)^2}$$

and

$$\begin{aligned}\operatorname{Var}(Z) &= \sum_{k=1}^N \frac{2\rho_k^4+2\rho_k^4+4\rho_k^2(1+\rho_k^2)+2\rho_k^4\cdot 2\rho_k^2 - 2\cdot 4\rho_k^3\cdot 2\rho_k}{4(1-\rho_k^2)^2} \\ &= \sum_{k=1}^N \frac{\rho_k^2(1-\rho_k^2)^2}{(1-\rho_k^2)^2} = \|P_{HK}\|_{HS}^2.\end{aligned}$$

Turning to the term $\operatorname{E}_0|Z|$, we first approximate Z by $Z_1 = \sum_1^N \rho_k\xi_k\eta_k$ (obtained by ignoring higher powers of ρ_k). Then

$$\begin{aligned}
\mathrm{E}_0 |Z - Z_1| &= \mathrm{E}_0 \Big| \sum_{k=1}^{N} \Big(-\frac{\rho_k^2 \xi_k^2 + \rho_k^2 \eta_k^2 - 2\rho_k^3 \xi_k \eta_k}{2(1-\rho_k^2)} - \tfrac{1}{2} \ln(1-\rho_k^2) \Big) \Big| \\
&\le \sum_{k=1}^{N} \Big(\frac{\rho_k^2 + \rho_k^2 + 2\rho_k^3}{2(1-\rho_k^2)} + \tfrac{1}{2} |\ln(1-\rho_k^2)| \Big) = \sum_{k=1}^{N} \Big(\frac{\rho_k^2}{1-\rho_k} + \tfrac{1}{2} |\ln(1-\rho_k^2)| \Big) \\
&\le \tfrac{3}{2} \frac{\|P_{HK}\|_{HS}^2}{1-\rho_1}.
\end{aligned}$$

Moreover, for the measure P_0, the random variable Z_1 is of the type studied in Example 3.52, with H_1 spanned by $\{\xi_k\}$ and H_2 spanned by $\{\eta_l\}$. Thus, by (3.28), recalling that $\kappa(1) = \sqrt{2/\pi}$ and observing $\mathrm{E}_0 |Z_1|^2 = \sum_k \rho_k^2 = \|P_{HK}\|_{HS}^2$,

$$\tfrac{2}{\pi} \|P_{HK}\|_{HS} \le \mathrm{E}_0 |Z_1| \le \sqrt{\tfrac{2}{\pi}} \|P_{HK}\|_{HS}.$$

Using the triangle inequality

$$\mathrm{E}_0 |Z_1| - \mathrm{E}_0 |Z - Z_1| \le \mathrm{E}_0 |Z| \le \mathrm{E}_0 |Z_1| + \mathrm{E}_0 |Z - Z_1|$$

and collecting the estimates above, we obtain (10.3).

In particular, this yields (10.4) for $\|P_{HK}\|_{HS} \le \frac{1}{2}$; moreover, (10.4) is trivial for larger $\|P_{HK}\|_{HS}$.

Since $|s - t| = |\sqrt{s} - \sqrt{t}|(\sqrt{s} + \sqrt{t}) \ge (\sqrt{s} - \sqrt{t})^2$ when $s, t \ge 0$, (10.6) implies

$$\begin{aligned}
\beta(H,K) &\ge \tfrac{1}{2} \int_{\mathbb{R}^{2N}} \big(\sqrt{p(\boldsymbol{x},\boldsymbol{y})} - \sqrt{p_0(\boldsymbol{x},\boldsymbol{y})} \big)^2 \, d\boldsymbol{x}\, d\boldsymbol{y} \\
&= 1 - \int_{\mathbb{R}^{2N}} \sqrt{p(\boldsymbol{x},\boldsymbol{y}) p_0(\boldsymbol{x},\boldsymbol{y})} \, d\boldsymbol{x}\, d\boldsymbol{y} \\
&= 1 - \prod_{k=1}^{N} \int_{\mathbb{R}^2} \big(p(x_k, y_k; \rho_k) p(x_k, y_k; 0) \big)^{1/2} \, dx_k \, dy_k \\
&= 1 - \prod_k (1-\rho_k^2)^{-1/4+1/2} (1 - \rho_k^2/4)^{-1/2},
\end{aligned}$$

by a standard evaluation of the two-dimensional Gaussian integrals in the last but one line. A Taylor expansion yields

$$\ln\big((1-\rho_k^2)^{1/4} / (1-\rho_k^2/4)^{1/2} \big) = \tfrac{1}{4} \ln(1-\rho_k^2) - \tfrac{1}{2} \ln(1-\rho_k^2/4) \le -\tfrac{1}{8} \rho_k^2,$$

and thus $\ln\big(1 - \beta(H,K)\big) \le -\frac{1}{8} \sum_k \rho_k^2$, which is (10.5). □

COROLLARY 10.15. *If $(H_n)_1^\infty$ and $(K_n)_1^\infty$ are sequences of subspaces of a Gaussian Hilbert space, then $\beta(H_n, K_n) \to 0 \iff \|P_{H_n K_n}\|_{HS} \to 0$.* □

For applications of the results above to mixing properties of Gaussian stochastic processes, see e.g. Ibragimov and Rozanov (1970).

XI

Limit theorems for generalized U-statistics

In this chapter we use the theory of Gaussian Hilbert spaces to obtain the asymptotic distributions of some important random variables. In the first section, we study U-statistics. In the second section we extend the results to asymmetric statistics. In the third section we extend the results further; as a special case we obtain results for random graphs.

Note that the original variables are defined without any reference to normal variables or Gaussian Hilbert spaces; the Gaussian Hilbert space is introduced as a convenient tool to treat the asymptotic distribution.

A common theme in these results is that 'typically' the asymptotic distribution is normal, but in some 'degenerate' cases other limits occur; these other limits can be represented as variables in a Wiener chaos $H^{:n:}$ for some Gaussian Hilbert space H, and can for example be expressed as multiple Gaussian stochastic integrals. The explanation for this phenomenon that emerges from the proofs below is that the variable in question may be expanded as a sum, where each term converges in distribution to some chaos, and the first term is asymptotically normal. Typically, the first term dominates the sum and all others are asymptotically negligible, but in degenerate cases the first term vanishes and the sum is dominated by one or several later terms, which may converge to a higher order chaos.

1. U-statistics

Let $X_1, X_2, \ldots$ be independent, identically distributed (i.i.d.) random variables with values in an arbitrary measurable space $\mathcal{S}$. We denote the common distribution of X_i by μ; thus μ is a probability measure on $\mathcal{S}$.

For any $k \geq 0$ and any (real or complex) measurable symmetric function f on $\mathcal{S}^k$ we define the sequence of random variables

$$S_n^k(f) = \sum_{1 \leq i_1 < \cdots < i_k \leq n} f(X_{i_1}, \ldots, X_{i_k}), \qquad n \geq 1; \tag{11.1}$$

if $k = 0$, f is a constant and (11.1) is interpreted as $S_n^0(f) = f$. Note that $S_n^k(f) = 0$ if $n < k$. We consider only the square integrable case so we assume that $f \in L^2(\mathcal{S}^k, \mu^k)$, which implies $S_n^k(f) \in L^2$.

We will later prefer to use a different normalization and define also

$$U_n^k(f) = \frac{1}{\binom{n}{k}} S_n^k(f). \tag{11.2}$$

Variables of the form (11.1) or (11.2) are known as *U-statistics.* In this section we find the asymptotic distributions of U-statistics, essentially following Dynkin and Mandelbaum (1983); we use the formalism of tensor products and Fock space given in Chapter 4 and Appendix E.

We use Proposition E.16 and identify the space of symmetric functions in $L^2(S^k, \mu^k)$ with the symmetric tensor product $L^2(\mathcal{S}, \mu)^{\odot k}$; observe that we define the norm by $\left(\int |f|^2 d\mu^k/k!\right)^{1/2}$. Consider now the subspace

$$L_0^2(\mathcal{S}, \mu) = \{f \in L^2(\mathcal{S}, \mu) : \int f d\mu = 0\}$$

of functions in $L^2(\mathcal{S}, \mu)$ orthogonal to the constants. It is easily seen, e.g. by considering tensor powers of the embedding $L_0^2 \to L^2$ and the projection $L^2 \to L_0^2$, that $L_0^2(\mathcal{S}, \mu)^{\odot k}$ equals the subspace of $L^2(\mathcal{S}, \mu)^{\odot k}$ consisting of all functions f such that

$$\int_S f(x_1, \dots, x_k)\, d\mu(x_k) = 0 \qquad \text{for a.e. } x_1, \dots, x_{k-1}. \tag{11.3}$$

(Since every $f \in L^2(\mathcal{S}, \mu)^{\odot k}$ is symmetric, we may as well integrate over any x_i, $i \le k$, in (11.3).) It follows that if $f \in L_0^2(\mathcal{S}, \mu)^{\odot k}$, then the $\binom{n}{k}$ terms in (11.1) are orthogonal and thus, letting $(n)_k = n(n-1)\cdots(n-k+1)$,

$$\mathrm{E}\,|S_n^k(f)|^2 = \binom{n}{k} \int_{S^k} |f|^2 d\mu^k = (n)_k \|f\|^2_{L_0^2(\mathcal{S},\mu)^{\odot k}}.$$

Moreover, if $g \in L_0^2(\mathcal{S}, \mu)^{\odot l}$, with $l \neq k$, then $S_n^k(f)$ and $S_n^l(g)$ are orthogonal.

Our main goal is to prove a limit theorem for $S_n^k(f)$ for some fixed k and f (Theorem 11.3 below), but in order to do this it turns out to be convenient to treat all k together. Thus we define, for any $f = (f_k)_0^\infty \in \Gamma\big(L_0^2(\mathcal{S}, \mu)\big)$,

$$Z_n(f) = \sum_{k=0}^{\infty} n^{-k/2} S_n^k(f_k)$$

(the sum is actually finite, since all terms with $k > n$ vanish). Then

$$\begin{aligned} \mathrm{E}\,|Z_n(f)|^2 &= \sum_{k=0}^{\infty} n^{-k}\, \mathrm{E}\,|S_n^k(f_k)|^2 = \sum_{k=0}^{\infty} n^{-k} (n)_k \|f_k\|^2_{L_0^2(\mathcal{S},\mu)^{\odot k}} \\ &\le \sum_{k=0}^{\infty} \|f_k\|^2_{L_0^2(\mathcal{S},\mu)^{\odot k}} = \|f\|^2_{\Gamma(L_0^2(\mathcal{S},\mu))}. \end{aligned} \tag{11.4}$$

This may be summarized as follows.

LEMMA 11.1. *For every $n \ge 1$, the mapping $f \to Z_n(f)$ is a linear contraction of $\Gamma\big(L_0^2(\mathcal{S}, \mu)\big)$ into $L^2(\Omega, \mathcal{F}, \mathrm{P})$.* □

We may now prove a limit theorem for these variables.

THEOREM 11.2. *Let H be a Gaussian Hilbert space (defined on some probability space $(\Omega, \mathcal{F}, \mathrm{P})$) and let $I\colon L^2_{0\mathbb{R}}(\mathcal{S}, \mu) \to H$ be an isometry into. Then, for every $f \in \Gamma(L^2_0(\mathcal{S}, \mu))$,*

$$Z_n(f) \xrightarrow{\mathrm{d}} \Gamma(I)(f) \in \Gamma(H) = L^2(\Omega, \mathcal{F}(H), \mathrm{P}) \qquad \text{as } n \to \infty;$$

moreover, the convergence holds jointly for any finite number of such f.

PROOF. We begin with the special case when $f = \exp_\odot(g) = \sum_0^\infty g^{\odot k}/k!$ for some $g \in L^2_{0\mathbb{R}}(\mathcal{S}, \mu)$. By (E.7), $g^{\odot k}/k!$ is the function $g(x_1)\cdots g(x_k)$ on $\mathcal{S}^k$. Thus

$$S_n^k(g^{\odot k}/k!) = \sum_{1 \le i_1 < \cdots < i_k \le n} g(X_{i_1}) \cdots g(X_{i_k})$$

and

$$\begin{aligned} Z_n(\exp_\odot(g)) &= \sum_{k=0}^n n^{-k/2} \sum_{1 \le i_1 < \cdots < i_k \le n} g(X_{i_1}) \cdots g(X_{i_k}) \\ &= \prod_1^n \bigl(1 + n^{-1/2} g(X_i)\bigr). \end{aligned}$$

If furthermore g is bounded, we may (for large n at least) take the logarithm and obtain by a Taylor expansion

$$\ln Z_n(\exp_\odot(g)) = n^{-1/2} \sum_1^n g(X_i) - \tfrac{1}{2} n^{-1} \sum_1^n g(X_i)^2 + O(n^{-1/2}).$$

The first term on the right hand side converges in distribution to $\mathrm{N}(0, \|g\|^2)$ by the central limit theorem, since $\mathrm{E}\, g(X_i) = \int g\, d\mu = 0$ for $g \in L^2_0(\mathcal{S}, \mu)$. The second term converges to $\frac{1}{2} \mathrm{E}\, g(X_1)^2 = \frac{1}{2}\|g\|^2$ a.s. by the law of large numbers. Consequently, using the assumption that $I(g) \sim \mathrm{N}(0, \|g\|^2)$,

$$\ln Z_n(\exp_\odot(g)) \xrightarrow{\mathrm{d}} I(g) - \tfrac{1}{2} \mathrm{E}\bigl(I(g)\bigr)^2 \tag{11.5}$$

and thus, using Theorem 3.33 also,

$$Z_n(\exp_\odot(g)) \xrightarrow{\mathrm{d}} \exp\bigl(I(g) - \tfrac{1}{2} \mathrm{E}\, I(g)^2\bigr) = {:}\exp(I(g)){:} = \Gamma(I)(\exp_\odot(g)). \tag{11.6}$$

This proves the result for $f = \exp_\odot(g)$, $g \in L^2_{0\mathbb{R}} \cap L^\infty$. Moreover, the same argument shows that (11.5) and (11.6) hold with joint convergence for several such f; hence

$$Z_n(f) \xrightarrow{\mathrm{d}} \Gamma(I)(f)$$

for any f that is a finite linear combination $\sum a_i \exp_\odot(g_i)$, $g_i \in L^2_{0\mathbb{R}} \cap L^\infty$. The set, V say, of such f is dense in $\Gamma(L^2_{0\mathbb{R}}(\mathcal{S}, \mu))$ by an analogue of Corollary 3.40, cf. Example 4.3, and the rest is now routine. Given any $f \in \Gamma(L^2_{0\mathbb{R}})$, we take a sequence $f_N \to f$ with every $f_N \in V$. Then $Z_n(f_N) \xrightarrow{\mathrm{d}} \Gamma(I)(f_N)$ as $n \to \infty$

for each N; moreover, as $N \to \infty$, $\Gamma(I)(f_N) \to \Gamma(I)f$ in L^2, and thus in distribution, and $Z_n(f_N) \to Z_n(f)$ in L^2 uniformly in n by Lemma 11.1. Consequently, $Z_n(f) \xrightarrow{d} \Gamma(I)f$ follows; see e.g. Billingsley (1968, Theorem 4.2).

Joint convergence follows similarly, or by the Cramér–Wold device. This implies, in particular, the result for complex f, considering the real and imaginary parts separately. □

The most important application of Theorem 11.2 is to the U-statistics, defined by (11.1) and (11.2) (not necessarily assuming (11.3)). We observe that if $f \in L^2(\mathcal{S}^k)$ is symmetric, then there exist $f_j^0 \in L_0^2(\mathcal{S},\mu)^{\odot j}$, $j = 0,\dots,k$, such that

$$f(x_1,\dots,x_k) = \sum_{j=0}^{k} \sum_{1\le i_1<\dots<i_j\le k} f_j^0(x_{i_1},\dots,x_{i_j}),$$

and thus

$$U_n^k(f) = \sum_{j=0}^{k} \binom{k}{j} U_n^j(f_j^0), \tag{11.7}$$

see Hoeffding (1961) or Dynkin and Mandelbaum (1983); in fact, from an abstract point of view, this reflects the fact that $L^2(\mathcal{S},\mu) = L_0^2(\mathcal{S},\mu) \oplus \mathbb{R}$ (or $\mathbb{C}$), and thus every $f \in (L^2)^{\odot k}$ equals $\sum_{j=0}^{k} f_j' \odot 1^{\odot(k-j)}$ for some $f_j' \in L_0^2(\mathcal{S},\mu)^{\odot j}$ (with $f_j^0 = (k-j)!\, f_j'$).

Note also that if f_j is the projection of f onto $L^2(\mathcal{S}^j)$ given by

$$f_j(x_1,\dots,x_j) = \int_{\mathcal{S}^{k-j}} f(x_1,\dots,x_k)\, d\mu(x_{j+1})\cdots d\mu(x_k),$$

so that $f_j(X_1,\dots,X_j) = \mathrm{E}(f(X_1,\dots,X_k) \mid X_1,\dots,X_j)$, then

$$f_l(x_1,\dots,x_l) = \sum_{j=0}^{l} \sum_{1\le i_1<\dots<i_j\le l} f_j^0(x_{i_1},\dots,x_{i_j}),$$

which can be solved for f_j^0, yielding

$$f_j^0(x_1,\dots,x_j) = \sum_{l=0}^{j} (-1)^{j-l} \sum_{1\le i_1<\dots<i_l\le j} f_l(x_{i_1},\dots,x_{i_l}).$$

We state the result using the notations of Section 7.2 for stochastic integrals. Recall that a Gaussian stochastic integral exists on every measure space, see Remark 7.21.

THEOREM 11.3. *Let I be a Gaussian stochastic integral on $(\mathcal{S},\mu)$, i.e. an isometry $I\colon L_{\mathbb{R}}^2(\mathcal{S},\mu) \to H$ into a Gaussian Hilbert space H, and let Z be the Gaussian stochastic measure corresponding to I. If f, f_j and f_j^0 are as above,* *and $m \geq 0$ is any integer such that $f_0^0 = \cdots = f_{m-1}^0 = 0$ (or, equivalently, $f_0 = \cdots = f_{m-1} = 0$), then*

$$n^{m/2} U_n^k(f) \xrightarrow{\mathrm{d}} \tbinom{k}{m} I_m(f_m^0) = \tbinom{k}{m} \int_{\mathcal{S}^m} f_m^0 \, dZ^m \in H^{:m:} \qquad \text{as } n \to \infty.$$

PROOF. By Theorem 11.2 and (11.7),

$$\begin{aligned} n^{m/2} U_n^k(f) &= \sum_{j=m}^{k} \tbinom{k}{j} n^{m/2} U_n^j(f_j^0) = \sum_{j=m}^{k} \tbinom{k}{j} n^{(m-j)/2} \frac{n^j}{\binom{n}{j}} n^{-j/2} S_n^j(f_j^0) \\ &= \sum_{j=m}^{k} \tbinom{k}{j} n^{(m-j)/2} j! \, \frac{n^j}{(n)_j} Z_n(f_j^0) \\ &\xrightarrow{\mathrm{d}} \tbinom{k}{m} m! \, \Gamma(I)(f_m^0) = (k)_m \hat{I}_m(f_m^0) = \tbinom{k}{m} \int_{\mathcal{S}^m} f_m^0 \, dZ^m, \end{aligned}$$

because all terms with $j > m$ tend to 0 in L^2 and thus in probability. □

REMARK 11.4. Under the assumptions of the theorem, $f_m^0 = f_m$. Moreover, it is natural to take m as the smallest integer for which $f_m^0 \neq 0$; then the limit given by the theorem is non-trivial.

The most important cases are the ones with small $m = 0, 1, 2$, which can be written as follows. The case $m = 0$ is rather trivial, $m = 1$ is the standard case with a normal limit (first found by a simpler method by Hoeffding (1948)), and $m = 2$ is a degenerate case with a non-normal limit that sometimes occurs in applications (typically when $f_1^0 = 0$ because of some symmetry property). Higher degeneracies, with $m \geq 3$, are rare in applications.

COROLLARY 11.5. *Let f, f_j and f_j^0 be as above, with f real, and let $\lambda = \int f \, d\mu^k$.*

(i) *The weak law of large numbers holds:*

$$U_n^k(f) \xrightarrow{\mathrm{p}} \lambda.$$

(ii) *The central limit theorem holds in the form*

$$n^{1/2}(U_n^k(f) - \lambda) \xrightarrow{\mathrm{d}} \mathrm{N}(0, \sigma^2),$$

with $\sigma^2 = k^2 \|f_1^0\|_{L^2(\mathcal{S},\mu)}^2 = k^2 \operatorname{Var} f_1(X_1) \geq 0$.

(iii) *If $\sigma^2 = 0$, then*

$$n(U_n^k(f) - \lambda) \xrightarrow{\mathrm{d}} \sum_j \tbinom{k}{2} \alpha_j (\xi_j^2 - 1),$$

where ξ_j are independent standard normal variables and the numbers α_j are the non-zero eigenvalues (counted with multiplicities) of the compact symmetric bilinear form

$$\begin{aligned}B(g,h) &= \iint f_2^0(x,y)g(x)h(y)\,d\mu(x)d\mu(y)\\ &= \mathrm{E}\big(f_2^0(X_1,X_2)g(X_1)h(X_2)\big)\\ &= \mathrm{E}\big((f(X_1,X_2,\dots,X_k)-\lambda)g(X_1)h(X_2)\big), \qquad g,h\in L^2(\mathcal{S},\mu).\end{aligned}$$

PROOF. Begin by observing $f_0^0 = f_0 = \mathrm{E}\,U_n^k(f) = \lambda$. Thus (i) is the case $m = 0$ of Theorem 11.3. Replacing f by $f - \lambda$, we may now assume $\lambda = 0$. Then (ii) is the case $m = 1$ of Theorem 11.3, while (iii) follows easily from the case $m = 2$ and Theorem 6.1, transferring the bilinear form (6.3) from H to $L^2(\mathcal{S},\mu)$ by I and observing that by Theorem 7.33 (there are in this case two permitted Feynman diagrams),

$$\begin{aligned}&\tfrac{1}{2}\mathrm{E}\Big(\int f_2^0\,dZ^2 I(g)I(h)\Big)\\ &\qquad = \tfrac{1}{2}\int_{\mathcal{S}^2}\big(f_2^0(x,y)g(x)h(y)+f_2^0(x,y)g(y)h(x)\big)\,d\mu(x)d\mu(y)\\ &\qquad = \int_{\mathcal{S}^2} f_2^0(x,y)g(x)h(y)\,d\mu(x)d\mu(y).\end{aligned}$$

□

REMARK 11.6. Similar limit theorems for U-statistics have been given by several authors, for example Rubin and Vitale (1980).

REMARK 11.7. Some similar results were also given already by von Mises (1947), who considered sums of the type $\sum_{1\le i_j\le n} f(X_{i_1},\dots,X_{i_k})$. (Such sums, where repeated indices are allowed, are called *V-statistics.*) It is easy to express sums of this form as a combination of several sums of the form (11.1) (with different k) and conversely; hence results for the two types of sums are essentially equivalent.

It may be shown that, for suitable f, this sum, normalized by $n^{-k/2}$, converges to the multiple Stratonovich integral $\mathring{I}_k(f)$ defined in Remark 7.38, see e.g. Budhiraja and Kallianpur (1995).

REMARK 11.8. The limit theorems for U-statistics may be extended to triangular schemes, with f depending on n, under suitable conditions; see Rubin and Vitale (1980). Note, however, that for triangular schemes, under other conditions, the limit may be normal in degenerate cases also (Jammalamadaka and Janson 1986).

REMARK 11.9. Dynkin and Mandelbaum (1983) consider instead of $Z_n(f)$ the Poissonized analogue defined by

$$\widetilde{Z}_\lambda(f) = \sum_{k=0}^{\infty} \lambda^{-k/2} S_N^k(f_k), \qquad \lambda > 0,$$

where $N \sim \mathrm{Po}(\lambda)$ is a Poisson variable independent of everything else. This has the added beauty that $f \mapsto \widetilde{Z}_\lambda(f)$ is an *isometry* of $\Gamma(L_0^2(\mathcal{S},\mu))$ into $L^2(\Omega,\mathcal{F},\mathrm{P})$.

REMARK 11.10. It follows from the calculation (11.4), by dominated convergence, that

$$\mathrm{E}\,|Z_n(f)|^2 \to \|f\|^2_{\Gamma(L_0^2(\mathcal{S},\mu))} = \|\Gamma(I)(f)\|^2_{\Gamma(H)} \qquad \text{as } n \to \infty.$$

Moreover, $\mathrm{E}\,Z_n(f) = f_0 = \mathrm{E}(\Gamma(I)(f))$.

Consequently, the mean and variance of $Z_n(f)$ converge to the mean and variance of $\Gamma(I)(f)$, i.e. to the mean and variance of the limit distribution. Similarly, the covariance of two variables $Z_n(f_1)$ and $Z_n(f_2)$ converges to the covariance of the joint limit distribution (this follows e.g. by considering $Z_n(f_1+f_2)$).

It follows easily that in the limit results given in Theorem 11.3 and Corollary 11.5 also, the means and variances (and covariances if we consider several f) converge to the corresponding quantities for the limits. Furthermore, the reader can easily verify that the same holds for every limit result proved in the remaining sections of this chapter.

In particular, Corollary 11.5(ii) implies that if $\operatorname{Var} f_1(X_1) \neq 0$, then $U_n^k(f)$ is asymptotically normal with the standard normalization:

$$(U_n^k(f) - \mathrm{E}\,U_n^k(f))/(\operatorname{Var} U_n^k(f))^{1/2} \xrightarrow{\mathrm{d}} \mathrm{N}(0,1).$$

Corresponding results hold for Corollary 11.20(ii) and Corollary 11.36.

REMARK 11.11. The results above may be extended to functional limit theorems. For example, if f, m and f_m^0 are as in Theorem 11.3, but now I is a Gaussian stochastic integral on $(\mathcal{S}\times[0,1],\mu\,dt)$ and Z is the corresponding Gaussian stochastic measure on $\mathcal{S}\times[0,1]$, then as $n\to\infty$,

$$n^{m/2} U^k_{[tn]}(f) \xrightarrow{\mathrm{d}} \binom{k}{m} I_m(f_m^0 \otimes \mathbf{1}_{[0,t]}^{\otimes m}) = \binom{k}{m} \int_{\mathcal{S}^m\times[0,t]^m} f_m^0\,dZ^m, \qquad 0 \le t \le 1, \tag{11.8}$$

with convergence in the Skorohod topology on $D[0,1]$. (See Billingsley (1968) for definitions.) This follows by simple modifications of the proofs above, using Donsker's theorem as a starting point and Doob's martingale inequality to show that the L^2-approximations hold uniformly in t. It is, further, easily seen that the limit in (11.8) is a continuous stochastic process. We omit the details.

Such theorems have been given by several authors, for example Hall (1979).

The results extend immediately to $0 \le t < \infty$, with the Skorohod topology on $D[0,\infty)$, see e.g. Lindvall (1973) for definitions. (When the limit is a continuous stochastic process, as it is in this case, convergence in $D[0,\infty)$ is equivalent to convergence in $D[0,T]$ for every finite T; in our case this is by a simple change of variable equivalent to convergence in $D[0,1]$.)

2. Asymmetric statistics

In this section we continue to assume that $X_1, X_2, \dots$ are i.i.d. random variables with distribution μ in some space $\mathcal{S}$ and study sums of the type (11.1) or (11.2) with $f \in L^2(\mathcal{S}^k, \mu^k)$, but we no longer require f to be symmetric. Note that it is now important that the sum in (11.1) is taken over increasing sequences $i_1 < \dots < i_k$ only; if we instead sum over all sequences of distinct i_j, we might as well symmetrize f and reduce to the symmetric case. (In the symmetric case, the sum over all sequences of distinct i_j differs from (11.1) only by the numerical factor $k!$.)

It turns out to be advantageous to study a somewhat more general sum with weights. Thus we define, for $f \in L^2(\mathcal{S}^k, \mu^k)$ and any function g on $[0,1]^k$, with $k \ge 1$,

$$S_n^k(f;g) = \sum_{i_1,\dots,i_k}^{*} g(\tfrac{i_1}{n},\dots,\tfrac{i_k}{n}) f(X_{i_1},\dots,X_{i_k}), \tag{11.9}$$

where $\sum^*_{i_1,\dots,i_k}$ denotes the sum over the $(n)_k$ sequences $i_1,\dots,i_k$ with distinct elements $i_j \in \{1,\dots,n\}$. In particular, $S_n^k(f) = S_n^k(f;\mathbf{1}_{D_k})$, where

$$D_k = \{(t_1,\dots,t_k) : 0 < t_1 < \dots < t_k < 1\} \tag{11.10}$$

(cf. (7.3), where we considered functions on $[0,\infty)$).

Since $S_n^k(f;g)$ depends on the values of g on a finite set only, it is not adapted to L^2-theory; therefore we also introduce a modification.

Let $\mathcal{F}_n$ be the collection of the n^k subcubes of $[0,1]^k$ of the form $(\frac{i_1-1}{n},\frac{i_1}{n}] \times \dots \times (\frac{i_k-1}{n},\frac{i_k}{n}]$, with $i_1,\dots,i_k \in \{1,\dots,n\}$, and define, for $g \in L^2([0,1]^k, dt)$ (where dt denotes Lebesgue measure in any dimension),

$$\overline{g}_n(t) = \frac{1}{|Q|}\int_Q g(s)\,ds, \qquad t \in Q \in \mathcal{F}_n, \tag{11.11}$$

and

$$\overline{S}_n^k(f;g) = \sum_{i_1,\dots,i_k}^{*} \overline{g}_n(\tfrac{i_1}{n},\dots,\tfrac{i_k}{n}) f(X_{i_1},\dots,X_{i_k}). \tag{11.12}$$

We consider as in Section 1 the subspace $L_0^2(\mathcal{S},\mu)$ of all functions with mean 0 in $L^2(\mathcal{S},\mu)$, and now take general tensor powers $L_0^2(\mathcal{S},\mu)^{\otimes k}$. The tensor power $L^2(\mathcal{S},\mu)^{\otimes k}$ may be identified with $L^2(\mathcal{S}^k,\mu^k)$, cf. Example E.10, and it is easily seen that then

$$L_0^2(\mathcal{S},\mu)^{\otimes k} = \Big\{ f \in L^2(\mathcal{S}^k,\mu^k) : \int_{\mathcal{S}} f(x_1,\dots,x_k)\,d\mu(x_j) = 0 \text{ a.e., } 1 \le j \le k \Big\}$$

(cf. (11.3) for the symmetric case).

We begin with variance estimates.

LEMMA 11.12. *Suppose that* $f \in L_0^2(\mathcal{S},\mu)^{\otimes k} \subset L^2(\mathcal{S}^k,\mu^k)$. *Then*

$$\mathrm{E}\,|S_n^k(f;g)|^2 \le k!\|f\|_2^2 \sum_{i_1,\dots,i_k}^{*} |g(\tfrac{i_1}{n},\dots,\tfrac{i_k}{n})|^2 \le n^k k!\|f\|_2^2 (\sup|g|)^2.$$

PROOF. It follows easily from $f \in L_0^2(\mathcal{S},\mu)^{\otimes k}$ that $f(X_{i_1},\dots,X_{i_k})$ and $f(X_{j_1},\dots,X_{j_k})$ are orthogonal unless the sequences $i_1,\dots,i_k$ and $j_1,\dots,j_k$ are permutations of each other. Consequently, letting π range over the $k!$ permutations of $\{1,\dots,k\}$ and using the Minkowski and Cauchy–Schwarz inequalities,

$$\begin{aligned}
\mathrm{E}\,|S_n^k(f;g)|^2 &= \sum_{i_1<\dots<i_k} \Big\| \sum_\pi g(\tfrac{i_{\pi(1)}}{n},\dots,\tfrac{i_{\pi(k)}}{n}) f(X_{i_{\pi(1)}},\dots,X_{i_{\pi(k)}}) \Big\|_2^2 \\
&\le \sum_{i_1<\dots<i_k} \Big(\sum_\pi |g(\tfrac{i_{\pi(1)}}{n},\dots,\tfrac{i_{\pi(k)}}{n})|\, \|f\|_2 \Big)^2 \\
&\le \sum_{i_1<\dots<i_k} k! \sum_\pi |g(\tfrac{i_{\pi(1)}}{n},\dots,\tfrac{i_{\pi(k)}}{n})|^2 \|f\|_2^2 \\
&= k! \sum_{i_1,\dots,i_k}^{*} |g(\tfrac{i_1}{n},\dots,\tfrac{i_k}{n})|^2 \|f\|_2^2.
\end{aligned}$$

□

LEMMA 11.13. *Suppose that* $f \in L_0^2(\mathcal{S},\mu)^{\otimes k} \subset L^2(\mathcal{S}^k,\mu^k)$ *and that* $g \in L^2([0,1]^k)$. *Then*

$$\mathrm{E}\,|\overline{S}_n^k(f;g)|^2 \le n^k k! \|f\|_2^2 \|g\|_2^2.$$

PROOF. Since $\overline{S}_n^k(f;g) = S_n^k(f;\overline{g}_n)$, this is a consequence of Lemma 11.12 and

$$\sum_{i_1,\dots,i_k}^{*} |\overline{g}_n(\tfrac{i_1}{n},\dots,\tfrac{i_k}{n})|^2 \le \sum_{i_1,\dots,i_k=1}^{n} |\overline{g}_n(\tfrac{i_1}{n},\dots,\tfrac{i_k}{n})|^2 = n^k \|\overline{g}_n\|_2^2 \le n^k \|g\|_2^2,$$

where the last inequality follows by the Cauchy–Schwarz inequality (or because $\overline{g}_n$ can be regarded as a conditional expectation of g). □

LEMMA 11.14. *Assume that g is bounded and a.e. continuous on* $[0,1]^k$. *Then, for every* $f \in L_0^2(\mathcal{S},\mu)^{\otimes k} \subset L^2(\mathcal{S}^k,\mu^k)$,

$$n^{-k}\,\mathrm{E}\,|S_n^k(f;g) - \overline{S}_n^k(f;g)|^2 \to 0 \qquad \text{as } n \to \infty.$$

In particular, if one of $n^{-k/2}S_n^k(f;g)$ and $n^{-k/2}\overline{S}_n^k(f;g)$ converges in distribution as $n\to\infty$, then so does the other, and to the same limit.

PROOF. Define g_n on $(0,1]^k$ by $g_n(t_1,\dots,t_k)=g(\frac{i_1}{n},\dots,\frac{i_k}{n})$ when $\frac{i_j-1}{n}<t_j\le\frac{i_j}{n}$, $j=1,\dots,k$. Then, by Lemma 11.12,

$$\begin{aligned} n^{-k}\operatorname{E}|S_n^k(f;g)-\overline{S}_n^k(f;g)|^2 &= n^{-k}\operatorname{E}|S_n^k(f;g-\overline{g}_n)|^2 \\ &\le k!\|f\|_2^2\sum_{i_1,\dots,i_k} n^{-k}|(g-\overline{g}_n)(\tfrac{i_1}{n},\dots,\tfrac{i_k}{n})|^2 = k!\|f\|_2^2\|(g_n-\overline{g}_n)\|^2_{L^2([0,1]^k)}. \end{aligned}$$

Since g is a.e. continuous, $g_n(t)\to g(t)$ and $\overline{g}_n(t)\to g(t)$ a.e. on $[0,1]^k$ as $n\to\infty$, and thus $g_n-\overline{g}_n\to 0$ a.e. Consequently, by dominated convergence, $\int_{[0,1]^k}|g_n-\overline{g}_n|^2\,dt\to 0$ as $n\to\infty$, and the result follows. □

REMARK 11.15. A function g on $[0,1]^k$ is bounded and a.e. continuous if and only if it is Riemann integrable. Moreover, every such function is Lebesgue measurable, for example because it is a.e. the limit of the sequence g_n defined in the proof above; in particular, every such g belongs to $L^2([0,1]^k)$.

We are now prepared for our limit theorem for $S_n^k(f;g)$, again using the notations of Section 7.2 for stochastic integrals.

THEOREM 11.16. *Let $I\colon L^2_{\mathbb{R}}(\mathcal{S},\mu)\otimes L^2_{\mathbb{R}}([0,1],dt)=L^2_{\mathbb{R}}(\mathcal{S}\times[0,1],\mu\times dt)\to H$ be an isometry into a Gaussian Hilbert space H, and let Z denote the corresponding Gaussian stochastic measure on $\mathcal{S}\times[0,1]$.*

(i) *If $k\ge 1$, $f\in L^2_0(\mathcal{S},\mu)^{\otimes k}$ and $g\in L^2([0,1]^k)$, then*

$$n^{-k/2}\overline{S}_n^k(f;g)\xrightarrow{\mathrm{d}} I_k(f\otimes g)=\int_{\mathcal{S}^k\times[0,1]^k} f(x)g(t)\,dZ^k\in H^{:k:}.$$

(ii) *If $k\ge 1$, $f\in L^2_0(\mathcal{S},\mu)^{\otimes k}$ and g is bounded and a.e. continuous on $[0,1]^k$, then*

$$n^{-k/2}S_n^k(f;g)\xrightarrow{\mathrm{d}} I_k(f\otimes g)=\int_{\mathcal{S}^k\times[0,1]^k} f(x)g(t)\,dZ^k\in H^{:k:}.$$

In both cases, the convergence holds jointly for any finite number of such pairs (f,g) (possibly with different k).

PROOF. Note that by Lemma 11.14, the two parts are equivalent when g is bounded and a.e. continuous. We prove the result in several steps.

Step 1. If $k=1$, $f\in L^2_{0\mathbb{R}}(\mathcal{S},\mu)$ and $g=\mathbf{1}_{(a,b]}$, $0\le a<b\le 1$, then

$$S_n^k(f;g)=\sum_{na<i\le nb} f(X_i). \tag{11.13}$$

The number of terms in this sum is $n(b-a)+O(1)$ and the terms are i.i.d. with mean 0 and variance $\int f^2\,d\mu=\|f\|_2^2$; thus it follows from the central limit theorem that

$$(b-a)^{-1/2}n^{-1/2}S_n^1(f;g)\xrightarrow{\mathrm{d}} \mathrm{N}(0,\|f\|_2^2)$$

or

$$n^{-1/2}S^1_n(f;g) \xrightarrow{\mathrm{d}} \mathrm{N}(0,(b-a)\|f\|_2^2).$$

Since $I_1(f\otimes g) \sim \mathrm{N}(0,\|f\otimes g\|_2^2) = \mathrm{N}(0,(b-a)\|f\|_2^2)$, this yields

$$n^{-1/2}S^1_n(f;g) \xrightarrow{\mathrm{d}} I_1(f\otimes g) = \int f\otimes g\,dZ, \tag{11.14}$$

which verifies the result in (ii) for a single pair (f,g) with g of this type.

Step 2. Consider a partition $0 = a_0 < a_1 < \cdots < a_m = 1$ and suppose that $\{(f_j,g_j)\}$ is a finite set of pairs with $f_j \in L^2_{0\mathbb{R}}(\mathcal{S},\mu)$ and each g_j equal to some $\mathbf{1}_{(a_{l-1},a_l]}$ (repetitions allowed). Let $E_l = \{j : g_j = \mathbf{1}_{(a_{l-1},a_l]}\}$.

The convergence in (11.14) extends, e.g. by the Cramér–Wold device, to the joint distribution for each set $\{(f_j,g_j) : j\in E_l\}$ with a fixed g_j.

On the other hand, the families $\{S^1_n(f_j;g_j) : j \in E_l\}$, $l = 1,\dots,m$, of random variables are independent of each other, because they by (11.13) are functions of disjoint subsets of $\{X_i\}$. Moreover, the corresponding families $\{f_j\otimes g_j : j\in E_l\}$, $l=1,\dots,m$, are orthogonal to each other, and hence the families $\{I_1(f_j\otimes g_j) : j \in E_l\}$ are independent of each other.

It follows that (11.14) holds jointly for all (f_j,g_j).

Step 3. Given any finite set of pairs (f_j,g_j) with $f_j \in L^2_{0\mathbb{R}}(\mathcal{S},\mu)$ and $g_j = \mathbf{1}_{(a_j,b_j]}$ for some intervals $(a_j,b_j] \subseteq (0,1]$, we consider the partition of $(0,1]$ defined by the set $\{a_j,b_j\}_j$ of all endpoints. Since each g_j is a sum of indicator functions of intervals in the partition, it is an easy consequence of Step 2 (and linearity in g) that (11.14) holds jointly for all (f_j,g_j).

Step 4. Suppose that $f_1,\dots,f_k \in L^2_{0\mathbb{R}}(\mathcal{S},\mu)$ and $g_j = \mathbf{1}_{(a_j,b_j]}$, where $(a_j,b_j]$, $j=1,\dots,k$, are disjoint intervals. Then, by Step 3,

$$\begin{aligned} n^{-k/2}S^k_n(f_1\otimes\cdots\otimes f_k; g_1\otimes\cdots\otimes g_k) &= n^{-k/2}\sum_{a_j<i_j/n\le b_j}\prod_{j=1}^k f_j(X_{i_j}) \\ &= \prod_{j=1}^k n^{-1/2}S^1_n(f_j;g_j), \end{aligned}$$

which by Step 3 converges in distribution to $\prod_{j=1}^k \int f_j\otimes g_j\,dZ$. However, the Gaussian variables $\int f_j\otimes g_j\,dZ$ are orthogonal, and thus, letting $f = f_1\otimes\cdots\otimes f_k$ and $g = g_1\otimes\cdots\otimes g_k$,

$$\prod_{j=1}^k \int f_j\otimes g_j\,dZ = :\int f_1\otimes g_1\,dZ\cdots\int f_k\otimes g_k\,dZ: = \int f\otimes g\,dZ^k.$$

Consequently,

$$n^{-k/2}S^k_n(f;g) \xrightarrow{\mathrm{d}} \int f\otimes g\,dZ^k$$

for any f and g of this type; moreover, the argument shows that this holds jointly for any finite set of such f and g.

Step 5. It is time to transfer attention to $\overline{S}_n^k$; we observe that by Lemma 11.14 we also have

$$n^{-k/2}\overline{S}_n^k(f;g) \xrightarrow{\mathrm{d}} \int f \otimes g \, dZ^k \tag{11.15}$$

when f and g are as in Step 4, with joint convergence for any finite set of such f and g.

Step 6. Let V_k be the subspace of $L^2(\mathcal{S}^k, \mu^k)$ consisting of all finite linear combinations of functions of the type $f_1 \otimes \cdots \otimes f_k$, $f_j \in L^2_{0\mathbb{R}}(\mathcal{S}, \mu)$, and let similarly W_k be the subspace of $L^2([0,1]^k)$ consisting of all finite linear combinations of functions of the type $g_1 \otimes \cdots \otimes g_k$, where $g_1, \dots, g_k$ are indicator functions for disjoint half-open intervals. By Step 5 and linearity in f and g separately, (11.15) holds for any $f \in V_k$ and $g \in W_k$, with joint convergence for any finite set of such f and g.

Step 7. Now V_k is dense in $L^2_0(\mathcal{S}, \mu)^{\otimes k}$ while W_k is dense in $L^2([0,1]^k)$. It follows by Lemma 11.13 and Billingsley (1968, Theorem 4.2), similarly as in the proof of Theorem 11.2, that (11.15) extends to all $f \in L^2_0(\mathcal{S}, \mu)^{\otimes k}$ and $g \in L^2([0,1]^k)$, still with joint convergence for finite sets, which completes the proof of (i).

Step 8. Finally, (ii) follows by (i) and Lemma 11.14. □

As in Section 1 we extend this result to arbitrary $f \in L^2(\mathcal{S}^k)$ by a decomposition; however, the lack of symmetry makes the decomposition (and the notation) somewhat more complicated this time.

For any subset A of $\{1, \dots, k\}$, let M_A be the subspace of $L^2(\mathcal{S}^k, \mu^k)$ consisting of all functions $f(x_1, \dots, x_k)$ that actually only depend on $\{x_i : i \in A\}$. In particular, $M_{\{1,\dots,k\}} = L^2(\mathcal{S}^k)$ while $M_\varnothing$ consists of the constants only. Note that $M_A \cong L^2(\mathcal{S}^{|A|}, \mu^{|A|})$, where $|A|$ is the cardinality of A; we denote the natural isometry (preserving the ordering of the coordinates) $M_A \to L^2(\mathcal{S}^{|A|}, \mu^{|A|})$ by $f \mapsto f \circ \pi_A$, using notation that will be fully explained in Section 3.

Further, let

$$M_A^0 = \{f \in M_A : f \perp M_B \text{ for every } B \subsetneq A\}.$$

LEMMA 11.17. *With notation as above,* $L^2(\mathcal{S}^k, \mu^k) = \bigoplus_{A \subseteq \{1,\dots,k\}} M_A^0$.

PROOF. Select an orthonormal basis $\{\varphi_i\}_{i \in \mathcal{I}}$ in $L^2(\mathcal{S}, \mu)$ with $\varphi_0 = 1$, and thus $\int_{\mathcal{S}} \varphi_i \, d\mu = 0$ for $i \neq 0$. The functions $\varphi_{i_1} \otimes \cdots \otimes \varphi_{i_k}$, $i_1, \dots, i_k \in \mathcal{I}$, constitute an orthonormal basis in $L^2(\mathcal{S}^k, \mu^k)$; it is easily seen that the subset of all such functions with $i_j = 0$ for $j \notin A$, or equivalently $\{j : i_j \neq 0\} \subseteq A$, constitutes an orthonormal basis in M_A; moreover, the subset with $\{j : i_j \neq 0\} = A$ consitutes an orthonormal basis in M_A^0. This yields a partition of the orthonormal basis in $L^2(\mathcal{S}^k)$, and the result follows. □

REMARK 11.18. More generally, for $A \subseteq \{1, \dots, k\}$, $M_A = \bigoplus_{B \subseteq A} M_B^0$.

Let $f \in L^2(\mathcal{S}^k, \mu^k)$, and let, for $A \subseteq \{1, \dots, k\}$, with P_M denoting the orthogonal projection onto a subspace M of $L^2(\mathcal{S}^k)$,

$$f_A = P_{M_A}(f) \in M_A,$$
$$f_A^0 = P_{M_A^0}(f) \in M_A^0,$$
$$\hat{f}_A^0 = f_A^0 \circ \pi_A \in L_0^2(\mathcal{S}, \mu)^{\otimes |A|} \subset L^2(\mathcal{S}^{|A|}, \mu^{|A|}).$$

Note that f_A is a conditional expectation,

$$f_A(X_1, \dots, X_k) = \mathrm{E}\big(f(X_1, \dots, X_k) \mid X_i,\, i \in A\big);$$

in particular, $f_\varnothing = \mathrm{E}\, f(X_1, \dots, X_k)$. Moreover, by Lemma 11.17,

$$f = \sum_A f_A^0. \tag{11.16}$$

More generally, by Remark 11.18,

$$f_B = \sum_{A \subseteq B} f_A^0, \qquad B \subseteq \{1, \dots, k\},$$

which can be inverted to

$$f_A^0 = \sum_{B \subseteq A} (-1)^{|A|-|B|} f_B, \qquad A \subseteq \{1, \dots, k\}.$$

Let the elements of A be $\alpha_1 < \cdots < \alpha_a$, with $a = |A|$, and let $\delta_i = \alpha_{i+1} - \alpha_i - 1$, $i = 0, \dots, a$, with $\alpha_0 = 0$ and $\alpha_{a+1} = k + 1$; thus $\{\delta_i\}_0^a$ are the successive lengths of the gaps before, between and after the indices in A. Note that each $\delta_i \geq 0$, and that $\sum_0^a \delta_i = k - a$.

Now suppose that $f \in M_A$. Then $S_n^k(f)$ is a sum of $f(X_{i_1}, \dots, X_{i_k})$, $i_1 < \cdots < i_k$, where each term depends only on $\{X_{i_l} : l \in A\}$; in fact, it equals $f \circ \pi_A(X_{j_1}, \dots, X_{j_l})$ for some $j_1 < \cdots < j_a \leq n$ ($j_l = i_{\alpha_l}$). Moreover, each such term occurs $\prod_0^a \binom{j_{l+1} - j_l - 1}{\delta_l}$ times, with $j_0 = 0$ and $j_{a+1} = n+1$, since this is the number of possible choices of the remaining indices $\{i_l : l \notin A\}$. Thus, defining

$$g_{A,n}(\tfrac{j_1}{n}, \dots, \tfrac{j_a}{n}) = \prod_{l=0}^{a} \binom{j_{l+1} - j_l - 1}{\delta_l} \mathbf{1}[1 \leq j_1 < \cdots < j_a \leq n],$$

it follows that

$$S_n^k(f) = S_n^a(f \circ \pi_A; g_{A,n}), \qquad f \in M_A.$$

Now, if $0 = j_0 < j_1 < \cdots < j_a < j_{a+1} = n + 1$, then

$$\begin{aligned} g_{A,n}(\tfrac{j_1}{n}, \dots, \tfrac{j_a}{n}) &= \prod_{l=0}^{a} \frac{(j_{l+1} - j_l)^{\delta_l}}{\delta_l!} + O(n^{\sum_l \delta_l - 1}) \\ &= n^{k-a} \prod_{l=0}^{a} \frac{(j_{l+1}/n - j_l/n)^{\delta_l}}{\delta_l!} + O(n^{k-a-1}). \end{aligned}$$

Consequently, if we define

$$g_A(t_1,\dots,t_a) = \begin{cases} \prod_{l=0}^{a}(t_{l+1}-t_l)^{\delta_l}/\delta_l!, & 0=t_0<t_1<\dots<t_a<t_{a+1}=1,\\ 0, & \text{otherwise,} \end{cases} \tag{11.17}$$

then, by Lemma 11.12,

$$\begin{aligned} \mathrm{E}\,|S_n^k(f)-n^{k-a}S_n^a(f\circ\pi_A;g_A)|^2 &= \mathrm{E}\,|S_n^a(f\circ\pi_A;g_{A,n}-n^{k-a}g_A)|^2\\ &= O(n^a\|f\circ\pi_A\|_2^2 n^{2(k-a-1)}) = O(n^{2k-a-2}\|f\|_2^2), \qquad f\in M_A. \end{aligned}$$

In particular, as $n\to\infty$,

$$\mathrm{E}\,|n^{a/2-k}S_n^k(f)-n^{-a/2}S_n^a(f\circ\pi_A;g_A)|^2 \to 0. \tag{11.18}$$

If now furthermore $f\in M_A^0$, then $f\circ\pi_A\in M_{\{1,\dots,a\}}^0 = L_0^2(\mathcal{S})^{\otimes a}$, and by Theorem 11.16(ii) and (11.18),

$$n^{a/2-k}S_n^k(f) \xrightarrow{\mathrm{d}} I_a(f\circ\pi_A\otimes g_A), \qquad f\in M_A^0. \tag{11.19}$$

Moreover, (11.19) holds jointly for any finite number of different f, possibly in different spaces M_A^0.

Finally, if $f\in L^2(\mathcal{S}^k)$, we have by the decomposition (11.16)

$$S_n^k(f) = \sum_A S_n^k(f_A^0),$$

which together with (11.19) yields our main result.

THEOREM 11.19. *Let, as above, $I\colon L_{\mathbb{R}}^2(\mathcal{S},\mu)\otimes L_{\mathbb{R}}^2([0,1],dt) = L_{\mathbb{R}}^2(\mathcal{S}\times[0,1],\mu\times dt)\to H$ be an isometry into a Gaussian Hilbert space H, and let Z denote the corresponding Gaussian stochastic measure on $\mathcal{S}\times[0,1]$. If f, f_A and f_A^0 are as above, and $m\ge 0$ is any integer such that $f_A^0=0$ whenever $|A|<m$ (or, equivalently, $f_A=0$ when $|A|<m$), then, as $n\to\infty$,*

$$n^{m/2-k}S_n^k(f) \xrightarrow{\mathrm{d}} I_m\Big(\sum_{|A|=m}\hat f_A^0\otimes g_A\Big) = \int_{\mathcal{S}^m\times[0,1]^m}\sum_{|A|=m}\hat f_A^0\otimes g_A\,dZ^m \in H^{:m:}.$$

□

For $m=0,1$ the result can be expressed as follows, using U_n^k defined by (11.2). (For $m=2$ too it is possible to give an explicit result, using Theorem 6.1, cf. Corollary 11.5(iii), but we omit it since the formula becomes rather complicated.)

COROLLARY 11.20. *Let f, f_A and f_A^0 be as above, with f real, and let $\lambda = f_\varnothing = \int f\,d\mu^k$.*

(i) *The weak law of large numbers holds:*

$$U_n^k(f)\xrightarrow{\mathrm{P}}\lambda.$$

(ii) *The central limit theorem holds in the form*

$$n^{1/2}(U_n^k(f) - \lambda) \xrightarrow{\text{d}} \mathrm{N}(0, \sigma^2),$$

with

$$\sigma^2 = k^2 \Big\| \sum_{i=1}^{k} \hat{f}^0_{\{i\}} \otimes \tbinom{k-1}{i-1} t^{i-1}(1-t)^{k-i} \Big\|^2_{L^2(\mathcal{S}\times[0,1])}$$
$$= \frac{k^2}{2k-1} \sum_{i,j=1}^{k} \tbinom{k-1}{i-1}\tbinom{k-1}{j-1}\tbinom{2k-2}{i+j-2}^{-1} \operatorname{Cov}\big(\hat{f}^0_{\{i\}}(X_1), \hat{f}^0_{\{j\}}(X_1)\big) \geq 0.$$

PROOF. By first subtracting λ from f, we may assume that $\lambda = f^0_0 = 0$. Taking $m = 1$ in Theorem 11.19, we then obtain

$$n^{1/2} U_n^k(f) = \frac{n^k}{\binom{n}{k}} n^{1/2-k} S_n^k(f) \xrightarrow{\text{d}} k!\, I_1\Big(\sum_{i=1}^{k} \hat{f}^0_{\{i\}} \otimes g_{\{i\}}\Big) \sim \mathrm{N}(0, \sigma^2),$$

where $\sigma^2 = \|k! \sum_{i=1}^{k} \hat{f}^0_{\{i\}} \otimes g_{\{i\}}\|_2^2$. Moreover, the definition (11.17) yields

$$g_{\{i\}}(t) = \frac{t^{i-1}(1-t)^{k-i}}{(i-1)!\,(k-i)!}, \qquad 0 \leq t \leq 1,$$

and the first expression for σ^2 follows. The second follows because a Beta integral evaluation yields

$$\langle g_{\{i\}}, g_{\{j\}} \rangle_{L^2([0,1])} = (k-1)!^{-2} \tbinom{k-1}{i-1}\tbinom{k-1}{j-1} \tfrac{(i+j-2)!\,(2k-i-j)!}{(2k-1)!}.$$

Finally, (i) follows by (ii), or by taking $m = 0$ in the theorem. □

REMARK 11.21. The results of Section 1 are easily obtained as special cases of the present results. In particular, note that if f is symmetric, then $\hat{f}^0_A = f^0_m$ when $|A| = m$; moreover, it is easily seen that

$$\sum_{|A|=m} g_A = \frac{1}{(k-m)!} \mathbf{1}_{D_m},$$

and thus

$$\sum_{|A|=m} \hat{f}^0_A \otimes g_A = \frac{1}{(k-m)!} f^0_m \otimes \mathbf{1}_{D_m}.$$

It is now easy to derive Theorem 11.3 from Theorem 11.19, which leads to another proof of the results in Section 1.

Conversely, Theorem 11.19 can be obtained from Theorem 11.3 as follows. Let $Y_1, Y_2, \dots$ be i.i.d. random variables, independent of $\{X_i\}$, with a uniform distribution on $[0, 1]$. Then

$$\sum_{i_1,\dots,i_k}^{*} f(X_{i_1}, \dots, X_{i_k}) \mathbf{1}[Y_{i_1} < \dots < Y_{i_k}] \tag{11.20}$$

has the same distribution as $S^k_n(f)$, as is seen by conditioning on (Y_i). The sum (11.20), however, equals $S^k_n(F)$, based on the i.i.d. sequence $\{(X_i, Y_i)\}_{i=1}^{\infty}$ of random variables in $\mathcal{S} \times [0,1]$, with

$$F = \sum_{\pi \in \mathfrak{S}_k} f(X_{\pi(1)}, \dots, X_{\pi(k)}) \mathbf{1}[Y_{\pi(1)} < \dots < Y_{\pi(k)}],$$

and Theorem 11.3 applies.

REMARK 11.22. For any fixed $a \le k$, the functions g_A, where A ranges over all subsets of $\{1, \dots, k\}$ with $|A| = a$, are linearly independent. This is easily seen by the change of coordinates $u_1 = t_1$ and $u_i = t_i - t_{i-1}$, $2 \le i \le a$; we omit the details.

EXAMPLE 11.23. Suppose that $f \in L^2_{\mathbb{R}}(\mathcal{S}^2)$ is antisymmetric; $f(x,y) = -f(y,x)$. Then $f^0_{\varnothing} = 0$, $f^0_{\{1\}}(x) = f_{\{1\}}(x) = \int_{\mathcal{S}} f(x,y)\, d\mu(y)$ and $f^0_{\{2\}} = -f^0_{\{1\}}$. Theorem 11.19 and Corollary 11.20 yield

$$n^{-3/2} S^2_n(f) \xrightarrow{\mathrm{d}} \mathrm{N}\Big(0, \tfrac{1}{3} \int (f^0_{\{1\}})^2 \, d\mu\Big),$$

and if $f^0_{\{1\}} = 0$ a.e., then $n^{-1} S^2_n(f) \xrightarrow{\mathrm{d}} \int f \, dZ^2$, where Z is a Gaussian stochastic measure on $(\mathcal{S}, \mu)$.

In the latter, degenerate, case, the limit distribution may also be expressed as a (generally infinite) linear combination of independent variables with the distribution of Lévy's stochastic area (Janson and Wichura 1983).

EXAMPLE 11.24. A *multi-sample U-statistic* is a sum of the type

$$\sum_{1 \le i_{j1} < \dots < i_{jk_j} \le n_j} f(X_{1i_{11}}, \dots, X_{1i_{1k}}, \dots, X_{mi_{m1}}, \dots, X_{mi_{mk}}),$$

where $n_1, \dots, n_m$ are given numbers and X_{ji}, $j = 1, \dots, m$, $1 \le i < \infty$ are independent random variables with a distribution depending on j only. If we let $k = k_1 + \dots + k_m$, $n = n_1 + \dots + n_m$, $\alpha_j = (n_1 + \dots + n_j)/n$, $X_i = (X_{ji})_{j=1}^m$ and $g(t_{1i_{11}}, \dots, t_{1i_{1k}}, \dots, t_{mi_{m1}}, \dots, t_{mi_{mk}}) = \prod_j \mathbf{1}[\alpha_{j-1} < t_{j1} < \dots < t_{jk_j} \le \alpha_j]$, then the sum may, after relabelling, be written $S^k_n(f; g)$. Hence the results above apply to multi-sample U-statistics, at least in the case when the ratios n_k/n_1 are fixed; we leave the details to the reader.

REMARK 11.25. As in Remark 11.11, it is possible to obtain functional limit theorems extending the results above. (For a special case, see Janson and Wichura (1983).)

3. Random graphs

The random variables studied in the preceding two sections were based on an i.i.d. sequence (X_i). In the theory of random graphs, it is common to study similar sums that instead are based on a family $(X_{ij})_{i<j}$ of i.i.d. doubly indexed variables.

EXAMPLE 11.26. Let X_{ij}, $1 \leq i < j \leq n$, be independent Bernoulli variables with a common distribution $\mathrm{Be}(p)$, i.e. $\mathrm{P}(X_{ij} = 1) = p$ and $\mathrm{P}(X_{ij} = 0) = 1 - p$, and interpret X_{ij} as the indicator of the event that there is an edge between i and j; this yields a random graph on $\{1, \ldots, n\}$ denoted by $G(n,p)$. The (random) number of triangles in $G(n,p)$ equals

$$\sum_{1 \leq i < j < k \leq n} X_{ij} X_{ik} X_{jk}.$$

We return to this example later.

For random hypergraphs, similar sums based on variables $X_{i_1 i_2 \cdots i_l}$ with $l \geq 3$ appear.

In order to be general, we allow all these possibilities at once and assume in this section the following (k is some given number):

For each l, $1 \leq l \leq k$, $\{X^{(l)}_{i_1 \cdots i_l}\}_{1 \leq i_1 < \cdots < i_l < \infty}$ is a family of i.i.d. random variables with values in some space $\mathcal{S}_l$; furthermore, the different families are independent.

We denote the common distribution of $\{X^{(l)}_{i_1 \cdots i_l}\}$ by μ_l and let $(\mathcal{S}, \mu)$ be the product space $\prod_{l=1}^{k} (\mathcal{S}_l, \mu_l)^{\binom{k}{l}}$. If f is any measurable function on $\mathcal{S}$, we let

$$Y(f; i_1, \ldots, i_k) = f(X^{(1)}_{i_1}, \ldots, X^{(1)}_{i_k}, X^{(2)}_{i_1 i_2}, X^{(2)}_{i_1 i_3}, \ldots, X^{(3)}_{i_1 i_2 i_3}, \ldots X^{(k)}_{i_1 \ldots i_k}) \tag{11.21}$$

and

$$S^k_n(f) = \sum_{1 \leq i_1 < \cdots < i_k \leq n} Y(f; i_1, \ldots, i_k).$$

We also introduce, generalizing (11.9) and (11.12), the weighted sums

$$S^k_n(f; g) = \sum_{i_1, \ldots, i_k}^{*} Y(f; i_1, \ldots, i_k) g(\tfrac{i_1}{n}, \ldots, \tfrac{i_k}{n}) \tag{11.22}$$

and $\overline{S}^k_n(f; g) = S^k_n(f; \overline{g}_n)$, where $\overline{g}_n$ is defined by (11.11).

REMARK 11.27. It is not really necessary to have families $\{X^{(l)}_{i_1 \cdots i_l}\}$ for all $l \leq k$; the results below extend immediately to the case when we have such families for only one or several values of l, for example by formally adding trivial families (with all variables equal to 0, say) for the missing values. In this way, the results of the preceding section become special cases of the results below. Further examples are given (without comment) later.

Before stating the results, we analyse the space $L^2(\mathcal{S}, \mu)$ and introduce more notations. We will again use the theory of tensor products in Appendix E.

We write the elements of $\mathcal{S}$ as $(x_1, \ldots, x_k, x_{12}, \ldots, x_{12 \ldots k})$ or as $(x_B)_B$, where B ranges over the non-empty subsets of $\{1, \ldots, k\}$.

For $A \subseteq \{1,\dots,k\}$, let M_A be the subspace of $L^2(\mathcal{S},\mu)$ consisting of all functions that depend only on the variables x_B with $B \subseteq A$.

If M and N are two subsets of $L^2(\mathcal{S})$, let MN be the set $\{gh : g \in M,\, h \in N\}$ of products, and define similarly the product of several subsets. Note that if $M_i \subseteq M_{A_i}$, $i = 1,\dots,p$, where the sets A_i are disjoint, then $M_1 \cdots M_p \subset L^2(\mathcal{S})$.

As in Section 2 we let

$$M_A^0 = \{f \in M_A : f \perp M_B \text{ if } B \subsetneq A\}.$$

In the present setting, we further define

$$\widetilde{M}_A^0 = \{f \in M_A^0 : f \perp M_B M_{A\setminus B} \text{ if } \varnothing \neq B \subsetneq A\}.$$

Finally, if $A_1,\dots,A_p$ are disjoint non-empty subsets of $\{1,\dots,k\}$, let $\widetilde{M}^0_{A_1,\dots,A_p}$ be the closed linear span of the product $\widetilde{M}^0_{A_1} \cdots \widetilde{M}^0_{A_p}$.

In order to study these subspaces further, we introduce bases. Thus, for each l, let $\{\varphi_i^l\}_{i\in\mathcal{I}_l}$ be an orthonormal basis in $L^2(\mathcal{S}_l,\mu_l)$ such that $\varphi_0^l = 1$. (For convenience, we use the same symbol 0 for every l for this special basis element, although the index sets $\mathcal{I}_l$ may be quite different.)

An orthonormal basis of $L^2(\mathcal{S},\mu)$ then is given by the set of all functions

$$\varphi_{\boldsymbol{i}}(\boldsymbol{x}) = \prod_{\varnothing\neq B\subseteq\{1,\dots,k\}} \varphi_{i_B}^{|B|}(x_B), \qquad \boldsymbol{x} = (x_B)_B \in \mathcal{S}, \tag{11.23}$$

where

$$\boldsymbol{i} = \{i_B\}_{\varnothing\neq B\subseteq\{1,\dots,k\}} = (i_1,\dots,i_{1\dots k}) \in \mathcal{I} = \mathcal{I}_1^k \times \mathcal{I}_2^{\binom{k}{2}} \times \cdots \times \mathcal{I}_k.$$

Furthermore, suitable subsets of the basis $\{\varphi_{\boldsymbol{i}}\}$ provide bases in the subspaces of $L^2(\mathcal{S},\mu)$ defined above. Define $\operatorname{supp}(\boldsymbol{i})$, the support graph of an index $\boldsymbol{i} = (i_B) \in \mathcal{I}$, to be the graph with the vertex set $\{j : i_B \neq 0 \text{ for some } B \text{ with } j \in B\}$ and the edge set $\{jj' : i_B \neq 0 \text{ for some } B \text{ with } j,j' \in B\}$. In other words, $\operatorname{supp}(\boldsymbol{i})$ is the union of the complete graphs on B for all $B \subseteq \{1,\dots,k\}$ such that $i_B \neq 0$. It is then easily seen that:

(i) An orthonormal basis in M_A is given by

$$\{\varphi_{\boldsymbol{i}} : \operatorname{supp}(\boldsymbol{i}) \text{ has vertex set } \subseteq A\}.$$

(ii) An orthonormal basis in M_A^0 is given by

$$\{\varphi_{\boldsymbol{i}} : \operatorname{supp}(\boldsymbol{i}) \text{ has vertex set } = A\}.$$

(iii) An orthonormal basis in $\widetilde{M}_A^0$ is given by

$$\{\varphi_{\boldsymbol{i}} : \operatorname{supp}(\boldsymbol{i}) \text{ is a connected graph with vertex set } = A\}.$$

(iv) An orthonormal basis in $\widetilde{M}^0_{A_1,\dots,A_p}$ is given by

$$\{\varphi_{\boldsymbol{i}} : \operatorname{supp}(\boldsymbol{i}) \text{ is a graph with connected components } A_1,\dots,A_p\}.$$

This yields the two orthogonal decompositions

$$L^2(\mathcal{S},\mu) = \bigoplus_{A\subseteq\{1,\dots,k\}} M^0_A$$

and

$$L^2(\mathcal{S},\mu) = \bigoplus \widetilde{M}^0_{A_1,\dots,A_p}, \tag{11.24}$$

with summation over all collections $\{A_1,\dots,A_p\}$, $p\geq 0$, of disjoint non-empty subsets of $\{1,\dots,k\}$. (The case $p=0$ gives the space $\widetilde{M}^0_\varnothing$ of constant functions.) Similarly, for every $A\subseteq\{1,\dots,k\}$,

$$M^0_A = \bigoplus \widetilde{M}^0_{A_1,\dots,A_p},$$

with summation over all such collections with $\bigcup A_j = A$.

Moreover, it follows easily from Example E.9 or E.11 that if $A_1,\dots,A_p$ are disjoint, then there is a natural identification

$$\widetilde{M}^0_{A_1,\dots,A_p} = \widetilde{M}^0_{A_1}\otimes\dots\otimes\widetilde{M}^0_{A_p}.$$

As in the preceding sections, we express the limit distributions in our main theorem using stochastic integrals, but now the definition is more complex and we proceed step by step as follows.

Let us temporarily consider only real function spaces, and let

$$J_l\colon \widetilde{M}^0_{\{1,\dots,l\}}\otimes L^2([0,1]^l)\to H_l\subseteq H, \qquad 1\leq l\leq k,$$

be Gaussian fields on the Hilbert spaces $\widetilde{M}^0_{\{1,\dots,l\}}\otimes L^2([0,1]^l)$ such that their ranges H_l are orthogonal (and thus independent) subspaces of a Gaussian Hilbert space H. This is equivalent to taking a Gaussian field on the (Hilbert) direct sum $\bigoplus_1^k \widetilde{M}^0_{\{1,\dots,l\}}\otimes L^2([0,1]^l)$; hence such fields J_l exist by Theorem 1.23.

If we want to use complex functions, we extend each J_l to a complex linear map into the complexification $H_{l\mathbb{C}}$; for simplicity we nevertheless continue to write H_l.

A permutation $\pi\in\mathfrak{S}_k$ acts on $\mathcal{S}$ and $[0,1]^k$ by permuting the coordinates: $(x_B)\mapsto(x_{\pi^{-1}(B)})$ and $(t_i)\mapsto(t_{\pi^{-1}(i)})$; this induces corresponding mappings of functions defined on these spaces, which we in both cases denote by $f\mapsto f\circ\pi$.

Observe that for any functions f on $\mathcal{S}$ and g on $[0,1]^k$, and $\pi\in\mathfrak{S}_k$,

$$S^k_n(f;g) = S^k_n(f\circ\pi; g\circ\pi), \tag{11.25}$$

and similarly for $\overline{S}^k_n$; of course, it is essential that the same permutation is used for both f and g.

Let, for $A\subseteq\{1,\dots,k\}$, $L^2([0,1]^A)$ denote the subspace of $L^2([0,1]^k)$ consisting of the functions that depend on $\{t_i : i\in A\}$ only. Moreover, let π_A be any permutation in $\mathfrak{S}_k$ that maps $\{1,\dots,|A|\}$ onto A and is increasing on $\{1,\dots,|A|\}$. Then $f\mapsto f\circ\pi_A$ maps $\widetilde{M}^0_A$ onto $\widetilde{M}^0_{\{1,\dots,|A|\}}$ and $L^2([0,1]^A)$

onto $L^2([0,1]^{|A|})$, which we regard as a subspace of $L^2([0,1]^k)$, and we define $J_A\colon \widetilde{M}^0_A \otimes L^2([0,1]^A) \to H_{|A|} \subseteq H$ by

$$J_A(f \otimes g) = J_{|A|}\big((f \circ \pi_A) \otimes (g \circ \pi_A)\big), \qquad f \in \widetilde{M}^0_A,\ g \in L^2([0,1]^A). \tag{11.26}$$

(Here, as several times below, we define a map on a tensor product by giving its values on elements of the form $f \otimes g$ only. We leave it to the reader to verify that in each case, our definition is indeed bilinear in f and g, which implies that it yields a well-defined linear map on the algebraic tensor product, and that this further extends to a bounded linear map on the Hilbert space tensor product, cf. Appendix E.)

If $A_1, \dots, A_p$ are disjoint, and $A = \bigcup A_j$, then $L^2([0,1]^A)$ can be identified with the tensor product $\bigotimes_{j=1}^p L^2([0,1]^{A_j})$, see Example E.10, and we define $J_{A_1,\dots,A_p}\colon \widetilde{M}^0_{A_1,\dots,A_p} \otimes L^2([0,1]^A) = \widetilde{M}^0_{A_1} \otimes \cdots \otimes \widetilde{M}^0_{A_p} \otimes L^2([0,1]^{A_1}) \otimes \cdots \otimes L^2([0,1]^{A_p}) \to H^{\odot p} = H^{:p:}$ by taking the tensor product of the operators $J_{A_1}, \dots, J_{A_p}$, which maps into $\bigotimes_1^p H_{|A_i|} \subseteq H^{\otimes p}$, followed by the natural map $H^{\otimes p} \to H^{\odot p} = H^{:p:}$ given by $h_1 \otimes \cdots \otimes h_p \mapsto h_1 \odot \cdots \odot h_p = {:}h_1 \cdots h_p{:}$. Thus

$$J_{A_1,\dots,A_p}(f_1 \otimes \cdots \otimes f_p \otimes g_1 \otimes \cdots \otimes g_p) = {:}J_{A_1}(f_1 \otimes g_1) \cdots J_{A_p}(f_p \otimes g_p){:}, \qquad f_j \in \widetilde{M}^0_{A_j},\ g_j \in L^2([0,1]^{A_j}). \tag{11.27}$$

Finally, we use the orthogonal decomposition (11.24) and write

$$f = \sum_{A_1,\dots,A_p} \widetilde{f}^0_{A_1,\dots,A_p}, \qquad \text{with } \widetilde{f}^0_{A_1,\dots,A_p} \in \widetilde{M}^0_{A_1,\dots,A_p}. \tag{11.28}$$

We also let $P_A\colon L^2([0,1]^k) \to L^2([0,1]^A)$ denote the orthogonal projection given by averaging over the coordinates $\{t_i : i \notin A\}$, and define $J\colon L^2(\mathcal{S}) \otimes L^2([0,1]^k) \to \Gamma(H)$ by

$$J(f \otimes g) = \sum_{A_1,\dots,A_p} J_{A_1,\dots,A_p}\big(\widetilde{f}^0_{A_1,\dots,A_p} \otimes P_{\bigcup A_j}(g)\big), \quad f \in L^2(\mathcal{S}),\ g \in L^2([0,1]^k), \tag{11.29}$$

with summation over all families $\{A_1, \dots, A_p\}$ of disjoint non-empty subsets of $\{1, \dots, k\}$.

Using these notations, we can now state our main theorem. We identify functions on D_k (defined in (11.10)) with functions on $[0,1]^k$ that vanish off D_k.

THEOREM 11.28. *Suppose that $f \in L^2(\mathcal{S})$ has the decomposition $f = \sum_A f^0_A$ with $f^0_A \in M^0_A$, $A \subseteq \{1, \dots, k\}$, and let $f^0_j = \sum_{|A|=j} f^0_A$, $0 \le j \le k$. If $m \ge 0$ is an integer such that $f^0_j = 0$ for every j with $j < m$, and g is bounded and a.s. continuous on D_k, then*

$$n^{m/2-k} S^k_n(f, g) \xrightarrow{\mathrm{d}} J(f^0_m \otimes g) \qquad \text{as } n \to \infty,$$

with joint convergence for any finite number of such f, g and corresponding m. The same holds for $\overline{S}^k_n$, with arbitrary $g \in L^2(D_k)$.

In particular, for such m,

$$n^{m/2-k} S^k_n(f) \xrightarrow{\mathrm{d}} J(f^0_m \otimes \mathbf{1}_{D_k}) \qquad \text{as } n \to \infty.$$

We postpone the proof, and give first some consequences and examples.

REMARK 11.29. The case $m = 0$ yields a weak law of large numbers as in Corollaries 11.5 and 11.20. The main interest, however, is the case $m > 0$. Note that we may always assume $f^0_\varnothing = 0$, and thus use $m \geq 1$, by applying the theorem to $f - \int_S f \, d\mu$.

REMARK 11.30. We may assume that the Gaussian fields J_l are defined on the (larger) spaces $M_{\{1,\dots,l\}} \otimes L^2([0,1]^l) \cong L^2(\mathcal{S}^{(l)} \times [0,1]^l)$, with $\mathcal{S}^{(l)} = \prod_{j=1}^l \mathcal{S}^{(l)}_j$. Thus J_l and their direct sum may be regarded as Gaussian stochastic integrals on suitable spaces, and the operator J defined above can be expressed as a sum of multiple stochastic integrals.

Note that if $f \in \widetilde{M}^0_{A_1,\dots,A_p}$, then $J(f \otimes g) \in H^{:p:}$. In particular, terms in (11.29) with different p are orthogonal. In general, all terms in (11.29) are not orthogonal, however, and there may be cancellations. The following lemma, which will be proved later, shows that complete cancellation is not possible when $g = \mathbf{1}_{D_k}$.

LEMMA 11.31. *There exist constants c_k, $C_k > 0$ such that for every $f \in L^2(\mathcal{S})$,*

$$c_k \|f\|_2 \leq \|J(f \otimes \mathbf{1}_{D_k})\|_2 \leq C_k \|f\|_2.$$

Theorem 11.28 and Lemma 11.31 yield the following simple qualitative limit theorem. Observe that $\mathrm{E}\, S^k_n(f) = \binom{n}{k} \int_S f \, d\mu$.

COROLLARY 11.32. *Let $f \in L^2_{\mathbb{R}}(\mathcal{S})$ have the decomposition* (11.28) *and suppose that $m \geq 1$ is such that $\widetilde{f}^0_{A_1,\dots,A_p} = 0$ whenever $0 < |A_1| + \dots + |A_p| < m$. Then $n^{m/2-k}(S^k_n(f) - \mathrm{E}\, S^k_n(f))$ converges in distribution to a limit. Furthermore:*

(i) *The limit is non-degenerate if and only if $\widetilde{f}^0_{A_1,\dots,A_p} \neq 0$ for some family $A_1, \dots, A_p$ with $|A_1| + \dots + |A_p| = m$.*

(ii) *The limit distribution is normal if and only if $\widetilde{f}^0_{A_1,\dots,A_p} = 0$ for every family $A_1, \dots, A_p$ with $p \geq 2$ and $|A_1| + \dots + |A_p| = m$.*

PROOF. We may assume that $\widetilde{f}^0_\varnothing = \int_S f \, d\mu = 0$. Then Theorem 11.28 yields convergence in distribution to $J(f^0_m \otimes \mathbf{1}_{D_k}) = \sum_p J(f^0_{m,p} \otimes \mathbf{1}_{D_k})$, where $f^0_{m,p} = \sum_{|A_1|+\dots+|A_p|=m} \widetilde{f}^0_{A_1,\dots,A_p}$.

By Lemma 11.31, $J(f^0_m \otimes \mathbf{1}_{D_k}) = 0$ if and only if $f^0_m = 0$, which yields (i). Furthermore, $J(f^0_{m,p} \otimes \mathbf{1}_{D_k}) \in H^{:p:}$ and thus by e.g. Theorem 6.12, $J(f^0_m \otimes \mathbf{1}_{D_k})$

is normal if and only if $J(f^0_{m,p} \otimes \mathbf{1}_{D_k}) = 0$ for every $p \geq 2$, which yields (ii) by Lemma 11.31. $\square$

Using a basis $\{\varphi_{\boldsymbol{i}}\}$ of the type (11.23), the corollary can be reformulated as follows. Recall the definition above of the support graph $\operatorname{supp}(\boldsymbol{i})$; we write $|\operatorname{supp}(\boldsymbol{i})|$ for the number of vertices in this graph.

COROLLARY 11.33. *Let $f \in L^2_{\mathbb{R}}(\mathcal{S})$ have the expansion $f = \sum_{\boldsymbol{i}} a_{\boldsymbol{i}} \varphi_{\boldsymbol{i}}$ in an orthonormal basis $\{\varphi_{\boldsymbol{i}}\}$ given by (11.23), and suppose that $m \geq 1$ is such that $a_{\boldsymbol{i}} = 0$ whenever $0 < |\operatorname{supp}(\boldsymbol{i})| < m$. Then $n^{m/2-k}(S^k_n(f) - \operatorname{E} S^k_n(f))$ converges in distribution to a limit. Furthermore:*

(i) *The limit is non-degenerate if and only if $a_{\boldsymbol{i}} \neq 0$ for some $\boldsymbol{i}$ with $|\operatorname{supp}(\boldsymbol{i})| = m$.*
(ii) *The limit distribution is normal if and only if $a_{\boldsymbol{i}} = 0$ for every $\boldsymbol{i}$ with $\operatorname{supp}(\boldsymbol{i})$ disconnected and $|\operatorname{supp}(\boldsymbol{i})| = m$.*

$\square$

REMARK 11.34. Corollary 11.33 shows that the asymptotic behaviour is governed by the terms $a_{\boldsymbol{i}}\varphi_{\boldsymbol{i}}$ such that $|\operatorname{supp}(\boldsymbol{i})|$ is as small as possible with $a_{\boldsymbol{i}}$ non-zero; in particular, the limit is normal if and only if every such $\boldsymbol{i}$ has connected support graph.

For U-statistics, as well as the asymmetric statistics studied in Section 2, every support graph is discrete (i.e., has no edges), so $\operatorname{supp}(\boldsymbol{i})$ is disconnected as soon as $|\operatorname{supp}(\boldsymbol{i})| \geq 2$; this explains the fact observed above that we obtain non-normal limits when we apply Theorem 11.3 or 11.19 with $m \geq 2$. In the present, more general, situation, the limit may also be normal in degenerate cases with higher values of m, due to the existence of larger connected support graphs.

If f is symmetric and there is no weight function g, Theorem 11.28 may be simplified by using the following alternative to the operator J.

Let

$$\widetilde{J}^1_l : \widetilde{M}^0_{\{1,\dots,l\}} \to H_l \subseteq H, \qquad 1 \leq l \leq k,$$

be Gaussian fields on the Hilbert spaces $\widetilde{M}^0_{\{1,\dots,l\}}$ such that their ranges H_l are orthogonal subspaces of a Gaussian Hilbert space H. (Again, we begin with the real case and complexify if we want to study complex functions.)

It turns out that a symmetry factor enters; thus define

$$J^1_l(f) = \sqrt{l!}\,\widetilde{J}^1_l(f), \qquad f \in \widetilde{M}^0_{\{1,\dots,l\}}. \tag{11.30}$$

We now proceed as in (11.26)–(11.29) and define $J^1_A : \widetilde{M}^0_A \to H_{|A|} \subseteq H$ by

$$J^1_A(f) = J^1_{|A|}(f \circ \pi_A), \tag{11.31}$$

and $J^1_{A_1,\dots,A_p}\colon \widetilde{M}^0_{A_1,\dots,A_p} \to H^{:p:}$ by

$$J^1_{A_1,\dots,A_p}(f_1 \otimes \dots \otimes f_p) = :J^1_{A_1}(f_1)\cdots J^1_{A_p}(f_p):, \qquad f_j \in \widetilde{M}^0_{A_j}, \tag{11.32}$$

and finally $J^1\colon L^2(\mathcal{S}) \to \Gamma(H)$ by, assuming the decomposition (11.28),

$$J^1(f) = \sum_{A_1,\dots,A_p} J^1_{A_1,\dots,A_p}(\widetilde{f}^0_{A_1,\dots,A_p}), \qquad f \in L^2(\mathcal{S}). \tag{11.33}$$

THEOREM 11.35. *If $f \in L^2(\mathcal{S})$ is symmetric, then with further assumptions and notations as in Theorem* 11.28,

$$n^{m/2-k} S^k_n(f) \xrightarrow{\mathrm{d}} \tfrac{1}{k!} J^1(f^0_m) \qquad \textit{as } n \to \infty,$$

with joint convergence for any finite number of such f and m.

We postpone the proof of this theorem too. In the normal case, it leads to a simple explicit formula.

COROLLARY 11.36. *Let $f \in L^2_{\mathbb{R}}(\mathcal{S})$ be symmetric with the decomposition* (11.28), *and suppose that $m \geq 1$ is such that $\widetilde{f}^0_{A_1,\dots,A_p} = 0$ when $0 < |A_1| + \dots + |A_p| < m$ or $p \geq 2$ and $|A_1| + \dots + |A_p| = m$. Then*

$$n^{m/2-k}\big(S^k_n(f) - \mathrm{E}\, S^k_n(f)\big) \xrightarrow{\mathrm{d}} \mathrm{N}(0, \sigma^2),$$

with

$$\sigma^2 = \frac{1}{m!\,(k-m)!^2} \|\widetilde{f}^0_A\|^2_2$$

for any A with $|A| = m$.

PROOF. We can apply Theorem 11.35, where furthermore, by our assumption, $f^0_m = \sum_{|A|=m} \widetilde{f}^0_A$. Hence (11.30)–(11.32) yield, by the symmetry,

$$J^1(f^0_m) = \sum_{|A|=m} J^1_A(\widetilde{f}^0_A) = \tbinom{k}{m} J^1_m(\widetilde{f}^0_{\{1,\dots,m\}}) = \tbinom{k}{m}\sqrt{m!}\,\widetilde{J}^1_m(\widetilde{f}^0_{\{1,\dots,m\}})$$

and thus

$$\tfrac{1}{k!} J^1(f^0_m) = m!^{-1/2}(k-m)!^{-1} \widetilde{J}^1_m(\widetilde{f}^0_{\{1,\dots,m\}}) \sim \mathrm{N}(0, \sigma^2),$$

with $\sigma^2 = \|m!^{-1/2}(k-m)!^{-1}\widetilde{f}^0_{\{1,\dots,m\}}\|^2_2$ because $\widetilde{J}^1_m$ is an isometry into a Gaussian Hilbert space. □

We give some applications of the results above to random graphs. For details not given here and further results, see Barbour, Karoński and Ruciński (1989), Janson and Nowicki (1991) and Janson (1990, 1994, 1995).

EXAMPLE 11.37. Consider the random graph $G(n,p)$ with i.i.d. edge indicators $X_{ij} \sim \mathrm{Be}(p)$, with $0 < p < 1$, see Example 11.26. Let T_n be the number of triangles in the graph; then $T_n = S^3_n(f)$, with $f(x_{12}, x_{13}, x_{23}) = x_{12}x_{13}x_{23}$.

We may identify $L^2(\mathcal{S})$ with the space of random variables of the form $h(X_{12}, X_{13}, X_{23})$ for some function h. Since each X_{ij} assumes only two values, this space is eight-dimensional. Furthermore, if we define $X'_{ij} = X_{ij} - \operatorname{E} X_{ij} = X_{ij} - p$, then X'_{12}, X'_{13}, X'_{23} are independent random variables with expectations zero, and it is easily seen that the eight variables 1, X'_{12}, X'_{13}, X'_{23}, $X'_{12}X'_{13}$, $X'_{12}X'_{23}$, $X'_{13}X'_{23}$, $X'_{12}X'_{13}X'_{23}$ constitute an orthogonal (unnormalized) basis in $L^2(\mathcal{S})$. In fact, apart from normalization this equals the basis φ_i constructed (more generally) in (11.23) using $\varphi_0^2(x) = 1$ and $\varphi_1^2(x) = (p(1-p))^{-1/2}(x-p)$, and it follows that $1 \in \widetilde{M}^0_{\varnothing}$, $X'_{12} \in \widetilde{M}^0_{\{1,2\}}$, $X'_{13} \in \widetilde{M}^0_{\{1,3\}}$, $X'_{23} \in \widetilde{M}^0_{\{2,3\}}$, while the remaining four basis elements belong to $\widetilde{M}^0_{\{1,2,3\}}$.

Our special f is easily decomposed by the substitution $X_{ij} = p + X'_{ij}$, which yields

$$f(X_{12}, X_{13}, X_{23}) = (p + X'_{12})(p + X'_{13})(p + X'_{23})$$
$$= p^3 + p^2X'_{12} + p^2X'_{13} + p^2X'_{23} + pX'_{12}X'_{13} + pX'_{12}X'_{23} + pX'_{13}X'_{23} + X'_{12}X'_{13}X'_{23},$$

and thus the decomposition (11.28) with

$$\widetilde{f}^0_{\varnothing} = p^3,$$
$$\widetilde{f}^0_{\{1,2\}} = p^2X'_{12}, \qquad \widetilde{f}^0_{\{1,3\}} = p^2X'_{13}, \qquad \widetilde{f}^0_{\{2,3\}} = p^2X'_{23},$$
$$\widetilde{f}^0_{\{1,2,3\}} = pX'_{12}X'_{13} + pX'_{12}X'_{23} + pX'_{13}X'_{23} + X'_{12}X'_{13}X'_{23}.$$

Consequently Corollary 11.36 applies with $m = 2$, and yields (for fixed p)

$$n^{-2}\left(T_n - \tbinom{n}{3}p^3\right) \xrightarrow{\mathrm{d}} \mathrm{N}(0, \sigma^2) \qquad \text{as } n \to \infty,$$

where $\sigma^2 = \frac{1}{2}\|\widetilde{f}^0_{\{1,2\}}\|_2^2 = \frac{1}{2}p^5(1-p)$. The asymptotic variance σ^2 can also be computed by $\sigma^2 = \lim_{n\to\infty} n^{-4} \operatorname{Var} T_n$, cf. Remark 11.10.

EXAMPLE 11.38. More generally, let K be a given graph on k vertices, and let N_K be the number of copies of K in $G(n,p)$. In order to apply the results above, we define for a graph K on $\{1, \dots, k\}$,

$$f_K(\boldsymbol{x}) = \sum_{F \cong K} \prod_{ij \in F} x_{ij}, \qquad \boldsymbol{x} \in \mathcal{S},$$

with summation over all graphs F on $\{1, \dots, k\}$ that are isomorphic to K; then $N_K = S_n^k(f_K)$.

If $\{\varphi_i\}$ is the orthonormal basis given by (11.23), then the support graphs of the elements of the basis have no isolated vertices. In particular, there is no φ_i such that $|\operatorname{supp}(\varphi_i)| = 1$; moreover, every support graph with two or three vertices is connected. Hence we may apply Corollary 11.33 with $m = 2$ and obtain

$$n^{1-k}(N_K - \operatorname{E} N_K) \xrightarrow{\mathrm{d}} \mathrm{N}(0, \sigma^2),$$

for some $\sigma^2 \geq 0$. (This can also be seen from Corollary 11.32 or 11.36, observing that $\widetilde{M}_A^0 = \{0\}$ whenever $|A| = 1$.)

In this case, it is easily seen that $\sigma^2 > 0$, which by Corollary 11.36 is equivalent to $P_{\widetilde{M}^0_{\{1,2\}}}(f_K) \neq 0$, where we let P_M denote the orthogonal projection onto a subspace M of $L^2(\mathcal{S}^k)$. In fact, an expansion as in Example 11.37 shows that if $\operatorname{aut}(K)$ is the number of automorhisms of K, $\alpha = k!/\operatorname{aut}(K)$ is the number of copies of K on $\{1,\dots,k\}$ and e is the number of edges in K, then $P_{\widetilde{M}^0_{\{1,2\}}}(f_K)(X_{12}) = \alpha\frac{2e}{k(k-1)}p^{e-1}(X_{12}-p)$; hence Corollary 11.36 yields $\sigma^2 = \frac{2e^2}{\operatorname{aut}(K)^2}p^{2e-1}(1-p)$.

EXAMPLE 11.39. Continuing the preceding example, let us now study the number of *induced* copies of K in $G(n,p)$. This number, N_K^* say, equals $S_n^k(f_K^*)$ with

$$f_K^*(\boldsymbol{x}) = \sum_{F \cong K} \prod_{ij \in F} x_{ij} \prod_{ij \notin F} (1 - x_{ij}), \qquad \boldsymbol{x} \in \mathcal{S},$$

again with summation over all graphs F on $\{1,\dots,k\}$ that are isomorphic to K. Hence Corollary 11.32 or 11.33 with $m = 2$ yields

$$n^{1-k}(N_K^* - \operatorname{E} N_K^*) \xrightarrow{\mathrm{d}} \mathrm{N}(0,\sigma^2),$$

for some $\sigma^2 \geq 0$.

In this case, however, it is possible that $P_{\widetilde{M}^0_{\{1,2\}}}(f_K^*) = 0$, and thus $\sigma^2 = 0$; an expansion as in Example 11.37 and a simple calculation show that $P_{\widetilde{M}^0_{\{1,2\}}}(f_K^*) = 0$ if and only if $e = p\binom{k}{2}$; thus this happens for exactly one value of $p \in (0,1)$ for every K that is neither discrete nor complete.

In the exceptional case when $\sigma^2 = 0$, we can apply Corollary 11.33 with $m = 3$. Since, as remarked above, all support graphs with three vertices are connected, this yields (in the exceptional case)

$$n^{3/2-k}(N_K^* - \operatorname{E} N_K^*) \xrightarrow{\mathrm{d}} \mathrm{N}(0,\sigma_1^2),$$

for some $\sigma_1^2 \geq 0$.

It is possible to calculate σ_1^2, and it turns out that for most graphs K, $\sigma_1^2 > 0$. There are, however, some graphs K such that also $\sigma_1^2 = 0$. (This happens, by Corollary 11.36, when $P_{\widetilde{M}^0_{\{1,2,3\}}}(f_K^*) = 0$.) We may then apply Corollary 11.32 or 11.33 with $m = 4$, and obtain

$$n^{2-k}(N_K^* - \operatorname{E} N_K^*) \xrightarrow{\mathrm{d}} W,$$

for some limit variable W. This time, however, there are disconnected support graphs with m vertices, viz. support graphs consisting of two disjoint edges, and it turns out that they (in this case) always have non-zero coefficients in the expansion of f_K^*; equivalently, $P_{\widetilde{M}^0_{\{1,2\},\{3,4\}}}(f_K^*) \neq 0$. Consequently the limit W is *not* normal. Theorem 11.35 shows that W can be expressed as an element of $H + H^{:2:}$, i.e. a chaos of order 2, and a calculation shows that in fact W can be written as $a\xi - b{:}\eta^2{:}$, where ξ and η are independent standard normal variables, $a \geq 0$ and $b > 0$ (Janson 1990).

The smallest graphs K for which such non-normal limits appear have eight vertices; one example is the wheel consisting of a cycle with seven vertices and a central vertex connected to all the others.

EXAMPLE 11.40. Let X_{ij} be independent $\mathrm{Be}(p)$ variables as in the preceding examples, but this time construct a directed graph such that, for $1 \leq i < j \leq n$, there is a directed edge from i to j if $X_{ij} = 1$ and a directed edge in the opposite direction if $X_{ij} = 0$. This yields a random tournament. (A tournament is a directed graph such that every pair of vertices is connected by exactly one edge, in some direction.)

The number T'_n of cyclically oriented triangles in this random tournament equals $S^3_n(f)$, with

$$f(X_{12}, X_{13}, X_{23}) = X_{12}X_{23}(1 - X_{13}) + (1 - X_{12})(1 - X_{23})X_{13}.$$

Substituting $X_{ij} = p + X'_{ij}$, we obtain after some cancellation

$$f(X_{12}, X_{13}, X_{23}) = p(1-p) + (1-2p)X'_{13} + X'_{12}X'_{23} - X'_{12}X'_{13} - X'_{23}X'_{13};$$

hence $f^0_1 = 0$ and $f^0_2 = (1-2p)X'_{13}$. If $p \neq 1/2$, we use Corollary 11.32 with $m = 2$ and obtain

$$n^{-2}\Big(T'_n - \tbinom{n}{3}p(1-p)\Big) \xrightarrow{\mathrm{d}} \mathrm{N}(0, \sigma^2); \qquad (11.34)$$

more precisely, Theorem 11.28 yields convergence to $J(f^0_2 \otimes \mathbf{1}_{D_k}) = J_{\{1,3\}}((1-2p)X'_{13} \otimes g_{\{1,3\}})$, with $g_{\{1,3\}}(t_1, t_2) = (t_2 - t_1)\mathbf{1}[0 < t_1 < t_2 < 1]$, which is (11.34) with

$$\begin{aligned}\sigma^2 &= \|(1-2p)X'_{13} \otimes g_{\{1,3\}}\|^2_2 = (1-2p)^2\|X'_{13}\|^2_2\|g_{\{1,3\}}\|^2_2 \\ &= \tfrac{1}{12}p(1-p)(1-2p)^2.\end{aligned}$$

If $p = 1/2$, this is still valid, but $\sigma^2 = 0$; if we instead apply Corollary 11.32 with $m = 3$, we obtain

$$n^{-3/2}\big(T'_n - \tfrac{1}{4}\tbinom{n}{3}\big) \xrightarrow{\mathrm{d}} \mathrm{N}(0, \sigma^2_1),$$

where a calculation based on Theorem 11.28 or Remark 11.10 yields $\sigma^2_1 = 1/32$.

EXAMPLE 11.41. Another way to obtain the random graph $G(n, p)$ is to let X_{ij} be i.i.d. and uniformly distributed on $[0, 1]$, and then let there be an edge between i and j if $X_{ij} \leq t$. This gives us the possibility of studying several values of p at once, and if we let p increase from 0 to 1, we may regard $G(n, p)$ as a random graph process increasing from a discrete graph to a complete; X_{ij} is the time edge ij is added.

In this setting, the results above may be extended to functional limit theorems for this random graph process, see Janson (1990, 1994).

Further examples are given in Janson and Nowicki (1991), including some examples with randomly coloured graphs where we in the notation above have variables of the types $X_i^{(1)}$ and $X_{ij}^{(2)}$ together.

REMARK 11.42. The theorems given above apply to the random graphs $G(n,p)$ with fixed p only, but they may be extended to the case $p = p(n) \to 0$ as $n \to \infty$ (at a suitable rate), see Janson (1994).

REMARK 11.43. Similar examples for random uniform hypergraphs are easily obtained; let for example $X^{(s)}_{i_1 \cdots i_s} \sim \mathrm{Be}(p)$ for a fixed $s \geq 3$ and interpret these variables as indicators for the hyperedges (a hyperedge here being a subset of order s of $\{1, \dots, n\}$). If we, for example, count the number of sub-hypergraphs isomorphic to some given hypergraph, we obtain an asymptotic normal distribution as in Example 11.38; if we instead count induced copies, we expect a sequence of more and more degenerate cases as in Example 11.39, culminating in a non-normal limit, but no one seems to have worked out the details.

We return to proving the results above, and first observe that Lemmas 11.12–11.14 extend, with the same proofs as before. In the present situation they can be stated as follows.

LEMMA 11.44. *Suppose that* $f \in \widetilde{M}^0_{\{1,\dots,k\}} \subset L^2(\mathcal{S}, \mu)$.

(i) *For every* g,

$$\mathrm{E}\,|S_n^k(f;g)|^2 \leq k!\|f\|_2^2 \sum_{i_1,\dots,i_k}^{*} |g(\tfrac{i_1}{n}, \dots, \tfrac{i_k}{n})|^2 \leq n^k k! \|f\|_2^2 (\sup |g|)^2.$$

(ii) *If* $g \in L^2([0,1]^k)$, *then*

$$\mathrm{E}\,|\overline{S}_n^k(f;g)|^2 \leq n^k k! \|f\|_2^2 \|g\|_2^2.$$

(iii) *Assume that* g *is bounded and a.e. continuous on* $[0,1]^k$. *Then,*

$$n^{-k}\,\mathrm{E}\,|S_n^k(f;g) - \overline{S}_n^k(f;g)|^2 \to 0 \qquad \text{as } n \to \infty.$$

□

In order to prove Theorems 11.28 and 11.35, we use yet another version of the operator J, which works also for weight functions g that do not vanish off D_k.

This time we start with Gaussian fields

$$\widetilde{J}_l^2 \colon \widetilde{M}^0_{\{1,\dots,l\}} \otimes L^2(D_l) \to H_l \subseteq H, \qquad l \leq k,$$

on the Hilbert spaces $\widetilde{M}^0_{\{1,\dots,l\}} \otimes L^2(D_l)$, assuming as before that their ranges H_l are orthogonal subspaces of a Gaussian Hilbert space H (and complexifying if necessary).

We next symmetrize. We regard $\mathfrak{S}_l$ as a subset of $\mathfrak{S}_k$ when $l \leq k$, and define a bounded linear map $J_l^2 \colon \widetilde{M}^0_{\{1,\dots,l\}} \otimes L^2([0,1]^l) \to H_l$ by

$$J_l^2(f \otimes g) = \sum_{\pi \in \mathfrak{S}_l} \widetilde{J}_l^2\big((f \circ \pi) \otimes (g \circ \pi)\mathbf{1}_{D_l}\big), \qquad f \in \widetilde{M}^0_{\{1,\dots,l\}},\ g \in L^2([0,1]^l). \tag{11.35}$$

We then proceed as in (11.26)–(11.29) and (11.31)–(11.33). We define $J_A^2 \colon \widetilde{M}^0_A \otimes L^2([0,1]^A) \to H_{|A|} \subseteq H$ by

$$J_A^2(f \otimes g) = J_{|A|}^2\big((f \circ \pi_A) \otimes (g \circ \pi_A)\big), \qquad f \in \widetilde{M}^0_A,\ g \in L^2([0,1]^A),$$

and $J^2_{A_1,\dots,A_p} \colon \widetilde{M}^0_{A_1,\dots,A_p} \otimes L^2([0,1]^A) \to H^{:p:}$ by

$$J^2_{A_1,\dots,A_p}(f_1 \otimes \cdots \otimes f_p \otimes g_1 \otimes \cdots \otimes g_p) = {:}J^2_{A_1}(f_1 \otimes g_1) \cdots J^2_{A_p}(f_p \otimes g_p){:},$$

and finally $J^2 \colon L^2(\mathcal{S}) \otimes L^2([0,1]^k) \to \Gamma(H)$ by, assuming (11.28),

$$J^2(f \otimes g) = \sum_{A_1,\dots,A_p} J^2_{A_1,\dots,A_p}\big(f^0_{A_1,\dots,A_p} \otimes P_{\bigcup A_j}(g)\big), \quad f \in L^2(\mathcal{S}),\ g \in L^2([0,1]^k).$$

THEOREM 11.45. *If $f \in L^2(\mathcal{S})$ and g is bounded and a.s. continuous on $[0,1]^k$, then with notation as in Theorem 11.28,*

$$n^{m/2-k} S_n^k(f,g) \xrightarrow{\mathrm{d}} J^2(f_m^0 \otimes g), \qquad \text{as } n \to \infty,$$

with joint convergence for any finite number of such f, g and m. The same holds for $\overline{S}_n^k$, with arbitrary $g \in L^2([0,1]^k)$.

We prove Theorem 11.45 by the method used for Theorem 11.16; the main difference is in the first step where we can no longer just appeal to the central limit theorem. We formulate that step as a lemma.

LEMMA 11.46. *If $f \in \widetilde{M}^0_{\{1,\dots,l\}}$ and $g = \mathbf{1}_{(a_1,b_1]} \otimes \cdots \otimes \mathbf{1}_{(a_l,b_l]}$, where $1 \leq l \leq k$ and $0 \leq a_1 < b_1 \leq a_2 < b_2 \leq \cdots \leq a_l < b_l \leq 1$, then*

$$n^{-l/2} S_n^l(f,g) \xrightarrow{\mathrm{d}} \widetilde{J}_l^2(f \otimes g),$$

with joint convergence for any finite number of l, f and g of this type.

PROOF. Assume first that f is real and bounded. We use the method of moments, so we fix $m \geq 1$ and expand

$$\mathrm{E}\big(n^{-l/2} S_n^l(f;g)\big)^m = n^{-ml/2} \sum_{(i_{sj})} \mathrm{E}\, Y(f; i_{11}, \dots, i_{1l}) \cdots Y(f; i_{m1}, \dots, i_{ml}), \tag{11.36}$$

where we sum over all arrays (i_{sj}) with $a_j < i_{sj}/n \leq b_j$, $1 \leq s \leq m$, $1 \leq j \leq l$. (Y is defined as in (11.21), with k replaced by l.)

Consider first a term in this sum where some index i_{sj} appears only once; say for definiteness that i_{11} differs from all other i_{sj}. Let $Y_s = Y(f; i_{s1}, \dots, i_{sl})$ and observe that if we condition on all variables $X^{(r)}_{j_1 \dots j_r}$ with $\{j_1, \dots, j_r\} \not\subseteq \{i_{11}, \dots, i_{1l}\}$, then $\prod_2^m Y_s$ can be written as $Y(g; i_{11}, \dots, i_{1l})$ for some function g such that $g \in M_B$ for the proper subset $B = \{2, \dots, l\}$ of $\{1, \dots, l\}$. The conditional expectation of $\prod_1^m Y_s$ then equals $\int_S fg \, d\mu$, but this integral vanishes because, by assumption, $f \in \widetilde{M}^0_{\{1,\dots,l\}}$ and thus $f \perp M_B$. Consequently also the unconditional expectation $\operatorname{E} \prod_1^m Y_s = 0$. Thus, we can delete every term in the sum in (11.36) where some index i_{sj} appears only once.

In the remaining terms, each index appears at least twice, so there are at most $ml/2$ different indices. In the terms where some index appears three times or more, the number of different indices is strictly less than $ml/2$, and thus at most $(ml-1)/2$. Consequently the number of such terms is $O(n^{(ml-1)/2})$, and since each term is bounded by the constant $\|f\|_\infty^m$, their total contribution to the sum is $O(n^{(ml-1)/2})$, which can be neglected.

It remains to consider the terms where each index appears exactly twice. For each such term, construct a graph G on $\{1, \dots, m\}$ by drawing an edge between p and q if $i_{pj} = i_{qj}$ for some j. (Note that $i_{pj} = i_{qs}$ with $j \neq s$ is impossible by the condition on (i_{sj}).)

We separate two cases. First, consider a term such that some vertex in G is joined to at least two other vertices, say that G contains the edges 12 and 13. Let $B = \{s : i_{1s} = i_{2s}\}$ and observe that with $A = \{1, \dots, l\}$, $\varnothing \neq B \subsetneq A$. Further, let Y_s be as above and condition again on all variables $X^{(r)}_{j_1 \dots j_r}$ with $\{j_1, \dots, j_r\} \not\subseteq \{i_{11}, \dots, i_{1l}\}$. Then $Y_2 = Y(g; i_{11}, \dots, i_{1l})$ and $\prod_3^m Y_s = Y(h; i_{11}, \dots, i_{1l})$, for some $g \in M_B$ and $h \in M_{A \setminus B}$. By assumption, $f \perp gh$, and thus the conditional expectation of $\prod_1^m Y_s$, which equals $\int_S fgh \, d\mu$, vanishes; consequently $\operatorname{E} \prod_1^m Y_s = 0$ for each such term.

It remains to consider only the terms where each vertex in G is joined to exactly one other vertex; this can happen only if m is even and the graph G is a complete Feynman diagram of order m, cf. Section 1.5. Now suppose that G is one of these Feynman diagrams. Then $Y_1, \dots, Y_m$ coincide two by two, and if we select one Y_s from each pair, the resulting $m/2$ random variables are independent and identically distributed, with $\operatorname{E} |Y_s|^2 = \int f^2$. Consequently, $\operatorname{E} \prod_1^m Y_s = (\int f^2)^{m/2} = \|f\|_2^m$. Moreover, for each such G, the number of corresponding terms in (11.36) is

$$\Big(\prod_1^l \big(n(b_j - a_j) + O(1)\big) \Big)^{m/2} = \Big(n^l \prod_1^l (b_j - a_j)\Big)^{m/2} + O(n^{ml/2-1})$$
$$= n^{ml/2} \big(\|g\|_2^m + o(1) \big),$$

and thus the total contribution to the right hand side of (11.36) for each G is $(\|f\|_2 \|g\|_2)^m + o(1)$.

Summing up, if c_m is the number of complete Feynman diagrams of order m,

$$\mathrm{E}\big(n^{-l/2}S_n^l(f;g)\big)^m \to c_m(\|f\|_2\|g\|_2)^m \qquad \text{as } n\to\infty.$$

By Theorem 1.36, the right hand side equals $\mathrm{E}(\widetilde{J}_l^2(f\otimes g))^m$, because $\widetilde{J}_l^2(f\otimes g)\sim \mathrm{N}(0,\|f\|_2^2\|g\|_2^2)$ by definition. Since here m is any natural number, the method of moments yields the sought-for result $n^{-l/2}S_n^l(f,g)\xrightarrow{\mathrm{d}}\widetilde{J}_l^2(f\otimes g)$.

Moreover, the same argument applies also (with more complicated notation) to mixed moments of several $S_n^l(f,g)$, which verifies joint convergence.

This completes the proof for bounded, real f. The extension to arbitrary real $f\in\widetilde{M}^0_{\{1,\dots,l\}}$ follows by the usual argument with Billingsley (1968, Theorem 4.2) and Lemma 11.44(i) (with k replaced by l), and the extension to complex f (if desired) by taking the real and imaginary parts separately. □

Proof of Theorem 11.45. We argue step by step as in the proof of Theorem 11.16, using the lemma as a starting point. It is important to note that at all steps below, joint convergence holds for any finite set of f and g of the type treated, although we do not repeat that further.

Step 1. Suppose that $f\in\widetilde{M}^0_{\{1,\dots,l\}}$ and $g=\mathbf{1}_{(a_1,b_1]}\otimes\cdots\otimes\mathbf{1}_{(a_l,b_l]}$, where the intervals $(a_i,b_i]$ are disjoint. This differs from the case studied in Lemma 11.46 only by a permutation of the coordinates, and it follows by the lemma, (11.25) and the definition (11.35) that

$$n^{-l/2}S_n^l(f,g)\xrightarrow{\mathrm{d}} J_l^2(f\otimes g).$$

Step 2. Let $A_1,\dots,A_p$ be a collection of non-empty disjoint subsets of $\{1,\dots,k\}$, and assume that $f=\bigotimes_1^p f_j$ with $f_j\in\widetilde{M}^0_{A_j}$. Assume further that $g=\mathbf{1}_{(a_1,b_1]}\otimes\cdots\otimes\mathbf{1}_{(a_k,b_k]}$, where the intervals $(a_i,b_i]$ are disjoint. Write $g_j=\bigotimes_{i\in A_j}\mathbf{1}_{(a_i,b_i]}\in L^2([0,1]^{A_j})$ and $A=\bigcup_j A_j$. Then, by the definition (11.22),

$$S_n^k(f;g)=\prod_{j=1}^p S_n^{|A_j|}(f_j\circ\pi_{A_j};g_j\circ\pi_{A_j})\prod_{i\notin A}(\lfloor nb_i\rfloor-\lfloor na_i\rfloor),$$

where the random variables $S_n^{|A_j|}(f_j\circ\pi_{A_j};g_j\circ\pi_{A_j})$ are independent because of our assumption on g. Moreover, by Step 1,

$$n^{-|A_j|/2}S_n^{|A_j|}(f_j\circ\pi_{A_j};g_j\circ\pi_{A_j})\xrightarrow{\mathrm{d}} J^2_{|A_j|}(f_j\circ\pi_{A_j}\otimes g_j\circ\pi_{A_j})=J^2_{A_j}(f_j\otimes g_j),$$

and these variables too are independent by the assumption on g. Hence,

$$\begin{aligned} n^{|A|/2-k} S_n^k(f;g) &\xrightarrow{\mathrm{d}} \prod_{i\notin A}(b_i - a_i) \prod_{j=1}^{p} J^2_{A_j}(f_j \otimes g_j) \\ &= \prod_{i\notin A}(b_i - a_i) :J^2_{A_1}(f_1\otimes g_1)\cdots J^2_{A_p}(f_p \otimes g_p): \\ &= J^2_{A_1,\dots,A_p}(f\otimes P_A(g)), \end{aligned}$$

because $P_A(g) = \prod_{i\notin A}(b_i - a_i) \bigotimes_1^p g_j$.

Step 3. Lemma 11.44(iii) shows that under the hypotheses of Step 2, we also have

$$n^{|A|/2-k}\overline{S}_n^k(f;g) \xrightarrow{\mathrm{d}} J^2_{A_1,\dots,A_p}(f\otimes P_A(g)) = J^2(f\otimes g). \tag{11.37}$$

Step 4. By the joint convergence, (11.37) holds also for f and g that are linear combinations of functions of the type studied in Steps 2 and 3.

Step 5. Such f and g are dense in $\widetilde{M}^0_{A_1,\dots,A_p}$ and $L^2([0,1]^k)$, respectively, and it follows from Lemma 11.44(ii), using Billingsley (1968, Theorem 4.2), that (11.37) holds for all $f\in \widetilde{M}^0_{A_1,\dots,A_p}$ and $g\in L^2([0,1]^k)$.

Step 6. Now let $f\in L^2(S)$ be arbitrary, and use the decomposition $f = \sum_{A_1,\dots,A_p} \widetilde{f}^0_{A_1,\dots,A_p}$ with $\widetilde{f}^0_{A_1,\dots,A_p} \in \widetilde{M}^0_{A_1,\dots,A_p}$; then $f^0_A = \sum_{\bigcup A_i = A} \widetilde{f}^0_{A_1,\dots,A_p}$ and thus $f^0_j = \sum_{|\bigcup A_i| = j} \widetilde{f}^0_{A_1,\dots,A_p}$.

Consequently (11.37) yields, by summing over all $A_1,\dots,A_p$ with total cardinality j,

$$n^{j/2-k}\overline{S}_n^k(f_j^0;g) \xrightarrow{\mathrm{d}} J^2(f^0_j\otimes g). \tag{11.38}$$

By assumption, $\overline{S}_n^k(f_j^0;g) = 0$ when $j<m$, while (11.38) implies that when $j>m$, $n^{m/2-k}\overline{S}_n^k(f_j^0;g) \xrightarrow{\mathrm{P}} 0$. Consequently, since $f = \sum_{j=0}^k f_j^0$,

$$n^{m/2-k}\overline{S}_n^k(f;g) \xrightarrow{\mathrm{d}} J^2(f^0_m\otimes g),$$

which proves the result for $\overline{S}_n^k$.

Step 7. The result for S_n^k now follows by Lemma 11.44(iii). $\square$

PROOF OF THEOREM 11.28. We may assume that $J_l = \widetilde{J}_l^2$ on $\widetilde{M}^0_{\{1,\dots,l\}}\otimes L^2(D_l)$. Then (11.35) yields, because $(\mathbf{1}_{D_l}\circ\pi)\mathbf{1}_{D_l} = 0$ if $\pi\in\mathfrak{S}_l$ is not the identity,

$$J_l^2(f\otimes(g\mathbf{1}_{D_l})) = \widetilde{J}_l^2(f\otimes(g\mathbf{1}_{D_l})) = J_l(f\otimes(g\mathbf{1}_{D_l}))$$

for every $f\in\widetilde{M}^0_{\{1,\dots,l\}}$ and $g\in L^2([0,1]^l)$.

Let, for $A\subseteq\{1,\dots,k\}$,

$$D_A = \{(t_i)_1^k \in [0,1]^k : t_i < t_j \text{ if } i<j \text{ and } i,j\in A\}.$$

Then $\mathbf{1}_{D_A} \circ \pi_A = \mathbf{1}_{D_l}$. Consequently,

$$J_A^2(f \otimes g\mathbf{1}_{D_A}) = J_{|A|}^2\big((f \circ \pi_A) \otimes (g \circ \pi_A)\mathbf{1}_{D_{|A|}}\big)$$
$$= J_{|A|}\big((f \circ \pi_A) \otimes (g \circ \pi_A)\mathbf{1}_{D_{|A|}}\big) = J_A(f \otimes g\mathbf{1}_{D_A}),$$

for $f \in \widetilde{M}_A^0$, $g \in L^2([0,1]^A)$, and hence

$$J_{A_1,\dots,A_p}^2\big(f \otimes g(\mathbf{1}_{D_{A_1}} \otimes \cdots \otimes \mathbf{1}_{D_{A_p}})\big) = J_{A_1,\dots,A_p}\big(f \otimes g(\mathbf{1}_{D_{A_1}} \otimes \cdots \otimes \mathbf{1}_{D_{A_p}})\big)$$

for $f \in \widetilde{M}_{A_1,\dots,A_p}^0$ and $g \in L^2([0,1]^A)$, $A = \bigcup_j A_j$.

If now g vanishes off D_k, then $P_A(g)$ vanishes off D_A, and on D_A each $\mathbf{1}_{D_{A_j}} = 1$; hence $P_A(g) = P_A(g)(\mathbf{1}_{D_{A_1}} \otimes \cdots \otimes \mathbf{1}_{D_{A_p}})$. Consequently, for every $f \in L^2(\mathcal{S})$ and $g \in L^2(D_k)$,

$$J^2(f \otimes g) = J(f \otimes g),$$

and the result follows by Theorem 11.45. □

PROOF OF THEOREM 11.35. We may assume that

$$\widetilde{J}_l^1 = \widetilde{J}_l^2(f \otimes \sqrt{l!}\mathbf{1}_{D_l}), \qquad f \in \widetilde{M}_{\{1,\dots,l\}}^0;$$

note that this defines a Gaussian field because $\|\sqrt{l!}\mathbf{1}_{D_l}\|_2 = 1$.

Let, for $A \subseteq \{1,\dots,k\}$, $\mathfrak{S}_A$ be the subgroup of $\mathfrak{S}_k$ fixing all elements of $\{1,\dots,k\} \setminus A$, and let $T_A f = \frac{1}{|A|!}\sum_{\pi \in \mathfrak{S}_A} f \circ \pi$ be the symmetrization of a function f over the coordinates in A. Further, let $T_l = T_{\{1,\dots,l\}}$.

Then, by (11.35),

$$J_l^2(f \otimes 1) = J_l(l!\,T_l f \otimes \mathbf{1}_{D_l}) = \sqrt{l!}\,\widetilde{J}_l^1(T_l f) = J_l^1(T_l f)$$

for every $f \in \widetilde{M}_{\{1,\dots,l\}}^0$. Consequently,

$$J_A^2(f \otimes 1) = J_A^1(T_A f), \qquad f \in \widetilde{M}_A^0,$$

and

$$J_{A_1,\dots,A_p}^2(f \otimes 1) = J_{A_1,\dots,A_p}^1(T_{A_1} \otimes \cdots \otimes T_{A_p} f), \qquad f \in \widetilde{M}_{A_1,\dots,A_p}^0.$$

If f is symmetric, and is decomposed as in (11.28), then $T_{A_1} \otimes \cdots \otimes T_{A_p}(\widetilde{f}_{A_1,\dots,A_p}^0) = \widetilde{f}_{A_1,\dots,A_p}^0$ for every $A_1,\dots,A_p$, and thus

$$J_{A_1,\dots,A_p}^2(\widetilde{f}_{A_1,\dots,A_p}^0 \otimes 1) = J_{A_1,\dots,A_p}^1(\widetilde{f}_{A_1,\dots,A_p}^0).$$

Moreover, $P_A(1) = 1$. Consequently, for every symmetric $f \in L^2(\mathcal{S})$,

$$J^2(f \otimes 1) = J^1(f).$$

The result now follows by taking $g = 1$ in Theorem 11.45, because $S_n^k(f) = \frac{1}{k!}S_n^k(f;1)$ when f is symmetric. □

PROOF OF LEMMA 11.31. Let $\{A_1, \dots, A_p\}$ and $\{A'_1, \dots, A'_{p'}\}$ be two collections, each consisting of disjoint non-empty subsets of $\{1, \dots, k\}$, and write $A = \bigcup A_j$, $A' = \bigcup A'_j$.

Note first that if $p \neq p'$, then $J_{A_1,\dots,A_p}$ and $J_{A'_1,\dots,A'_{p'}}$ map into $H^{:p:}$ and $H^{:p':}$, respectively, and these spaces are orthogonal.

Now assume $p' = p$. If $f_j \in \widetilde{M}^0_{A_j}$, $f'_j \in \widetilde{M}^0_{A'_j}$, $g_j \in L^2([0,1]^{A_j})$ and $g'_j \in L^2([0,1]^{A'_j})$, then by (11.27) and Theorem 3.9,

$$\langle J_{A_1,\dots,A_p}(f_1 \otimes \dots \otimes f_p \otimes g_1 \otimes \dots \otimes g_p), J_{A'_1,\dots,A'_p}(f'_1 \otimes \dots \otimes f'_p \otimes g'_1 \otimes \dots \otimes g'_p)\rangle$$
$$= \sum_{\sigma \in \mathfrak{S}_p} \prod_i \langle J_{A_i}(f_i \otimes g_i), J_{A'_{\sigma(i)}}(f'_{\sigma(i)} \otimes g'_{\sigma(i)})\rangle. \quad (11.39)$$

The right hand side vanishes unless $|A'_{\sigma(i)}| = |A_i|$ for each i. In this case, we choose a permutation $\pi \in \mathfrak{S}_k$ such that π maps A_i onto $A'_{\sigma(i)}$, and is increasing on A_i for every i; we then obtain, from (11.26) and the assumption that each J_l is an isometry into,

$$\langle J_{A_i}(f_i \otimes g_i), J_{A'_{\sigma(i)}}(f'_{\sigma(i)} \otimes g'_{\sigma(i)})\rangle = \langle f_i \otimes g_i, (f'_{\sigma(i)} \circ \pi) \otimes (g'_{\sigma(i)} \circ \pi)\rangle$$
$$= \langle f_i, f'_{\sigma(i)} \circ \pi\rangle \langle g_i, g'_{\sigma(i)} \circ \pi\rangle.$$

It thus follows from (11.39) that, letting V be the (possibly empty) set of permutations obtained by choosing a π for each such σ,

$$\langle J_{A_1,\dots,A_p}(f \otimes g), J_{A'_1,\dots,A'_p}(f' \otimes g')\rangle = \sum_{\pi \in V} \langle f, f' \circ \pi\rangle \langle g, g' \circ \pi\rangle,$$

where by multilinearity and continuity we may take any $f \in \widetilde{M}^0_{A_1,\dots,A_p} = \widetilde{M}^0_{A_1} \otimes \dots \otimes \widetilde{M}^0_{A_p}$, $f' \in \widetilde{M}^0_{A'_1,\dots,A'_p}$, $g \in L^2([0,1]^A)$, $g' \in L^2([0,1]^{A'})$.

In particular, for such f and f',

$$\langle J(f \otimes \mathbf{1}_{D_k}), J(f' \otimes \mathbf{1}_{D_k})\rangle = \sum_{\pi \in V} \langle f, f' \circ \pi\rangle \langle P_A(\mathbf{1}_{D_k}), P_{A'}(\mathbf{1}_{D_k}) \circ \pi\rangle,$$

where, however, $\langle P_A(\mathbf{1}_{D_k}), P_{A'}(\mathbf{1}_{D_k}) \circ \pi\rangle = 0$ unless π is increasing on A. There is at most one such π in our set V, and we can summarize the result as follows, using the functions g_A defined in (11.17):

Let $f \in \widetilde{M}^0_{A_1,\dots,A_p}$ and $f' \in \widetilde{M}^0_{A'_1,\dots,A'_{p'}}$. If $|A| = |A'|$, let $\pi \in \mathfrak{S}_k$ map A onto A' and be increasing on A. If further the sets $\pi(A_1), \dots, \pi(A_p)$ equal $A'_1, \dots, A'_p$ (in some order), then

$$\langle J(f \otimes \mathbf{1}_{D_k}), J(f' \otimes \mathbf{1}_{D_k})\rangle = \langle f, f' \circ \pi\rangle \langle g_A, g_{A'}\rangle = \langle f \otimes g_A, f' \circ \pi \otimes g_{A'}\rangle.$$

This result partitions the family of all collections $\{A_1,\dots,A_p\}$ into classes such that different classes have orthogonal ranges for $J(\widetilde{f}^0_{A_1,\dots,A_p}\otimes \mathbf{1}_{D_k})$; consequently, it suffices to consider terms in one such class. If the class contains $\{A_1,\dots,A_p\}$, it contains exactly one collection $\{A'_1,\dots,A'_p\}$ for every $A'\subseteq\{1,\dots,k\}$ with $|A'|=|\bigcup A_j|$; moreover, for each such collection, $f\mapsto f\circ\pi$, with π as above, is an isometry of $\widetilde{M}^0_{A'_1,\dots,A'_p}$ onto $\widetilde{M}^0_{A_1,\dots,A_p}$. Hence the result follows if we can show that if $l\le k$ and M is a Hilbert space, then for any elements $f_A\in M$, where A ranges over the subsets of $\{1,\dots,k\}$ with $|A|=l$,

$$c_k\Bigl(\sum_A\|f_A\|^2_M\Bigr)^{1/2}\le\Bigl\|\sum_A f_A\otimes g_A\Bigr\|_{M\otimes L^2([0,1]^l)}\le C_k\Bigl(\sum_A\|f_A\|^2_M\Bigr)^{1/2}. \tag{11.40}$$

Using an orthonormal basis in M, this reduces to the case when M is one-dimensional, but then (11.40) is equivalent to the fact that the functions $g_A\in L^2([0,1]^l)$ are linearly independent, see Remark 11.22. □

XII

Applications to operator theory

We will here give some general results on extensions to vector-valued functions of operators on L^p-spaces. More precisely, we consider functions with values in a Hilbert space H. Although the statements are purely functional analytical, and do not involve any probability, our proofs use Gaussian Hilbert spaces. For further results, see Gasch and Maligranda (1994) and the references cited there.

Some of the results below will be used in Chapter 15.

We consider in this chapter either real-valued scalar functions and real Hilbert spaces or complex-valued scalar functions and complex Hilbert spaces; unless otherwise indicated, all results are valid for both cases. (In the complex case, the Gaussian Hilbert spaces used in the arguments below should be replaced by complex Gaussian Hilbert spaces, see Section 1.4; we usually do not mention this further.)

We use the notation of Appendix C; in particular, if f is a real or complex function on a space M and $h \in H$, then $f \otimes h$ is the H-valued function $x \mapsto f(x)h$, and if $(M, \mathcal{M}, \mu)$ is a measure space, the spaces $L^p(M, \mathcal{M}, \mu; H)$ of H-valued functions are defined as for real or complex functions. If $1 \leq p < \infty$, then the functions that are finite sums $\sum_1^n f_i \otimes h_i$ with $f_i \in L^p(M, \mathcal{M}, \mu)$ and $h_i \in H$ are dense in $L^p(M, \mathcal{M}, \mu; H)$ (these are the functions that map into some finite-dimensional subspace of H); moreover, so is the set of (integrable) simple functions, i.e. the functions that may be written as $\sum_1^n \mathbf{1}_{A_i} \otimes h_i$ with $A_i \in \mathcal{M}$ disjoint with finite measures and $h_i \in H$. For $p = \infty$, however, this is not true, unless H or $L^\infty(M, \mathcal{M}, \mu)$ has finite dimension. We let $L_0^\infty(M, \mathcal{M}, \mu; H)$ denote the closed hull of $\{\sum_1^n f_i \otimes h_i : n \geq 0,\ f_i \in L^\infty(M, \mathcal{M}, \mu),\ h_i \in H\}$, which equals the closed hull of the simple functions $\{\sum_1^n \mathbf{1}_{A_i} \otimes h_i : n \geq 0,\ A_i \in \mathcal{M},\ h_i \in H\}$. It is easily seen that $L_0^\infty(M, \mathcal{M}, \mu; H)$ is the space of all measurable functions $M \to H$ that (a.e.) map into some compact subset of H.

THEOREM 12.1. *Let $T: L^p(M_1, \mathcal{M}_1, \mu_1) \to L^p(M_2, \mathcal{M}_2, \mu_2)$ be a bounded linear operator, where $(M_1, \mathcal{M}_1, \mu_1)$ and $(M_2, \mathcal{M}_2, \mu_2)$ are two measure spaces and $1 \leq p < \infty$. If H is a Hilbert space, then the following hold.*

(i) *There exists a unique bounded linear operator $\widetilde{T}: L^p(M_1, \mathcal{M}_1, \mu_1; H) \to L^p(M_2, \mathcal{M}_2, \mu_2; H)$ such that $\widetilde{T}(f \otimes h) = Tf \otimes h$ for $f \in L^p(M_1, \mathcal{M}_1, \mu_1)$ and $h \in H$. Moreover, $\|\widetilde{T}\| \leq \|T\|$.*

(ii) *If* $\|Tf\| \geq \delta\|f\|$ *for all* $f \in L^p(M_1, \mathcal{M}_1, \mu_1)$ *and some* $\delta > 0$, *then* $\|\widetilde{T}F\| \geq \delta\|F\|$ *for all* $F \in L^p(M_1, \mathcal{M}_1, \mu_1; H)$.

For $p = \infty$, *the conclusions remain valid with* L^p *replaced by* L_0^∞.

PROOF. Since elements of the type $\sum_1^n f_i \otimes h_i$ are dense in $L^p(M_1; H)$ ($L_0^\infty(M_1; H)$ if $p = \infty$), (i) follows if we show that

$$\Big\|\sum_1^n Tf_i \otimes h_i\Big\|_{L^p(M_2;H)} \leq \|T\| \, \Big\|\sum_1^n f_i \otimes h_i\Big\|_{L^p(M_1;H)}, \tag{12.1}$$

when $f_1, \dots, f_n \in L^p(M_1)$ and $h_1, \dots, h_n \in H$.

Suppose first that $1 \leq p < \infty$. Since all functions f_i and Tf_i have σ-finite support, we may, by restricting M_1 and M_2 if necessary, assume that M_1 and M_2 are σ-finite.

Evidently, (12.1) is not affected if we replace H by an isomorphic Hilbert space; hence we may assume that H is a Gaussian Hilbert space, defined on some probability space $(\Omega, \mathcal{F}, \mathrm{P})$. (In the complex case, we assume instead that H is a complex Gaussian Hilbert space, see Section 1.4.) By (1.1), the norm in H is then proportional to the L^p-norm, and thus (12.1) is equivalent to

$$\Big\|\sum_1^n Tf_i \otimes h_i\Big\|_{L^p(M_2;\, L^p(\Omega))} \leq \|T\| \, \Big\|\sum_1^n f_i \otimes h_i\Big\|_{L^p(M_1;\, L^p(\Omega))}.$$

By Fubini's theorem, this equation can also be written

$$\Big\|\sum_1^n Tf_i \otimes h_i\Big\|_{L^p(\Omega;\, L^p(M_2))} \leq \|T\| \, \Big\|\sum_1^n f_i \otimes h_i\Big\|_{L^p(\Omega;\, L^p(M_1))},$$

which is evident by applying the assumption for each fixed ω, because

$$\sum_1^n Tf_i \otimes h_i(x, \omega) = \sum_1^n h_i(\omega) Tf_i(x) = T\Big(\sum_1^n h_i(\omega) f_i\Big)(x). \tag{12.2}$$

This completes the proof of (12.1), and thus of (i), when $p < \infty$; (ii) is proved in the same way, reversing the inequalities above.

In the case $p = \infty$, we use another argument (not involving Gaussian Hilbert spaces). Let H_0 be the subspace of H spanned by $h_1, \dots, h_n$, and take a countable dense subset $\{h_j'\}_1^\infty$ of the unit ball of H_0. For each j,

$$\begin{aligned}\Big\|\Big\langle\sum_{i=1}^n Tf_i \otimes h_i, h_j'\Big\rangle_H\Big\|_{L^\infty(M_2)} &= \Big\|T\sum_{i=1}^n f_i \langle h_i, h_j'\rangle_H\Big\|_{L^\infty(M_2)} \\ &\leq \|T\| \, \Big\|\Big\langle\sum_i f_i \otimes h_i, h_j'\Big\rangle_H\Big\|_{L^\infty(M_1)}.\end{aligned}$$

In other words,

$$\Big|\Big\langle \sum_{i=1}^{n} Tf_i \otimes h_i, h'_j \Big\rangle_H \Big| \le \|T\| \, \Big\| \Big\langle \sum_i f_i \otimes h_i, h'_j \Big\rangle_H \Big\|_{L^\infty(M_1)}$$

$$\le \|T\| \, \Big\| \sum_i f_i \otimes h_i \Big\|_{L^\infty(M_1;H)} \quad \text{a.e.}$$

for each j, which implies

$$\Big\| \sum_{i=1}^{n} Tf_i \otimes h_i \Big\|_H \le \|T\| \, \Big\| \sum_i f_i \otimes h_i \Big\|_{L^\infty(M_1;H)} \quad \text{a.e.,}$$

i.e. (12.1). This proves (i) in the case $p = \infty$. Again, (ii) is proved in the same way, reversing the inequalities. □

Theorem 12.1(i) can be generalized to operators $L^p \to L^q$ with $p \le q$ as follows.

THEOREM 12.2. *Let $T\colon L^p(M_1, \mathcal{M}_1, \mu_1) \to L^q(M_2, \mathcal{M}_2, \mu_2)$ be a bounded linear operator, where $(M_1, \mathcal{M}_1, \mu_1)$ and $(M_2, \mathcal{M}_2, \mu_2)$ are two measure spaces, $1 \le p \le q \le \infty$ and $p < \infty$. If H is a Hilbert space, then there exists a unique bounded linear operator $\widetilde{T}\colon L^p(M_1, \mathcal{M}_1, \mu_1; H) \to L^q(M_2, \mathcal{M}_2, \mu_2; H)$ such that $\widetilde{T}(f \otimes h) = Tf \otimes h$ for $f \in L^p(M_1, \mathcal{M}_1, \mu_1)$ and $h \in H$. Moreover, $\|\widetilde{T}\| \le \|T\|$.*

PROOF. Arguing as in the proof of Theorem 12.1, it suffices to show that if H is a Gaussian Hilbert space, then

$$\Big\| \sum_{1}^{n} Tf_i \otimes h_i \Big\|_{L^q(M_2;\, L^p(\Omega))} \le \|T\| \, \Big\| \sum_{1}^{n} f_i \otimes h_i \Big\|_{L^p(M_1;\, L^p(\Omega))}. \tag{12.3}$$

In this case, (12.2) implies

$$\Big\| \sum_{1}^{n} Tf_i \otimes h_i \Big\|_{L^p(\Omega;\, L^q(M_2))} \le \|T\| \, \Big\| \sum_{1}^{n} f_i \otimes h_i \Big\|_{L^p(\Omega;\, L^p(M_1))}. \tag{12.4}$$

Moreover, since $p \le q$, Proposition C.4 yields

$$\Big\| \sum_{1}^{n} Tf_i \otimes h_i \Big\|_{L^q(M_2;\, L^p(\Omega))} \le \Big\| \sum_{1}^{n} Tf_i \otimes h_i \Big\|_{L^p(\Omega;\, L^q(M_2))}.$$

This and (12.4) yield (12.3). □

COROLLARY 12.3. *Let $(M_1, \mathcal{M}_1, \mu_1)$ and $(M_2, \mathcal{M}_2, \mu_2)$ be two measure spaces. If $1 \le p \le q \le \infty$ and $T\colon L^p_{\mathbb{R}}(M_1, \mathcal{M}_1, \mu_1) \to L^q_{\mathbb{R}}(M_2, \mathcal{M}_2, \mu_2)$ is a bounded linear operator, then its complexification $T_{\mathbb{C}}\colon L^p_{\mathbb{C}}(M_1, \mathcal{M}_1, \mu_1; H) \to L^q_{\mathbb{C}}(M_2, \mathcal{M}_2, \mu_2; H)$ has the same norm.*

PROOF. $\mathbb{C}$ can be regarded as a two-dimensional real Hilbert space, and the extension provided by Theorems 12.1 and 12.2 equals the complexification. $\square$

We give two applications of these results to Gaussian Hilbert spaces, showing that the operators π_n and M_r defined in Chapters 2 and 4 extend to the vector-valued case. (We use the same notations for the operators as in the scalar case.)

EXAMPLE 12.4. If H is a Gaussian Hilbert space, $1 < p < \infty$ and H_0 is a real or complex Hilbert space, then Theorems 5.14 and 12.1 show that π_n is a bounded linear operator in $L^p(\Omega, \mathcal{F}(H), \mathrm{P}; H_0)$, which satisfies $\pi_n(X \otimes h_0) = \pi_n(X) \otimes h_0$ for $X \in L^p(\Omega, \mathcal{F}(H), \mathrm{P})$ and $h_0 \in H_0$. It is easily seen that the extensions for different p are compatible, which justifies our use of a single notation π_n.

It follows that $\pi_n^2 = \pi_n$, i.e. π_n is a projection in $L^p(\Omega, \mathcal{F}(H), \mathrm{P}; H_0)$. For $p = 2$, this is the orthogonal projection of $L^2(H_0)$ onto the subspace $H^{:n:} \otimes H_0$ spanned by $X \otimes h$ with $X \in H^{:n:}$, $h \in H_0$. (This is the Hilbert space tensor product of $H^{:n:}$ and H_0, and $L^2(H_0)$ is the Hilbert space tensor product of $L^2 \otimes H_0$, see Example E.12.)

Moreover, Theorems 5.14, 3.50 and 12.2 show that π_n maps $L^p(H_0)$ into $L^q(H_0)$ when $1 < p < q < \infty$. Since π_n is a projection, it follows easily that π_n maps $L^p(H_0)$ onto $H^{:n:} \otimes H_0$ for every p with $1 < p < \infty$; it follows also that, as in the scalar case, $H^{:n:} \otimes H_0 \subset L^p(H_0)$ for $p < \infty$ and that all L^p-norms with $1 < p < \infty$ are equivalent on $H^{:n:} \otimes H_0$. (This also holds a fortiori for $0 < p < \infty$ by Hölder's inequality as in the proof of Theorem 3.50.)

EXAMPLE 12.5. If H is a Gaussian Hilbert space, H_0 is a real or complex Hilbert space and $-1 \leq r \leq 1$, then Theorems 4.20 and 12.1 show that M_r is a contraction in $L^p(\Omega, \mathcal{F}(H), \mathrm{P}; H_0)$ for every p with $1 \leq p \leq \infty$; it satisfies $M_r(X \otimes h_0) = M_r(X) \otimes h_0$ for $X \in L^1(\Omega, \mathcal{F}(H), \mathrm{P})$ and $h_0 \in H_0$.

Moreover, Theorems 5.8 and 12.2 show that if $1 < p < q < \infty$ and $0 \leq r^2 \leq (p-1)/(q-1)$, then M_r actually is a contraction of $L^p(\Omega, \mathcal{F}(H), \mathrm{P}; H_0)$ into $L^q(\Omega, \mathcal{F}(H), \mathrm{P}; H_0)$.

If H_0 is a complex Hilbert space, then Theorems 5.28 and 12.2 show that for $1 < p \leq q < \infty$, M_r is also defined for complex r in the set D_{pq} given by Theorem 5.28, and then M_r is a contraction of $L^p(\Omega, \mathcal{F}(H), \mathrm{P}; H_0)$ into $L^q(\Omega, \mathcal{F}(H), \mathrm{P}; H_0)$. Furthermore, if $X \in L^p(\Omega, \mathcal{F}(H), \mathrm{P}; H_0)$, then $z \mapsto M_z X$ is a continuous function $D_{pq} \to L^q(\Omega, \mathcal{F}(H), \mathrm{P}; H_0)$ which is analytic in the interior D_{pq}°; this follows from Theorem 5.28(v) for variables of the form $X \otimes h_0$, by linearity for sums of such variables, and in general by uniform convergence (as in the proof of Theorem 5.28).

Similarly (and for $p > 1$ as a special case), if $1 \leq p < \infty$ and $X \in L^p(H_0)$, then $M_r X \to X$ in $L^p(H_0)$ as $r \to 1$.

Theorem 12.2 and Corollary 12.3 are not valid for $p > q$, see Example 12.9 below. It is true, however, that T always extends to a bounded linear operator $\widetilde{T}$ in this case too, but the norm of $\widetilde{T}$ may be somewhat larger than the norm of T. This (in particular the extreme case $p = \infty$, $q = 1$) is a version of Grothendieck's theorem, also known as Grothendieck's inequality (Grothendieck 1956). We give a proof (also using Gaussian Hilbert spaces) based on Krivine (1979). For other versions of Grothendieck's theorem, and various applications, see Pisier (1986).

REMARK 12.6. The lemma below is another, equivalent, version of Grothendieck's theorem, and it can be shown that the best constants in the lemma and the theorem below are the same; the best constant is known as *Grothendieck's constant.* The exact value of Grothendieck's constant is not known, but it has been conjectured that the value of the constant C given in the lemma and theorem below is the best possible in the real case. (It can be improved in the complex case, see Haagerup (1987) who uses the method below with complex Gaussian variables, which turns out to be much more complicated.)

LEMMA 12.7. *Let $C = \pi/2\ln(1+\sqrt{2}) \approx 1.782$. If H is a Hilbert space, then there exist two families $\{X(h)\}_{h\in H}$ and $\{X'(h)\}_{h\in H}$ of bounded random variables defined on some probability space $(\Omega, \mathcal{F}, \mathrm{P})$, i.e. two (non-linear) mappings $X, X' \colon H \to L^\infty(\Omega, \mathcal{F}, \mathrm{P})$, such that*

$$\|X(h)\|_\infty = \|X'(h)\|_\infty = C^{1/2}\|h\|_H, \qquad h \in H$$

(and more precisely $|X(h)| = |X'(h)| = C^{1/2}\|h\|_H$ a.s.), and

$$\langle X(h), X'(h')\rangle_{L^2} = \mathrm{E}\big(X(h)\overline{X'(h')}\big) = \langle h, h'\rangle_H, \qquad h, h' \in H.$$

PROOF. Consider first the real case. If H_1 is any real Hilbert space, there exist by Theorem 1.23 a Gaussian Hilbert space H_0 and an isometry $h \mapsto \xi_h$ mapping H_1 onto H_0. Define, for $h \in H_1$ with $\|h\|_{H_1} = 1$, $\varphi(h) = \operatorname{sign}\xi_h \in L^\infty(\Omega, \mathcal{F}, \mathrm{P})$. Then $|\varphi(h)| = 1$ a.s., and for every $h, h' \in H_1$ with $\|h\|_{H_1} = \|h'\|_{H_1} = 1$, by Lemma 10.12,

$$\begin{aligned}\langle \varphi(h), \varphi(h')\rangle_{L^2} &= \operatorname{Cov}(2\cdot \mathbf{1}[\xi_h > 0] - 1, 2\cdot\mathbf{1}[\xi_{h'} > 0] - 1)\\ &= 4\operatorname{Cov}(\mathbf{1}[\xi_h > 0], \mathbf{1}[\xi_{h'} > 0]) = \tfrac{2}{\pi}\arcsin\langle h, h'\rangle_{H_1}.\end{aligned}$$

The next step is to construct, given a real Hilbert space H, another real Hilbert space H_1 and two mappings $\psi, \psi' \colon H \to H_1$ such that $\langle \psi(h), \psi'(h')\rangle_{H_1} = \sin\langle h, h'\rangle_H$. This can easily be done with $H_1 = \Gamma(H)$, the Fock space defined in Appendix E, but we prefer to be more concrete and use Gaussian Hilbert spaces again (cf. Theorem 4.1). We thus assume, as we may by Theorem 1.23, that H is a Gaussian Hilbert space, and let $H_1 = L^2_{\mathbb{R}}(\Omega, \mathcal{F}(H), \mathrm{P})$.

We define, for $h \in H$,

$$\psi(h) = {:}\sinh(h){:} = \tfrac{1}{2}({:}e^h{:} - {:}e^{-h}{:}) = \sum_{k=0}^{\infty} \frac{{:}h^{2k+1}{:}}{(2k+1)!},$$

$$\psi'(h) = {:}\sin(h){:} = \tfrac{1}{2i}({:}e^{ih}{:} - {:}e^{-ih}{:}) = \sum_{k=0}^{\infty} (-1)^k \frac{{:}h^{2k+1}{:}}{(2k+1)!}.$$

It then follows from Corollary 3.37 or Theorem 3.9 that

$$\|\psi(h)\|_2^2 = \|\psi'(h)\|_2^2 = \sinh(\|h\|_H^2), \qquad h \in H,$$

and

$$\langle \psi(h), \psi'(h')\rangle_{H_1} = \sin\langle h, h'\rangle_H, \qquad h, h' \in H.$$

With this choice of H_1, let φ be the mapping into $L^\infty(\Omega, \mathcal{F}, \mathrm{P})$ (for some new probability space $(\Omega, \mathcal{F}, \mathrm{P})$) constructed in the first part of the proof. Now suppose that $h, h' \in H$ with $\|h\| = \|h'\| = a$, where $a = (\ln(1+\sqrt{2}))^{1/2}$ is chosen such that $\sinh(a^2) = 1$. Then $\|\psi(h)\|_{H_1} = \|\psi'(h')\|_{H_1} = 1$, and thus $\varphi(\psi(h)), \varphi(\psi'(h')) \in L^\infty$ with

$$\langle \varphi(\psi(h)), \varphi(\psi'(h'))\rangle_{L^2} = \tfrac{2}{\pi} \arcsin\langle \psi(h), \psi'(h')\rangle_{H_1} = \tfrac{2}{\pi}\langle h, h'\rangle_H.$$

We may thus define, for $h \in H$ with $\|h\| = a$, $X(h) = (\pi/2)^{1/2}\varphi(\psi(h))$, $X'(h) = (\pi/2)^{1/2}\varphi(\psi'(h))$ and obtain for such h, h'

$$|X(h)| = |X'(h)| = (\tfrac{\pi}{2})^{1/2} = C^{1/2}\|h\|_H \quad \text{a.s.}$$

and

$$\langle X(h), X'(h')\rangle_{L^2} = \langle h, h'\rangle_H.$$

The construction is extended to all $h \in H$ by homogeneity: let $X(0) = X'(0) = 0$ and for $h \neq 0$,

$$X(h) = a^{-1}\|h\| X(ah/\|h\|) = C^{1/2}\|h\|\varphi(\psi(ah/\|h\|)),$$

$$X'(h) = a^{-1}\|h\| X'(ah/\|h\|) = C^{1/2}\|h\|\varphi(\psi'(ah/\|h\|)).$$

This completes the proof in the real case. If H is a complex Hilbert space, select an orthonormal basis $\{e_\alpha\}$ in H and let H_0 be the real Hilbert space spanned by $\{e_\alpha\}$; thus $H = \{h_1 + ih_2 : h_1, h_2 \in H_0\}$. Define X and X' as above for H_0, and extend them to H by $X(h_1 + ih_2) = X(h_1) + iX(h_2)$, $X'(h_1 + ih_2) = X'(h_1) + iX'(h_2)$. □

THEOREM 12.8. *Let $T\colon L^p(M_1, \mathcal{M}_1, \mu_1) \to L^q(M_2, \mathcal{M}_2, \mu_2)$ be a bounded linear operator, where $(M_1, \mathcal{M}_1, \mu_1)$ and $(M_2, \mathcal{M}_2, \mu_2)$ are two measure spaces and $1 \le q < p < \infty$. If H is a Hilbert space, then there exists a unique bounded linear operator $\widetilde{T}\colon L^p(M_1, \mathcal{M}_1, \mu_1; H) \to L^q(M_2, \mathcal{M}_2, \mu_2; H)$ such that $\widetilde{T}(f \otimes h) = Tf \otimes h$ for $f \in L^p(M_1, \mathcal{M}_1, \mu_1)$ and $h \in H$. Moreover, $\|\widetilde{T}\| \le C\|T\|$, where $C = \pi/2\ln(1+\sqrt{2}) \approx 1.782$.*

For $1 \le q < p = \infty$, the conclusions remain valid with L^p replaced by L^∞_0.

PROOF. It suffices to consider simple functions in $L^p(M_1; H)$ and show that

$$\Big\|\sum_i T\mathbf{1}_{A_i} \otimes h_i\Big\|_{L^q(M_2;H)} \le C\|T\|\,\Big\|\sum_i \mathbf{1}_{A_i} \otimes h_i\Big\|_{L^p(M_1;H)}$$

when $A_1, \ldots, A_n \in \mathcal{M}_1$ are disjoint and $h_1, \ldots, h_n \in H$. Moreover, by duality, it suffices to show that

$$\Big|\int_{M_2} \Big\langle \sum_i T\mathbf{1}_{A_i} \otimes h_i, \sum_j \mathbf{1}_{B_j} \otimes h'_j \Big\rangle_H \, d\mu_2\Big| \le C\|T\|\,\Big\|\sum_i \mathbf{1}_{A_i} \otimes h_i\Big\|_{L^p(M_1;H)} \Big\|\sum_j \mathbf{1}_{B_j} \otimes h'_j\Big\|_{L^{q'}(M_2;H)}, \tag{12.5}$$

where $B_1, \ldots, B_m \in \mathcal{M}_2$ are disjoint, $h'_1, \ldots, h'_m \in H$ and q' is the conjugate exponent. Let $X(h)$ and $X'(h)$ be as in Lemma 12.7. Then, writing $\langle f, g\rangle_{L^q} = \int_{M_2} f\bar{g}\, d\mu_2$, we have

$$\begin{aligned}\int_{M_2} \Big\langle \sum_i T\mathbf{1}_{A_i} \otimes h_i, \sum_j \mathbf{1}_{B_j} \otimes h'_j \Big\rangle_H \, d\mu_2 &= \sum_{i,j} \langle T\mathbf{1}_{A_i}, \mathbf{1}_{B_j}\rangle_{L^q} \langle h_i, h'_j\rangle_H \\ &= \sum_{i,j} \langle T\mathbf{1}_{A_i}, \mathbf{1}_{B_j}\rangle_{L^q}\, \mathrm{E}\big(X(h_i)\overline{X'(h'_j)}\big) = \mathrm{E}\sum_{i,j} \langle X(h_i) T\mathbf{1}_{A_i}, X'(h'_j)\mathbf{1}_{B_j}\rangle_{L^q} \\ &= \mathrm{E}\Big\langle T\big(\sum_i X(h_i)\mathbf{1}_{A_i}\big), \sum_j X'(h'_j)\mathbf{1}_{B_j}\Big\rangle_{L^q}.\end{aligned} \tag{12.6}$$

The random function $\sum_i X(h_i)\mathbf{1}_{A_i} \in L^p(M_1, \mathcal{M}_1, \mu_1)$ for each $\omega \in \Omega$, and, for a.e. ω and every point in M_1,

$$\Big|\sum_i X(h_i)\mathbf{1}_{A_i}\Big| = \sum_i |X(h_i)|\mathbf{1}_{A_i} = C^{1/2} \sum_i \|h_i\|_H \mathbf{1}_{A_i} = C^{1/2} \Big\|\sum_i \mathbf{1}_{A_i} \otimes h_i\Big\|_H$$

and thus, for a.e. ω,

$$\begin{aligned}\Big\|T\sum_i X(h_i)\mathbf{1}_{A_i}\Big\|_{L^q(M_2)} &\le \|T\|\,\Big\|\sum_i X(h_i)\mathbf{1}_{A_i}\Big\|_{L^p(M_1)} \\ &= C^{1/2}\|T\|\,\Big\|\sum_i \mathbf{1}_{A_i} \otimes h_i\Big\|_{L^p(M_1;H)}.\end{aligned}$$

Similarly, for a.e. ω,

$$\Big\|\sum_j X'(h'_j)\mathbf{1}_{B_j}\Big\|_{L^{q'}(M_2)} = C^{1/2}\Big\|\sum_j \mathbf{1}_{B_j} \otimes h'_j\Big\|_{L^{q'}(M_2;H)}.$$

Consequently, (12.5) follows from (12.6) and Hölder's inequality. □

EXAMPLE 12.9. Let $M_1 = M_2 = M$ be a set with two points, each of unit measure. Then $L^p_{\mathbb{R}}(M) = \mathbb{R}^2$ with the ℓ^p-norm.

Define $T\colon \mathbb{R}^2 \to \mathbb{R}^2$ by $T(x,y) = (x+y, x-y)$. Regarding T as an operator $L^\infty_\mathbb{R}(M) \to L^1_\mathbb{R}(M)$, it is easily seen that $\|T\| = 2$, while its complexification $L^\infty_\mathbb{C}(M) \to L^1_\mathbb{C}(M)$ has norm $2\sqrt{2}$. (Use $\|T(1,i)\|_{L^1} = \|(1+i, 1-i)\|_{L^1} = 2\sqrt{2}$.)

Hence, as claimed above, Theorem 12.8 is not valid without the constant C (at least not in the real case and for $p = \infty$, $q = 1$); moreover, this example shows that necessarily $C \geq \sqrt{2}$.

EXAMPLE 12.10. An example which shows that Corollary 12.3, and thus also Theorem 12.2, fails for every pair (p,q) with $1 \leq q < p \leq \infty$ was given by Riesz (1926, §36). For this example we let $M_1 = M_2 = M$ be a set with three points with unit measures; equivalently, we consider complexifications of a linear operator on $\mathbb{R}^3$.

Define, for $\varepsilon \geq 0$, $T_\varepsilon(x,y,z) = (x - \varepsilon(x+y+z), y - \varepsilon(x+y+z), z - \varepsilon(x+y+z))$. Riesz showed that for every p and q with $p > q$, the (ℓ^p, ℓ^q)-operator norm of $T_{\varepsilon\mathbb{C}}\colon \mathbb{C}^3 \to \mathbb{C}^3$ is larger than the norm of $T_\varepsilon\colon \mathbb{R}^3 \to \mathbb{R}^3$ for every sufficiently small $\varepsilon > 0$.

We sketch a proof, leaving the details to the reader. The norm of $T_{\varepsilon\mathbb{C}}$ is at least $3^{1/q-1/p}$, since $T_\varepsilon(1, e^{2\pi i/3}, e^{4\pi i/3}) = (1, e^{2\pi i/3}, e^{4\pi i/3})$. Next consider $x = (x_1, x_2, x_3)$ belonging to the unit ball B_p of $\mathbb{R}^3$. First, since $T_0 = I$, Hölder's inequality yields $\|T_0 x\|_q \leq 3^{1/q-1/p}$, with equality only for the eight points with $|x_1| = |x_2| = |x_3| = 3^{-1/p}$. At these eight points, $\frac{d}{d\varepsilon}\|T_\varepsilon x\|_q < 0$ at $\varepsilon = 0$; by continuity, this holds in some neighbourhoods, and there are an open subset U of B_p containing these eight points and an $\varepsilon_0 > 0$ such that $\|T_\varepsilon x\|_q < \|T_0 x\|_q \leq 3^{1/q-1/p}$ for $x \in U$ and $0 < \varepsilon < \varepsilon_0$. On the other hand, for $x \in B_p \setminus U$, $\|T_0 x\|_q < 3^{1/q-1/p}$, and by compactness and continuity, $\|T_\varepsilon x\|_q < 3^{1/q-1/p}$ for $x \in B_p \setminus U$ and $0 \leq \varepsilon < \varepsilon_1$, for some $\varepsilon_1 > 0$. Hence, for $0 < \varepsilon < \min(\varepsilon_0, \varepsilon_1)$, $\|T_\varepsilon\| < 3^{1/q-1/p} \leq \|T_{\varepsilon\mathbb{C}}\|$.

EXAMPLE 12.11. Let H be a Gaussian Hilbert space and let $(M, \mathcal{M}, \mu) = (\Omega, \mathcal{F}(H), \mathrm{P})$. Consider the projection π_1 onto H, first as a mapping $L^2_\mathbb{R} \to L^1_\mathbb{R}$.

If $X \in L^2_\mathbb{R}(M) = L^2_\mathbb{R}(\Omega)$, then by (1.1), $\|\pi_1(X)\|_1 = \kappa(1)\|\pi_1(X)\|_2 \leq \kappa(1)\|X\|_2$, with equality when $X \in H$; thus $\|\pi_1\|_{L^2_\mathbb{R}, L^1_\mathbb{R}} = \kappa(1) = \sqrt{\frac{2}{\pi}}$.

Now, assume that $\dim H = \infty$ and let H_0 be another infinite-dimensional Gaussian Hilbert space defined on a probability space $(\Omega_0, \mathcal{F}_0, \mathrm{P}_0)$; we may thus regard elements of $L^0(H_0)$ as functions on $\Omega \times \Omega_0$.

Let $\varepsilon > 0$. By Example 3.52, there exists a finite sum $X = \sum_1^n \xi_i \otimes \eta_i$, with $\xi_i \in H$ and $\eta_i \in H_0$, $i = 1, \dots, n$, such that $\|X\|_{L^1(\Omega\times\Omega_0)} > (1-\varepsilon)\kappa(1)\|X\|_{L^2(\Omega\times\Omega_0)}$. Thus,

$$\begin{aligned}\|X\|_{L^1(H_0)} &= \|X\|_{L^1(\Omega;\, L^2(\Omega_0))} = \kappa(1)^{-1}\|X\|_{L^1(\Omega;\, L^1(\Omega_0))} = \kappa(1)^{-1}\|X\|_{L^1(\Omega\times\Omega_0)} \\ &> (1-\varepsilon)\|X\|_{L^2(\Omega\times\Omega_0)} = (1-\varepsilon)\|X\|_{L^2(H_0)}.\end{aligned}$$

Since $\pi_1(X) = X$, this shows that $\|\pi_1\|_{L^2(H_0),L^1(H_0)} > 1 - \varepsilon$, and thus

$$\|\pi_1\|_{L^2(H_0),L^1(H_0)} = 1 = \kappa(1)^{-1}\|\pi_1\|_{L^2_{\mathbb{R}},L^1_{\mathbb{R}}}.$$

Moreover, π_1 is a self-adjoint projection, and thus $\pi_1 = \pi_1^2 = \pi_1\pi_1^*$. Consequently,

$$\|\pi_1\|_{L^\infty_{\mathbb{R}},L^1_{\mathbb{R}}} = \|\pi_1\pi_1^*\|_{L^\infty_{\mathbb{R}},L^1_{\mathbb{R}}} \le \|\pi_1\|_{L^2_{\mathbb{R}},L^1_{\mathbb{R}}}\|\pi_1^*\|_{L^\infty_{\mathbb{R}},L^2_{\mathbb{R}}} = \|\pi_1\|^2_{L^2_{\mathbb{R}},L^1_{\mathbb{R}}} = \kappa(1)^2 = \tfrac{2}{\pi}.$$

On the other hand, duality also yields

$$\|\pi_1\|_{L^\infty(H_0),L^2(H_0)} = \|\pi_1\|_{L^2(H_0),L^1(H_0)} = 1.$$

Hence, if $0 < \varepsilon < 1$, there exists $X \in L^\infty(H_0)$ with $\|X\|_{L^\infty(H_0)} = 1$ and $\|\pi_1(X)\|_{L^2(H_0)} > 1 - \varepsilon$, and thus

$$\begin{aligned}(1-\varepsilon)^2 < \|\pi_1(X)\|^2_{L^2(H_0)} &= \mathrm{E}\langle \pi_1(X), \pi_1(X)\rangle_{H_0} = \mathrm{E}\langle \pi_1(X), X\rangle_{H_0}\\ &\le \|\pi_1(X)\|_{L^1(H_0)}\|X\|_{L^\infty(H_0)} = \|\pi_1(X)\|_{L^1(H_0)}.\end{aligned}$$

Consequently,

$$\|\pi_1\|_{L^\infty(H_0),L^1(H_0)} = 1 = \kappa(1)^{-2}\|\pi_1\|_{L^\infty_{\mathbb{R}},L^1_{\mathbb{R}}} = \tfrac{\pi}{2}\|\pi_1\|_{L^\infty_{\mathbb{R}},L^1_{\mathbb{R}}},$$

which shows that Grothendieck's constant is at least $\pi/2 \approx 1.57$, cf. Grothendieck (1956).

In the complex case, the same argument with complex Gaussian Hilbert spaces shows that the complex Grothendieck's constant is at least $\kappa(1)_{\mathbb{C}}^{-2} = 4/\pi \approx 1.27$. (These bounds are not sharp, cf. Pisier (1986, §5.e).)

The results above may be extended to vector-valued functions as follows. Suppose that H, H_1, H_2 are three given Hilbert spaces (all real or all complex) and that T maps H_1-valued functions into H_2-valued. We then want a mapping $\widetilde{T}$ with formally the same defining condition $\widetilde{T}(f \otimes h) = Tf \otimes h$ as before; we thus want $\widetilde{T}$ to map $(H_1 \otimes H)$-valued functions into $(H_2 \otimes H)$-valued. (We consider only the Hilbert space tensor products of the spaces, see Appendix E.)

Theorem 12.1 and its proof extend as follows.

THEOREM 12.12. *Let $T\colon L^p(M_1, \mathcal{M}_1, \mu_1; H_1) \to L^p(M_2, \mathcal{M}_2, \mu_2; H_2)$ be a bounded linear operator, where $(M_1, \mathcal{M}_1, \mu_1)$ and $(M_2, \mathcal{M}_2, \mu_2)$ are two measure spaces, H_1 and H_2 are Hilbert spaces and $1 \le p < \infty$. If H is a third Hilbert space, then following hold, for some constants C_p, c_p.*

(i) *There exists a unique bounded linear operator $\widetilde{T}\colon L^p(M_1, \mathcal{M}_1, \mu_1; H_1 \otimes H) \to L^p(M_2, \mathcal{M}_2, \mu_2; H_2 \otimes H)$ such that $\widetilde{T}(f \otimes h) = Tf \otimes h$ for $f \in L^p(M_1, \mathcal{M}_1, \mu_1; H_1)$ and $h \in H$. Moreover, $\|\widetilde{T}\| \le C_p\|T\|$.*

(ii) *If $\|Tf\| \ge \delta\|f\|$ for all $f \in L^p(M_1, \mathcal{M}_1, \mu_1; H_1)$ and some $\delta > 0$, then $\|\widetilde{T}F\| \ge c_p\delta\|F\|$ for all $F \in L^p(M_1, \mathcal{M}_1, \mu_1; H_1 \otimes H)$.*

PROOF. We may assume that H_1, H_2 and H are three Gaussian Hilbert spaces, defined on probability spaces Ω_1, Ω_2 and Ω. Let $f_1, \dots, f_n \in L^p(M_1)$, $h'_1, \dots, h'_n \in H_1$ and $h_1, \dots, h_n \in H$. We obtain by Fubini's theorem and (1.1),

$$
\begin{aligned}
\Big\|\sum_{i=1}^n T(f_i \otimes h'_i) \otimes h_i\Big\|_{L^p(M_2;L^p(\Omega_2\times\Omega))} &= \Big\|\sum_{i=1}^n T(f_i \otimes h'_i) \otimes h_i\Big\|_{L^p(\Omega;L^p(M_2;L^p(\Omega_2)))} \\
&= \kappa(p)\Big\|\sum_{i=1}^n T(f_i \otimes h'_i) \otimes h_i\Big\|_{L^p(\Omega;\, L^p(M_2;H_2))} \\
&\le \kappa(p)\|T\|\,\Big\|\sum_{i=1}^n f_i \otimes h'_i \otimes h_i\Big\|_{L^p(\Omega;\, L^p(M_1;H_1))} \\
&= \|T\|\,\Big\|\sum_{i=1}^n f_i \otimes h'_i \otimes h_i\Big\|_{L^p(\Omega;\, L^p(M_1;\, L^p(\Omega_1)))} \\
&= \|T\|\,\Big\|\sum_{i=1}^n f_i \otimes h'_i \otimes h_i\Big\|_{L^p(M_1;\, L^p(\Omega_1\times\Omega))}
\end{aligned}
\tag{12.7}
$$

We may regard $H_1 \oplus H$ as a Gaussian Hilbert space defined on the product space $\Omega_1 \times \Omega$. Now $H_1 \otimes H \subset P_2(H_1 \oplus H)$; in fact, it is contained in $(H_1 \oplus H)^{:2:}$, because it is spanned by elements of the type $h_1 h = {:}h_1 h{:}$. (It equals the space ${:}H_1 H{:}$ in the notation used in Example 3.52.) Thus, by Theorem 3.50, the norm in $H_1 \otimes H$, which equals the L^2-norm on $\Omega_1 \times \Omega$, is equivalent to the L^p-norm, and similarly the norm in $H_2 \otimes H$ is equivalent to the norm in $L^p(\Omega_2 \times \Omega)$. Consequently, (12.7) yields

$$
\Big\|\sum_{i=1}^n T(f_i \otimes h'_i) \otimes h_i\Big\|_{L^p(M_2;\, H_2\otimes H)} \le C_p\|T\|\,\Big\|\sum_{i=1}^n f_i \otimes h'_i \otimes h_i\Big\|_{L^p(M_1;\, H_1\otimes H)}.
$$

This implies (i), and (ii) follows similarly by reversing the inequalities. □

Note that, unlike the special case $H_1 = H_2 = \mathbb{R}$ or $\mathbb{C}$ in Theorem 12.1, we do not claim that the norm of $\widetilde{T}$ equals $\|T\|$; this is not an artefact of the proof as is seen by Example 12.15 below.

Note also that we assumed $p < \infty$ above. Part (i) holds also for $p = \infty$ (with L^p replaced by L^∞_0); this follows by a duality argument or by Theorem 12.14 below. On the other hand, part (ii) does not extend to $p = \infty$, as is seen by Example 12.16 below.

REMARK 12.13. The constants C_p and c_p come (in the proof above) from the constants in the equivalence of the L^2- and L^p-norms on $H_1 \otimes H = {:}H_1 H{:}$ and $H_2 \otimes H = {:}H_2 H{:}$. It follows from the proof above and Example 3.52 that for $1 \le p \le 2$, we can take $C_p = \kappa(p)^{-1}$ and $c_p = \kappa(p)$, where $\kappa(p)$ is as in (1.1), and indeed Example 12.15 below shows that this value for C_p is the best possible. (Recall that $\kappa(p) = \sqrt{2}\big(\Gamma(\frac{p+1}{2})/\sqrt{\pi}\big)^{1/p}$ in the real case; in the complex case, this is replaced by $\kappa(p)_{\mathbb{C}} = \Gamma(\frac{p}{2}+1)^{1/p}$.)

For $p \geq 2$, the same argument yields $C_p = \kappa(p)$ and $c_p = \kappa(p)^{-1}$. This value of C_p is not the best possible; a simple duality argument shows that we can take $C_p = C_{p'} = \kappa(p')^{-1}$, and again Example 12.15 shows that this is the best possible value.

For c_p, we do not know the best possible value for $p < 2$, but for $p \geq 2$, the argument above and Example 12.15 show that $c_p = \kappa(p)^{-1}$ is best possible. Note that $c_p \to 0$ as $p \to \infty$, while C_p can be chosen bounded. (Note also that the proof yields $c_p = \kappa(p) \geq \kappa(1)$ for $1 \leq p \leq 2$; hence there is no duality argument relating c_p and $c_{p'}$ as there is for C_p.)

Also Theorems 12.2 and 12.8 extend to the present setting; since we may lose a constant in any case, there is no reason to separate them.

THEOREM 12.14. *Let* $T\colon L^p(M_1,\mathcal{M}_1,\mu_1;H_1) \to L^p(M_2,\mathcal{M}_2,\mu_2;H_2)$ *be a bounded linear operator, where* $(M_1,\mathcal{M}_1,\mu_1)$ *and* $(M_2,\mathcal{M}_2,\mu_2)$ *are two measure spaces,* H_1 *and* H_2 *are Hilbert spaces and* $1 \leq p < \infty$, $1 \leq q \leq \infty$. *If* H *is a third Hilbert space, then there exists a unique bounded linear operator* $\widetilde{T}\colon L^p(M_1,\mathcal{M}_1,\mu_1;H_1\otimes H) \to L^q(M_2,\mathcal{M}_2,\mu_2;H_2\otimes H)$ *such that* $\widetilde{T}(f\otimes h) = Tf\otimes h$ *for* $f \in L^p(M_1,\mathcal{M}_1,\mu_1;H_1)$ *and* $h \in H$. *Moreover,* $\|\widetilde{T}\| \leq C\|T\|$, *where* $C = \pi/2\ln(1+\sqrt{2}) \approx 1.782$.

For $1 \leq q \leq p = \infty$, *the conclusions remain valid with* L^p *replaced by* L_0^∞.

PROOF. We follow the proof of Theorem 12.8. It suffices to consider simple functions in $L^p(M_1;H_1\otimes H)$; moreover, choosing an orthonormal basis $\{e_\alpha\}$ in H_1 and approximating elements in $H_1\otimes H$ by finite sums $\sum_j e_{\alpha_j}\otimes h_j$, we see that it suffices to consider simple functions of the special form $\sum_{i,k}\mathbf{1}_{A_i}e_k\otimes h_{ik}$, where $A_i \in \mathcal{M}_1$ are disjoint, $\{e_k\}_1^N$ are orthonormal in H_1, and $h_{ik} \in H$. If, similarly, $B_j \in \mathcal{M}_2$ are disjoint, $\{e'_l\}_1^M$ are orthonormal in H_2, and $h'_{jl} \in H$, then as in (12.6), again letting $X(h)$ and $X'(h)$ be as in Lemma 12.7,

$$\int_{M_2}\Big\langle\sum_{i,k}T(\mathbf{1}_{A_i}\otimes e_k)\otimes h_{ik},\sum_{j,l}\mathbf{1}_{B_j}\otimes e'_l\otimes h'_{jl}\Big\rangle_{H_2\otimes H}d\mu_2$$
$$=\mathrm{E}\Big\langle T\Big(\sum_{i,k}X(h_{ik})\mathbf{1}_{A_i}\otimes e_k\Big),\sum_{j,l}X'(h'_{jl})\mathbf{1}_{B_j}\otimes e'_l\Big\rangle_{L^q(M_2;H_2)}. \tag{12.8}$$

Moreover, for a.e. ω and every $x \in M_1$,

$$\Big\|\sum_{i,k}X(h_{ik})\mathbf{1}_{A_i}\otimes e_k\Big\|_{H_1}^2 = \sum_i\mathbf{1}_{A_i}\sum_k|X(h_{ik})|^2 = \sum_i\mathbf{1}_{A_i}\sum_k C\|h_{ik}\|_H^2$$
$$= C\sum_i\mathbf{1}_{A_i}\Big\|\sum_k e_k\otimes h_{ik}\Big\|_{H_1\otimes H}^2 = C\Big\|\sum_{i,k}\mathbf{1}_{A_i}e_k\otimes h_{ik}\Big\|_{H_1\otimes H}^2$$

and thus, for a.e. ω,

$$\|T\sum_{i,k} X(h_{ik})\mathbf{1}_{A_i}\otimes e_k\|_{L^q(M_2;H_2)} \le \|T\|\,\|\sum_{i,k} X(h_{ik})\mathbf{1}_{A_i}\otimes e_k\|_{L^p(M_1;H_1)}$$
$$\le \|T\|\,C^{1/2}\|\sum_{i,k} \mathbf{1}_{A_i}e_k\otimes h_{ik}\|_{L^p(M_1;H_1\otimes H)}.$$

Similarly, for a.e. ω,

$$\|\sum_{j,l} X'(h'_{jl})\mathbf{1}_{B_j}\otimes e'_l\|_{L^{q'}(M_2;H_2)} = C^{1/2}\|\sum_{j,l} \mathbf{1}_{B_j}e'_l\otimes h'_{jl}\|_{L^{q'}(M_2;H_2\otimes H)}.$$

Consequently, (12.8) and Hölder's inequality yield

$$\left|\int_{M_2}\Big\langle\sum_{i,k} T(\mathbf{1}_{A_i}\otimes e_k)\otimes h_{ik},\sum_{j,l}\mathbf{1}_{B_j}\otimes e'_l\otimes h'_{jl}\Big\rangle_{H_2\otimes H}d\mu_2\right|$$
$$\le C\|T\|\,\|\sum_{i,k}\mathbf{1}_{A_i}e_k\otimes h_{ik}\|_{L^p(M_1;H_1\otimes H)}\,\|\sum_{j,l}\mathbf{1}_{B_j}e'_l\otimes h'_{jl}\|_{L^{q'}(M_2;H_2\otimes H)},$$

which implies the result. □

EXAMPLE 12.15. It is not possible to take C_p and c_p in Theorem 12.12 as 1 (except when $p = 2$).

For an example, let H_1 be a Gaussian Hilbert space on a probability space $(\Omega_1,\mathcal{F}_1,\mathrm{P}_1)$ and let M_1 consist of a single point with $\mu_1(M_1) = 1$; thus $L^p(M_1;H_1) = H_1$ isometrically, for every $p \le \infty$, if we identify $1\otimes\xi \in L^p(H_1)$ with $\xi \in H_1$. Further, let $H_2 = \mathbb{R}$ and $(M_2,\mathcal{M}_2,\mu_2) = (\Omega_1,\mathcal{F}_1,\mathrm{P}_1)$, and let T be the inclusion map $H_1 \to L^p(M_2,\mathcal{M}_2,\mu_2)$, regarded as a map $L^p(M_1;H_1) \to L^p(M_2,\mathcal{M}_2,\mu_2)$. Thus, in our usual notation, $T(1\otimes\xi) = \xi$, $\xi \in H_1$.

It follows that

$$\|T(1\otimes\xi)\|_{L^p(M_2)} = \|\xi\|_p = \kappa(p)\|\xi\|_{H_1} = \kappa(p)\|1\otimes\xi\|_{L^p(M_1;H_1)},$$

and thus $\|T\| = \kappa(p)$. (In fact, $\kappa(p)^{-1}T$ is an isometry of $L^p(M_1;H_1)$ into $L^p_{\mathbb{R}}(M_2)$.) Let H be another Gaussian Hilbert space, defined on a probability space $(\Omega,\mathcal{F},\mathrm{P})$. Then $\widetilde{T}\colon L^p(M_1;H_1\otimes H) \cong H_1\otimes H \to L^p(M_2;H_2\otimes H) = L^p(M_2;H)$ may be identified with the identity map $H_1\otimes H \to L^p(M_2;H) = L^p(\Omega_1;H)$, regarding $H_1\otimes H$ as a space of functions on $\Omega_1\times\Omega$ in the standard way (cf. Example E.10). If $X \in H_1\otimes H$, then

$$\|\widetilde{T}X\|_{L^p(\Omega_1;H)} = \kappa(p)^{-1}\|X\|_{L^p(\Omega_1;\,L^p(\Omega))} = \kappa(p)^{-1}\|X\|_p$$

and thus

$$\|\widetilde{T}\| = \sup_{X\in H_1\otimes H}\big(\kappa(p)^{-1}\|X\|_p/\|X\|_2\big).$$

If now $1 \le p < 2$ and $\dim H_1 = \dim H = \infty$, then $\sup_{X \in H_1 \otimes H}(\|X\|_p/\|X\|_2) = \kappa(p)$ by Example 3.52, and thus $\|\widetilde{T}\| = 1 = \kappa(p)^{-1}\|T\| > \|T\|$.

For $2 < p \le \infty$, we consider the adjoint map $T^*\colon L^p(M_2) \to L^p(M_1; H) \cong H$; this is the projection $\pi_1\colon L^p(\Omega, \mathcal{F}, \mathrm{P}) \to H$. It is easily seen that the extension $\widetilde{T}^*$ is the adjoint of $\widetilde{T}\colon L^{p'}(M_1; H) \to L^{p'}(M_2)$, where $1/p + 1/p' = 1$, and it follows that $\|\widetilde{T}^*\| = \|\widetilde{T}\| = \kappa(p')^{-1}\|T^*\| > \|T^*\|$.

For the lower bound c_p, we similarly observe that for $p \ge 2$, the operator T yields, again by Example 3.52 and assuming $\dim H_1 = \dim H = \infty$,

$$\inf_{X \in H_1 \otimes H}\bigl(\|\widetilde{T}X\|_{L^p(\Omega_1; H)}/\|X\|_{H_1 \otimes H}\bigr) = \inf_{X \in H_1 \otimes H}\bigl(\kappa(p)^{-1}\|X\|_p/\|X\|_2\bigr) = 1.$$

Since we have $\|TX\|_p = \kappa(p)\|X\|_{L^p(M_1; H_1)}$ for every $X \in L^p(M_1; H_1)$, this shows that we must take $c_p \le \kappa(p)^{-1}$ when $p \ge 2$.

EXAMPLE 12.16. The preceding example shows that $c_p \to 0$ in part (ii) of Theorem 12.12 as $p \to \infty$, which at least suggests that this part of the theorem does not extend to $p = \infty$. An example verifying this is obtained by taking M_1 to be a one-point space as in Example 12.15 and H_1 any separable infinite-dimensional Hilbert space. Moreover, let $M_2 = \mathbb{N}$ (with counting measure) and let $H_2 = \mathbb{R}$ (or $\mathbb{C}$); thus $L^p(M_2; H_2) = \ell^\infty$. Finally, let $\{h_k\}_1^\infty$ be a countable dense subset of the unit ball of H_1, and define $T\colon L^\infty(M_1; H_1) \cong H_1 \to \ell^\infty$ by $Th = (\langle h, h_k\rangle)_1^\infty$; then T is an isometry.

Now let $H = H_1$, and let $\{e_n\}_1^\infty$ be an orthonormal basis in H. If $N \ge 1$, then $X_N = \sum_{n=1}^N e_n \otimes e_n \in H_1 \otimes H \cong L_0^\infty(M_1; H_1 \otimes H)$ with $\|X_N\| = N^{1/2}$ and

$$\begin{aligned}\|\widetilde{T}X_N\|_{\ell^\infty(H)} &= \Bigl\|\Bigl(\sum_{n=1}^N \langle e_n, h_k\rangle e_n\Bigr)_k\Bigr\|_{\ell^\infty(H)} = \sup_k \Bigl\|\sum_{n=1}^N \langle e_n, h_k\rangle e_n\Bigr\|_H \\ &\le \sup_k \|h_k\|_H = 1.\end{aligned}$$

Hence there is no constant c such that $\|\widetilde{T}X_N\| \ge c\|X_N\|$ for all N.

REMARK 12.17. Example 12.15 gives yet another proof that the constant C in Lemma 12.7, and thus in Theorem 12.8 (cf. Remark 12.6) cannot be taken as 1. (The example gives the explicit estimate $C \ge \kappa(1)^{-1} = \sqrt{\pi/2}$ for Grothendieck's constant, which is inferior to the bound obtained in Example 12.11.)

REMARK 12.18. We have here only considered vector-valued functions with values in a Hilbert space. If we instead consider functions with values in a Banach space, the situation is more complicated. For example, if we replace the Hilbert space H by a Banach space B in Theorem 12.1, then (i) holds for $p = 1$ and $p = \infty$, as is easily shown, but for $1 < p < \infty$ there is in general no bounded linear operator $\widetilde{T} : L^p(M_1; B) \to L^p(M_2; B)$ with $\widetilde{T}(f \otimes b) = Tf \otimes b$ for $f \in L^p(M_1, \mathcal{M}_1, \mu_1)$ and $b \in B$.

XIII

Some operators from quantum physics

In this chapter we assume that H is a Gaussian Hilbert space. We will use H to define and study several operators on $L^2(\Omega, \mathcal{F}(H), \mathrm{P})$. These operators are important in quantum physics, but we will not go into any such applications here; cf. for example Segal (1956), Glimm and Jaffe (1981, Chapter 6) and Meyer (1993).

There are also other applications of these operators, and we will use some of the results below in Chapter 15.

Some of the operators will be studied again in Chapters 15 and 16, where we also consider actions on L^p for $p \neq 2$.

The reader may note that the operators treated here can be defined for any abstract Fock space based on an arbitrary Hilbert space, and that many of the results make sense in this generality, cf. for example Baez, Segal and Zhou (1992), Meyer (1993) and Parthasaraty (1992). Nevertheless, we will exclusively consider the Gaussian case here, where the extra structure is both helpful and, we hope, illuminating. (Of course, any result that can be stated for an abstract Fock space is valid in general as long as it is valid for the concrete realization treated here for Gaussian spaces.)

We warn the reader that several of the operators defined below will be unbounded and defined only on a dense subset of L^2. A general background on such operators can be found in for example Conway (1990) or Rudin (1991). The reader who is not interested in these technical details, such as the properties of being closed, symmetric or self-adjoint, may ignore all such comments and just consider the operators as defined on the space $\mathcal{P}(H)$ or $\overline{\mathcal{P}}_*(H)$.

For convenience, we write L^2 for $L^2(\Omega, \mathcal{F}(H), \mathrm{P})$ in this chapter. (Alternatively, we assume that the σ-field $\mathcal{F}$ is generated by H.) We use $\mathrm{Dom}(T)$ to denote the domain of an operator T.

1. The Fourier–Wiener transform

This is the operator $F_W = \Gamma(iI) = M_i$ on $L^2_{\mathbb{C}}(\Omega, \mathcal{F}(H), \mathrm{P})$ given by Theorem 4.5 (Cameron and Martin 1947). We obtain immediately

THEOREM 13.1. *The Fourier–Wiener transform F_W is a unitary operator on $L^2_{\mathbb{C}}$, given by*

$$F_W\Big(\sum_0^\infty X_n\Big) = \sum_0^\infty i^n X_n, \qquad X_n \in H^{:n:}.$$

F_W is an algebra automorphism on $\overline{\mathcal{P}}_*$:

$$F_W(X \odot Y) = F_W(X) \odot F_W(Y), \qquad X, Y \in \overline{\mathcal{P}}_*. \tag{13.1}$$

Moreover, F_W^4 is the identity map. □

Clearly, the operator is characterized, as a bounded linear operator on $L^2_{\mathbb{C}}$, by (13.1) and

$$F_W(\xi) = i\xi, \qquad \xi \in H.$$

See Example 4.23 for another formula for the Fourier–Wiener transform.

EXAMPLE 13.2. If $\xi, \eta \in H$, then

$$F_W(\xi\eta) = F_W(:\xi\eta: + \operatorname{E}\xi\eta) = -:\xi\eta: + \operatorname{E}\xi\eta = -\xi\eta + 2\operatorname{E}\xi\eta.$$

This shows that ordinary products do not behave as well as Wick products under the Fourier–Wiener transform.

Similarly, using Corollary 3.17 and Theorem 3.4, with $\sigma^2 = \operatorname{E}\xi^2$,

$$F_W(\xi^3) = -i\xi^3 + 6i\sigma^2\xi,$$
$$F_W(\xi^4) = \xi^4 - 12\sigma^2\xi^2 + 12\sigma^4,$$

and more generally

$$F_W(\xi^n) = \sum_{r=0}^{n/2} \frac{n!}{r!(n-2r)!}(i\xi)^{n-2r}\sigma^{2r}.$$

We leave the details to the reader.

EXAMPLE 13.3. If $\zeta \in H_{\mathbb{C}}$, then, as in Example 4.8,

$$F_W(:e^{\zeta}:) = F_W\Big(\sum_0^\infty \frac{1}{n!} :\zeta^n:\Big) = \sum_0^\infty \frac{i^n}{n!} :\zeta^n: = :e^{i\zeta}:$$

and thus, by Theorem 3.33,

$$F_W(e^{\zeta}) = F_W(e^{\operatorname{E}\zeta^2/2} :e^{\zeta}:) = e^{\operatorname{E}\zeta^2/2} :e^{i\zeta}: = e^{i\zeta + \operatorname{E}\zeta^2}.$$

Because of Theorem 2.12 and Corollary 3.40, each of these formulae characterizes F_W.

EXAMPLE 13.4. Consider the one-dimensional Gaussian space spanned by the function x on the probability space $(\mathbb{R}, \mathcal{B}, \gamma_\sigma)$, where γ_σ is the Gaussian measure

$$d\gamma_\sigma = (2\pi\sigma^2)^{-1/2} e^{-x^2/2\sigma^2}\,dx,$$

with $\sigma^2 > 0$. (When $\sigma^2 = 1$, this is the space considered in Example 1.15; the general case differs only by a scaling.)

We showed in Examples 4.18 and 4.22 that if $\sigma^2 = 1$, then

$$F_W(f(x)) = \mathcal{M}_i f(x) = (4\pi)^{-1/2} e^{x^2/4} \int_{-\infty}^{\infty} e^{ixy/2} e^{-y^2/4} f(y)\, dy, \tag{13.2}$$

at least if f has compact support. By continuity, (13.2) is valid for every $f \in L^2(\gamma_\sigma)$ if we interpret the integral as the L^2 limit of $\int_{-N}^{N}$ as $N \to \infty$; in particular (13.2) is valid whenever the integral converges absolutely for some, and thus all, x. These restrictions apply to all integrals that follow in this example, but we will not mention them again.

In the general case, $\xi = x/\sigma$ is a standard normal variable. Hence, if $f \in L^2(\gamma_\sigma)$ and $g(t) = f(\sigma t)$,

$$\begin{aligned} F_W(f(x)) &= F_W(g(\xi)) \\ &= (4\pi)^{-1/2} e^{\xi^2/4} \int_{-\infty}^{\infty} e^{i\xi\eta/2} e^{-\eta^2/4} g(\eta) d\eta \\ &= (4\pi\sigma^2)^{-1/2} e^{x^2/4\sigma^2} \int_{-\infty}^{\infty} e^{ixy/2\sigma^2} e^{-y^2/4\sigma^2} f(y) dy. \end{aligned} \tag{13.3}$$

If we define

$$V_\sigma f(x) = (2\pi\sigma^2)^{-1/4} e^{-x^2/4\sigma^2} f(x) \tag{13.4}$$

so that V_σ is an isometry $L^2(\gamma_\sigma) \to L^2(dx)$, then (13.3) may be written

$$F_W f(x) = V_\sigma^{-1} (4\pi\sigma^2)^{-1/2} \int_{-\infty}^{\infty} e^{ixy/2\sigma^2} V_\sigma f(y) dy;$$

in other words, $V_\sigma F_W V_\sigma^{-1}$ is the unitary operator on $L^2(\mathbb{R}, dx)$ given by

$$f \mapsto (4\pi\sigma^2)^{-1/2} \int_{-\infty}^{\infty} e^{ixy/2\sigma^2} f(y) dy,$$

which is the usual Fourier transform up to a rescaling. Two common versions of the Fourier transform are given by the choices $\sigma^2 = 1/2$ and $\sigma^2 = 1/4\pi$.

EXAMPLE 13.5. The preceding example can be generalized to several dimensions by considering the d-dimensional Gaussian space consisting of the linear functions on the probability space $(\mathbb{R}^d, \gamma_\sigma^d)$, $d \geq 1$. There are only notational changes; in particular, if

$$V_\sigma f(x) = (2\pi\sigma^2)^{-d/4} e^{-|x|^2/4\sigma^2} f(x),$$

then $V_\sigma F_W V_\sigma^{-1}$ is the version of the Fourier transform on $L^2(\mathbb{R}^d, dx)$ given by

$$f \mapsto (4\pi\sigma^2)^{-d/2} \int_{\mathbb{R}^d} e^{ix\cdot y/2\sigma^2} f(y) dy$$

(at least when $f \in L^1(\mathbb{R}^d) \cap L^2(\mathbb{R}^d)$).

We observe also that $F_W^2 = \Gamma(i^2 I) = \Gamma(-I)$ is given by changing the sign of every variable in H. More precisely:

THEOREM 13.6.

$$F_W^2 f(\xi_1, \xi_2, \dots) = f(-\xi_1, -\xi_2, \dots) \tag{13.5}$$

for any sequence $\xi_1, \xi_2, \dots$ *in* H *and square integrable function* $f(\xi_1, \xi_2, \dots)$.

PROOF. The formula (13.5) is easily verified if f is a polynomial in a finite number of variables $\xi_1, \dots, \xi_n \in H$, using for example (13.1) and Corollary 3.17. The general case follows by standard approximation arguments. □

2. Annihilation and creation operators

The *annihilation operator* $A(\xi)$ and the *creation operator* $A^*(\xi)$ are, for every $\xi \in H$, two unbounded, closed, densely defined linear operators on $L^2(\Omega, \mathcal{F}(H), \mathrm{P})$. (Except that both operators are bounded, because they vanish, in the trivial case $\xi = 0$.)

In order to define the operators, we first restrict attention to $\overline{\mathcal{P}}_*$ and define

$$A^*(\xi)(X) = \xi \odot X, \qquad X \in \overline{\mathcal{P}}_*. \tag{13.6}$$

Recall that this operation is well-defined, and bounded on each $H^{:n:}$, by Theorem 3.47.

Before proceeding, recall that the *adjoint* T^* of a densely defined linear operator T in a Hilbert space is defined as follows: $x \in \mathrm{Dom}(T^*)$ if there exists a z such that $\langle x, Ty \rangle = \langle z, y \rangle$ for all $y \in \mathrm{Dom}(T)$, and then T^*x is defined to be this z (which is unique because T is assumed to be densely defined). This yields a linear operator defined on some subspace of the Hilbert space; $\mathrm{Dom}(T^*)$ is dense if and only if T is closable, i.e. if the closure of the graph of T is the graph of a linear operator $\overline{T}$ (called the closure of T), and then T^{**} is defined and equals $\overline{T}$. In particular, if T is closed and densely defined, then so is T^*, and $T^{**} = T$. (Proofs and more details may be found in Conway (1990, Chapter X).)

We will soon extend $A^*(\xi)$ to a larger domain, but to be careful, let $A_0^*(\xi)$ temporarily denote the operator with domain $\overline{\mathcal{P}}_*$ just defined by (13.6). We then define $A(\xi)$ as the adjoint of $A_0^*(\xi)$; this is legitimate because $\overline{\mathcal{P}}_*$ is dense in L^2 by Theorem 2.6. Moreover, let $Y \in H^{:n:}$ and $X \in H^{:m:}$. If $m = n-1$, then

$$\begin{aligned}\langle Y, A_0^*(\xi)X \rangle &= \langle Y, \xi \odot X \rangle = \langle Y, \pi_{m+1}(\xi X) \rangle \\ &= \langle Y, \xi X \rangle = \langle \xi Y, X \rangle = \langle \pi_{n-1}(\xi Y), X \rangle,\end{aligned}$$

while if $m \neq n-1$, $A_0^*(\xi)X \in H^{:m+1:}$ is orthogonal to $Y \in H^{:n:}$ and $\pi_{n-1}(\xi Y) \in H^{:n-1:}$ is orthogonal to $X \in H^{:m:}$, whence

$$\langle Y, A_0^*(\xi)X \rangle = 0 = \langle \pi_{n-1}(\xi Y), X \rangle.$$

Hence, by linearity, if $Y \in H^{:n:}$, then $\langle Y, A_0^*(\xi)X\rangle = \langle \pi_{n-1}(\xi Y), X\rangle$ for any $X \in \overline{\mathcal{P}}_*$, and thus the adjoint $A(\xi)Y = A_0^*(\xi)^*(Y)$ is defined and

$$A(\xi)Y = \pi_{n-1}(\xi Y).$$

This shows that $A(\xi)$ is defined on each $H^{:n:}$, and thus on $\overline{\mathcal{P}}_*$. In particular, $A(\xi)$ is densely defined so we can define $A^*(\xi)$ to be its adjoint $A(\xi)^*$. Thus $A^*(\xi) = A_0^*(\xi)^{**}$, which by the general facts quoted above equals the closure of $A_0^*(\xi)$; in particular, $A^*(\xi)$ is an extension of $A_0^*(\xi)$, so on $\overline{\mathcal{P}}_*$ it is given by (13.6). Since an operator and its closure have the same adjoint, this also yields $A(\xi) = A^*(\xi)^*$.

We collect some basic properties.

THEOREM 13.7. *The following hold for any $\xi, \eta \in H$.*

(i) *$A(\xi)$ and $A^*(\xi)$ are closed, densely defined operators in L^2 which are the adjoints of each other; in particular,*

$$\langle A(\xi)X, Y\rangle = \langle X, A^*(\xi)Y\rangle, \qquad X \in \mathrm{Dom}(A(\xi)),\ Y \in \mathrm{Dom}(A^*(\xi)). \tag{13.7}$$

(ii) *The domains of $A(\xi)$ and $A^*(\xi)$ contain $\overline{\mathcal{P}}_*$, and both operators map $\overline{\mathcal{P}}_*$ into itself; moreover*

$$A^*(\xi)X = \xi \odot X, \qquad X \in \overline{\mathcal{P}}_*.$$

(iii) *If $n \geq 0$, then*

$$A(\xi)(H^{:n:}) \subseteq H^{:n-1:} \qquad \text{and} \qquad A^*(\xi)(H^{:n:}) \subseteq H^{:n+1:}; \tag{13.8}$$

and the operators satisfy

$$A(\xi)X = \pi_{n-1}(\xi X), \qquad X \in H^{:n:},$$

and

$$A^*(\xi)X = \xi \odot X = \pi_{n+1}(\xi X), \qquad X \in H^{:n:}.$$

In particular, for any $\xi_1, \dots, \xi_n \in H$,

$$A(\xi)(\,{:}\xi_1 \cdots \xi_n{:}\,) = \sum_{j=1}^{n} \mathrm{E}(\xi\xi_j)\,{:}\xi_1 \cdots \xi_{j-1}\xi_{j+1} \cdots \xi_n{:}\,, \tag{13.9}$$

$$A^*(\xi)(\,{:}\xi_1 \cdots \xi_n{:}\,) = \,{:}\xi\xi_1 \cdots \xi_n{:}\,. \tag{13.10}$$

(iv) *The restrictions to $H^{:n:}$ have the norms*

$$\|A(\xi)\|_{H^{:n:}} = \sqrt{n}\|\xi\|_2$$

and

$$\|A^*(\xi)\|_{H^{:n:}} = \sqrt{n+1}\|\xi\|_2.$$

Hence $A(\xi)$ and $A^*(\xi)$ are unbounded operators in L^2 unless $\xi = 0$.

(v) *$A(\xi)$ and $A^*(\xi)$ have the same domain*

$$\begin{aligned}\operatorname{Dom}(A(\xi)) &= \operatorname{Dom}(A^*(\xi)) \\ &= \Big\{\sum_0^\infty X_n : X_n \in H^{:n:} \text{ and } \sum_0^\infty (\|X_n\|_2^2 + \|A(\xi)X_n\|_2^2) < \infty\Big\}. \qquad (13.11)\end{aligned}$$

(vi) *$A(\xi)$ and $A^*(\xi)$ equal the closures of their restrictions to $\overline{\mathcal{P}}_*$ or $\mathcal{P}$.*

(vii) *Whenever $X \in \operatorname{Dom}(A(\xi)) = \operatorname{Dom}(A^*(\xi))$,*

$$\|A^*(\xi)X\|_2^2 = \|A(\xi)X\|_2^2 + \|\xi\|_2^2\|X\|_2^2$$

and

$$A(\xi)X + A^*(\xi)X = \xi X. \qquad (13.12)$$

(viii) *The operators satisfy the commutation relations*

$$[A(\xi), A(\eta)] = [A^*(\xi), A^*(\eta)] = 0, \qquad (13.13)$$

$$[A(\xi), A^*(\eta)] = \langle \xi, \eta \rangle I. \qquad (13.14)$$

REMARK 13.8. The relations (13.13)–(13.14) are known as the *canonical commutation relations*, or *CCR* for short. Since the operators are not everywhere defined, the commutation relations have to be interpreted with some care. The left hand sides of (13.13) and (13.14) are only densely defined, while the right hand sides are bounded, everywhere defined operators, so the domains are different. A more precise notation is $[A(\xi), A^*(\eta)] \subseteq \langle \xi, \eta \rangle I$, etc.

We will see other forms of the CCR in Theorems 13.21 and 13.28.

PROOF. The assertions in (i), (ii) and the first half of (iii) follow by the definitions and calculations before the theorem. The second half of (iii) follows from the first part and Theorems 3.15 and 3.47.

In order to prove (iv), we may without loss of generality suppose $\|\xi\| = 1$. If we then select an orthonormal basis in H containing ξ, it follows from Theorem 3.21, by collecting terms with the same Wick power of ξ, that if $H_1 = \{\xi\}^\perp \subset H$, then every element of $H^{:n:}$ may be written

$$X = \sum_{m=0}^{n} {:}\xi^m{:}\, Y_m,$$

where $Y_m \in H_1^{:n-m:}$. Moreover, we then have

$$\|X\|_2^2 = \sum_0^n \|{:}\xi^m{:}\|_2^2 \|Y_m\|_2^2 = \sum_0^n m! \|Y_m\|_2^2$$

and, using (13.9) and (13.10), it is easily seen that

$$\|A(\xi)X\|_2^2 = \|\sum_{m=0}^{n} m :\xi^{m-1}: Y_m\|_2^2 = \sum_{m=0}^{n} m^2(m-1)!\|Y_m\|_2^2 \le n\|X\|_2^2,$$

$$\|A^*(\xi)X\|_2^2 = \|\sum_{m=0}^{n} :\xi^{m+1}: Y_m\|_2^2 = \sum_{m=0}^{n}(m+1)!\|Y_m\|_2^2 \le (n+1)\|X\|_2^2.$$

Since the choice $X = :\xi^n:$ yields equalities, (iv) follows.

If $X \in \overline{\mathcal{P}}_*$, then

$$A^*(\xi)A^*(\eta)X = \xi \odot \eta \odot X = \eta \odot \xi \odot X = A^*(\eta)A^*(\xi)X;$$

hence $[A^*(\xi), A^*(\eta)] = A^*(\xi)A^*(\eta) - A^*(\eta)A^*(\xi) = 0$ on $\overline{\mathcal{P}}_*$. By the duality (13.7), or by (13.9), similarly $[A(\xi), A(\eta)] = 0$ on $\overline{\mathcal{P}}_*$. Further, (13.9) and (13.10) imply

$$A(\xi)A^*(\eta) :\xi_1 \cdots \xi_n: = \mathrm{E}(\xi\eta) :\xi_1 \cdots \xi_n: + A^*(\eta)A(\xi) :\xi_1 \cdots \xi_n:.$$

Hence, by linearity and continuity on each $H^{:n:}$, using Corollary 3.27,

$$[A(\xi), A^*(\eta)]X = \mathrm{E}(\xi\eta)X$$

for every $X \in \overline{\mathcal{P}}_*$.

This proves (13.13) and (13.14) on $\overline{\mathcal{P}}_*$. Moreover, if $X \in \overline{\mathcal{P}}_*$, then

$$\begin{aligned}\|A^*(\xi)X\|_2^2 &= \langle A^*(\xi)X, A^*(\xi)X\rangle \\ &= \langle A(\xi)A^*(\xi)X, X\rangle = \langle A^*(\xi)A(\xi)X, X\rangle + \langle \|\xi\|_2^2 X, X\rangle \\ &= \|A(\xi)X\|_2^2 + \|\xi\|_2^2\|X\|_2^2. \end{aligned} \tag{13.15}$$

Furthermore, if $X \in H^{:n:}$, then

$$A(\xi)X + A^*(\xi)X = \pi_{n-1}(\xi X) + \pi_{n+1}(\xi X) = \xi X,$$

since $\pi_m(\xi X) = 0$ whenever $m \ne n \pm 1$ as follows from Theorem 3.15. Hence (vii) holds for $X \in \overline{\mathcal{P}}_*$.

Next, suppose that $X = \sum_n X_n \in \mathrm{Dom}(A(\xi))$ and let $A(\xi)X = \sum_n Y_n$ with $X_n, Y_n \in H^{:n:}$. Then, for every $Z \in H^{:n:}$, since $A^*(\xi)Z \in H^{:n+1:}$ by (iii),

$$\langle Y_n, Z\rangle = \langle A(\xi)X, Z\rangle = \langle X, A^*(\xi)Z\rangle = \langle X_{n+1}, A^*(\xi)Z\rangle = \langle A(\xi)X_{n+1}, Z\rangle.$$

Since Y_n and $A(\xi)X_{n+1}$ belong to $H^{:n:}$, this implies $Y_n = A(\xi)X_{n+1}$, and thus

$$\begin{aligned}\sum_0^\infty(\|X_n\|_2^2 + \|A(\xi)X_n\|_2^2) &= \sum_0^\infty(\|X_n\|^2 + \|Y_{n-1}\|^2) \\ &= \|X\|_2^2 + \|A(\xi)X\|_2^2 < \infty.\end{aligned}$$

Conversely, if $\sum_0^\infty(\|X_n\|_2^2 + \|A(\xi)X_n\|_2^2) < \infty$, with $X_n \in H^{:n:}$, then $\sum_0^N X_n \to X = \sum_0^\infty X_n$ and $A(\xi)(\sum_0^N X_n) \to Y = \sum_0^\infty A(\xi)X_n$ as $N \to \infty$. Since $A(\xi)$ is a closed operator, $X \in \mathrm{Dom}(A(\xi))$.

We have proved that $\mathrm{Dom}(A(\xi))$ is given by the set in (13.11). A similar argument shows that $\mathrm{Dom}(A^*(\xi))$ is given by the corresponding set with $A(\xi)$

replaced by $A^*(\xi)$. Hence (13.15) implies $\operatorname{Dom}(A(\xi)) = \operatorname{Dom}(A^*(\xi))$, and (v) is proved.

This argument shows also that if $X \in \operatorname{Dom}(A(\xi)) = \operatorname{Dom}(A^*(\xi))$, and $X_n = \pi_{\leq n} X$, then $X_n \to X$, $A(\xi)X_n \to A(\xi)X$ and $A^*(\xi)X_n \to A^*(\xi)X$ as $n \to \infty$. Hence (vi) and the extensions of (vii) and (viii) from $\overline{\mathcal{P}}_*$ to the domains of the operators follow, and the proof is complete. □

EXAMPLE 13.9. If ξ is standard Gaussian, then by (13.9) and (13.10),

$$A(\xi) :\xi^n: = n :\xi^{n-1}:,$$
$$A^*(\xi) :\xi^n: = :\xi^{n+1}:.$$

In particular, if H is a one-dimensional Gaussian space spanned by a standard Gaussian variable ξ, then the variables $e_n = (n!)^{-1/2} :\xi^n:$, $n \geq 0$, form an orthonormal basis in L^2 by Theorem 3.21, and in terms of this basis,

$$A(\xi)e_n = \sqrt{n}\, e_{n-1},$$
$$A^*(\xi)e_n = \sqrt{n+1}\, e_{n+1}.$$

EXAMPLE 13.10. If $\xi, \eta \in H$ (or, more generally, if $\xi \in H$ and $\eta \in H_{\mathbb{C}}$), then, similarly, since $A(\xi)$ is a closed operator,

$$A(\xi) :e^\eta: = \sum_{n=0}^{\infty} \frac{1}{n!} A(\xi) :\eta^n: = \sum_{n=1}^{\infty} \frac{1}{(n-1)!} \operatorname{E}(\xi\eta) :\eta^{n-1}: = \operatorname{E}(\xi\eta) :e^\eta:$$

and thus by (13.12)

$$A^*(\xi) :e^\eta: = \xi :e^\eta: - A(\xi) :e^\eta: = (\xi - \operatorname{E}\xi\eta) :e^\eta:.$$

Since $e^\eta = e^{\operatorname{E}\eta^2/2} :e^\eta:$ by Theorem 3.33, we also have

$$A(\xi)(e^\eta) = \operatorname{E}(\xi\eta)e^\eta,$$
$$A^*(\xi)(e^\eta) = (\xi - \operatorname{E}(\xi\eta))e^\eta.$$

EXAMPLE 13.11. Consider again the one-dimensional Gaussian Hilbert space spanned by the function x on $(\mathbb{R}, \mathcal{B}, \gamma_\sigma)$, see Example 13.4.

Assume first $\sigma^2 = 1$. By Example 3.35,

$$\sum_{n=0}^{\infty} \frac{t^n}{n!} h_n(x) = :e^{tx}: = e^{tx - t^2/2}.$$

This function is an analytic function in $(t, x) \in \mathbb{C}^2$; hence we may differentiate termwise and obtain

$$\sum_{n=0}^{\infty} \frac{t^n}{n!} \frac{d}{dx} h_n(x) = \frac{\partial}{\partial x} e^{tx - t^2/2} = t e^{tx - t^2/2} = \sum_{n=1}^{\infty} \frac{t^n}{(n-1)!} h_{n-1}(x).$$

Hence we obtain the well-known formula (which also follows from (3.11))

$$\frac{d}{dx} h_n(x) = n h_{n-1}(x), \qquad n \geq 0. \tag{13.16}$$

Since $:x^n: = h_n(x)$ by Theorem 3.19, (13.9) and (13.16) yield

$$A(x) :x^n: = n :x^{n-1}: = \frac{d}{dx} :x^n:.$$

Consequently,

$$A(x) = \frac{d}{dx}$$

on every polynomial; by Theorem 13.7(vi) and continuity, this holds also on $\mathrm{Dom}(A(x))$, with the derivative interpreted in the sense of the theory of distributions (Rudin 1991, Chapters 6 and 7). We now obtain, from (13.12),

$$A^*(x) = x - \frac{d}{dx}. \tag{13.17}$$

In particular, letting f' denote the distributional derivative of f, this shows that if $f(x) \in \mathrm{Dom}(A(x))$, then $f' \in L^2(\gamma_1)$. Using elementary distribution theory, it is also easy to identify the domain of the operators and show that

$$\mathrm{Dom}\big(A(x)\big) = \mathrm{Dom}\big(A^*(x)\big) = \{f \in L^2(\gamma_1) : f' \in L^2(\gamma_1)\}, \tag{13.18}$$

where f' denotes the distributional derivative of f. Indeed, we have already shown that if $f(x) \in \mathrm{Dom}(A(x))$, then $f' \in L^2(\gamma_1)$. Conversely, suppose that $f,\ f' \in L^2(\gamma_1)$. Let $p(x)$ be a polynomial. Then, since f is a tempered distribution and $p(x)e^{-x^2/2}$ is a rapidly decreasing function, by the definition of derivatives in distribution theory and using (13.17),

$$\begin{aligned}\int f'(x)p(x)e^{-x^2/2}\,dx &= -\int f(x)\frac{d}{dx}\big(p(x)e^{-x^2/2}\big)\,dx \\ &= -\int f(x)(p'(x) - xp(x))e^{-x^2/2}\,dx = \int f(x)\big(A^*(x)p\big)(x)e^{-x^2/2}\,dx,\end{aligned}$$

and thus in $L^2(\gamma_1)$,

$$\langle f', p\rangle = \langle f, A^*(x)p\rangle, \qquad p \in \mathcal{P},$$

which shows that $f \in \mathrm{Dom}(A(x))$ by the definition of $A(x)$.

It is here possible to avoid distributional derivatives; $\mathrm{Dom}(A(x))$ also equals the space of all $f \in L^2(\gamma_1)$ that are absolutely continuous with $f' \in L^2(\gamma_1)$, where f' is the usual derivative (which then exists a.e.). This is easily proved from (13.18), or by Theorem 15.98 below.

Note that Theorem 13.7(vii) implies that if $f(x),\ f'(x) \in L^2(\gamma_1)$, then also $xf(x) \in L^2(\gamma_1)$.

For a general $\sigma^2 > 0$, we have $:x^n: = \sigma^n h_n(x/\sigma)$, and it follows again that

$$\frac{d}{dx} :x^n: = n :x^{n-1}:$$

which yields, by (13.9),

$$A(x) :x^n: = n\sigma^2 :x^{n-1}: = \sigma^2 \frac{d}{dx} :x^n:.$$

Hence we now have, on the natural domains,

$$A(x) = \sigma^2 \frac{d}{dx},$$
$$A^*(x) = x - \sigma^2 \frac{d}{dx}.$$

EXAMPLE 13.12. For the d-dimensional Gaussian space consisting of the linear functions on $(\mathbb{R}^d, \gamma_\sigma^d)$ we similarly obtain, checking first on the basis given by Theorem 3.21,

$$A(x_i) = \sigma^2 \frac{\partial}{\partial x_i},$$
$$A^*(x_i) = x_i - \sigma^2 \frac{\partial}{\partial x_i}.$$

Hence, for any $a \in \mathbb{R}^d$,

$$A(a \cdot x) = \sigma^2 \partial_a, \tag{13.19}$$

where ∂_a denotes the directional derivative along a. As in Example 13.11, the domains of these operators consist of all functions in L^2 such that these derivatives belong to L^2 (interpreting the derivatives as distributions, for example). Cf. also Theorem 15.98 and Example 15.107 below.

EXAMPLE 13.13. *Stein's method* to show results on normal approximation of random variables can be described as follows, cf. Stein (1986) and Diaconis and Zabell (1991). Consider again the Gaussian space spanned by $\xi = x$ on $(\mathbb{R}, \mathcal{B}, \gamma)$. By Example 13.9 the creation operator $A^*(\xi)$ is given by

$$A^*(\xi)\Big(\sum_0^\infty a_n e_n\Big) = \sum_1^\infty \sqrt{n} a_{n-1} e_n$$

where $e_n \in H^{:n:}$ is as in Example 13.9.

It follows that $A^*(\xi)$ maps its domain onto

$$\bigoplus_1^\infty H^{:n:} = L^2 \ominus H^{:0:} = \{X \in L^2 : \mathrm{E}\, X = 0\} = \Big\{f \in L^2(\gamma) : \int f\, d\gamma = 0\Big\}$$

with an inverse $A^*(\xi)^{-1}\colon \sum_1^\infty a_n e_n \mapsto \sum_0^\infty (n+1)^{-1/2} a_{n+1} e_n$.

Now, let $f \in L^2(\gamma)$. Then $f - \mathrm{E}\, f(\xi)$ belongs to the range of $A^*(\xi)$, and if we define $g = A^*(\xi)^{-1}(f - \mathrm{E}\, f(\xi))$, then by (13.17)

$$f(x) - \mathrm{E}\, f(\xi) = A^*(x)g(x) = xg(x) - g'(x).$$

Hence if W is any random variable such that the expectations exist,

$$\mathrm{E}\, f(W) - \mathrm{E}\, f(\xi) = \mathrm{E}\big(Wg(W) - g'(W)\big). \tag{13.20}$$

The left hand side of (13.20) vanishes if $W \sim \mathrm{N}(0, 1)$, and the supremum of the left hand side over a suitable class F of f yields a measure of the distance from the distribution of W to the standard normal distribution.

(One attractive choice is $F = \{f : |f(x) - f(y)| \le |x - y|, x, y \in \mathbb{R}\}$, which gives the Wasserstein distance, also known as the Dudley, Fortet–Mourier or Kantorovich distance.) Stein's method consists of estimating this distance by estimating the right hand side of (13.20), using properties of $A^*(x)^{-1}$ and specific properties of the random variable W under study.

Note that $A^*(x)^{-1}$ is given by the explicit formula

$$A^*(x)^{-1}f(x) = -e^{x^2/2}\int_{-\infty}^{x} f(t)e^{-t^2/2}\,dt.$$

We give some further results for the annihilation and creation operators.

THEOREM 13.14. *For any $\xi \in H$,*

$$F_W^{-1}A(\xi)F_W = iA(\xi)$$
$$F_W^{-1}A^*(\xi)F_W = -iA^*(\xi).$$

In particular, $\mathrm{Dom}(A(\xi)) = \mathrm{Dom}(A^*(\xi))$ *is invariant under F_W.*

PROOF. It suffices to verify the equalities on each $H^{:n:}$, but then they follow immediately by (13.8). □

In the next two results, $\mathcal{N}$ denotes the number (or Ornstein–Uhlenbeck) operator defined in Example 4.7.

THEOREM 13.15. *If $\{\xi_i\}_{i\in I}$ is an orthonormal basis in H, then*

$$\sum_{i\in I} A^*(\xi_i)A(\xi_i) = \mathcal{N}.$$

More precisely, the sum $\sum_i A^*(\xi_i)A(\xi_i)X$ is defined and converges in L^2 if and only if $X \in \mathrm{Dom}(\mathcal{N}) = \{X \in L^2 : \sum_0^\infty n^2\|\pi_n(X)\|_2^2 < \infty\}$, and then the sum equals $\mathcal{N}X$.

PROOF. Consider the orthonormal basis $e_\alpha = (\alpha!)^{-1/2} :\xi^\alpha:$ given by Theorem 3.21. Since $A(\xi_i) :\xi^\alpha: = \alpha_i :\xi^{\alpha'}:$, with $\alpha'_j = \alpha_j - \delta_{ij}$ (provided $\alpha_i > 0$), and thus $A^*(\xi_i)A(\xi_i) :\xi^\alpha: = \alpha_i :\xi^\alpha:$, it follows that if $X = \sum_\alpha a_\alpha e_\alpha$, then

$$A^*(\xi_i)A(\xi_i)X = \sum_\alpha \alpha_i a_\alpha e_\alpha.$$

On the other hand, $\mathcal{N}X = \sum_\alpha |\alpha| a_\alpha e_\alpha$, with $|\alpha| = \sum_i \alpha_i$, and the result follows. □

THEOREM 13.16. *If $X \in \mathrm{Dom}(\mathcal{N}^{1/2}) = \{X \in L^2 : \sum_0^\infty n\|\pi_n(X)\|_2^2 < \infty\}$, then $X \in \mathrm{Dom}(A(\xi))$ for every $\xi \in H$, and*

$$\|A(\xi)X\|_2 \le \|\xi\|_2\|\mathcal{N}^{1/2}X\|_2 = \|\xi\|_2\Big(\sum_0^\infty n\|\pi_n(X)\|_2\Big)^{1/2},$$

$$\|A^*(\xi)X\|_2 \le \|\xi\|_2\|(\mathcal{N}+1)^{1/2}X\|_2 = \|\xi\|_2\Big(\sum_0^\infty (n+1)\|\pi_n(X)\|_2\Big)^{1/2}.$$

PROOF. A simple consequence of Theorem 13.7(iii),(iv),(v). □

So far there has been a symmetry between properties of the annihilation and the creation operators, but the final result is special for annihilation operators. (Further results follow from Theorem 15.98 below.)

THEOREM 13.17. *The annihilation operator $A(\xi)$ is a derivation in the algebras $\mathcal{P}$ and $\overline{\mathcal{P}}_*$, both for the ordinary multiplication and for the Wick multiplication. In other words, for any $X, Y \in \overline{\mathcal{P}}_*$,*

$$A(\xi)(XY) = (A(\xi)X) \cdot Y + X \cdot A(\xi)Y, \tag{13.21}$$

$$A(\xi)(X \odot Y) = (A(\xi)X) \odot Y + X \odot A(\xi)Y. \tag{13.22}$$

PROOF. Since $A(\xi)$ is continuous on every $\overline{\mathcal{P}}_n$, it suffices to show (13.21) and (13.22) for polynomial variables X and Y. By linearity, it suffices to verify (13.22) for Wick monomials $X = {:}\xi_1 \cdots \xi_n{:}$ and $Y = {:}\eta_1 \cdots \eta_m{:}$, but the result then follows by (13.9).

Similarly, (13.21) is equivalent to

$$A(\xi)(\xi_1 \cdots \xi_n) = \sum_{j=1}^{n} \mathrm{E}(\xi\xi_j)\xi_1 \cdots \xi_{j-1}\xi_{j+1} \cdots \xi_n, \tag{13.23}$$

which follows easily by (13.9) and Corollary 3.17. Alternatively, by polarization (Theorem D.2), if suffices to verify (13.23) when $\xi_1 = \cdots = \xi_n$, i.e.

$$A(\xi)(\eta^n) = n\,\mathrm{E}(\xi\eta)\eta^{n-1},$$

which follows by identifying the coefficients of t^n in the formula

$$A(\xi)e^{t\eta} = t\,\mathrm{E}(\xi\eta)e^{t\eta},$$

obtained by substituting $t\eta$ in Example 13.10. Some care has to be taken since $A(\xi)$ is not continuous, but for this argument it suffices that it is a closed operator.

A third proof of (13.21) may be given by observing that the explicit formula (13.19) in Example 13.12 implies that (13.21) holds for polynomial variables in the special d-dimensional Gaussian Hilbert spaces considered there. Since all Gaussian Hilbert spaces of the same dimension are isomorphic, (13.21) holds for polynomial variables in any finite-dimensional Gaussian space. The general result follows. □

REMARK 13.18. Conversely, $A(\xi)$ is characterized on $\mathcal{P}$ by either (13.21) or (13.22), together with $A(\xi)(\eta) = \mathrm{E}\,\xi\eta$ when $\eta \in H$.

REMARK 13.19. The operator $A(\xi)$ generates a one-parameter group of operators on $\overline{\mathcal{P}}_*$ given by $\exp(tA(\xi)) = \exp(A(t\xi)) = \sum_{n=0}^{\infty} \frac{t^n}{n!} A(\xi)^n$, $t \in \mathbb{R}$. (Note that the sum $\sum_{n=0}^{\infty} \frac{1}{n!} A(\xi)^n X$ is actually finite for every $X \in \overline{\mathcal{P}}_*$.) The derivation properties in Theorem 13.17 are, by simple algebraic manipulations, equivalent to the fact that these operators are algebra isomorphisms, for both the ordinary multiplication and the Wick multiplication. We will study these operators in greater generality in Chapter 14.

EXAMPLE 13.20. (*Wick ordering.*) We are now able to describe the connection between our definition of the Wick product and Wick's (1950) original definition. Wick worked in a quite different framework, viz. products of operators in quantum field theory instead of Gaussian random variables, but the link is provided by the definition above of annihilation and creation operators for Gaussian Hilbert spaces. (This corresponds physically to considering bosons only; Wick's definition is somewhat more general and allows also for fermions, but they have no counterpart in the present context.)

Hence, let H be a Gaussian Hilbert space, and consider the algebra of operators generated by the operators $A(\xi)$ and $A^*(\xi)$, $\xi \in H$. Since we are only interested in formal, algebraic properties, we regard all operators as defined on $\overline{\mathcal{P}}_*(H)$. We also define, for $X \in \overline{\mathcal{P}}_*$, the multiplication operator M_X on $\overline{\mathcal{P}}_*$ by $M_X(Y) = XY$, and note that by Theorem 13.7(vii), $M_\xi = A(\xi) + A^*(\xi)$ when $\xi \in H$.

The definition by Wick (1950) of his product (which he called *S-product*) of a number of such operators can be phrased as follows for operators of the form M_ξ: Express each operator as a sum of annihilation and creation operators, expand the product into a sum of products of such operators, and reorder each such product so that all creation operators come to the left and the annihilation operators to the right. (This is known as *Wick ordering*; note that the creation and annihilation operators commute among themselves, so their order does not matter.) We use Wick's notation $:UV\cdots Z:$ for this Wick product of the operators $U, V, \dots, Z$.

For example, for $\xi, \eta \in H$,

$$\begin{aligned} :M_\xi M_\eta: &= :(A(\xi) + A^*(\xi))(A(\eta) + A^*(\eta)): \\ &= A(\xi)A(\eta) + A^*(\eta)A(\xi) + A^*(\xi)A(\eta) + A^*(\xi)A^*(\eta). \end{aligned}$$

This is easily evaluated using the commutation relations Theorem 13.7(viii), which yield

$$:M_\xi M_\eta: - M_\xi M_\eta = A^*(\eta)A(\xi) - A(\xi)A^*(\eta) = -\langle \xi, \eta \rangle I$$

and thus, for any $X \in \overline{\mathcal{P}}_*$,

$$:M_\xi M_\eta:(X) = M_\xi M_\eta(X) - \langle \xi, \eta \rangle X = (\xi\eta - \langle \xi, \eta \rangle)X = :\xi\eta: X,$$

where $:\xi\eta:$ (as always for us) denotes the Wick product defined in Chapter 3 of the Gaussian random variables ξ and η. Consequently, the Wick product (in Wick's sense) of the multiplication operators M_ξ and M_η equals the multiplication operator corresponding to the Wick product (in our sense) of ξ and η; in other words, the two different Wick products coincide (at least for two elements of H) if we identify random variables in $\overline{\mathcal{P}}_*$ with the corresponding multiplication operators.

This extends to products of several variables in H as follows.

CLAIM. *If* $\xi_1, \dots, \xi_n \in H$, *then* $:M_{\xi_1} \cdots M_{\xi_n}: = M_{:\xi_1 \cdots \xi_n:}$.

We can prove this claim by induction on n; the case $n = 1$ is trivial (and $n = 2$ is verified above). Hence assume that the claim is true for some n and let $\xi_1, \dots, \xi_{n+1} \in H$. Write, for convenience, $\xi = \xi_{n+1}$ and $Y = :\xi_1 \cdots \xi_n:$. Then, by Wick's definition and induction,

$$\begin{aligned} :M_{\xi_1} \cdots M_{\xi_{n+1}}: &= A^*(\xi) :M_{\xi_1} \cdots M_{\xi_n}: + :M_{\xi_1} \cdots M_{\xi_n}: A(\xi) \\ &= A^*(\xi) M_Y + M_Y A(\xi). \end{aligned}$$

Hence, for any $X \in \overline{\mathcal{P}}_*$, using Theorem 13.17 and (13.12),

$$\begin{aligned} :M_{\xi_1} \cdots M_{\xi_{n+1}}: (X) &= A^*(\xi)(YX) + YA(\xi)(X) \\ &= A^*(\xi)(YX) + A(\xi)(YX) - XA(\xi)(Y) \\ &= \xi YX - XA(\xi)(Y) = (\xi Y - A(\xi)(Y))X \\ &= (A^*(\xi)(Y))X. \end{aligned}$$

Since $A^*(\xi)(Y) = \xi \odot :\xi_1 \cdots \xi_n: = :\xi_1 \cdots \xi_{n+1}:$ by our definitions of ξ and Y and Theorem 3.47, this verifies the induction step.

3. Position and momentum operators

In this section we use complex scalars and consider operators on $L^2 = L^2_{\mathbb{C}}$. Define, for $\xi \in H$, the *position operator* $Q(\xi)$ and the *momentum operator* $P(\xi)$ by

$$Q(\xi) = A(\xi) + A^*(\xi), \tag{13.24}$$

$$P(\xi) = -i(A(\xi) - A^*(\xi)). \tag{13.25}$$

These are symmetric operators on $\operatorname{Dom}(A(\xi)) = \operatorname{Dom}(A^*(\xi))$; apart from a factor 2 they are the real and imaginary parts (in the operator theory sense) of the operator $A(\xi)$. We extend these operators (keeping the same notations) by taking their closures. These are still symmetric, and a part of the following theorem is that they are self-adjoint.

THEOREM 13.21. *The following hold for any* $\xi, \eta \in H$.

(i) $P(\xi)$ *and* $Q(\xi)$ *are unbounded self-adjoint operators on* $L^2_{\mathbb{C}}(\Omega, \mathcal{F}(H), \mathrm{P})$ (*except that they vanish, and thus are bounded, when* $\xi = 0$).

(ii) *The position operator is just the multiplication operator*

$$Q(\xi)X = \xi X$$

defined on $\{X \in L^2 : \xi X \in L^2\}$.

(iii) *The operators are related by*

$$P(\xi) = F_W Q(\xi) F_W^{-1}, \tag{13.26}$$

$$Q(\xi) = -F_W P(\xi) F_W^{-1}. \tag{13.27}$$

(iv) *The position and momentum operators satisfy the canonical commutation relations*

$$[P(\xi), P(\eta)] = [Q(\xi), Q(\eta)] = 0, \tag{13.28}$$

$$[P(\xi), Q(\eta)] = -2i\langle \xi, \eta \rangle I. \tag{13.29}$$

Again, the commutation relations have to be interpreted with care, since the domains differ.

PROOF. By (13.12), $Q(\xi)X = \xi X$ for $X \in \operatorname{Dom} A(\xi)$. It is easily seen that taking the closure yields a self-adjoint operator, for example as follows. Let T denote the operator $X \mapsto \xi X$ defined on $\operatorname{Dom}(A(\xi))$, and suppose that $Y \in \operatorname{Dom}(T^*)$. Then Y, $T^*Y \in L^2$ and $\langle T^*Y, X\rangle = \langle Y, TX\rangle = \langle Y, \xi X\rangle$ for all $X \in \mathcal{P}$, whence

$$\langle T^*Y - \xi Y, X\rangle = 0, \qquad X \in \mathcal{P}.$$

Since $\xi \in L^4$, $\xi Y \in L^{4/3}$ by Hölder's inequality and thus $T^*Y - \xi Y \in L^{4/3}$. Moreover, $\mathcal{P}$ is dense in the dual space L^4 by Theorem 2.11, and thus $T^*Y = \xi Y$. Conversely, if $Y, \xi Y \in L^2$ then $\langle \xi Y, X\rangle = \langle Y, TX\rangle$, $X \in \operatorname{Dom}(T)$, so $Y \in \operatorname{Dom}(T^*)$.

Hence $\operatorname{Dom}(T^*) = \{Y \in L^2 : \xi Y \in L^2\}$ and $T^*Y = \xi Y$. Clearly T^* is symmetric, so $T^{**} \supseteq T^*$. On the other hand, $T \subseteq T^*$ and thus $T^* \supseteq T^{**}$. Hence $\overline{T} = T^{**} = T^*$, and T^* is self-adjoint. Since by definition $Q(\xi) = \overline{T}$, this proves (ii); it also proves that $Q(\xi)$ is self-adjoint.

Since the Gaussian random variable ξ is unbounded, unless $\xi = 0$, the multiplication operator $Q(\xi)$ is an unbounded operator.

Next, by Theorem 13.14, on $\operatorname{Dom}(A(\xi))$,

$$F_W^{-1}Q(\xi)F_W = F_W^{-1}A(\xi)F_W + F_W^{-1}A^*(\xi)F_W = iA(\xi) - iA^*(\xi) = -P(\xi)$$

and similarly

$$F_W^{-1}P(\xi)F_W = Q(\xi).$$

Clearly, these identities extend to the closures of the operators, which proves (iii).

In particular, $P(\xi)$ and $Q(\xi)$ are unitarily equivalent. Hence also $P(\xi)$ is self-adjoint and unbounded, which completes the proof of (i).

Finally, the canonical commutation relations (13.28) and (13.29) follow easily for variables in $\overline{\mathcal{P}}_*$ from the the definitions (13.24), (13.25) and corresponding relations (13.13) and (13.14) for the annihilation and creation operators. The general case follows by duality; if $X \in \operatorname{Dom}([P(\xi), Q(\eta)])$, then for every $Y \in \overline{\mathcal{P}}_*$,

$$\begin{aligned}\langle [P(\xi), Q(\eta)]X, Y\rangle &= \langle (P(\xi)Q(\eta) - Q(\eta)P(\xi))X, Y\rangle \\ &= \langle X, (Q(\eta)P(\xi) - P(\xi)Q(\eta))Y\rangle = \langle X, -[P(\xi), Q(\eta)]Y\rangle \\ &= \langle X, 2i\langle \xi, \eta\rangle Y\rangle = \langle -2i\langle \xi, \eta\rangle X, Y\rangle,\end{aligned}$$

which proves (13.29), and similarly for (13.28). □

COROLLARY 13.22. *For every* $\xi \in H$,

$$\operatorname{Dom}(A(\xi)) = \operatorname{Dom}(A^*(\xi)) = \operatorname{Dom}(P(\xi)) \cap \operatorname{Dom}(Q(\xi)) \tag{13.30}$$

and

$$A(\xi) = \tfrac{1}{2}(Q(\xi) + iP(\xi)), \tag{13.31}$$

$$A^*(\xi) = \tfrac{1}{2}(Q(\xi) - iP(\xi)). \tag{13.32}$$

(*By* (13.30), *the left and right hand sides have the same domain.*)

PROOF. The formulae (13.31) and (13.32) follow directly from (13.24) and (13.25) once we have verified (13.30).

Suppose that $X \in \operatorname{Dom}(P(\xi)) \cap \operatorname{Dom}(Q(\xi))$. For any $Y \in \operatorname{Dom}(A(\xi)) \subseteq \operatorname{Dom}(P(\xi)) \cap \operatorname{Dom}(Q(\xi))$, since $Q(\xi)$ and $P(\xi)$ are symmetric,

$$\begin{aligned}\langle X, A(\xi)Y\rangle &= \tfrac{1}{2}\langle X, Q(\xi)Y + iP(\xi)Y\rangle \\ &= \tfrac{1}{2}\langle X, Q(\xi)Y\rangle - \tfrac{i}{2}\langle X, P(\xi)Y\rangle \\ &= \tfrac{1}{2}\langle Q(\xi)X, Y\rangle - \tfrac{i}{2}\langle P(\xi)X, Y\rangle \\ &= \langle \tfrac{1}{2}Q(\xi)X - \tfrac{i}{2}P(\xi)X, Y\rangle.\end{aligned}$$

Hence $X \in \operatorname{Dom}(A^*(\xi))$, which equals $\operatorname{Dom}(A(\xi))$ by Theorem 13.7. The result follows. □

EXAMPLE 13.23. Continuing Example 13.11, we find in $L^2(\mathbb{R}, \gamma_\sigma)$,

$$\begin{aligned}P(x) &= -2i\sigma^2\frac{d}{dx} + ix, \\ Q(x) &= x.\end{aligned}$$

Transferring these by the isometry $V_\sigma\colon L^2(\gamma_\sigma) \to L^2(dx)$ defined by (13.4), we obtain the unitarily equivalent pair of operators on $L^2(dx)$

$$\begin{aligned}P &= V_\sigma P(x) V_\sigma^{-1} = -2i\sigma^2\frac{d}{dx}, \\ Q &= V_\sigma Q(x) V_\sigma^{-1} = x.\end{aligned}$$

These operators form the standard Schrödinger representation of the canonical commutation rules. We see that $2\sigma^2$ plays the role of Planck's constant $\hbar$, and the CCR reduce to the single formula

$$[P, Q] = -2i\sigma^2 I = -i\hbar I.$$

Similarly, for the d-dimensional Gaussian space in Examples 13.5 and 13.12, we obtain the operators on $L^2(\mathbb{R}^d, dx)$

$$\begin{aligned}P_k &= V_\sigma P(x_k) V_\sigma^{-1} = -i\hbar\frac{\partial}{\partial x_k}, \\ Q_k &= V_\sigma Q(x_k) V_\sigma^{-1} = x_k.\end{aligned}$$

The domains are in all cases the natural ones.

EXAMPLE 13.24. By the definitions (13.24) and (13.25), and the commutation relation (13.14), we have, at least on $\overline{\mathcal{P}}_*$,

$$P(\xi)^2 + Q(\xi)^2 = 2A(\xi)A^*(\xi) + 2A^*(\xi)A(\xi)$$
$$= 4A^*(\xi)A(\xi) + 2\|\xi\|_2^2 I.$$

If $\dim(H) < \infty$ and $\{\xi_k\}$ is an orthonormal basis in H, it thus follows by Theorem 13.15 that (at least on $\overline{\mathcal{P}}_*$)

$$\sum_k P(\xi_k)^2 + \sum_k Q(\xi_k)^2 = 4\mathcal{N} + 2\dim(H)I.$$

For the d-dimensional Gaussian space considered in Examples 13.5, 13.12, 13.23, we thus obtain on $L^2(\mathbb{R}^d, \gamma_\sigma)$, since $(x_k/\sigma)_{k=1}^d$ is an orthonormal basis in H,

$$\tfrac{1}{2}\sum_k P(x_k)^2 + \tfrac{1}{2}\sum_k Q(x_k)^2 = 2\sigma^2\mathcal{N} + d\sigma^2 I = \hbar(\mathcal{N} + \tfrac{d}{2}I),$$

where again $\hbar = 2\sigma^2$.

The corresponding operator $\frac{1}{2}\sum P_k^2 + \frac{1}{2}\sum Q_k^2$ on $L^2(\mathbb{R}^d, dx)$ is the Schrödinger representation of the Hamilton operator for a d-dimensional harmonic oscillator in quantum mechanics. We thus recover, using Theorem 3.21, the well-known fact that the eigenvalues of this Hamilton operator are $(n+d/2)\hbar$, $n \geq 0$, and that an orthonormal basis of $L^2(\mathbb{R}^d, dx)$ consisting of eigenvectors is given by

$$V_\sigma\Big(\prod_{k=1}^d (n_k!)^{-1/2} h_{n_k}(x_k/\sigma)\Big), \qquad n_1, \dots, n_k \geq 0.$$

In this connection, the annihilation and creation operators, which operate on the eigenvectors changing the eigenvalue by $\pm\hbar$, are known as *ladder operators*.

Recall that if A is a self-adjoint operator in a complex Hilbert space, then e^{iA} is a unitary operator (in particular, e^{iA} is defined everywhere). Moreover, the unitary operators e^{itA}, $-\infty < t < \infty$, constitute a strongly continuous group and, conversely, every strongly continuous group of unitary operators $U(t)$ arises in this way from a unique self-adjoint A. (See e.g. Conway (1990) or Dunford and Schwartz (1958).) It is thus natural to study the unitary groups $t \mapsto e^{itP(\xi)} = e^{iP(t\xi)}$ and $t \mapsto e^{itQ(\xi)} = e^{iQ(t\xi)}$ in $L^2_{\mathbb{C}}(\Omega, \mathcal{F}(H), \mathrm{P})$.

THEOREM 13.25. *Let $\xi \in H$. Then $e^{iP(\xi)}$ and $e^{iQ(\xi)}$ are unitary operators on $L^2_{\mathbb{C}} = L^2_{\mathbb{C}}(\Omega, \mathcal{F}(H), \mathrm{P})$; they satisfy*

$$e^{iQ(\xi)}X = e^{i\xi}X, \qquad X \in L^2_{\mathbb{C}}, \tag{13.33}$$

and

$$e^{iP(\xi)}e^{\zeta} = e^{2\mathrm{E}\zeta\xi - \mathrm{E}\xi^2}e^{\zeta-\xi}, \qquad \zeta \in H_{\mathbb{C}}. \tag{13.34}$$

PROOF. Since $Q(\xi)$ is just the operation of multiplication by ξ, $e^{iQ(\xi)}$ is multiplication by $e^{i\xi}$, which gives (13.33).

In particular, for $\xi \in H$ and $\zeta \in H_{\mathbf{C}}$,

$$e^{iQ(\xi)}e^{\zeta} = e^{i\xi+\zeta}.$$

Hence, using Theorem 13.21(iii) and Example 13.3,

$$\begin{aligned} e^{iP(\xi)}e^{\zeta} &= F_W e^{iQ(\xi)} F_W^{-1} e^{\zeta} = F_W e^{iQ(\xi)} e^{-i\zeta + \mathrm{E}\,\zeta^2} \\ &= e^{\mathrm{E}\,\zeta^2} F_W(e^{i\xi - i\zeta}) = e^{\mathrm{E}\,\zeta^2 + \zeta - \xi - \mathrm{E}(\zeta-\xi)^2} \\ &= e^{2\,\mathrm{E}\,\zeta\xi - \mathrm{E}\,\xi^2} e^{\zeta-\xi}. \end{aligned}$$

□

EXAMPLE 13.26. Using Wick exponentials $:e^{\zeta}: = e^{\zeta - \mathrm{E}\,\zeta^2/2}$, cf. Theorem 3.33, we obtain from Theorem 13.25 by simple calculations

$$e^{iQ(\xi)} :e^{\zeta}: = e^{i\,\mathrm{E}(\xi\zeta) - \frac{1}{2}\mathrm{E}\,\xi^2} :e^{\zeta + i\xi}:,$$
$$e^{iP(\xi)} :e^{\zeta}: = e^{\mathrm{E}\,\zeta\xi - \frac{1}{2}\mathrm{E}\,\xi^2} :e^{\zeta-\xi}:.$$

EXAMPLE 13.27. Suppose that $\xi \in H$ and $\zeta \in H_{\mathbf{C}}$. Then, by (13.34),

$$e^{iP(\xi)}e^{t\zeta} = e^{-\xi - \mathrm{E}\,\xi^2} e^{t(\zeta + 2\,\mathrm{E}\,\zeta\xi)}, \qquad t \in \mathbf{C}, \tag{13.35}$$

and an expansion in powers of t yields

$$e^{iP(\xi)}(\zeta^n) = e^{-\xi - \mathrm{E}\,\xi^2}(\zeta + 2\,\mathrm{E}\,\xi\zeta)^n. \tag{13.36}$$

Similarly, if $\zeta_1, \dots, \zeta_n \in H_{\mathbf{C}}$, we obtain by considering $\zeta = \sum t_k \zeta_k$ in (13.35), or by polarization in (13.36),

$$e^{iP(\xi)}\Big(\prod_1^n \zeta_k\Big) = e^{-\xi - \mathrm{E}\,\xi^2} \prod_1^n (\zeta_k + 2\,\mathrm{E}\,\xi\zeta_k)$$

and thus, for every polynomial p,

$$e^{iP(\zeta)} p(\zeta_1, \dots, \zeta_n) = e^{-\xi - \mathrm{E}\,\xi^2} p(\zeta_1 + 2\,\mathrm{E}\,\xi\xi_1, \dots, \zeta_n + 2\,\mathrm{E}\,\xi\zeta_n).$$

This is generalized in Remark 14.4 below.

We have given two versions of the CCR, viz. (13.13)–(13.14) for the annihilation and creation operators and (13.28)–(13.29) for the position and momentum operators. A third version, known as the *Weyl commutation relations*, is obtained by considering the unitary operators $e^{iP(\xi)}$ and $e^{iQ(\eta)}$. Note that in this version all operators are everywhere defined, so there is no problem with the domains. (This form of the CCR is perhaps best understood in terms of representations of the Heisenberg group, see Folland (1989).)

THEOREM 13.28. *Let $\xi, \eta \in H$. Then*

$$e^{iP(\xi)}e^{iP(\eta)} = e^{iP(\xi+\eta)} = e^{iP(\eta)}e^{iP(\xi)}, \tag{13.37}$$

$$e^{iQ(\xi)}e^{iQ(\eta)} = e^{iQ(\xi+\eta)} = e^{iQ(\eta)}e^{iQ(\xi)}, \tag{13.38}$$

$$e^{iP(\xi)}e^{iQ(\eta)} = e^{2i\langle\xi,\eta\rangle}e^{iQ(\eta)}e^{iP(\xi)}. \tag{13.39}$$

PROOF. Formula (13.38) follows immediately by (13.33), and (13.37) follows most easily by (13.38) and (13.26).

For (13.39), let $\zeta \in H_{\mathbb{C}}$. Then, by Theorem 13.25,

$$e^{iQ(\eta)}e^{\zeta} = e^{\zeta+i\eta}$$

and

$$e^{iP(\xi)}e^{iQ(\eta)}e^{\zeta} = e^{2\,\mathrm{E}(\zeta+i\eta)\xi-\mathrm{E}\,\xi^2}e^{\zeta+i\eta-\xi}.$$

Similarly,

$$e^{iQ(\eta)}e^{iP(\xi)}e^{\zeta} = e^{2\,\mathrm{E}\,\zeta\xi-\mathrm{E}\,\xi^2}e^{\zeta-\xi+i\eta}.$$

Hence the two sides of (13.39) give the same result when applied to an element e^{ζ}. Since linear combinations of such elements are dense in $L^2_{\mathbb{C}}$ by Theorem 2.12, and the operators are bounded, (13.39) follows. □

EXAMPLE 13.29. Continuing Examples 13.11 and 13.23, we find by Theorem 13.25

$$\begin{aligned} e^{itP(x)}e^{sx} &= e^{2st\sigma^2-t^2\sigma^2+sx-tx} \\ &= e^{-tx-\sigma^2t^2}e^{s(x+2\sigma^2t)}. \end{aligned}$$

It is easily checked that $f(x) \mapsto e^{-tx-\sigma^2t^2}f(x+2\sigma^2t)$ is an isometry in $L^2(\mathbb{R},\gamma_\sigma)$. Hence, by Theorem 2.12,

$$e^{itP(x)}f(x) = e^{-tx-\sigma^2t^2}f(x+2\sigma^2t)$$

for all $f \in L^2(\mathbb{R},\gamma_\sigma)$.

This becomes even simpler if we transfer to $L^2(dx)$ again. Then

$$e^{itP}f(x) = V_\sigma e^{itP(x)}V_\sigma^{-1}f(x) = f(x+2\sigma^2t) = f(x+\hbar t),$$

an obviously unitary operator. Moreover,

$$e^{itQ}f(x) = V_\sigma e^{itx}V_\sigma^{-1}f(x) = e^{itx}f(x)$$

and the Weyl commutation relations become (beside the group properties)

$$e^{isP}e^{itQ} = e^{i\hbar st}e^{itQ}e^{isP}.$$

XIV

The Cameron–Martin shift

Cameron and Martin (1944) studied the effects of a shift on Brownian motion. This has been generalized to other Gaussian processes and Gaussian spaces. Traditionally, the shift is described by a transformation of the underlying probability space. We will, however, in accordance with our general philosophy of concentrating on the random variables and ignoring the probability space as much as possible, instead define the shift as a transformation of the random variables, mapping each random variable to another. In Section 2, we show that this transformation in many natural examples is indeed given by a shift on the probability space, which shows that the approaches are essentially equivalent. We will then also see the connection with the Cameron–Martin space defined in Chapter 8.

We assume in this chapter that H is a fixed Gaussian Hilbert space. For convenience we write L^p for $L^p(\Omega, \mathcal{F}(H), \mathrm{P})$, $0 \leq p \leq \infty$, (or assume that $\mathcal{F}(H) = \mathcal{F}$) in this chapter.

We also write $L^{p+} = \bigcup_{q>p} L^q$, $0 \leq p < \infty$, and $L^{p-} = \bigcap_{q<p} L^q$, $0 < p \leq \infty$.

1. The Cameron–Martin shift

THEOREM 14.1. *Let $\xi \in H$. Then there exists a unique continuous linear map ρ_ξ of L^0 into itself such that*

$$\rho_\xi(1) = 1, \tag{14.1}$$

$$\rho_\xi(XY) = \rho_\xi(X)\rho_\xi(Y) \tag{14.2}$$

(i.e., ρ_ξ is an algebra homomorphism) and

$$\rho_\xi(\eta) = \eta + \mathrm{E}\,\xi\eta, \qquad \eta \in H. \tag{14.3}$$

Moreover, ρ_ξ has the following properties.

(i) *ρ_ξ is invertible (i.e., an automorphism), and*

$$\rho_\xi^{-1} = \rho_{-\xi}.$$

(ii) *More generally,*

$$\rho_{\xi_1}\rho_{\xi_2} = \rho_{\xi_1+\xi_2}, \qquad \xi_1, \xi_2 \in H, \tag{14.4}$$

and ρ_0 is the identity map. Hence $\{\rho_\xi\}_{\xi \in H}$ is a group of automorphisms.

(iii) *If* $(X_i)_{i=1}^N$, $N \le \infty$, *is a finite or infinite sequence of random variables in* L^0, *and* $f\colon \mathbb{R}^N \to \mathbb{R}$ (*or* $f\colon \mathbb{C}^N \to \mathbb{C}$) *is Borel measurable, then*

$$\rho_\xi\big(f(X_1, X_2, \dots)\big) = f\big(\rho_\xi(X_1), \rho_\xi(X_2), \dots\big). \tag{14.5}$$

(iv) *If* $X \in L^0$, *then*

$$\mathrm{E}\,\rho_\xi(X) = \mathrm{E}(:e^\xi:X), \tag{14.6}$$

provided one of the expectations is finite or $X \ge 0$.

(v) *Similarly, if* $X \in L^0$, *then*

$$\mathrm{E}\,X = \mathrm{E}(:e^{-\xi}:\rho_\xi(X)),$$

provided one of the expectations is finite or $X \ge 0$.

(vi) *If* $0 < q < p \le \infty$, *then* ρ_ξ *is a continuous map of* L^p *into* L^q; *more precisely*

$$\|\rho_\xi(X)\|_q \le e^{\mathrm{E}\,\xi^2/2(p-q)}\|X\|_p. \tag{14.7}$$

Consequently, ρ_ξ *preserves each of the spaces* L^{p+}, $0 \le p < \infty$, *and* L^{p-}, $0 < p \le \infty$.

(vii) *If* $X_n \to X$ *a.s. as* $n \to \infty$, *then*

$$\rho_\xi(X_n) \to \rho_\xi(X) \quad \text{a.s.}$$

(viii) *The map* $\xi \to \rho_\xi(X)$ *is a continuous map* $H \to L^0$ *for every* $X \in L^0$. (*In other words,* $\xi \mapsto \rho_\xi$ *is strongly continuous.*) *Moreover, if* $X \in L^p$, *with* $0 < p \le \infty$, *then* $\xi \mapsto \rho_\xi(X)$ *is a continuous map* $H \to L^q$ *for every* $q < p$.

(ix) *More generally, the map* $(\xi, X) \mapsto \rho_\xi(X)$ *is continuous* $H \times L^0 \to L^0$ *and* $H \times L^p \to L^q$ *for* $0 < q < p \le \infty$.

(x) *For any* $X \in L^0$, *there exists a jointly measurable version of the stochastic process* $\xi \mapsto \rho_\xi(X)$ *indexed by* $\xi \in H$, *i.e. a measurable function* $Y(\xi, \omega)$ *on* $H \times \Omega$ (*where* H *is equipped with the Borel* σ*-field*) *such that* $Y(\xi, \omega) = \rho_\xi(X)(\omega)$ *a.s. for every* ξ.

(xi) ρ_ξ *preserves* $\mathcal{P}(H)$ *and* $\overline{\mathcal{P}}_*(H)$ *and is an algebra automorphism for the Wick multiplication too:*

$$\rho_\xi(X \odot Y) = \rho_\xi(X) \odot \rho_\xi(Y), \qquad X, Y \in \overline{\mathcal{P}}_*(H). \tag{14.8}$$

PROOF. We begin by verifying that if $\xi, \xi_1, \dots, \xi_m \in H$ and f is a bounded measurable function on $\mathbb{R}^m$, then

$$\mathrm{E}\,f(\xi_1 + \mathrm{E}\,\xi_1\xi, \dots, \xi_m + \mathrm{E}\,\xi_m\xi) = \mathrm{E}\big(:e^\xi: f(\xi_1, \dots, \xi_m)\big). \tag{14.9}$$

This is trivial if $\xi = 0$. Otherwise we consider the linear span of $\{\xi\} \cup \{\xi_i\}_{i=1}^m$ and choose there an orthogonal basis $\{\xi_j'\}_{j=1}^{m'}$, with $\xi_1' = \xi$. By expressing each ξ_i as a linear combination of the ξ_j', (14.9) becomes a formula of the same type for a bounded measurable function f' and the variables ξ_j'; dropping the primes from the notations, we thus see that it suffices to show (14.9) when the variables ξ_j are orthogonal, and thus independent, and $\xi_1 = \xi$. In this case, (14.9) becomes

$$\mathrm{E}\, f(\xi_1 + \mathrm{E}\,\xi_1^2, \xi_2, \dots, \xi_m) = \mathrm{E}\big(:e^{\xi_1}: f(\xi_1, \xi_2, \dots, \xi_n) \big),$$

which, by the independence of $\xi_1, \dots, \xi_m$ and Fubini's theorem, follows from the one-dimensional formula

$$\mathrm{E}\, f(\xi + \mathrm{E}\,\xi^2) = \mathrm{E}\big(:e^{\xi}: f(\xi) \big),$$

for every bounded measurable f. This formula is easily proved directly by a computation of the expectations and a simple change of variable: if $\sigma^2 = \mathrm{E}\,\xi^2$, then, using Theorem 3.33,

$$\begin{aligned}
\mathrm{E}\, f(\xi + \mathrm{E}\,\xi^2) &= (2\pi\sigma^2)^{-1/2} \int_{-\infty}^{\infty} f(x + \sigma^2) e^{-x^2/2\sigma^2}\, dx \\
&= (2\pi\sigma^2)^{-1/2} \int_{-\infty}^{\infty} f(y) e^{-(y-\sigma^2)^2/2\sigma^2}\, dy \\
&= (2\pi\sigma^2)^{-1/2} \int_{-\infty}^{\infty} f(y) e^{y - \sigma^2/2} e^{-y^2/2\sigma^2}\, dy \\
&= \mathrm{E}\big(f(\xi) :e^{\xi}: \big).
\end{aligned}$$

Having proved (14.9), we extend it to infinitely many variables as follows. Consider a set $\{\xi_i\}_1^\infty$ in H, and let $\mathcal{A}$ be the set of all bounded Borel functions $f\colon \mathbb{R}^\infty \to \mathbb{R}$ such that

$$\mathrm{E}\, f(\xi_1 + \mathrm{E}\,\xi_1\xi,\ \xi_2 + \mathrm{E}\,\xi_2\xi, \dots) = \mathrm{E}\big(:e^{\xi}: f(\xi_1, \xi_2, \dots) \big). \tag{14.10}$$

(The more abstractly minded reader may prefer to consider Borel functions $f\colon \mathbb{R}^H \to \mathbb{R}$ instead; the argument will be the same.) Clearly, $\mathcal{A}$ is a linear space that is closed under bounded monotone convergence, and by (14.9), $\mathcal{A}$ contains the algebra of bounded functions of finitely many variables. By the monotone class theorem, Theorem A.1, $\mathcal{A}$ is the set of all bounded Borel functions.

We have thus proved that (14.10) holds for every sequence $\{\xi_i\}$ and every bounded measurable f; by monotone convergence, (14.10) holds also for unbounded $f \geq 0$.

It is easily seen that every random variable on $(\Omega, \mathcal{F}(H), \mathrm{P})$ can be written as $f(\xi_1, \xi_2, \dots)$ for some infinite sequence $(\xi_i)_1^\infty$ in H and some Borel function f on $\mathbb{R}^\infty$. If $f(\xi_1, \xi_2, \dots)$ and $f'(\xi_1', \xi_2', \dots)$ are two such representations of the same random variable, meaning that they are a.s. equal, then (14.10) applied to the positive function $|f(x_1, x_2, \dots) - f'(x_1', x_2', \dots)|^2$ and the variables $\{\xi_i\}_i \cup \{\xi_j'\}_j$ yields

$$\begin{aligned}
\mathrm{E}\, \big| f(\xi_1 + \mathrm{E}\,\xi\xi_1, \xi_2 + \mathrm{E}\,\xi\xi_2, \dots) - f'(\xi_1' + \mathrm{E}\,\xi\xi_1', \xi_2' + \mathrm{E}\,\xi\xi_2', \dots) \big|^2 \\
= \mathrm{E}\big(:e^{\xi}: |f(\xi_1, \xi_2, \dots) - f'(\xi_1', \xi_2', \dots)|^2 \big) = 0.
\end{aligned}$$

Hence $f(\xi_1 + \mathrm{E}\,\xi\xi_1, \xi_2 + \mathrm{E}\,\xi\xi_2, \dots) = f'(\xi_1' + \mathrm{E}\,\xi\xi_1', \xi_2' + \mathrm{E}\,\xi\xi_2', \dots)$ a.s. Since we identify random variables that are a.s. equal, this shows that ρ_ξ can be unambiguously defined on L^0 by

$$\rho_\xi\big(f(\xi_1, \xi_2, \dots)\big) = f(\xi_1 + \mathrm{E}\,\xi\xi_1, \xi_2 + \mathrm{E}\,\xi\xi_2, \dots), \tag{14.11}$$

whenever f is a (real or complex) Borel function on $\mathbb{R}^\infty$ and $\xi_1, \xi_2, \dots \in H$.

In order to prove (iii), let $X_i = f_i((\xi_{ij})_j)$ for some Borel functions f_i and $\xi_{ij} \in H$. Then (14.5) follows for any Borel function f by applying (14.11) to the composite function $f\big(f_1((x_{1j})_j), f_2((x_{2j})_j), \dots\big)$ and the variables $\{\xi_{ij}\}$.

In particular, choosing first $f(x_1, x_2) = ax_1 + bx_2$ and then $f(x_1, x_2) = x_1 x_2$, (14.5) implies that ρ_ξ is linear and that (14.2) holds.

We obtain (14.3) directly from the definition (14.11) with $f(x_1, x_2, \dots) = x_1$. Similarly, (14.1) holds. Furthermore, (14.6) holds by (14.10) when $X \geq 0$, and thus by linearity also whenever any of the expectations is finite. This proves (iv).

In order to verify (14.7), let $r = p/(p-q)$. Then $r^{-1} + (p/q)^{-1} = 1$, and thus, by (14.5) with $f(x) = |x|^q$, (14.6), Hölder's inequality and Corollary 3.38,

$$\mathrm{E}\,|\rho_\xi(X)|^q = \mathrm{E}\,\rho_\xi(|X|^q) = \mathrm{E}(\,{:}e^\xi{:}\,|X|^q) \leq \|\,{:}e^\xi{:}\,\|_r \||X|^q\|_{p/q} = e^{(r-1)\mathrm{E}\,\xi^2/2} \|X\|_p^q.$$

Taking the qth root, we obtain (14.7) because $(r-1)/q = 1/(p-q)$. The remaining assertions in (vi) follow directly.

If $\{X_n\}_1^\infty \in L^0$ and $X_n \xrightarrow{\mathrm{P}} X$ as $n \to \infty$, let f be the function

$$f(x) = x/(1+|x|). \tag{14.12}$$

Since f is continuous, $f(X_n) \xrightarrow{\mathrm{P}} f(X)$, and since f is bounded, this implies $f(X_n) \to f(X)$ in L^2, say, by uniform integrability.

By (iii) and (vi), as $n \to \infty$,

$$f\big(\rho_\xi(X_n)\big) = \rho_\xi\big(f(X_n)\big) \to \rho_\xi\big(f(X)\big) = f\big(\rho_\xi(X)\big) \qquad \text{in } L^1,$$

and thus in probability, and since f has a continuous inverse, this yields

$$\rho_\xi(X_n) \xrightarrow{\mathrm{P}} \rho_\xi(X).$$

Consequently, ρ_ξ is continuous as a map $L^0 \to L^0$.

Uniqueness follows because (14.1)–(14.3) determine ρ_ξ on the space of polynomial variables, which is dense in L^0 by Theorem 2.11.

We have proved that ρ_ξ is a continuous algebra homomorphism. If $\xi_1, \xi_2 \in H$, then $\rho_{\xi_1}\rho_{\xi_2}$ is thus a continuous algebra homomorphism, and by (14.3),

$$\rho_{\xi_1}\rho_{\xi_2}(\eta) = \rho_{\xi_1}(\eta + \mathrm{E}\,\xi_2\eta) = \eta + \mathrm{E}\,\xi_1\eta + \mathrm{E}\,\xi_2\eta = \eta + \mathrm{E}(\xi_1 + \xi_2)\eta.$$

Consequently, the uniqueness assertion implies (14.4). Similarly, ρ_0 equals the identity map, and (ii) is proved.

(i) is a special case of (ii). Furthermore, (ii) and (iv) yield $\mathrm{E}\,X = \mathrm{E}(\rho_{-\xi}(\rho_\xi(X))) = \mathrm{E}(\,{:}e^{-\xi}{:}\,\rho_\xi(X))$, i.e. (v).

In order to show (vii), define a Borel function f on $\mathbb{R}^\infty$ (or $\mathbb{C}^\infty$) by

$$f(x, x_1, x_2, \dots) = \begin{cases} 0, & \text{if } x_n \to x \text{ as } n \to \infty, \\ 1, & \text{otherwise.} \end{cases}$$

If $X_n \to X$ a.s., then $f(X, X_1, X_2, \dots) = 0$ a.s. and thus, by (iii),

$$f\big(\rho_\xi(X), \rho_\xi(X_1), \rho_\xi(X_2), \dots\big) = \rho_\xi\big(f(X, X_1, X_2, \dots)\big) = 0 \quad \text{a.s.},$$

i.e.

$$\rho_\xi(X_n) \to \rho_\xi(X) \quad \text{a.s.}$$

For (viii) and (ix), we may assume that $p < \infty$. Thus, let $0 < q < p < \infty$ and suppose that $\xi_n \to \xi$ in H and $X_n \to X$ in L^p as $n \to \infty$. For $\varepsilon > 0$, let Y be a polynomial variable in $\mathcal{P}$ with $\|X - Y\|_p < \varepsilon$; such a Y exists by Theorem 2.11. It follows by (14.1)–(14.3) that $\rho_\xi(Y)$ is a polynomial in some $\eta_1, \dots, \eta_N \in H$, with coefficients that are polynomials in $\mathrm{E}(\xi\eta_i)$. Hence $\xi \mapsto \rho_\xi(Y)$ is a continuous map $H \to L^q$; in particular, $\|\rho_{\xi_n}(Y) - \rho_\xi(Y)\|_q \to 0$ as $n \to \infty$. Furthermore, by the triangle inequality and (14.7), letting $s = 1$ for $q \geq 1$ and $s = q$ for $q < 1$, and $C = \sup_n e^{\mathrm{E}\xi_n^2/2(p-q)} < \infty$,

$$\begin{aligned} \|\rho_{\xi_n}(X_n) - \rho_\xi(X)\|_q^s &\leq \|\rho_{\xi_n}(X_n - Y)\|_q^s + \|\rho_{\xi_n}(Y) - \rho_\xi(Y)\|_q^s + \|\rho_\xi(Y - X)\|_q^s \\ &\leq C^s\|X_n - Y\|_p^s + \|\rho_{\xi_n}(Y) - \rho_\xi(Y)\|_q^s + C^s\|X - Y\|_p^s \\ &\leq (2C^s + 1)\varepsilon^s, \end{aligned}$$

provided n is large enough. Since ε is arbitrary, $\rho_{\xi_n}(X_n) \to \rho_\xi(X)$ in L^q, which proves the second claim in (ix).

For the first claim we assume that $\xi_n \to \xi$ in H and $X_n \to X$ in L^0. Using the function f in (14.12) again, $f(X_n) \to f(X)$ in probability and in L^2, and by what we have just proved,

$$f\big(\rho_{\xi_n}(X_n)\big) = \rho_{\xi_n}\big(f(X_n)\big) \to \rho_\xi\big(f(X)\big) = f\big(\rho_\xi(X)\big)$$

in L^1 and thus in probability. Hence

$$\rho_{\xi_n}(X_n) \xrightarrow{\mathrm{P}} \rho_\xi(X),$$

which completes the proof of (ix), and thus also of (viii).

The right hand side of (14.11), regarded as function of $\xi \in H$ and $\omega \in \Omega$, is measurable on $H \times \Omega$. Hence (x) follows by using the same f and $\xi_1, \xi_2, \dots$ for all ξ.

In order to prove (xi), we begin by establishing a translation formula for Hermite polynomials:

$$h_n(x + a) = \sum_{k=0}^{n} \binom{n}{k} a^k h_{n-k}(x).$$

This is just Taylor's formula, in view of the relation $h_n' = h_{n-1}$ (13.16); it also follows easily from (3.11), or from the generating function (3.19) by identifying the coefficients of t^n in

$$\sum_0^\infty \frac{t^n}{n!} h_n(x+a) = e^{t(x+a)-t^2/2} = \sum_0^\infty \frac{t^k}{k!} a^k \sum_0^\infty \frac{t^l}{l!} h_l(x).$$

If η is any standard normal variable in H, we thus obtain, with $a = \mathrm{E}\,\xi\eta$,

$$\begin{aligned}\rho_\xi(:\eta^n:) = \rho_\xi\big(h_n(\eta)\big) = h_n\big(\rho_\xi(\eta)\big) = h_n(\eta + a) &= \sum_{k=0}^n \binom{n}{k} a^k h_{n-k}(\eta) \\ = \sum_{k=0}^n \binom{n}{k} a^k \eta^{\odot(n-k)} = (\eta + a)^{\odot n} &= \rho_\xi(\eta)^{\odot n}.\end{aligned}$$

By homogeneity we obtain for any $\eta \in H$,

$$\rho_\xi(\eta^{\odot n}) = \rho_\xi(\eta)^{\odot n},$$

and by polarization (see Theorem D.2),

$$\rho_\xi(\eta_1 \odot \cdots \odot \eta_n) = \rho_\xi(\eta_1) \odot \cdots \odot \rho_\xi(\eta_n), \qquad \eta_1, \dots, \eta_n \in H.$$

It follows that (14.8) holds whenever X and Y are of the form $\eta_1 \odot \cdots \odot \eta_n = :\eta_1 \cdots \eta_n:$.

By linearity (and Theorem 3.15), the formula remains valid for any polynomial variables X and Y, and by continuity on $\overline{\mathcal{P}}_n \times \overline{\mathcal{P}}_m$ for any fixed $n, m \geq 0$, it holds for all $X, Y \in \overline{\mathcal{P}}_*$. □

REMARK 14.2. It follows also, e.g. by (14.11), that ρ_ξ maps L^∞ into itself, but it is important to note that ρ_ξ does not preserve L^p for any $p < \infty$ (unless $\xi = 0$). For example, if ξ is standard normal, then $X = (1 + |\xi|)^{-a} e^{\xi^2/2p} \in L^p$ when $ap > 1$, but for every $\varepsilon \neq 0$, $\rho_{\varepsilon\xi} X = (1 + |\xi + \varepsilon|)^{-a} e^{(\xi+\varepsilon)^2/2p} \notin L^p$.

We will therefore occasionally use the spaces L^{p+} and L^{p-}, which behave better by (vi) above.

REMARK 14.3. If L^{p+} ($p \geq 1$) and L^{p-} ($p > 1$) are equipped with the natural inductive and projective topologies, respectively (cf. e.g. Schaefer 1971, §§II.5–6), then ρ_ξ is a linear homeomorphism of each L^{p+} and L^{p-} onto itself. (The topology on L^{p-} is simply given by the family of norms $\|\cdot\|_q$, $1 \leq q < p$. The topology on L^{p+} is characterized by the property that a linear map of L^{p+} into any locally convex vector topological space is continuous if and only if its restriction to L^q is continuous for each $q > p$.) This extends also to $p < 1$ if we leave the realm of locally convex spaces.

Similarly, ρ_ξ is a linear homeomorphism on $\overline{\mathcal{P}}_*$, equipped with the natural inductive topology described in Remark 3.49.

REMARK 14.4. It follows easily that for any $p > 0$, the operator $X \mapsto {:}e^{-\xi}{:}^{1/p}\rho_\xi(X)$ is an isometry of L^p onto itself. In particular, taking $p = 2$, we obtain the unitary operator $X \mapsto e^{-\xi/2-\mathrm{E}\,\xi^2/4}\rho_\xi(X)$ on L^2, which maps e^η to $e^{-\xi/2-\mathrm{E}\,\xi^2/4+\eta+\mathrm{E}\,\xi\eta}$ for $\eta \in H$, and thus (for complex scalars) by Theorems 13.25 and 2.12 equals $e^{iP(\xi)/2}$.

Consequently, for any $\xi \in H$ and $X \in L^2_{\mathbb{C}}$,

$$e^{iP(\xi)}X = e^{-\xi-\mathrm{E}\,\xi^2}\rho_{2\xi}(X).$$

REMARK 14.5. It is easily seen that the operator $\exp(A(\xi))$ on $\overline{\mathcal{P}}_*$, defined in Remark 13.19, equals ρ_ξ.

REMARK 14.6. The theorem extends in all relevant parts to random variables with values in a Banach space B, simply by replacing L^p by $L^p(B)$ $(0 \le p \le \infty)$. (See Appendix C for definitions; note that we restrict ourselves to a.s. separably valued random variables.)

In fact, any a.s. separably valued random variable on $(\Omega, \mathcal{F}(H), \mathrm{P})$ with values in a Banach space B can be written as $f(\xi_1, \xi_2, \dots)$ for some infinite sequence $(\xi_i)_1^\infty$ in H and some Borel function $f\colon \mathbb{R}^\infty \to B$; thus we can still use (14.11) as a definition. Note that $\|\rho_\xi(X)\|_B = \rho_\xi(\|X\|_B)$, which for example means that the inequality (14.7) for the L^p-norms immediately extends to $L^p(B)$.

REMARK 14.7. Even if $\xi \in H$ and $X \in L^0$ are (stochastically) independent, it is possible that $\rho_\xi(X) \ne X$; an example is given by $X = \eta\,\mathrm{sign}(\xi)$, with ξ and η independent standard normal. Conversely, $\rho_\xi(X)$ may equal X even when ξ and X are dependent, for example when ξ is standard normal and $X = \sin(2\pi\xi)$. It is true, however, that $\rho_{t\xi}(X) = X$ for every real t if and only if X is measurable with respect to the σ-field generated by $\{\xi\}^\perp \subset H$; this is a special case of Corollary 16.55 below, and can also be shown directly.

EXAMPLE 14.8. Let $\xi \in H$. By linearity, (14.3) extends to $H_{\mathbb{C}}$, i.e. if $\zeta \in H_{\mathbb{C}}$, then $\rho_\xi(\zeta) = \zeta + \mathrm{E}(\zeta\xi)$. It follows by (14.5) that, for $\zeta \in H_{\mathbb{C}}$,

$$\rho_\xi(e^\zeta) = e^{\zeta+\mathrm{E}(\zeta\xi)} = e^{\mathrm{E}(\zeta\xi)}e^\zeta, \qquad \zeta \in H_{\mathbb{C}},$$

and, using Theorem 3.33,

$$\rho_\xi({:}e^\zeta{:}) = \rho_\xi(e^\zeta)e^{-\mathrm{E}(\zeta^2)/2} = e^{\mathrm{E}(\zeta\xi)}\,{:}e^\zeta{:}.$$

We give some further results that will be used in the next chapter.

THEOREM 14.9. *If $X \in L^0$ and $\rho_\xi X = X$ for every $\xi \in H$, then $X = C$ a.s., for some constant C.*

PROOF. Suppose first that $X \in L^2$. By Theorem 14.1(iv), for every $\xi \in H$,

$$\mathrm{E}(X\,{:}e^\xi{:}) = \mathrm{E}\,\rho_\xi(X) = \mathrm{E}\,X = \mathrm{E}\big((\mathrm{E}\,X)\,{:}e^\xi{:}\big).$$

Since the family $\{e^{\xi}\}$ is total in L^2 by Theorem 2.12, this implies $X = \mathrm{E}\,X$ a.s.

In general, let $f(x) = x/(1+|x|)$, and let $Y = f(X) \in L^\infty \subset L^2$. Then $\rho_\xi(Y) = f(\rho_\xi(X)) = f(X) = Y$ for each $\xi \in H$, and the first part shows that $Y = \mathrm{E}\,Y$ a.s. Since $f\colon \mathbb{C} \to \mathbb{C}$ is injective, this implies $X = f^{-1}(\mathrm{E}\,Y)$ a.s. □

A generalization of Theorem 14.9 is given in Corollary 16.55.

THEOREM 14.10. *Let* $X \in L^{1+}$, $\xi \in H$ *and* $-1 \le r \le 1$. *Then*

$$\rho_\xi(M_r X) = M_r(\rho_{r\xi} X). \tag{14.13}$$

PROOF. Fix p with $1 < p < \infty$ and consider $X \in L^p$. By Theorems 14.1(vi) and 4.20, both sides of (14.13) define continuous linear maps of $X \in L^p$ into L^1. Hence, it suffices to verify the formula X belonging to some total subset of L^p; we choose the set of Wick exponentials $:e^{\eta}:$, $\eta \in H$, which is total by Theorem 2.12.

Thus, let $X = :e^{\eta}:$ with $\eta \in H$. Then, by Examples 4.8 and 14.8, $M_r X = :e^{r\eta}:$, $\rho_{r\xi} X = e^{\mathrm{E}(r\xi\eta)} X$ and

$$\rho_\xi(M_r X) = e^{r\,\mathrm{E}(\xi\eta)} :e^{r\eta}: = M_r(\rho_{r\xi} X).$$

□

2. Shifts and the Cameron–Martin space

As stated above, the map ρ_ξ is in many important cases given by a shift on the probability space. Note that if τ is a measurable map of Ω into itself, and X is a random variable, i.e. a measurable function $\Omega \to \mathbb{R}$ (or $\mathbb{C}$), then $X \circ \tau\colon \omega \mapsto X(\tau(\omega))$ is another random variable on the same probability space. However, since we identify random variables that are a.s. equal, we have to further require that if $X = 0$ a.s., then $X \circ \tau = 0$ a.s., in order for the mapping $X \mapsto X \circ \tau$ to be well-defined on L^0; this condition can also be written

$$\text{If } A \subset \Omega \text{ with } \mathrm{P}(A) = 0, \text{ then } \mathrm{P}(\tau^{-1}(A)) = \mathrm{P}\{\omega : \tau(\omega) \in A\} = 0 \tag{14.14}$$

or

$$\text{The measure } \mathrm{P} \circ \tau^{-1} \text{ is absolutely continuous with respect to } \mathrm{P}. \tag{14.15}$$

EXAMPLE 14.11. Consider once more the one-dimensional Gaussian Hilbert space spanned by $\xi = x$ on the probability space $(\mathbb{R}, \mathcal{B}, \gamma)$ in Example 1.15. For any real t,

$$\rho_{t\xi}(\xi) = \xi + t$$

by (14.3), and thus by (14.5), for any measurable function f on $\mathbb{R}$,

$$\rho_{t\xi}(f(\xi)) = f(\xi + t).$$

Note that $f(\xi)$ is the random variable $\omega \mapsto f(\omega)$ on $\Omega = \mathbb{R}$, and that every random variable X on this probability space is of this form. Consequently, the result may be written

$$\rho_{t\xi}(X)(\omega) = X(\omega + t), \qquad \omega \in \mathbb{R},$$

or

$$\rho_{t\xi}(X) = X \circ \tau_t,$$

where τ_t is the translation $x \mapsto x + t$ of $\Omega = \mathbb{R}$ onto itself. (We know by Theorem 14.1 that the mapping $X \mapsto \rho_{t\xi}(X) = X \circ \tau_t$ is well-defined on L^0. In this example it is also immediate that (14.14) holds.)

EXAMPLE 14.12. Similarly, if H is the d-dimensional Gaussian Hilbert space consisting of all linear functions on $(\Omega, \mathcal{F}, \mathrm{P}) = (\mathbb{R}^d, \mathcal{B}^d, \gamma^d)$, where γ^d is the standard Gaussian measure on $\mathbb{R}^d$, and $\{\xi_i\}_1^d$ is the orthonormal basis in H consisting of the coordinate functions, then for every $a = (a_1, \dots, a_d) \in \mathbb{R}^d$ and $\xi = \sum_i a_i \xi_i \in H$,

$$\rho_\xi(X) = X \circ \tau_a,$$

where τ_a is the translation $x \mapsto x + a$ of $\mathbb{R}^d$ onto itself.

For infinite-dimensional spaces, the following result gives a convenient criterion for the transformation ρ_ξ to be given by a shift in Ω.

THEOREM 14.13. *Suppose that S is a set of random variables in H that generates the σ-field $\mathcal{F}$. If $\xi \in H$ and $\tau_\xi \colon \Omega \to \Omega$ is a measurable mapping such that*

$$\eta \circ \tau_\xi = \eta + \langle \eta, \xi \rangle \text{ a.s.}, \qquad \eta \in S, \tag{14.16}$$

then the measures $\mathrm{P} \circ \tau_\xi^{-1}$ and P are mutually absolutely continuous, and

$$\rho_\xi(X) = X \circ \tau_\xi \text{ a.s.}, \qquad X \in L^0. \tag{14.17}$$

REMARK 14.14. Here the random variables $\eta \in S$ should be regarded as specific measurable functions on Ω, and not (as usually in this book) some representatives modulo a.s. equality, since otherwise (14.16) does not make sense without a priori assuming $\mathrm{P} \circ \tau_\xi^{-1}$ to be absolutely continuous with respect to P. (That these measures are equivalent is part of the conclusion, so there is no problem *after* applying the theorem.)

Recall also that the equalities in (14.16) and (14.17) mean equality a.s.; indeed, $\rho_\xi(X)$ is defined by Theorem 14.1 as an element of L^0, but it is defined only up to a.s. equality. When Theorem 14.13 applies, (14.17) yields a possibility of defining $\rho_\xi(X)$ *everywhere* in a consistent manner.

PROOF. Consider the set $\mathcal{A}$ of all real measurable functions $X \colon \Omega \to \mathbb{R}$ such that (14.17) holds. It follows from (14.16) and Theorem 14.1 that $\mathcal{A}$ is a linear space which is closed under pointwise convergence and contains every function $f(\eta_1, \dots, \eta_n)$ with $\eta_1, \dots, \eta_n \in S$. By the monotone class theorem, Corollary A.2, with C chosen as the set of all bounded functions of the form $f(\eta_1,\dots,\eta_n)$ with $\eta_1,\dots,\eta_n\in S$, $\mathcal{A}$ contains every measurable function, which proves (14.17).

In particular, if $A\subseteq\Omega$ is measurable and $X=\mathbf{1}_A$, then

$$\mathrm{P}(A)=0 \iff \mathbf{1}_A=0 \text{ a.s.} \iff \rho_\xi(\mathbf{1}_A)=0 \text{ a.s.} \iff \mathbf{1}_A\circ\tau_\xi=0 \text{ a.s.} \\ \iff \mathrm{P}\big(\tau_\xi^{-1}(A)\big)=0.$$

□

EXAMPLE 14.15. Let $\Omega=C([0,1])$ be the space of continuous functions on $[0,1]$, let $\mathcal{F}$ be the Borel σ-field on Ω and let P be the Wiener measure on Ω. (There are no essential differences if we instead consider the infinite interval $[0,\infty)$ as in Section 7.1, with $\Omega=C([0,\infty))$.)

The coordinate mappings $B_t\colon\omega\mapsto\omega(t)$, $\omega\in\Omega$, form a Brownian motion; moreover, they generate the Borel σ-field, see e.g. Billingsley (1968) or use Blackwell's theorem (Cohn 1980, Section 8.6).

If $\xi\in H(B)$, the Gaussian Hilbert space generated by $\{B_t\}_{0\le t\le 1}$, let $g_\xi\in C([0,1])$ be given by $g_\xi(t)=\langle\xi,B_t\rangle$ and define

$$T_\xi(\omega)=\omega+g_\xi,\qquad \omega\in\Omega.$$

Then Theorem 14.13 applies with $S=\{B_t\}_t$ because

$$B_t\circ T_\xi(\omega)=T_\xi(\omega)(t)=\omega(t)+g_\xi(t)=B_t(\omega)+\langle B_t,\xi\rangle,\qquad \omega\in\Omega,$$

which is (14.16). Consequently,

$$\rho_\xi(X)(\omega)=X(\omega+g_\xi), \tag{14.18}$$

which is the classical definition of the Cameron–Martin shift.

Note that g_ξ is the function on $[0,1]$ denoted $R(\xi)$ in Section 8.4. By Example 8.19, the set of functions g_ξ that appear as shifts in (14.18) thus equals the Cameron–Martin space

$$H_1=R(H(B))=\Big\{\int_0^{\cdot} f(s)\,ds : f\in L^2_{\mathbb{R}}([0,1])\Big\} \\ =\Big\{g\in C_{\mathbb{R}}([0,1]) : g(0)=0,\ g \text{ is absolutely continuous},\ \int_0^1|g'|^2\,dt<\infty\Big\}.$$

Why did only shifts $\omega\mapsto\omega+g$ with $g\in H_1\subsetneq C_{\mathbb{R}}([0,1])$ appear in this example? The reason is that although $T\colon\omega\mapsto\omega+g$ is well-defined for any $g\in C_{\mathbb{R}}([0,1])$, (14.15) holds only for $g\in H_1$. Moreover, there is a nice dichotomy: $\mathrm{P}\circ T^{-1}$ and P are mutually absolutely continuous when $g\in H_1$ and mutually singular when $g\notin H_1$. This is known as the Cameron–Martin theorem and was proved in this form by Maruyama (1950).

We prove these assertions more generally, for a general Gaussian stochastic process. Recall (cf. Appendix B) that a stochastic process $(X_t)_{t\in T}$ (on some arbitrary index set T) can be regarded as a map $(\Omega, \mathcal{F}, \mathrm{P}) \to \mathbb{R}^T$; this map induces a probability measure μ on $(\mathbb{R}^T, \mathcal{B}^T)$, and the coordinate process $X_t'(\omega) = \omega(t)$, $t \in T$, is a stochastic process defined on $(\mathbb{R}^T, \mathcal{B}^T, \mu)$ that has the same distribution as $(X_t)_{t\in T}$. We call $(\mathbb{R}^T, \mathcal{B}^T, \mu)$ the *canonical model* for $(X_t)_{t\in T}$.

REMARK 14.16. If T is a topological space and (X_t) has a continuous version, another canonical model may be defined with $\Omega = C(T)$, the set of continuous functions on T. (We equip $C(T)$ with the σ-field generated by the coordinate mappings $\omega \to \omega(t)$; under some topological conditions, e.g. if $T = [0,1]$ or $[0,\infty)$, this equals the Borel σ-field.) For example, this yields for Brownian motion the Wiener space $(\Omega, \mathcal{F}, \mathrm{P})$ studied in the preceding example.

Since $C(T) \subset \mathbb{R}^T$ and the inclusion mapping connects the two models, the following theorem applies to this type of canonical model as well.

THEOREM 14.17. *Suppose that $\{\xi_t\}_{t\in T}$ is a centred Gaussian stochastic process with an arbitrary index set T. Let H be the Gaussian Hilbert space generated by $\{\xi_t\}_{t\in T}$ and let $H_1 = \{t \mapsto \langle \xi_t, \xi\rangle : \xi \in H\}$ be the Cameron–Martin space of the process. Let $(\mathbb{R}^T, \mathcal{B}^T, \mathrm{P})$ be the canonical model for the process, and let g be a real-valued function on T. Finally, let $\tau_g \colon \mathbb{R}^T \to \mathbb{R}^T$ be given by $\tau_g(\omega) = \omega + g$.*

(i) *If $g \in H_1$, then P and $\mathrm{P}\circ\tau_g^{-1}$ are mutually absolutely continuous.*
(ii) *If $g \notin H_1$, then P and $\mathrm{P}\circ\tau_g^{-1}$ are mutually singular.*

PROOF. We may assume that $\{\xi_t\}$ is given by the canonical model; thus $\Omega = \mathbb{R}^T$ and $\xi_t(\omega) = \omega(t)$.

(i) If $g \in H_1$, then $g(t) = \langle \xi_t, \xi\rangle$ for some $\xi \in H$, which yields

$$\xi_t \circ \tau_g(\omega) = \xi_t(\omega + g) = \omega(t) + g(t) = \xi_t(\omega) + \langle \xi_t, \xi\rangle.$$

Hence Theorem 14.13, with $S = \{\xi_t\}_{t\in T}$, shows that P and $\mathrm{P}\circ\tau_g^{-1}$ are mutually absolutely continuous.

(ii) Consider the real linear space V spanned by $\{\xi_t\}$: V consists of all functions $\Omega = \mathbb{R}^T \to \mathbb{R}$ that are finite linear combinations $\sum a_i \xi_{t_i}$ with $a_i \in \mathbb{R}$ and $t_i \in T$. Note that the elements of V are linear functions on Ω; hence, if $\eta = \sum a_i \xi_{t_i} \in V$, then

$$\eta(\tau_g(\omega)) = \eta(\omega + g) = \sum a_i \xi_{t_i}(\omega + g) = \sum a_i\big(\omega(t_i) + g(t_i)\big) = \eta(\omega) + \eta(g).$$

Moreover, $\eta \mapsto \eta(g)$ is a linear functional on V.

Let G be the Gaussian linear space G spanned by $\{\xi_t\}$; then G too consists of all finite linear combinations $\sum a_i \xi_{t_i}$, but now regarded as elements of L^2. Hence G equals the quotient space of V modulo the subspace $V_0 = \{\eta \in V : \eta = 0 \text{ P-a.s.}\}$.

Suppose first that the linear functional $\eta \mapsto \eta(g)$ is bounded on V in the sense that

$$|\eta(g)| \leq C\|\eta\|_2, \qquad \eta \in V, \tag{14.19}$$

for some $C < \infty$. In particular, $\eta(g) = 0$ if $\eta \in V_0$ so $\eta \mapsto \eta(g)$ defines a bounded linear functional on the Gaussian linear space G, which may be extended to the completion H. Hence there exists $\xi \in H$ such that

$$\eta(g) = \langle \eta, \xi \rangle$$

for every $\eta \in V$. In particular,

$$g(t) = \xi_t(g) = \langle \xi_t, \xi \rangle,$$

which shows that $g \in H_1$.

We may thus suppose that (14.19) does not hold for any C. By homogeneity, for every $k \geq 1$ there exists $\eta_k \in V$ such that $\|\eta_k\|_2 \leq 1$ and $\eta_k(g) \geq 2k$. Let $A_k = \{\omega : \eta_k < k\}$. Then

$$\begin{aligned}\tau_g^{-1}(A_k) &= \{\omega : \tau_g(\omega) \in A_k\} = \{\omega : \eta_k(\tau_g(\omega)) < k\} = \{\omega : \eta_k(\omega) + \eta_k(g) < k\} \\ &= \{\omega : \eta_k(\omega) < k - \eta_k(g)\} \subseteq \{\omega : \eta_k(\omega) < -k\},\end{aligned}$$

and thus

$$\mathrm{P}(A_k) = 1 - \mathrm{P}(\eta_k > k) \geq 1 - e^{-k^2/2}$$

while

$$\mathrm{P} \circ \tau_g^{-1}(A_k) = \mathrm{P}(\eta_k < k - \eta_k(g)) \leq \mathrm{P}(\eta_k < -k) \leq e^{-k^2/2}.$$

It follows easily, by the Borel–Cantelli lemma, that if $A = \limsup A_k = \bigcap_{n=1}^{\infty} \bigcup_{k=n}^{\infty} A_k$, then $\mathrm{P}(A) = 1$ and $\mathrm{P} \circ \tau_g^{-1}(A) = 0$, and thus the measures are mutually singular. □

XV

Malliavin calculus

The Malliavin calculus (also known as *stochastic calculus of variation*) is a differential calculus for functions (i.e. random variables) defined on a space with a Gaussian measure. (In applications, the space is usually some version of the Wiener space.) In accordance with our general principle, we present here a version concentrating on the random variables without explicit mention of the underlying space.

We define in Sections 1–3 the basic derivative operators ∂_ξ and ∇ for an arbitrary Gaussian Hilbert space, and in Section 9 the dual divergence operator. We also give a detailed treatment of the Sobolev spaces $\mathcal{D}^{k,p}$ in Sections 5–8; this includes a proof of the important Meyer inequalities in Section 8. Results on existence and smoothness of densities are given in Sections 4 and 10; these results are central in many applications. Finally, a connection with the Skorohod integral is established in Section 11.

The first application of Malliavin calculus (Malliavin 1978) was to study smoothness of solutions to partial differential operators. Many other applications have been developed later, for example to stochastic differential equations and stochastic integrals. We will not treat any of these applications here; for applications, other versions of the theory and further results on analysis on Wiener space we refer to for example Bell (1987), Bouleau and Hirsch (1991), Ikeda and Watanabe (1984), Malliavin (1993, 1997), Nualart (1995, 1997+), Nualart and Zakai (1986), Ocone (1987), Peters (1997+), Stroock (1981), Üstünel (1995), Watanabe (1984), Zakai (1985). Moreover, there are many related aspects of stochastic analysis which are not treated here, for example the white noise analysis studied in Hida, Kuo, Potthoff and Streit (1993), Holden, Øksendal, Ubøe and Zhang (1996), Kuo (1996), Obata (1994).

We use heavily the Cameron–Martin shift defined in Chapter 14, and continue with the notation used there. In particular, H is a fixed Gaussian Hilbert space and $L^p = L^p(\Omega, \mathcal{F}(H), \mathrm{P})$, $0 \le p \le \infty$. We also use the notation of Appendix C for Banach- (Hilbert-) space-valued random variables, and Appendix E for tensor products. Moreover, we let ∂_i denote $\partial/\partial x_i$ for functions on $\mathbb{R}^n$.

We also use some results from Chapters 12 and 13.

1. The directional derivative

DEFINITION 15.1. Let $X \in L^0$ and $\xi \in H$. We say that the *directional derivative* $\partial_\xi X$ *exists* (or, for emphasis, *exists in probability*) if the 'difference quotients' $(\rho_{t\xi}X - X)/t$ converge in probability as $t \to 0$; we then define

$$\partial_\xi X = \lim_{t\to 0} \frac{\rho_{t\xi}X - X}{t}.$$

REMARK 15.2. Since

$$\frac{\rho_{-t\xi}X - X}{-t} = \rho_{-t\xi}\Big(\frac{\rho_{t\xi}X - X}{t}\Big)$$

it follows from Theorem 14.1(ix) that the existence of $\lim_{t\searrow 0}\big((\rho_{t\xi}X - X)/t\big)$ suffices for the existence of $\partial_\xi X$.

EXAMPLE 15.3. If $\zeta \in H_{\mathbb{C}}$, then by (14.3) and linearity, cf. Example 14.8, $\rho_{t\xi}(\zeta) = \zeta + t\,\mathrm{E}(\zeta\xi)$, and thus $\partial_\xi\zeta = \mathrm{E}(\zeta\xi)$.

EXAMPLE 15.4. Let H be the one-dimensional Gaussian Hilbert space spanned by a standard normal variable ξ (for example, H could be the space in Example 14.11). If f is an a.e. differentiable function on $\mathbb{R}$, then

$$\frac{\rho_{t\xi}(f(\xi)) - f(\xi)}{t} = \frac{f(\xi+t) - f(\xi)}{t} \to f'(\xi) \quad \text{a.s.},$$

as $t \to 0$, and thus

$$\partial_\xi(f(\xi)) = f'(\xi).$$

EXAMPLE 15.5. Similarly, for a d-dimensional Gaussian Hilbert space with an orthonormal basis $(\xi_1, \dots, \xi_d)$, and any real $a_1, \dots, a_d$, if f is a.e. differentiable on $\mathbb{R}^d$, then

$$\partial_{\sum a_i\xi_i} f(\xi_1, \dots, \xi_d) = \sum_i a_i \frac{\partial f}{\partial x_i}(\xi_1, \dots, \xi_d).$$

For the d-dimensional Gaussian Hilbert space in Example 14.12, this shows that for a.e. differentiable functions on $\mathbb{R}^d$, $\partial_{\sum a_i\xi_i}$ equals the standard directional derivative.

EXAMPLE 15.6. For the Gaussian Hilbert space spanned by Brownian motion defined on the Wiener space as in Example 14.15,

$$\partial_\xi f(\omega) = \frac{d}{dt} f(\omega + t g_\xi)\Big|_{t=0} \quad \text{a.s.},$$

whenever the right hand side exists a.s. More generally, this holds for the canonical model of any Gaussian stochastic process.

THEOREM 15.7. *Suppose that* $\xi, \eta \in H$ *and* $X, Y, X_1, \dots, X_n \in L^0$.

(i) *If* $\partial_\xi X_1, \dots, \partial_\xi X_n$ *exist,* $X_1, \dots, X_n \in L^0_{\mathbb{R}}$ *and* $f\colon \mathbb{R}^n \to \mathbb{R}$ *(or* $\mathbb{R}^n \to \mathbb{C}$*) is continuously differentiable, then* $\partial_\xi f(X_1, \dots, X_n)$ *exists, with*

$$\partial_\xi f(X_1, \dots, X_n) = \sum_{i=1}^n \frac{\partial f}{\partial x_i}(X_1, \dots, X_n)\partial_\xi X_i.$$

(ii) *If* $\partial_\xi X_1, \dots, \partial_\xi X_n$ *exist, and* $f\colon \mathbb{C}^n \to \mathbb{C}$ *is continuously differentiable, then* $\partial_\xi f(X_1, \dots, X_n)$ *exists, with*

$$\partial_\xi f(X_1, \dots, X_n) = \sum_{i=1}^n \frac{\partial f}{\partial z_i}(X_1, \dots, X_n)\partial_\xi X_i + \sum_{i=1}^n \frac{\partial f}{\partial \bar{z}_i}(X_1, \dots, X_n)\overline{\partial_\xi X_i}. \tag{15.1}$$

(iii) *In particular, if* $\partial_\xi X$ *and* $\partial_\xi Y$ *exist then*

$$\partial_\xi(tX) = t\partial_\xi X, \qquad t \in \mathbb{C},$$
$$\partial_\xi(X + Y) = \partial_\xi X + \partial_\xi Y,$$
$$\partial_\xi(XY) = (\partial_\xi X)Y + X(\partial_\xi Y).$$

(iv) *If* $\partial_\xi X$ *exists and* $t \in \mathbb{R}$*, then* $\partial_{t\xi}$ *exists and*

$$\partial_{t\xi} X = t\partial_\xi X.$$

(v) *If* $\partial_\xi X$ *and* $\partial_\eta X$ *exist, then* $\partial_{\xi+\eta} X$ *exists and*

$$\partial_{\xi+\eta}(X) = \partial_\xi X + \partial_\eta X.$$

(vi) *If* $\partial_\xi X$ *exists, then* $\partial_\xi(\rho_\eta(X))$ *exists and*

$$\partial_\xi(\rho_\eta(X)) = \rho_\eta(\partial_\xi X).$$

PROOF. For (i) and (ii), consider a sequence $t_k \to 0$. Then, as $k \to \infty$, $(\rho_{t_k\xi}(X_i) - X_i)/t_k \to \partial_\xi X_i$ in probability for every i. It is well-known that this implies that there exists a subsequence of $(t_k)_{k=1}^\infty$ such that the convergence holds a.s. along the subsequence (for every i). Consequently, as $t \to 0$ along this subsequence, using Theorem 14.1 and standard calculus (for fixed $\omega \in \Omega$), a.s.

$$\begin{aligned} &\frac{\rho_{t\xi}(f(X_1, \dots, X_n)) - f(X_1, \dots, X_n)}{t} \\ &\qquad = \frac{f(\rho_{t\xi}(X_1), \dots, \rho_{t\xi}(X_n)) - f(X_1, \dots, X_n)}{t} \\ &\qquad \to Y = \sum_{i=1}^n \frac{\partial f}{\partial x_i}(X_1, \dots, X_n)\partial_\xi X_i \end{aligned} \tag{15.2}$$

in the real case; in the complex case we obtain the same result with Y equal to the right hand side of (15.1). Since each sequence $t_k \to 0$ thus has a subsequence such that (15.2) holds a.s., and thus in probability, and Y does not depend on the sequence, a standard argument shows that (15.2) holds in probability for $t \to 0$ without restriction, which proves (i) and (ii).

The formulae in (iii) are special cases of (ii), and (iv) follows directly from the definition.

For (v), it follows by Theorem 14.1(ii) and (ix) that, as $t \to 0$,

$$\frac{\rho_{t(\xi+\eta)}(X) - X}{t} = \rho_{t\xi}\Big(\frac{\rho_{t\eta}(X) - X}{t}\Big) + \frac{\rho_{t\xi}(X) - X}{t}$$
$$\to \rho_0(\partial_\eta X) + \partial_\xi X = \partial_\eta X + \partial_\xi X.$$

Similarly, (vi) follows by

$$\frac{\rho_{t\xi}(\rho_\eta(X)) - \rho_\eta(X)}{t} = \frac{\rho_{t\xi+\eta}(X) - \rho_\eta(X)}{t} = \rho_\eta\Big(\frac{\rho_{t\xi}(X) - X}{t}\Big) \to \rho_\eta(\partial_\xi X).$$

□

Example 15.3 and Theorem 15.7(i) yield the following result.

THEOREM 15.8. *If $\xi_1, \dots, \xi_n \in H$ and $f\colon \mathbb{R}^n \to \mathbb{C}$ is continuously differentiable, then*

$$\partial_\xi f(\xi_1, \dots, \xi_n) = \sum_{i=1}^n \langle \xi_i, \xi \rangle \frac{\partial f}{\partial x_i}(\xi_1, \dots, \xi_n).$$

□

EXAMPLE 15.9. In particular, this applies to every polynomial variable in $\mathcal{P}$. Theorem 15.8 yields the explicit formula

$$\partial_\xi(\xi_1 \cdots \xi_n) = \sum_{j=1}^n \langle \xi_j, \xi \rangle \xi_1 \cdots \xi_{j-1}\xi_{j+1} \cdots \xi_n.$$

Note that by (13.23) and linearity, this implies $\partial_\xi X = A(\xi)X$ for every $X \in \mathcal{P}$ and $\xi \in H$. By (13.9), or as a consequence of Theorem 14.1(xi), we also have

$$\partial_\xi(\,{:}\xi_1 \cdots \xi_n{:}\,) = \sum_{j=1}^n \langle \xi_j, \xi \rangle \,{:}\xi_1 \cdots \xi_{j-1}\xi_{j+1} \cdots \xi_n{:}\,.$$

EXAMPLE 15.10. If $\zeta \in H_{\mathbb{C}}$, then Theorem 15.7(ii) shows that

$$\partial_\xi e^\zeta = \langle \zeta, \xi \rangle e^\zeta.$$

Since ${:}e^\zeta{:}$ equals the constant $e^{-\mathrm{E}\zeta^2/2}$ times e^ζ, we also obtain

$$\partial_\xi \,{:}e^\zeta{:} = \langle \zeta, \xi \rangle \,{:}e^\zeta{:}\,.$$

Note that if $\partial_\xi X$ exists, then by Theorem 14.1, cf. Theorem 15.7(vi),

$$\lim_{h\to 0} \frac{\rho_{(t+h)\xi}(X) - \rho_{t\xi}(X)}{h} = \lim_{h\to 0} \rho_{t\xi}\Big(\frac{\rho_{h\xi}(X) - X}{t}\Big) = \rho_{t\xi}(\partial_\xi X), \tag{15.3}$$

and thus $t \mapsto \rho_{t\xi}(X)$ is differentiable for every t, as a function of $\mathbb{R}$ into L^0. The next example shows, however, that differentiability in L^0 is a rather weak property.

EXAMPLE 15.11. Let $\xi \in H$ be standard normal and let $X = \mathbf{1}[\xi > 0]$. Then,

$$\rho_{t\xi} X = \mathbf{1}[\rho_{t\xi}(\xi) > 0] = \mathbf{1}[\xi + t > 0] = \mathbf{1}[\xi > -t]$$

and a simple calculation shows that $\partial_\xi X = \lim_{t \to 0} t^{-1}(\mathbf{1}[\xi > -t] - \mathbf{1}[\xi > 0])$ exists in probability (and a.s.), with $\partial_\xi X = 0$. (Cf. Example 15.4, and note that $\mathbf{1}[x > 0]$ has derivative 0 a.e.) It follows similarly, see also (15.3), that $\frac{d}{dt}\rho_{t\xi}(X) = 0$ for every t, in probability. Hence the mapping $t \mapsto \rho_{t\xi}(X)$ has an everywhere vanishing derivative in L^0, without being constant.

Example 15.11 can be generalized to any indicator variable.

THEOREM 15.12. *If $A \in \mathcal{F}(H)$ is any event, and $\xi \in H$, then $\partial_\xi \mathbf{1}_A$ exists and $\partial_\xi \mathbf{1}_A = 0$.*

PROOF. By Theorem 14.1, for any real t, $\rho_{t\xi}(\mathbf{1}_A)^2 = \rho_{t\xi}(\mathbf{1}_A^2) = \rho_{t\xi}(\mathbf{1}_A)$, and thus $\rho_{t\xi}(\mathbf{1}_A) \in \{0, 1\}$ a.s.; consequently $\rho_{t\xi}(\mathbf{1}_A) - \mathbf{1}_A \in \{-1, 0, 1\}$ a.s. and

$$\mathrm{P}(\rho_{t\xi}(\mathbf{1}_A) - \mathbf{1}_A \neq 0) = \mathrm{P}(|\rho_{t\xi}(\mathbf{1}_A) - \mathbf{1}_A| = 1) = \|\rho_{t\xi}(\mathbf{1}_A) - \mathbf{1}_A\|_1.$$

Since $\mathbf{1}_A \in L^2$, Theorem 14.1(viii) yields $\rho_{t\xi}(\mathbf{1}_A) - \mathbf{1}_A \to 0$ in L^1 as $t \to 0$, and thus $\mathrm{P}(\rho_{t\xi}(\mathbf{1}_A) - \mathbf{1}_A \neq 0) \to 0$. Consequently, $(\rho_{t\xi}(\mathbf{1}_A) - \mathbf{1}_A)/t \to 0$ in probability. □

As a consequence, we obtain the following theorems which say that ∂_ξ is a local operator in a measure theoretic sense.

THEOREM 15.13. *If $\partial_\xi X$ and $\partial_\xi Y$ exist, and $A \in \mathcal{F}(H)$ is an event such that $X = Y$ a.s. on A, then $\partial_\xi X = \partial_\xi Y$ a.s. on A.*

PROOF. By Theorems 15.12 and 15.7(iii), $\partial_\xi(\mathbf{1}_A X)$ exists and

$$\partial_\xi(\mathbf{1}_A X) = \mathbf{1}_A \partial_\xi X + X \partial_\xi \mathbf{1}_A = \mathbf{1}_A \partial_\xi X.$$

Similarly, $\partial_\xi(\mathbf{1}_A Y) = \mathbf{1}_A \partial_\xi Y$. Since $\mathbf{1}_A X = \mathbf{1}_A Y$, this yields $\mathbf{1}_A \partial_\xi X = \mathbf{1}_A \partial_\xi Y$. □

THEOREM 15.14. *Let $X \in L^0$ be a random variable, and let $\xi \in H$. Suppose that there exist random variables $X_1, X_2, \ldots \in L^0$ and events $A_1, A_2, \ldots \in \mathcal{F}(H)$ such that $\partial_\xi X_i$ exists for every i, $X = X_i$ a.s. on A_i, and $\mathrm{P}(\bigcup_1^\infty A_i) = 1$. Then $\partial_\xi X$ exists, with $\partial_\xi X = \partial_\xi X_i$ a.s. on A_i.*

PROOF. By replacing A_i by $A_i \setminus \bigcup_1^{i-1} A_j$, we may assume that the sets A_i, $1 \le i < \infty$ are disjoint. Let $Y = \sum_1^\infty \mathbf{1}_{A_i} \partial_\xi X_i$ and $B_n = \bigcup_1^n A_i$, $n \ge 1$. Then $\mathbf{1}_{B_n} X = \sum_1^n \mathbf{1}_{A_i} X = \sum_1^n \mathbf{1}_{A_i} X_i$, and thus, by Theorems 15.12 and 15.7(iii),

$$\partial_\xi(\mathbf{1}_{B_n} X) = \sum_1^n \partial_\xi(\mathbf{1}_{A_i} X_i) = \sum_1^n \mathbf{1}_{A_i} \partial_\xi X_i = \mathbf{1}_{B_n} Y.$$

For every $\varepsilon > 0$, $t \neq 0$ and $n \geq 1$,

$$\begin{aligned}
&\mathrm{P}\big(|(\rho_{t\xi}(X) - X)/t - Y| > \varepsilon\big) \\
&\leq \mathrm{P}(\rho_{t\xi}(\mathbf{1}_{B_n}) \neq 1) + \mathrm{P}(\mathbf{1}_{B_n} \neq 1) + \mathrm{P}\big(|(\rho_{t\xi}(\mathbf{1}_{B_n}X) - \mathbf{1}_{B_n}X)/t - \mathbf{1}_{B_n}Y| > \varepsilon\big).
\end{aligned}$$

As $t \to 0$, the last term tends to 0 because $\partial_\xi(\mathbf{1}_{B_n}X) = \mathbf{1}_{B_n}Y$, and

$$\begin{aligned}
\mathrm{P}(\rho_{t\xi}(\mathbf{1}_{B_n}) \neq 1) &= \mathrm{P}(\rho_{t\xi}(\mathbf{1}_{B_n}) = 0) = \|\rho_{t\xi}(1 - \mathbf{1}_{B_n})\|_1 \\
&\to \|1 - \mathbf{1}_{B_n}\|_1 = \mathrm{P}(\mathbf{1}_{B_n} \neq 1).
\end{aligned}$$

Consequently,

$$\limsup_{t\to 0} \mathrm{P}\big(|(\rho_{t\xi}(X) - X)/t - Y| > \varepsilon\big) \leq 2\,\mathrm{P}(\mathbf{1}_{B_n} \neq 1) = 2(1 - \mathrm{P}(B_n)). \tag{15.4}$$

Now let $n \to \infty$; then $\mathrm{P}(B_n) \to 1$, and thus (15.4) yields $\lim_{t\to 0} \mathrm{P}\big(|(\rho_{t\xi}(X) - X)/t - Y| > \varepsilon\big) = 0$. Since $\varepsilon > 0$ is arbitrary, this proves $\partial_\xi X = Y$. Finally, $\partial_\xi X = \partial_\xi X_i$ on A_i follows from Theorem 15.13. □

COROLLARY 15.15. *If $X \in L^0$ is a discrete random variable, i.e. a random variable with only a finite or countable number of possible values, then $\partial_\xi X = 0$.*

PROOF. Apply Theorem 15.14 with each X_i constant. □

Example 15.11, and more generally Theorem 15.12 and Corollary 15.15, show that it is in general impossible to recover a variable from its derivative. We therefore introduce another definition. Recall that a function f of a real variable is *absolutely continuous* if there exists a locally integrable function g such that

$$f(t) = f(0) + \int_0^t g(s)\,ds, \qquad -\infty < t < \infty. \tag{15.5}$$

An absolutely continuous function is differentiable a.e., and g in (15.5) equals f' a.e.

REMARK 15.16. In this chapter we use the standard convention $\int_0^t = -\int_t^0$ for $t < 0$. Note also that the definition above is not the only one in use for functions on $\mathbb{R}$; many authors require g to be integrable on the entire line $\mathbb{R}$, and not (as we do) just on finite intervals, or impose another global condition, and the property above should in their terminology be called *locally absolutely continuous*, or *absolutely continuous on finite intervals*.

REMARK 15.17. It is well-known, cf. Cohn (1980, Section 4.4), that f is absolutely continuous if and only if for every $\varepsilon > 0$ and $T < \infty$, there exists $\delta > 0$ such that $\sum_i |f(b_i) - f(a_i)| < \varepsilon$ holds whenever $\{(a_i, b_i)\}$ is a finite sequence of disjoint intervals contained in $[-T, T]$ for which $\sum_i (b_i - a_i) < \delta$.

We also use the language of stochastic processes, cf. Appendix B, and say that a *version* of $\rho_{t\xi}(X)$ is a function Y on $\mathbb{R}\times\Omega$ such that $Y(t,\cdot)=\rho_{t\xi}(X)$ a.s. for every $t\in\mathbb{R}$. We observe that Theorem 14.1(x) implies that $\rho_{t\xi}(X)$ always has a jointly measurable version.

DEFINITION 15.18. Let $\xi\in H$. A random variable $X\in L^0$ is *absolutely continuous along* ξ, which we abbreviate to ξ-*a.c.*, if there exists an absolutely continuous version of $\rho_{t\xi}(X)$, i.e. a version that is an absolutely continuous function of t for every $\omega\in\Omega$.

We observe that an (absolutely) continuous version of a stochastic process on $\mathbb{R}$ is jointly measurable; if $U(t,\omega)$ is continuous in t for every $\omega\in\Omega$, then $U=\lim_{n\to\infty}U_n$ where $U_n(t,\omega)=U(\lfloor nt\rfloor/n,\omega)$ is jointly measurable.

LEMMA 15.19. *X is ξ-a.c. if and only if there exists a measurable function Y on $\mathbb{R}\times\Omega$ such that $\int_{-T}^{T}|Y(s,\omega)|\,ds<\infty$ a.s. for every $T>0$ and, for every $t\in\mathbb{R}$,*

$$\rho_{t\xi}X(\omega)-X(\omega)=\int_0^t Y(s,\omega)\,ds \qquad a.s. \tag{15.6}$$

PROOF. If X is ξ-a.c., and $U(t,\omega)$ is an absolutely continuous version of $\rho_{t\xi}(X)$, then

$$U(t,\omega)-U(0,\omega)=\int_0^t \frac{d}{ds}U(s,\omega)\,ds$$

and $\frac{d}{ds}U(s,\omega)=\lim_{n\to\infty}n(U(s+1/n,\omega)-U(s,\omega))$, defined e.g. as 0 when the limit does not exist, is measurable on $\mathbb{R}\times\Omega$.

Conversely, if Y is given as above, we may assume that $\int_{-T}^{T}|Y(s,\omega)|\,ds<\infty$ for every $T>0$ and every ω, by redefining $Y(t,\omega)$ to be 0 for every t when ω is in the null set $\bigcup_{N=1}^{\infty}\{\omega:\int_{-N}^{N}|Y(s,\omega)|\,ds=\infty\}$. Then $X(\omega)+\int_0^t Y(s,\omega)\,ds$ is an absolutely continuous version of $\rho_{t\xi}(X)$. □

LEMMA 15.20. *Let $\xi\in H$ and $X\in L^0$. If X is ξ-a.c., then $\partial_\xi X$ exists and $Y(s,\omega)$ in (15.6) equals a.s. $\rho_{s\xi}(\partial_\xi X)$, for a.e. s.*

PROOF. Suppose that X is ξ-a.c., and let Y be as in Lemma 15.19. Let

$$A=\{(t,\omega)\in\mathbb{R}\times\Omega:\frac{1}{h}\int_t^{t+h}Y(s,\omega)\,ds\to Y(t,\omega)\text{ as }h\to 0\}.$$

It is easy to see that A is measurable. Moreover, for a.e. $\omega\in\Omega$, $Y(\cdot,\omega)$ is locally integrable, and thus a.e. $t\in\mathbb{R}$ is such that $\frac{d}{dt}\int_0^t Y(s,\omega)\,ds=Y(t,\omega)$, i.e. $(t,\omega)\in A$ (Cohn 1980, Section 6.3). By Fubini's theorem applied to the product measure $dt\times\mathrm{P}$ and the complement of A, a.e. real t belongs to the set

$$B=\{t\in\mathbb{R}:(t,\omega)\in A\text{ for a.e. }\omega\}.$$

Suppose now that $t \in B$, and let $h \to 0$ along a sequence. Then

$$\frac{\rho_{h\xi}(\rho_{t\xi}(X)) - \rho_{t\xi}(X)}{h} = \frac{\rho_{(t+h)\xi}(X) - \rho_{t\xi}(X)}{h} = \frac{1}{h}\int_t^{t+h} Y(s,\omega)\,ds \to Y(t,\omega)$$

a.s., and thus

$$\partial_\xi(\rho_{t\xi}(X)) = Y(t,\cdot), \qquad t \in B.$$

Theorem 15.7(vi) now shows, using Theorem 14.1(i), that $\partial_\xi(X)$ exists, and

$$Y(t,\cdot) = \rho_{t\xi}(\partial_\xi X), \qquad t \in B.$$

□

THEOREM 15.21. *Let $\xi \in H$ and $X \in L^0$. Then X is ξ-a.c. if and only if $\partial_\xi X$ exists and, for every real t,*

$$\rho_{t\xi}X(\omega) - X(\omega) = \int_0^t \rho_{s\xi}(\partial_\xi X)(\omega)\,ds \qquad a.s.,$$

with the integral a.s. absolutely convergent. (Here $\rho_{s\xi}(\partial_\xi X)$ denotes any jointly measurable version.)

PROOF. A simple consequence of the lemmas. □

THEOREM 15.22. *If $X_1, \dots, X_n$ are ξ-a.c., and f is continuously differentiable on $\mathbb{R}^n$ (or $\mathbb{C}^n$), then $f(X_1, \dots, X_n)$ is ξ-a.c.*

PROOF. This follows, by fixing $\omega \in \Omega$, from Definition 15.18, Theorem 14.1(iii), and the corresponding fact in real analysis that if $f_1, \dots, f_n$ are absolutely continuous, then $f(f_1, \dots, f_n)$ is too (which easily follows from the characterization in Remark 15.17). □

REMARK 15.23. More generally, the result holds for any f that is locally Lipschitz, i.e. such that for every $R > 0$ there exists an $M < \infty$ such that $|f(x) - f(y)| \le M|x - y|$ if $|x|, |y| \le R$.

EXAMPLE 15.24. Let $\xi \in H$. By Definition 15.18 and (14.3), every $\eta \in H$ is ξ-a.c. Consequently, Theorem 15.22 shows that every variable $f(\xi_1, \dots, \xi_n)$ as in Theorem 15.8 is ξ-a.c. In particular, every polynomial variable in $\mathcal{P}$ is ξ-a.c. Similarly, if $\zeta \in H_{\mathbb{C}}$, then e^ζ and $:e^\zeta:$ are ξ-a.c.

Finally we give an example of a random variable that does *not* have a derivative. (We use a version of Weierstrass' construction of a non-differentiable function.)

EXAMPLE 15.25. Let $g(x) = \mathbf{1}[\lfloor x \rfloor \text{ is odd}]$, where $\lfloor x \rfloor$ denotes the integer part of $x \in \mathbb{R}$, and

$$f(x) = \sum_{k=1}^{\infty} 3^{-k} g(4^k x).$$

Given x and $n \geq 1$, let m be the smallest integer such that $g(4^m(x+4^{-n})) \neq g(4^m x)$. Then $m \leq n$ and

$$|f(x+4^{-n}) - f(x)| \geq 3^{-m} - \sum_{m+1}^{\infty} 3^{-k} = \tfrac{1}{2}3^{-m} \geq \tfrac{1}{2}3^{-n}.$$

Hence $|f(x+4^{-n}) - f(x)|/4^{-n} \geq \frac{1}{2}(\frac{4}{3})^n$ for every x and $n \geq 1$. It follows that if ξ is standard normal, then $\partial_\xi f(\xi)$ does not exist.

2. The gradient operator

In the preceding section we studied the directional derivative $\partial_\xi X$ for a fixed ξ. Let us now fix $X \in L^0$ and vary ξ, assuming that $\partial_\xi X$ exists for every $\xi \in H$. By Theorem 15.7(iv),(v), we then have $\partial_{t\xi} X = t\partial_\xi X$ and $\partial_{\xi+\eta}(X) = \partial_\xi X + \partial_\eta X$ for all $t \in \mathbb{R}$ and $\xi, \eta \in H$. These equalities hold only a.s., and it will be seen in Example 15.97 below that it is not necessarily the case that there exist versions of the variables such that the equalities hold for every $\omega \in \Omega$, with $\xi \mapsto \partial_\xi X(\omega)$ a continuous linear functional, and thus $\partial_\xi X(\omega) = \langle Y(\omega), \xi \rangle_H$ for some $Y(\omega) \in H$. Nevertheless, this holds in well-behaved cases, and we use this property as a definition.

DEFINITION 15.26. Let $X \in L^0_{\mathbb{R}}$. If $\partial_\xi X$ exists for every $\xi \in H$, and moreover there exists an H-valued random variable $\nabla X \in L^0(H)$ such that $\partial_\xi X(\omega) = \langle \nabla X(\omega), \xi \rangle_H$ a.s. for every $\xi \in H$, then we call ∇X the *gradient* of X. In this case we also say that the gradient of X exists (in probability). (Note that Proposition C.3 implies that ∇X is uniquely defined when it exists.)

If $X \in L^0_{\mathbb{C}}$, we define the gradient similarly, but now allow $\nabla X \in L^0(H_{\mathbb{C}})$.

REMARK 15.27. The gradient operator ∇ goes under several different names in the literature, including the *derivative*, *H-derivative* and *Malliavin derivative*, and it is often denoted by D.

REMARK 15.28. The gradient ∇X is thus, when it exists, a vector-valued random variable with values in the Hilbert space H itself (or its complexification). This may lead to some confusion, since the elements of H are themselves random variables; for example (in the real case) in the defining relation $\langle \nabla X(\omega), \xi \rangle_H = \partial_\xi X(\omega)$, we take in the inner product the value of ∇X at a particular $\omega \in \Omega$, which is an element of H, and the random variable ξ regarded as an element of H. The best way to avoid confusion is perhaps to regard the gradient as taking its values in an abstract Hilbert space H, which happens to equal our Gaussian Hilbert space.

If H is given as a Gaussian Hilbert space indexed by some other Hilbert space H_0, it is also possible, and often useful, to regard ∇X as a random variable taking its values in H_0 (or its complexification), using the defining isometry between H_0 and H. For example, given a Gaussian stochastic integral on a measure space $(M, \mathcal{M}, \mu)$, we can regard ∇X as a random variable in $L^2(M, \mathcal{M}, \mu)$, see further Section 11. (This definition is often used in the literature, and it is useful in many applications.)

Yet another possibility, which we will use in Section 8, is to regard ∇X as a random variable taking its values in an independent copy of H. (One version of this is to regard ∇X as a function on Ω^2.)

EXAMPLE 15.29. Let $(\Omega, \mathcal{F}, \mathrm{P})$ be $\mathbb{R}^d$ with the standard Gaussian measure, and let H be the d-dimensional Gaussian Hilbert space of linear functions, with the coordinate functions ξ_i, $i = 1, \dots, d$, as the canonical orthonormal basis.

It follows from Example 15.5 that if X is a continuously differentiable function on $\mathbb{R}^d$, then ∇X can be identified with the usual gradient.

We make another definition.

DEFINITION 15.30. A random variable $X \in L^0$ is *ray absolutely continuous*, abbreviated *r.a.c.*, if it is ξ-a.c. for every $\xi \in H$.

We let $\mathcal{D}^{1,0}$ denote the set of all random variables X in L^0 such that X is r.a.c. and ∇X exists.

EXAMPLE 15.31. By Example 15.9, every polynomial variable in $\mathcal{P}$ has a gradient, with the explicit formulae, for $\xi_1, \dots, \xi_n \in H_{\mathbb{C}}$,

$$\nabla(\xi_1 \cdots \xi_n) = \sum_{j=1}^{n} \xi_1 \cdots \xi_{j-1}\xi_{j+1} \cdots \xi_n \otimes \xi_j,$$

and

$$\nabla(:\xi_1 \cdots \xi_n:) = \sum_{j=1}^{n} :\xi_1 \cdots \xi_{j-1}\xi_{j+1} \cdots \xi_n: \otimes \xi_j.$$

In particular, $\nabla 1 = 0$ and $\nabla \xi = 1 \otimes \xi$, $\xi \in H$.

It is easily seen that every polynomial variable is r.a.c., directly, by Example 15.24 or by Lemma 15.33 below. In other words, $\mathcal{P} \subset \mathcal{D}^{1,0}$.

EXAMPLE 15.32. By Example 15.10, for every $\zeta \in H_{\mathbb{C}}$,

$$\nabla e^{\zeta} = e^{\zeta} \otimes \zeta$$

and

$$\nabla :e^{\zeta}: = :e^{\zeta}: \otimes \zeta.$$

These variables too are r.a.c., and thus they belong to $\mathcal{D}^{1,0}$.

LEMMA 15.33. *If $X_1, \dots, X_n$ are r.a.c., and f is continuously differentiable on $\mathbb{R}^n$ (or $\mathbb{C}^n$), then $f(X_1, \dots, X_n)$ is r.a.c.*

PROOF. Apply Theorem 15.22 for each $\xi \in H$. □

The chain rule holds for the gradient in the following form.

THEOREM 15.34. *Suppose that $\nabla X_1, \dots, \nabla X_n$ exist and that f is continuously differentiable on $\mathbb{R}^n$ (or $\mathbb{C}^n$). Then $\nabla f(X_1, \dots, X_n)$ exists. In the real case,*

$$\nabla f(X_1, \dots, X_n) = \sum_{i=1}^{n} \frac{\partial f}{\partial x_i}(X_1, \dots, X_n) \nabla X_i,$$

and in the complex case,

$$\nabla f(X_1, \dots, X_n) = \sum_{i=1}^{n} \frac{\partial f}{\partial z_i}(X_1, \dots, X_n) \nabla X_i + \sum_{i=1}^{n} \frac{\partial f}{\partial \bar{z}_i}(X_1, \dots, X_n) \overline{\nabla X_i}.$$

Furthermore, if $X_1, \dots, X_n \in \mathcal{D}^{1,0}$, then $f(X_1, \dots, X_n) \in \mathcal{D}^{1,0}$.

PROOF. This is a direct consequence of Theorem 15.7, Definition 15.26 and Lemma 15.33. □

In particular, this yields the following extension of Theorem 15.8.

THEOREM 15.35. *If $\xi_1, \dots, \xi_n \in H$ and $f \colon \mathbb{R}^n \to \mathbb{C}$ is continuously differentiable, then $f(\xi_1, \dots, \xi_n) \in \mathcal{D}^{1,0}$ and*

$$\nabla f(\xi_1, \dots, \xi_n) = \sum_{i=1}^{n} \frac{\partial f}{\partial x_i}(\xi_1, \dots, \xi_n) \otimes \xi_i.$$

□

Note the following uniqueness result.

THEOREM 15.36. *If $X \in \mathcal{D}^{1,0}$ and $\nabla X = 0$, then $X = C$ a.s., for some constant C.*

PROOF. Let $\xi \in H$. Since $\partial_\xi X = \langle \nabla X, \xi \rangle_{H_\mathbb{C}} = 0$ and X is ξ-a.c. by assumption, Theorem 15.21 yields $\rho_{t\xi} X - X = 0$ for each real t, and thus, in particular, $\rho_\xi X = X$.

The result follows by Theorem 14.9. □

This theorem fails without the assumption that $X \in \mathcal{D}^{1,0}$, i.e. that X is r.a.c.; this is seen by the next result which follows from Theorem 15.12 and Corollary 15.15 in the preceding section.

THEOREM 15.37. *If $A \in \mathcal{F}(H)$ is any event, and $\xi \in H$, then $\nabla \mathbf{1}_A = 0$. More generally, if $X \in L^0$ is a discrete random variable, then $\nabla X = 0$.* □

This theorem thus gives examples of random variables that are not r.a.c. (except in trivial cases), but nevertheless have gradients.

COROLLARY 15.38. *If $A \in \mathcal{F}(H)$ and $\mathbf{1}_A \in \mathcal{D}^{1,0}$, then $\mathrm{P}(A) = 0$ or 1.*

PROOF. By Theorem 15.37, $\nabla \mathbf{1}_A = 0$; hence $\mathbf{1}_A$ is a.s. constant by Theorem 15.36. □

Finally, we note that the localization theorems 15.13 and 15.14 too extend to ∇.

THEOREM 15.39. *If ∇X and ∇Y exist, and $A \in \mathcal{F}(H)$ is an event such that $X = Y$ a.s. on A, then $\nabla X = \nabla Y$ a.s. on A.*

PROOF. By Theorem 15.13, $\langle \nabla X, \xi \rangle_{H_\mathbf{C}} = \partial_\xi X = \partial_\xi Y = \langle \nabla Y, \xi \rangle_{H_\mathbf{C}}$ a.s. on A for every $\xi \in H$, and the result follows by Proposition C.3. □

THEOREM 15.40. *Let $X \in L^0$ be a random variable, and let $\xi \in H$. Suppose that there exist random variables $X_1, X_2, \ldots \in L^0$ and events $A_1, A_2, \ldots \in \mathcal{F}(H)$ such that ∇X_i exists for every i, $X = X_i$ on A_i, and $\mathrm{P}(\bigcup_1^\infty A_i) = 1$. Then ∇X exists, with $\nabla X = \nabla X_i$ on A_i.*

PROOF. We may assume that the sets A_i are disjoint. Define $Y = \sum_i \mathbf{1}_{A_i} \nabla X_i \in L^0(H_\mathbf{C})$. By Theorem 15.14, $\partial_\xi X$ exists for every $\xi \in H$, and $\partial_\xi X = \sum_i \mathbf{1}_{A_i} \partial_\xi X_i = \sum_i \mathbf{1}_{A_i} \langle \nabla X_i, \xi \rangle_{H_\mathbf{C}} = \langle Y, \xi \rangle_{H_\mathbf{C}}$. □

3. Higher derivatives

The gradient of a real or complex random variable is (when it exists) a random variable with values in the Hilbert space H or $H_\mathbf{C}$. In order to define higher gradients, we thus generalize the definitions above to random variables with values in an arbitrary Hilbert space. (See Appendix C for definitions.) For simplicity, and maximal generality, we use 'weak definitions', using linear functionals to reduce to the scalar case treated above.

DEFINITION 15.41. Let $X \in L^0(H_0)$ be an a.s. separably valued random variable with values in a real or complex Hilbert space H_0.

We say that X is ξ-a.c. or r.a.c. if the scalar random variable $\langle X, h \rangle_{H_0}$ is, for every $h \in H_0$. Further, $\partial_\xi X = Y$ if $Y \in L^0(H_0)$ is a random variable such that $\partial_\xi \langle X, h \rangle_{H_0} = \langle Y, h \rangle_{H_0}$ for every $h \in H_0$, and similarly $\nabla X = Z$ if $Z \in L^0(H \otimes H_0)$ is a random variable such that $\nabla \langle X, h \rangle_{H_0} = \langle Z, h \rangle_{H_0}$ for every $h \in H_0$. (Note that in this case $\langle Z, h \rangle_{H_0} \in L^0(H)$; the inner product acts by $\langle \eta \otimes h_0, h \rangle_{H_0} = \langle h_0, h \rangle_{H_0} \eta$, for any $\eta \in H$ and $h, h_0 \in H_0$.)

By Proposition C.3, $\partial_\xi X$ and ∇X are uniquely defined when they exist.

REMARK 15.42. Here the tensor product $H \otimes H_0$ is taken regarding the spaces as real vector spaces in the complex case too; a more precise notation is $H \otimes_\mathbf{R} H_0$. In the complex case $H \otimes_\mathbf{R} H_0 \cong H_\mathbf{C} \otimes_\mathbf{C} H_0$, and thus we may then also write $H_\mathbf{C} \otimes H_0$.

REMARK 15.43. By the definition, $\partial_\xi X = Y$ if and only if $(\rho_{t\xi} X - X)/t \xrightarrow{\mathrm{p}} Y$ weakly in H_0 as $t \to 0$; recall that $\rho_{t\xi}$ extends to $L^0(H_0)$ by Remark 14.6. We will here not consider the question whether these difference quotients converge strongly in H_0, or similar questions for the other definitions.

REMARK 15.44. Peters (1997+) has studied a weaker version where ∇X takes its values in the space $L(H, H_0)$ of bounded linear operators $H \to H_0$ rather than $H \otimes H_0$ (which can be identified with the subspace of Hilbert–Schmidt operators).

The definition above and Definition 15.26 yield immediately the following alternative characterizations.

PROPOSITION 15.45. *Let $X \in L^0(H_0)$ and $Z \in L^0(H \otimes H_0)$. Then the following are equivalent.*

(i) $\nabla X = Z$.
(ii) $\partial_\xi \langle X, h\rangle_{H_0} = \langle Z, \xi \otimes h\rangle_{H\otimes H_0}$ *for every* $\xi \in H$ *and* $h \in H_0$.
(iii) $\partial_\xi X = \langle Z, \xi\rangle_H$ *for every* $\xi \in H$.

□

We can now iterate, and for suitable X define $\nabla^2 X$ with values in $H \otimes H \otimes H_0$, $\nabla^3 X$ with values in $H \otimes H \otimes H \otimes H_0$, etc. In particular, if X is a real (complex) random variable, then $\nabla^k X$ is, if it exists, a random variable with values in $H^{\otimes k}$ ($H_{\mathbb{C}}^{\otimes k}$) such that

$$\langle \nabla^k X, \xi_1 \otimes \cdots \otimes \xi_k\rangle_{H_{\mathbb{C}}^{\otimes k}} = \partial_{\xi_1} \cdots \partial_{\xi_k} X, \qquad \xi_1, \ldots, \xi_k \in H.$$

EXAMPLE 15.46. It is easily seen that if $X \in L^0$ is a scalar-valued random variable, H_0 is a Hilbert space and $h_0 \in H_0$, then $X \otimes h_0$ is ξ-a.c. or r.a.c. if X is, and $\partial_\xi(X \otimes h_0) = \partial_\xi X \otimes h_0$ and $\nabla(X \otimes h_0) = \nabla X \otimes h_0$, provided $\partial_\xi X$ and ∇X exist.

In particular, by Theorem 15.35 and induction, if $\xi_1, \ldots, \xi_n \in H$ and $f \colon \mathbb{R}^n \to \mathbb{C}$ is k times continuously differentiable ($k \geq 1$), then $\nabla^k f(\xi_1, \ldots, \xi_n)$ exists and

$$\nabla^k f(\xi_1, \ldots, \xi_n) = \sum_{j_1, \ldots, j_k} \frac{\partial^k f}{\partial x_{j_1} \cdots \partial x_{j_k}}(\xi_1, \ldots, \xi_n) \otimes \xi_{j_1} \otimes \cdots \otimes \xi_{j_k}. \tag{15.7}$$

Moreover, $\nabla^{k-1} f(\xi_1, \ldots, \xi_n)$ is then r.a.c.

4. Absolute continuity

One of the main applications of the Malliavin calculus is to prove existence and smoothness of densities of certain random variables. We give here some results on the existence of densities; smoothness will be studied under more restrictive conditions in Section 10.

We begin with a version of the chain rule with a rather weak regularity condition.

LEMMA 15.47. *Suppose that $g \colon \mathbb{R} \to \mathbb{R}$ is absolutely continuous and that $f \colon \mathbb{R} \to \mathbb{R}$ is a bounded Borel function. Let $F(x) = \int_0^x f(t)\, dt$. Then $F \circ g$ is* *absolutely continuous, and* $(F \circ g)'(x) = \big(f \circ g(x)\big)g'(x)$ *a.e.; in other words,*

$$F(g(t)) - F(g(0)) = \int_0^t f(g(s))g'(s)\,ds, \qquad -\infty < t < \infty. \tag{15.8}$$

PROOF. Consider the class $\mathcal{A}$ of bounded real Borel functions f such that the conclusion holds. It is clear that $\mathcal{A}$ is a linear space, and it is easily seen that $\mathcal{A}$ contains every bounded continuous function.

Moreover, if $f_n \in \mathcal{A}$ are uniformly bounded, $|f_n| \le C$ say, and $f_n \to f$ pointwise as $n \to \infty$, then $F_n(x) = \int_0^x f_n(t)\,dt \to F(x)$ for each x and $\int_0^t f_n(g(s))g'(s)\,ds \to \int_0^t f(g(s))g'(s)\,ds$ by dominated convergence; consequently we can let $n \to \infty$ in (15.8) applied to f_n and obtain $f \in \mathcal{A}$.

The monotone class theorem, Theorem A.1, now shows that $\mathcal{A}$ contains every bounded Borel function. □

THEOREM 15.48. *Let* $1 \le p < \infty$. *Suppose that* $X \in \mathcal{D}^{1,0}_{\mathbb{R}}$ *and that* $f\colon \mathbb{R} \to \mathbb{R}$ *is a bounded Borel function. Let* $F(x) = \int_0^x f(t)\,dt$. *Then* $F(X) \in \mathcal{D}^{1,0}$, *and* $\nabla(F(X)) = f(X)\nabla X$.

PROOF. By Theorem 15.21 and Lemma 15.47, for every $\xi \in H$,

$$\begin{aligned}\rho_{t\xi}(F(X)) - F(X) = F(\rho_{t\xi}X) - F(X) &= \int_0^t f(\rho_{s\xi}X)\rho_{s\xi}(\partial_\xi X)\,ds \\ &= \int_0^t \rho_{s\xi}(f(X)\partial_\xi X)\,ds \qquad \text{a.s.}\end{aligned}$$

Hence, Lemmas 15.19 and 15.20 show that $F(X)$ is ξ-a.c. and $\partial_\xi F(X) = f(X)\partial_\xi X = \langle f(X)\nabla X, \xi\rangle_{H_C}$. The result follows by the definitions in Section 2. □

COROLLARY 15.49. *If* $X \in \mathcal{D}^{1,0}_{\mathbb{R}}$ *and* $E \subset \mathbb{R}$ *is a Borel set with Lebesgue measure* 0, *then* $\mathbf{1}[X \in E]\nabla X = 0$ *a.s.*

PROOF. Apply Theorem 15.48 with $f(x) = \mathbf{1}[x \in E]$. Then $F(x) = 0$ for each x, and thus $f(X)\nabla X = \nabla F(X) = \nabla 0 = 0$. □

As a simple consequence we obtain the following result on absolute continuity. See also the related Theorem 4.24, which has somewhat different conditions. (Neither of the theorems is contained in the other.)

THEOREM 15.50. *If* $X \in \mathcal{D}^{1,0}_{\mathbb{R}}$ *and* $\nabla X \ne 0$ *a.s., then the distribution of* X *is absolutely continuous.*

PROOF. Corollary 15.49 shows that if E is a Borel set with Lebesgue measure 0, then $\mathbf{1}[X \in E] = 0$ a.s., and thus $\mathrm{P}(X \in E) = 0$. □

Another result on the distribution of variables in $\mathcal{D}^{1,0}_{\mathbb{R}}$ is the following.

THEOREM 15.51. *If* $X_1, \dots, X_n \in \mathcal{D}^{1,0}_{\mathbb{R}}$, *then the support of the distribution of* $(X_1, \dots, X_n)$ *in* $\mathbb{R}^n$ *is connected.*

PROOF. Suppose that the support is disconnected. Then there exist two disjoint closed sets F_1 and F_2 in $\mathbb{R}^n$ such that $(X_1, \dots, X_n) \in F_1 \cup F_2$ a.s. and $0 < \mathrm{P}((X_1, \dots, X_n) \in F_1) < 1$.

There exists a continuously differentiable function ψ on $\mathbb{R}^n$ that is 1 on F_1 and 0 on F_2. By Theorem 15.34, $\psi(X_1, \dots, X_n) \in \mathcal{D}^{1,0}$. Moreover, $\psi(X_1, \dots, X_n) = \mathbf{1}_A$ a.s., where A is the event $\{(X_1, \dots, X_n) \in F_1\}$. Consequently $\mathbf{1}_A \in \mathcal{D}^{1,0}$, and Corollary 15.38 shows that $\mathrm{P}(A) = 0$ or 1, a contradiction. □

COROLLARY 15.52. *The support of the distribution of a variable in* $\mathcal{D}^{1,0}_{\mathbb{R}}$ *is a closed interval (finite or infinite, and possibly the whole real line or a single point).* □

REMARK 15.53. Theorem 15.51 extends immediately to complex-valued variables in $\mathcal{D}^{1,0}$, because $\mathbb{C}^m$ can be identified with $\mathbb{R}^{2m}$.

Finally we come to the multi-variate generalization of Theorem 15.50, showing the existence of a density for the joint distribution of several variables.

THEOREM 15.54. *Suppose that* $X_1, \dots, X_n \in \mathcal{D}^{1,0}_{\mathbb{R}}$. *If the vectors* $\nabla X_1, \dots, \nabla X_n$ *in* H *are a.s. linearly independent, then* $X = (X_1, \dots, X_n)$ *has an absolutely continuous distribution in* $\mathbb{R}^n$.

REMARK 15.55. The condition that $\nabla X_1, \dots, \nabla X_n$ in H are a.s. linearly independent is often expressed by the equivalent condition that the matrix $(\langle \nabla X_i, \nabla X_j \rangle_H)_{i,j=1}^n$ is a.s. non-singular. This matrix is often called the *Malliavin matrix*.

Cf. Theorem 15.151 below where a further integrability condition implies a stronger result.

PROOF. We use the method of Bouleau and Hirsch (1986), which is based on results from real analysis for a change of variables in integrals by maps $\mathbb{R}^m \to \mathbb{R}^n$ with rather little regularity.

We say, see e.g. Federer (1969, 3.1.2), that a function $f \colon \mathbb{R} \to \mathbb{R}$ has *approximate derivative* a at $x \in \mathbb{R}$ if for every $\varepsilon > 0$, x is a density point of $\{y : |f(y) - f(x) - a(y - x)| \le \varepsilon |y - x|\}$; we then write $\operatorname{ap} \partial f(x) = a$. For a function $f \colon \mathbb{R}^m \to \mathbb{R}$, we define the approximate partial derivative $\operatorname{ap} \partial_i f(x)$, $i = 1, \dots, m$, by keeping x_j fixed for $j \neq i$ and applying the definition just given to $x_i \mapsto f(x)$.

We use the real analysis results in the following version. We let λ_k denote the Lebesgue measure in $\mathbb{R}^k$.

LEMMA 15.56. *Suppose that* $m > n \ge 1$ *and that* $f \colon \mathbb{R}^m \to \mathbb{R}^n$ *has approximate partial derivatives* $\operatorname{ap} \partial_j f_k(x)$ *with* $j = 1, \dots, m$ *and* $k = 1, \dots, n$, *for a.e.* $x \in \mathbb{R}^m$, *and let* A *be the set of* $x \in \mathbb{R}^m$ *such that the* $m \times n$ *matrix* $(\operatorname{ap} \partial_j f_k(x))_{jk}$ *has rank* n. *If* $E \subset \mathbb{R}^n$ *with* $\lambda_n(E) = 0$, *then* $\lambda_m(A \cap f^{-1}(E)) = 0$.

PROOF. By Federer (1969, Theorem 3.1.4), f is approximate differentiable a.e., and thus (Federer 1969, Theorem 3.2.11 and the comments in 3.2.1),

$$\int_{f^{-1}(E)} \operatorname{ap} J_n f(x)\, d\lambda_m(x) = \int_{\mathbb{R}^n} g(y)\, d\lambda_n(y),$$

where g is a function that vanishes off E and $\operatorname{ap} J_n f$ is the approximate n-dimensional Jacobian, which satisfies $\operatorname{ap} J_n f \geq 0$ everywhere and $\operatorname{ap} J_n f > 0$ on A. Since $\int_{\mathbb{R}^n} g(y)\, d\lambda_n(y) = \int_E g(y)\, d\lambda_n(y) = 0$, the conclusion follows. □

We will use two further lemmas.

LEMMA 15.57. *Suppose that $\xi \in H$ and that X is ξ-a.c. If $\rho_{t\xi}(X)$ and $\rho_{t\xi}(\partial_\xi X)$ denote jointly measurable versions, then for a.e. $\omega \in \Omega$, the function $t \mapsto \rho_{t\xi}(X)(\omega)$ has for a.e. $t \in \mathbb{R}$ the approximate derivative $\rho_{t\xi}(\partial_\xi X)$.*

PROOF. For every $t \in \mathbb{R}$ and a.e. $\omega \in \Omega$, by Theorem 15.21,

$$\rho_{t\xi}(X)(\omega) = X(\omega) + \int_0^t \rho_{s\xi}(\partial_\xi X)(\omega)\, ds. \tag{15.9}$$

By Fubini's theorem for $\lambda_1 \times \mathrm{P}$ on $\mathbb{R} \times \Omega$, for a.e. $\omega \in \Omega$, (15.9) holds for a.e. real t; moreover, for a.e. ω, $\int_{-N}^{N} |\rho_{s\xi}(\partial_\xi X)(\omega)|\, ds < \infty$ for every integer $N \geq 1$, and thus the right hand side of (15.9) is a.e. differentiable with derivative $\rho_{t\xi}(\partial_\xi X)(\omega)$. It follows that the left hand side of (15.9) for a.e. ω has the approximate derivative $\rho_{t\xi}(\partial_\xi X)(\omega)$ for a.e. t. □

LEMMA 15.58. *Let $\xi, \eta \in H$. If X is ξ-a.c., then so is $\rho_\eta X$.*

PROOF. We may assume that X is real. Then there exist a sequence $\xi_1, \xi_2, \dots \in H$ and a Borel function $f\colon \mathbb{R}^\infty \to \mathbb{R}$ such that $\partial_\xi X = f(\xi_1, \xi_2, \dots)$ a.s.; write $f = f_+ - f_-$ with $f_+ = \max(f, 0)$, $f_- = \max(-f, 0)$, and thus $f_+, f_- \geq 0$ and $f_+ + f_- = |f|$. Fix $t \in \mathbb{R}$ and define g_+ and g_- on $\mathbb{R}^\infty$ by

$$g_\pm(x_1, x_2, \dots) = \int_0^t f_\pm(x_1 + s\,\mathrm{E}(\xi\xi_1), x_2 + s\,\mathrm{E}(\xi\xi_2), \dots)\, ds \leq \infty;$$

further, let $Z_\pm = g_\pm(\xi_1, \xi_2, \dots)$. Then by (14.11) and Theorem 15.21,

$$Z_+ + Z_- = \int_0^t \rho_{s\xi}(|\partial_\xi X|)\, ds = \int_0^t |\rho_{s\xi}(\partial_\xi X)|\, ds < \infty \qquad \text{a.s.}$$

and thus $Z_+, Z_- \in L^0$, and

$$Z_+ - Z_- = \int_0^t \rho_{s\xi}(\partial_\xi X)\, ds = \rho_{t\xi}(X) - X \qquad \text{a.s.}$$

Furthermore, a.s.,

$$\rho_\eta(Z_+ + Z_-) = g_+(\xi_1 + \mathrm{E}(\eta\xi_1), \dots) + g_-(\xi_1 + \mathrm{E}(\eta\xi_1), \dots)$$
$$= \int_0^t |f(\xi_1 + \mathrm{E}(\eta\xi_1) + s\,\mathrm{E}(\xi\xi_1), \xi_2 + \mathrm{E}(\eta\xi_2) + s\,\mathrm{E}(\xi\xi_2), \dots)|\,ds,$$

so this integral is finite a.s., and similarly

$$\rho_\eta(Z_+ - Z_-) = \int_0^t f(\xi_1 + \mathrm{E}(\eta\xi_1) + s\,\mathrm{E}(\xi\xi_1), \xi_2 + \mathrm{E}(\eta\xi_2) + s\,\mathrm{E}(\xi\xi_2), \dots)\,ds.$$

Since

$$\rho_{t\xi}(\rho_\eta(X)) - \rho_\eta(X) = \rho_\eta(\rho_{t\xi}(X) - X) = \rho_\eta(Z_+ - Z_-),$$

this shows that $\rho_\eta(X)$ is ξ-a.c. by Lemma 15.19. □

PROOF OF THEOREM 15.54 (concluded). Let $E \subset \mathbb{R}^n$ with $\lambda_n(E) = 0$. We have to show that $\mathrm{P}(X \in E) = 0$.

There exists a separable subspace H_0 of H such that $\nabla X_i \in H_0$ a.s. for $i = 1, \dots, n$. Choose a dense (or just total) sequence $(\xi_j)_1^\infty$ in H_0, and choose jointly measurable versions of $\rho_\xi(X)$ and $\rho_\xi(\partial_{\xi_j} X)$, $\xi \in H$ and $j \geq 1$, cf. Theorem 14.1(x).

Fix $m > n$. Define, for $i = 1, \dots, n$ and $j = 1, \dots, m$,

$$F_i(t_1, \dots, t_m; \omega) = \rho_{t_1\xi_1 + \dots + t_m\xi_m}(X_i)(\omega),$$
$$G_{ij}(t_1, \dots, t_m; \omega) = \rho_{t_1\xi_1 + \dots + t_m\xi_m}(\partial_{\xi_j} X_i)(\omega).$$

We further let F denote the vector $(F_1, \dots, F_n)$ and G the matrix (G_{ij}).

If $1 \leq i \leq n$ and $t_2, \dots, t_n$ are fixed, let $\widetilde{X} = \rho_{t_2\xi_2 + \dots + t_m\xi_m}(X_i)$; thus $\rho_{t_1\xi_1}(\widetilde{X}) = F_i(t_1, \dots, t_m; \omega)$ and $\rho_{t_1\xi_1}(\partial_{\xi_1}\widetilde{X}) = G_{i1}(t_1, \dots, t_m; \omega)$ a.s., using Theorems 14.1(ii) and 15.7(vi). Furthermore, $\widetilde{X}$ is ξ-a.c. by Lemma 15.58, and thus Lemma 15.57 shows that, for a.e. $(\omega, t_1) \in \Omega \times \mathbb{R}$, $\operatorname{ap}\partial_1 F_i(t_1, \dots, t_m; \omega)$ exists and equals $G_{i1}(t_1, \dots, t_m; \omega)$. Hence, using Fubini's theorem on $\mathbb{R}^m \times \Omega$, for a.e. $(t_1, \dots, t_m, \omega)$, $\operatorname{ap}\partial_1 F_i = G_{i1}$. Similarly, for each j, $\operatorname{ap}\partial_j F_i = G_{ij}$ for a.e. $(t_1, \dots, t_m, \omega)$. It follows that for a.e. ω, F satisfies the assumption of Lemma 15.56 that $\operatorname{ap}\partial_j F_i$ exists a.e. and moreover, with A as in the lemma,

$$\mathbf{1}_A = \mathbf{1}[\operatorname{rank}(\operatorname{ap}\partial_j F_i)_{1,1}^{n,m} = n] = \mathbf{1}[\operatorname{rank}(G) = n] \qquad \text{a.e.}$$

The conclusion of Lemma 15.56 may be written

$$\mathbf{1}[t \in A \text{ and } F(t) \in E] = 0 \qquad \text{for a.e. } t \in \mathbb{R}^n;$$

consequently we obtain from this lemma, for a.e. $\omega \in \Omega$ and $t \in \mathbb{R}^m$,

$$\mathbf{1}[F(t;\omega) \in E \text{ and } \operatorname{rank}(G(t;\omega)) = n] = 0.$$

By the definitions of F and G, and Theorem 14.1(iii), this means that for a.e. $(t_1, \dots, t_m) \in \mathbb{R}^m$,

$$\rho_{t_1\xi_1+\dots+t_m\xi_m} \mathbf{1}[X \in E \text{ and } \operatorname{rank}((\partial_{\xi_j} X_i)_{1,1}^{n,m}) = n] = 0 \qquad \text{a.s.}$$

Choosing one such $(t_1, \dots, t_m)$ we obtain, recalling $\partial_{\xi_j} X_i = \langle \nabla X_i, \xi_j \rangle_H$ a.s.,

$$\mathrm{P}\big(X \in E \text{ and } \operatorname{rank}((\langle \nabla X_i, \xi_j \rangle_H)_{1,1}^{n,m}) = n\big) = 0. \tag{15.10}$$

Now let $m \to \infty$. It is easily seen that if $\nabla X_1(\omega), \dots, \nabla X_n(\omega)$ are linearly independent elements of H_0, then the matrix $(\langle \nabla X_i, \xi_j \rangle_H)_{1,1}^{n,m}$ has rank n for all large m. Thus (15.10) yields

$$\mathrm{P}(X \in E) = \mathrm{P}(X \in E \text{ and } \nabla X_1, \dots, \nabla X_n \text{ are linearly independent}) = 0.$$

□

5. Sobolev spaces

We define some fundamental spaces of random variables with certain degrees of differentiability as follows.

DEFINITION 15.59. If $1 \le p \le \infty$ and $\xi \in H$, let

$$\mathcal{D}^{\xi,p} = \{X \in L^p : X \text{ is } \xi\text{-a.c. and } \partial_\xi X \in L^p\}.$$

Similarly, if $1 \le p \le \infty$, let

$$\begin{aligned} \mathcal{D}^{1,p} &= \{X \in L^p : \nabla X \in L^p(H_{\mathbb{C}}) \text{ and } X \text{ is r.a.c.}\} \\ &= \{X \in \mathcal{D}^{1,0} : X \in L^p \text{ and } \nabla X \in L^p(H_{\mathbb{C}})\}. \end{aligned}$$

More generally, if $1 \le p \le \infty$ and $k \ge 0$, let

$$\mathcal{D}^{k,p} = \{X \in L^p : \nabla X \in L^p(H_{\mathbb{C}}), \dots, \nabla^k X \in L^p(H_{\mathbb{C}}^{\otimes k}) \\ \text{and } X, \dots, \nabla^{k-1} X \text{ are r.a.c.}\}.$$

The spaces $\mathcal{D}^{k,p}$ are called (*Gaussian*) *Sobolev spaces.* Note that $\mathcal{D}^{0,p} = L^p$. We are primarily interested in the spaces $\mathcal{D}^{k,p}$, and in particular $\mathcal{D}^{1,p}$; the spaces $\mathcal{D}^{\xi,p}$ are mainly technical tools.

REMARK 15.60. Here, as well as often below, we use $H_{\mathbb{C}}$ in order to cover both the real case and the more general complex case. In the real case $H_{\mathbb{C}}$ may be replaced by H, but we will not comment further upon that.

We equip $\mathcal{D}^{\xi,p}$ with the norm

$$\|X\|_{\xi,p} = (\|X\|_p^p + \|\partial_\xi X\|_p^p)^{1/p}$$

and $\mathcal{D}^{k,p}$ with the norm

$$\|X\|_{k,p} = \Big(\sum_{j=0}^{k} \|\nabla^j X\|_{L^p(H_{\mathbb{C}}^{\otimes j})}^p\Big)^{1/p} = \Big(\sum_{j=0}^{k} \mathrm{E}\,\|\nabla^j X\|_{H_{\mathbb{C}}^{\otimes j}}^p\Big)^{1/p},$$

with the usual interpretation for $p = \infty$. In particular, $\mathcal{D}^{1,p}$ has the norm

$$\|X\|_{1,p} = \bigl(\|X\|_{L^p}^p + \|\nabla X\|_{L^p(H_{\mathbb{C}})}^p\bigr)^{1/p} = \bigl(\mathrm{E}\,|X|^p + \mathrm{E}\,\|\nabla X\|_{H_{\mathbb{C}}}^p\bigr)^{1/p}.$$

We will see below that $\mathcal{D}^{\xi,p}$ and $\mathcal{D}^{k,p}$ are Banach spaces.

We further define $\mathcal{D}^{\infty,p} = \bigcap_{k=1}^{\infty} \mathcal{D}^{k,p}$, and

$$\mathcal{D}^{\infty-} = \bigcap_{p<\infty} \mathcal{D}^{\infty,p} = \bigcap_{k\geq 1} \bigcap_{p<\infty} \mathcal{D}^{k,p}.$$

These spaces are equipped with the topologies given by the families of norms $\|\cdot\|_{k,p}$, for the corresponding ranges of k and p.

REMARK 15.61. The spaces $\mathcal{D}^{k,\infty}$ are often too small to be useful, for example they do not even contain the Gaussian variables in H. The space $\mathcal{D}^{\infty-}$ is a useful substitute in many applications.

The definitions above cover both the real and the complex cases; as usual we use subscripts $\mathbb{R}$ or $\mathbb{C}$ to distinguish between these cases when necessary. More generally, we may also consider random variables with values in an arbitrary (real or complex) Hilbert space H_0; we define $\mathcal{D}^{\xi,p}(H_0)$ and $\mathcal{D}^{k,p}(H_0)$ as above, now requiring $X \in L^p(H_0)$, $\partial_\xi X \in L^p(H_0)$ and $\nabla^j X \in L^p(H^{\otimes j} \otimes H_0)$, and equip them with the norms

$$\|X\|_{\xi,p,H_0} = \bigl(\|X\|_{L^p(H_0)}^p + \|\partial_\xi X\|_{L^p(H_0)}^p\bigr)^{1/p}$$

and

$$\|X\|_{k,p,H_0} = \Bigl(\sum_{j=0}^{k} \|\nabla^j X\|_{L^p(H^{\otimes j}\otimes H_0)}^p\Bigr)^{1/p}.$$

Many results for $\mathcal{D}^{\xi,p}$ and $\mathcal{D}^{k,p}$ extend without difficulty to the Hilbert-space-valued case. Several examples of this are given in the results below; in other cases we, for simplicity, state only the scalar-valued case and leave the Hilbert space extension to the reader.

REMARK 15.62. There are several important reasons for the absolute continuity conditions in Definition 15.59; one of them is that without these conditions, the resulting (larger) spaces are *not* Banach spaces. For example, let $\xi \in H$ be standard normal and let $X_n = \lfloor n\xi \rfloor / n$, $n \geq 1$. Then, as $n \to \infty$, $X_n \to \xi \in L^p$ and, using Theorem 15.37, since each X_n is a discrete random variable, $\nabla X_n = 0 \to 0 \in L^p(H)$. Since $\nabla \xi = 1 \otimes \xi \neq 0$, it follows easily that $(X_n)_1^\infty$ is a Cauchy sequence (in $\|\cdot\|_{1,p}$) that does not converge.

Another reason for these conditions is that they are also required for the density results in Section 7.

REMARK 15.63. We define $\mathcal{D}^{k,p}$ by first defining ∇^j on a larger space and then imposing suitable conditions on the variables.

For $p < \infty$, an alternative, which is common in the literature, is to first define the norm $\|\cdot\|_{k,p}$ by the formula above for a suitable class of 'nice' random variables for which ∇^j is explicitly given by (15.7), for example polynomial variables, and then take the completion in this norm. This, indeed, defines the same space, which follows from the facts proved below that $\mathcal{D}^{k,p}$ is complete and that the polynomial variables are dense in it; cf. Sugita (1985).

We begin by observing some immediate consequences of the definitions just made.

THEOREM 15.64.

(i) *If $0 \le k \le l$ and $1 \le p \le q \le \infty$, then $\mathcal{D}^{l,q}$ is a subspace of $\mathcal{D}^{k,p}$.*
(ii) *If $\xi \in H$ and $1 \le p \le q \le \infty$, then $\mathcal{D}^{\xi,q}$ is a subspace of $\mathcal{D}^{\xi,p}$.*
(iii) *If $\xi \in H$ and $1 \le p \le \infty$, then $\mathcal{D}^{1,p}$ is a subspace of $\mathcal{D}^{\xi,p}$.*
(iv) *If $1 \le p < q \le \infty$ and $\xi \in H$, then $\mathcal{D}^{\xi,q} = \{X \in \mathcal{D}^{\xi,p} : X, \partial_\xi X \in L^q\}$.*
(v) *If $1 \le p < q \le \infty$ and $k \ge 0$, then $\mathcal{D}^{k,q} = \{X \in \mathcal{D}^{k,p} : X \in L^q, \dots, \nabla^k X \in L^q(H_{\mathbb{C}}^{\otimes k})\}$.*
(vi) *If $0 \le j \le k$ and $1 \le p \le \infty$, then ∇^j maps $\mathcal{D}^{k,p}$ continuously into $\mathcal{D}^{k-j,p}(H_{\mathbb{C}}^{\otimes j})$.*
(vii) *If $k, l \ge 0$, $1 \le p \le \infty$ and H_0 is a Hilbert space, then $X \in \mathcal{D}^{k+l,p}(H_0)$ if and only if $X \in \mathcal{D}^{k,p}(H_0)$ and $\nabla^k X \in \mathcal{D}^{l,p}(H^{\otimes k} \otimes H_0)$.*
(viii) *If $1 \le p \le \infty$, then $X \in \mathcal{D}^{1,p}$ if and only if not only $X \in \mathcal{D}^{\xi,p}$ for every $\xi \in H$ but also there exists $Y \in L^p(H_{\mathbb{C}})$ such that $\partial_\xi X = \langle Y, \xi \rangle_{H_{\mathbb{C}}}$ for all $\xi \in H$; in this case further $\nabla X = Y$.*
(ix) *If $1 \le p \le \infty$ and H_0 is a Hilbert space, then $X \in \mathcal{D}^{1,p}(H_0)$ if and only if there exists $Y \in L^p(H \otimes H_0)$ such that, for every $h \in H_0$, $\langle X, h \rangle_{H_0} \in \mathcal{D}^{1,p}$ and $\nabla \langle X, h \rangle_{H_0} = \langle Y, h \rangle_{H_0}$; in this case further $\nabla X = Y$.*
(x) *Let H_0 be a Hilbert space, $1 \le p \le \infty$ and $k \ge 0$. If $X \in \mathcal{D}^{k,p}$ and $h \in H_0$, then $X \otimes h \in \mathcal{D}^{k,p}(H_0)$.*
(xi) *Let H_1 and H_2 be Hilbert spaces, $T\colon H_1 \to H_2$ a linear operator, $1 \le p \le \infty$ and $k \ge 0$. If $X \in \mathcal{D}^{k,p}(H_1)$, then $T(X) \in \mathcal{D}^{k,p}(H_2)$; moreover, $\nabla^j(T(X)) = T(\nabla^j X)$, $j \le k$, and $\|T(X)\|_{k,p,H_2} \le \|T\| \|X\|_{k,p,H_1}$.*

□

REMARK 15.65. It follows from the results in Section 7 below that if $p < \infty$, then the subspaces in (i)–(iii) are dense.

EXAMPLE 15.66. If $\xi_1, \dots, \xi_n \in H$ and $f\colon \mathbb{R}^n \to \mathbb{C}$ is infinitely differentiable then $\nabla^k f(\xi_1, \dots, \xi_n)$ exists and is r.a.c. for every $k \ge 0$ by Example 15.46. If further every derivative $\partial^\alpha f$ is polynomially bounded on $\mathbb{R}^n$, then (15.7) implies that $\nabla^k f(\xi_1, \dots, \xi_n) \in L^p(H_{\mathbb{C}}^{\otimes k})$ for every $k \ge 0$ and $p < \infty$, and hence $f(\xi_1, \dots, \xi_n) \in \mathcal{D}^{k,p}$ for every $k \ge 0$, $p < \infty$; in other words, $f(\xi_1, \dots, \xi_n) \in \mathcal{D}^{\infty-}$.

In particular, this applies to polynomial variables f. Thus, $\mathcal{P} \subset \mathcal{D}^{\infty-}$.

EXAMPLE 15.67. If $\xi_1, \dots, \xi_n \in H$ and $f\colon \mathbb{R}^n \to \mathbb{C}$ is infinitely differentiable with compact support, then similarly $f(\xi_1, \dots, \xi_n) \in \mathcal{D}^{k,\infty}$, $k \ge 0$. We let $\mathcal{S}_0$ denote the set of such variables.

EXAMPLE 15.68. It also follows from Example 15.46 that if $\zeta \in H_{\mathbb{C}}$, then e^{ζ} and $:e^{\zeta}:$ belong to $\mathcal{D}^{k,p}$ for every $k \geq 0$ and $p < \infty$, and thus to $\mathcal{D}^{\infty-}$; cf. Example 15.32.

Moreover, if $\xi \in H$, then $e^{i\xi}$ and $:e^{i\xi}:$ belong to $\mathcal{D}^{k,\infty}$, $k \geq 0$.

EXAMPLE 15.69. If H_0 is a real or complex Hilbert space, let $\mathcal{P} \otimes H_0$ be the space of finite linear combinations $\sum_1^N X_i \otimes h_i$, with polynomial variables $X_i \in \mathcal{P}$ and $h_i \in H_0$. ($\mathcal{P} \otimes H_0$ is the algebraic tensor product of $\mathcal{P}$ and H_0, cf. Appendix E.)

It follows by Example 15.66 and Theorem 15.64(x) that $\mathcal{P} \otimes H_0 \subset \mathcal{D}^{k,p}(H_0)$ for every $k \geq 0$ and $p < \infty$, i.e. $\mathcal{P} \otimes H_0 \subset \mathcal{D}^{\infty-}(H_0)$.

Note also that $\mathcal{P} \otimes H_0$ is dense in $L^p(H_0)$ for every $p < \infty$, as a simple consequence of Theorem 2.12 and the fact that measurable simple functions are dense in $L^p(H_0)$.

We turn to proving the completeness of the Sobolev spaces, beginning with two lemmas which circumvent the technical problem that the Cameron–Martin shift is not bounded on L^1 by using a suitable weight. (We use the weight $e^{-\xi^2}$, but many alternatives are possible.)

LEMMA 15.70. *Let $\xi \in H$ and $\sigma^2 = \mathrm{E}\,\xi^2 \geq 0$. If $X \in L^1$, then $e^{-\xi^2} \rho_{t\xi}(X) \in L^1$ for each real t, with*

$$\|e^{-\xi^2} \rho_{t\xi}(X)\|_1 \leq e^{(1+2\sigma^2)t^2/4} \|X\|_1. \tag{15.11}$$

Moreover, the mapping $t \mapsto e^{-\xi^2} \rho_{t\xi}(X)$ is a continuous function $\mathbb{R} \to L^1$.

PROOF. By Theorem 14.1,

$$\begin{aligned}\|e^{-\xi^2} \rho_{t\xi}(X)\|_1 &= \mathrm{E}\big(e^{-\xi^2} \rho_{t\xi}(|X|)\big) = \mathrm{E}\big(:e^{t\xi}: \rho_{-t\xi}(e^{-\xi^2} \rho_{t\xi}(|X|))\big)\\ &= \mathrm{E}\big(:e^{t\xi}: \rho_{-t\xi}(e^{-\xi^2})|X|\big) = \mathrm{E}\big(e^{t\xi - t^2\sigma^2/2} e^{-(\xi - t\sigma^2)^2}|X|\big),\end{aligned}$$

and (15.11) follows because

$$\begin{aligned}t\xi - t^2\sigma^2/2 - (\xi - t\sigma^2)^2 &= (1+2\sigma^2)t\xi - \xi^2 - t^2\sigma^2/2 - t^2\sigma^4\\ &= -\big((\tfrac{1}{2} + \sigma^2)t - \xi\big)^2 + \big(\tfrac{1}{4} + \tfrac{\sigma^2}{2}\big)t^2 \leq (1+2\sigma^2)t^2/4.\end{aligned}$$

If $X = e^{\eta}$ for some $\eta \in H$, then $t \mapsto e^{-\xi^2} \rho_{t\xi}(e^{\eta}) = e^{-\xi^2} e^{\eta + t\,\mathrm{E}\,\xi\eta}$ is a continuous function $\mathbb{R} \to L^1$. In general, if $X \in L^1$, then by Theorem 2.12, there exists a sequence $X_n \to X$ in L^1 such that each X_n is a linear combination of exponentials e^{η}, $\eta \in H$, and thus $t \mapsto e^{-\xi^2} \rho_{t\xi}(X)$ is a continuous function into L^1. By (15.11), $e^{-\xi^2} \rho_{t\xi}(X_n) \to e^{-\xi^2} \rho_{t\xi}(X)$ in L^1, uniformly for t in any bounded interval, and the continuity of $t \mapsto e^{-\xi^2} \rho_{t\xi}(X)$ follows. ∎

LEMMA 15.71. *Let $X, Y \in L^1$ and $\xi \in H$. Then $X \in \mathcal{D}^{\xi,1}$ and $\partial_\xi X = Y$ if and only if, for every real t,*

$$e^{-\xi^2}(\rho_{t\xi}(X) - X) = \int_0^t e^{-\xi^2} \rho_{s\xi}(Y)\, ds, \tag{15.12}$$

where the integral converges in L^1.

PROOF. The integral in (15.12) converges as a Bochner integral in L^1 by Lemma 15.70. Thus, if we assume as we may that we have chosen a jointly measurable version of $\rho_{s\xi}(Y)$, the integral may be evaluated pointwise, for a.e. $\omega \in \Omega$, see Appendix C.

Suppose first that $X \in \mathcal{D}^{\xi,1}$ and $\partial_\xi X = Y$. Then X is ξ-a.c., and thus by Theorem 15.21, for a.e. ω,

$$\begin{aligned}\int_0^t e^{-\xi(\omega)^2} \rho_{s\xi}(Y)(\omega)\, ds &= e^{-\xi(\omega)^2} \int_0^t \rho_{s\xi}(\partial_\xi X)(\omega)\, ds \\ &= e^{-\xi(\omega)^2}(\rho_{t\xi}(X)(\omega) - X(\omega)).\end{aligned}$$

Consequently, (15.12) holds.

Conversely, if (15.12) holds, we may evaluate the integral pointwise and cancel $e^{-\xi(\omega)^2}$, which yields

$$\int_0^t \rho_{s\xi}(Y)(\omega)\, ds = \rho_{t\xi}(X)(\omega) - X(\omega),$$

for a.e. $\omega \in \Omega$. By Lemmas 15.19 and 15.20, X is then ξ-a.c., $\partial_\xi X$ exists and $\rho_{s\xi}(\partial_\xi X) = \rho_{s\xi}(Y)$ for some s. Theorem 14.1 yields $\partial_\xi X = Y$, and $X \in \mathcal{D}^{\xi,1}$ follows. □

THEOREM 15.72. *Let $1 \leq p \leq \infty$, $\xi \in H$ and $k \geq 1$.*

(i) *If $(X_n)_1^\infty$ is a sequence in $\mathcal{D}^{\xi,p}$ such that $X_n \to X$ and $\partial_\xi X_n \to Y$ in L^p as $n \to \infty$, for some $X, Y \in L^p$, then $X \in \mathcal{D}^{\xi,p}$, $\partial_\xi X = Y$ and $X_n \to X$ in $\mathcal{D}^{\xi,p}$ as $n \to \infty$.*

(ii) *If $(X_n)_1^\infty$ is a sequence in $\mathcal{D}^{k,p}$ such that $X_n \to X$ in L^p and, for $j = 1, \dots, k$, $\nabla^j X_n \to Y_j$ in $L^p(H_{\mathbb{C}}^{\otimes j})$ as $n \to \infty$, for some $X \in L^p$ and $Y_j \in L^p(H_{\mathbb{C}}^{\otimes j})$, then $X \in \mathcal{D}^{k,p}$, $\nabla^j X = Y_j$ for $j = 1, \dots, k$, and $X_n \to X$ in $\mathcal{D}^{k,p}$ as $n \to \infty$.*

The corresponding results for variables with values in a real or complex Hilbert space H_0 are also valid.

PROOF. Consider first $\mathcal{D}^{\xi,p}$. By Lemma 15.71, since $X_n \in \mathcal{D}^{\xi,p} \subseteq \mathcal{D}^{\xi,1}$,

$$e^{-\xi^2}(\rho_{t\xi}(X_n) - X_n) = \int_0^t e^{-\xi^2} \rho_{s\xi}(\partial_\xi X_n)\, ds. \tag{15.13}$$

Now let $n \to \infty$. It follows by Lemma 15.70 that $e^{-\xi^2}\rho_{s\xi}(\partial_\xi X_n) \to e^{-\xi^2}\rho_{s\xi}(Y)$ in L^1 uniformly for $0 \leq s \leq t$ (or $t \leq s \leq 0$), and $e^{-\xi^2}(\rho_{t\xi}(X_n) - X_n) \to$

$e^{-\xi^2}(\rho_{t\xi}(X) - X)$ in L^1. Thus (15.13) yields

$$e^{-\xi^2}(\rho_{t\xi}(X) - X) = \int_0^t e^{-\xi^2}\rho_{s\xi}(Y)\,ds.$$

The converse part of Lemma 15.71 thus yields $\partial_\xi X = Y$ and $X \in \mathcal{D}^{\xi,1}$; furthermore, since $X \in L^p$ and $\partial_\xi X = Y \in L^p$, $X \in \mathcal{D}^{\xi,p}$. Evidently $X_n \to X$ in $\mathcal{D}^{\xi,p}$, which completes the proof of (i).

Next, consider $\mathcal{D}^{1,p}$; thus assume $X_n \to X$ in L^p and $\nabla X_n \to Y$ in $L^p(H_\mathbb{C})$, for some $X \in L^p$ and $Y \in L^p(H_\mathbb{C})$.

It follows that, for every $\xi \in H$,

$$\partial_\xi X_n = \langle \nabla X_n, \xi\rangle_{H_\mathbb{C}} \to \langle Y, \xi\rangle_{H_\mathbb{C}} \in L^p,$$

and thus, by the just proved part (i), X is ξ-a.c. and $\partial_\xi X = \langle Y, \xi\rangle_{H_\mathbb{C}}$.

By Definitions 15.30 and 15.26, X is r.a.c. and $\nabla X = Y$, and thus $X \in \mathcal{D}^{1,p}$. It follows immediately that $X_n \to X$ in $\mathcal{D}^{1,p}$, which completes the proof for $\mathcal{D}^{1,p}$.

Next consider the Hilbert space version for $\mathcal{D}^{1,p}$ (the Hilbert space version of (i) follows similarly). Thus assume that $X_n \to X$ in $L^p(H_0)$ and $\nabla X_n \to Y$ in $L^p(H \otimes H_0)$, for some $X \in L^p(H_0)$ and $Y \in L^p(H \otimes H_0)$.

If $h \in H_0$, then $\langle X_n, h\rangle_{H_0} \to \langle X, h\rangle_{H_0}$ in L^p and $\nabla\langle X_n, h\rangle_{H_0} = \langle \nabla X_n, h\rangle_{H_0} \to \langle Y, h\rangle_{H_0}$ in $L^p(H_\mathbb{C})$, and thus the just proved scalar case yields that $\langle X, h\rangle_{H_0} \in \mathcal{D}^{1,p}$ with $\nabla\langle X, h\rangle_{H_0} = \langle Y, h\rangle_{H_0}$. Theorem 15.64(ix) shows that $X \in \mathcal{D}^{1,p}(H_0)$ and $\nabla X = Y$, and $X_n \to X$ in $\mathcal{D}^{1,p}(H_0)$ follows.

Finally, consider $\mathcal{D}^{k,p}$ or, more generally, $\mathcal{D}^{k,p}(H_0)$. Thus, writing $Y_0 = X$, let $X_n \in \mathcal{D}^{k,p}(H_0)$ be such that for $j = 0, \dots, k$, $\nabla^j X_n \to Y_j$ in $L^p(H^{\otimes j} \otimes H_0)$ as $n \to \infty$. If $0 \le j \le k-1$, thus also $\nabla(\nabla^j X_n) \to Y_{j+1}$ in $L^p(H \otimes H^{\otimes j} \otimes H_0)$, and the case just considered shows that that $Y_j \in \mathcal{D}^{1,p}(H^{\otimes j} \otimes H_0)$ with $\nabla Y_j = Y_{j+1}$.

By Theorem 15.64(vii) and induction on j, $X = Y_0 \in \mathcal{D}^{j,p}(H_0)$ with $\nabla^j X = Y_j$, $0 \le j \le k$. Consequently, $X \in \mathcal{D}^{k,p}(H_0)$, and $X_n \to X$ in $\mathcal{D}^{k,p}(H_0)$. □

COROLLARY 15.73. *Let $1 \le p \le \infty$, $\xi \in H$ and $k \ge 0$. Then the spaces $\mathcal{D}^{\xi,p}$ and $\mathcal{D}^{k,p}$, and more generally $\mathcal{D}^{\xi,p}(H_0)$ and $\mathcal{D}^{k,p}(H_0)$ for a real or complex Hilbert space H_0, are Banach spaces.*

PROOF. Consider for example $\mathcal{D}^{\xi,p}$; the other cases are similar. If $(X_n)_1^\infty$ is a Cauchy sequence in $\mathcal{D}^{\xi,p}$, then $(X_n)_1^\infty$ and $(\partial_\xi X_n)_1^\infty$ are Cauchy sequences in L^p. Since L^p is complete, this implies that $X_n \to X$ and $\partial_\xi X_n \to Y$ in L^p as $n \to \infty$, for some $X, Y \in L^p$, and Theorem 15.72 shows that $X_n \to X$ in $\mathcal{D}^{\xi,p}$. □

COROLLARY 15.74. *The spaces $\mathcal{D}^{\infty,p}$, $1 \le p \le \infty$, and $\mathcal{D}^{\infty-}$ are Fréchet spaces.* □

COROLLARY 15.75. *Let $1 \le p \le \infty$. If $\xi \in H$, then the set $\tilde{\mathcal{D}}^{\xi,p} = \{(X, \partial_\xi X) : X \in \mathcal{D}^{\xi,p}\}$ is a closed subspace of $L^p \times L^p$; similarly, if $k \ge 0$, then $\tilde{\mathcal{D}}^{k,p} = \{(X, \nabla X, \dots, \nabla^k X) : X \in \mathcal{D}^{k,p}\}$ is a closed subspace of $\bigoplus_0^k L^p(H^{\otimes j})$ or $\bigoplus_0^k L^p(H_{\mathbb{C}}^{\otimes j})$ (in the real and complex cases, respectively).* □

REMARK 15.76. In other words, the restriction of ∂_ξ to $\mathcal{D}^{\xi,p}$ is a closed operator $L^p \to L^p$, and (by the case $k = 1$) the restriction of ∇ to $\mathcal{D}^{1,p}$ is a closed operator $L^p_{\mathbb{R}} \to L^p(H^{\otimes j})$ or $L^p_{\mathbb{C}} \to L^p(H_{\mathbb{C}}^{\otimes j})$.

If $1 \le p < \infty$, these operators are densely defined, since $\mathcal{P} \subset \mathcal{D}^{1,p} \subset \mathcal{D}^{\xi,p}$ by Example 15.66.

REMARK 15.77. The spaces $\tilde{\mathcal{D}}^{\xi,p}$ and $\tilde{\mathcal{D}}^{k,p}$ in Corollary 15.75 are obviously isometric to $\mathcal{D}^{\xi,p}$ and $\mathcal{D}^{k,p}$ if we use the ℓ^p-sum-norm on $L^p \times L^p$ and $\bigoplus_0^k L^p(H^{\otimes j})$ or $\bigoplus_0^k L^p(H_{\mathbb{C}}^{\otimes j})$; for example for $\mathcal{D}^{1,p}$, the isometry $\mathcal{D}^{1,p} \to \tilde{\mathcal{D}}^{1,p}$ is given by $X \mapsto (X, \nabla X)$, and its inverse by the projection $(X, Y) \mapsto X$.

The chain rule holds for variable in Sobolev spaces, for example in the following version which follows directly from Definition 15.59 and Theorem 15.34. (There is also a similar result for complex variables and $f : \mathbb{C}^n \to \mathbb{C}$.)

THEOREM 15.78. *Let $1 \le p \le \infty$. Suppose that $X_1, \dots, X_m \in \mathcal{D}_{\mathbb{R}}^{1,0}$ and that $f : \mathbb{R}^m \to \mathbb{R}$ is a continuously differentiable function. If $f(X_1, \dots, X_m) \in L^p$ and*

$$Y = \sum_{i=1}^{m} \partial_i f(X_1, \dots, X_m) \nabla X_i \in L^p(H),$$

then $f(X_1, \dots, X_m) \in \mathcal{D}^{1,p}$ and $\nabla f(X_1, \dots, X_m) = Y$. □

EXAMPLE 15.79. Let $1 \le p \le \infty$. If $X_1, \dots, X_m \in \mathcal{D}_{\mathbb{R}}^{1,p}$ and f is a function on $\mathbb{R}^m$ with bounded continuous partial derivatives, then $f(X_1, \dots, X_m) \in \mathcal{D}^{1,p}$.

COROLLARY 15.80. *Let $1 \le p, q, r \le \infty$ with $1/p = 1/q + 1/r$. If $X \in \mathcal{D}^{1,q}$ and $Y \in \mathcal{D}^{1,r}$, then $XY \in \mathcal{D}^{1,p}$ and $\nabla(XY) = X\nabla Y + Y\nabla X$.*

PROOF. If X and Y are real, this follows from Theorem 15.78 with $f(x, y) = xy$ and Hölder's inequality; the complex case follows by linearity. □

This result can be extended to higher derivatives. We use two preliminary lemmas. The first is an improvement of Theorem 15.64(ix).

LEMMA 15.81. *Let H_0 be a Hilbert space and $1 \le p \le \infty$. If $X \in L^p(H_0)$ and $Y \in L^p(H \otimes H_0)$ are such that, for every h in some total subset of H_0, $\langle X, h \rangle_{H_0} \in \mathcal{D}^{1,p}$ and $\nabla \langle X, h \rangle_{H_0} = \langle Y, h \rangle_{H_0}$, then $X \in \mathcal{D}^{1,p}(H_0)$ and $\nabla X = Y$.*

PROOF. Let $V = \{h \in H_0 : \langle X,h\rangle_{H_0} \in \mathcal{D}^{1,p} \text{ and } \nabla\langle X,h\rangle_{H_0} = \langle Y,h\rangle_{H_0}\}$; by assumption, V is total. It is obvious that V is a subspace. Furthermore, let $h \in \overline{V}$; then there exist $h_n \in V$ such that $h_n \to h$ in H_0. Let $X_n = \langle X,h_n\rangle_{H_0} \in \mathcal{D}^{1,p}$; then $X_n \to \langle X,h\rangle_{H_0}$ in L^p and $\nabla X_n = \langle Y,h_n\rangle_{H_0} \to \langle Y,h\rangle_{H_0}$ in $L^p(H)$ as $n \to \infty$, and thus Theorem 15.72 shows that $\langle X,h\rangle_{H_0} \in \mathcal{D}^{1,p}$ and $\nabla\langle X,h\rangle_{H_0} = \langle Y,h\rangle_{H_0}$, i.e. $h \in V$.

Consequently, V is a closed total subspace, and thus $V = H_0$. The result follows by Theorem 15.64(ix). □

LEMMA 15.82. *Let H_1 and H_2 be two Hilbert spaces, and let $1 \le p,q,r \le \infty$ with $1/p = 1/q + 1/r$. If $X \in \mathcal{D}^{1,q}(H_1)$, and $Y \in \mathcal{D}^{1,r}(H_2)$, then $X \otimes Y \in \mathcal{D}^{1,p}(H_1 \otimes H_2)$. Moreover, if $\tau\colon H_1 \otimes H \otimes H_2 \to H \otimes H_1 \otimes H_2$ is the natural isomorphism given by $\tau(h_1 \otimes h \otimes h_2) = h \otimes h_1 \otimes h_2$, then $\nabla(X \otimes Y) = \nabla X \otimes Y + \tau(X \otimes \nabla Y)$.*

PROOF. A simple consequence of Corollary 15.80 and Lemma 15.81, using the total subset $\{h_1 \otimes h_2\}$ of $H_1 \otimes H_2$. □

THEOREM 15.83. *Let $k \ge 0$ and $1 \le p,q,r \le \infty$ with $1/p = 1/q + 1/r$. If $X \in \mathcal{D}^{k,q}$ and $Y \in \mathcal{D}^{k,r}$, then $XY \in \mathcal{D}^{k,p}$; moreover, $\|XY\|_{k,p} \le C_k\|X\|_{k,q}\|Y\|_{k,r}$ for some constant $C_k < \infty$. More generally, if $X \in \mathcal{D}^{k,q}(H_1)$ and $Y \in \mathcal{D}^{k,r}(H_2)$, where H_1 and H_2 are two Hilbert spaces, then $X \otimes Y \in \mathcal{D}^{k,p}(H_1 \otimes H_2)$ with $\|X \otimes Y\|_{k,p,H_1\otimes H_2} \le C_k\|X\|_{k,q,H_1}\|Y\|_{k,r,H_2}$.*

PROOF. The Hilbert space version follows by Lemma 15.82 and induction. □

COROLLARY 15.84. *$\mathcal{D}^{\infty-}$ is an algebra, with continuous multiplication.* □

COROLLARY 15.85. *If $X \in \mathcal{D}^{\infty-}$ and $Y \in \mathcal{D}^{\infty-}(H_0)$, where H_0 is a Hilbert space, then $XY \in \mathcal{D}^{\infty-}(H_0)$.* □

We continue with a similar but more technical result that will be used in Section 9.

LEMMA 15.86. *Let $1 \le p < \infty$, and let H_1 be a real Hilbert space. If $X \in \mathcal{D}^{1,p}(H_1)$ and φ is a continuously differentiable function on $\mathbb{R}$ such that $\varphi(x)$ and $|x|^{1/2}\varphi'(x)$ are bounded, then $\varphi(\|X\|^2_{H_1}) \in \mathcal{D}^{1,p}$ and*

$$\nabla\varphi(\|X\|^2_{H_1}) = 2\varphi'(\|X\|^2_{H_1})\langle\nabla X, X\rangle_{H_1}.$$

PROOF. Let $\{h_i\}_1^\infty$ be an orthonormal basis in H_1, or at least in a separable subspace that a.s. contains X. (The case when H_1 has finite dimension is similar but simpler.) Let $X_i = \langle X,h_i\rangle_{H_1}$, $Y_n = \sum_1^n X_i \otimes h_i$ and $Z_n = \varphi(\sum_1^n X_i^2)$. Then $X_i \in \mathcal{D}^{1,p}_{\mathbb{R}}$, and Theorem 15.78 yields, for every $n \ge 1$, that $Z_n \in \mathcal{D}^{1,p}$ with

$$\nabla Z_n = \sum_{i=1}^n 2\varphi'(\sum_{j=1}^n X_j^2)X_i\nabla X_i = 2\varphi'(\sum_{j=1}^n X_j^2)\langle Y_n, \nabla X\rangle_{H_1};$$

note that

$$\|\nabla Z_n\|_H \le 2\varphi'(\|Y_n\|_{H_1}^2)\|Y_n\|_{H_1}\|\nabla X\|_{H\otimes H_1} \le C\|\nabla X\|_{H\otimes H_1} \in L^p. \quad (15.14)$$

Let $n \to \infty$. Then $Z_n \to \varphi(\|X\|_{H_1}^2)$ and $\nabla Z_n \to 2\varphi'(\|X\|_{H_1}^2)\langle \nabla X, X\rangle_{H_1}$ a.s.; by dominated convergence, using (15.14), these limits hold in L^p and $L^p(H)$ too, and the result follows by Theorem 15.72. □

REMARK 15.87. Taking $\varphi(x) = \arctan x$ in this lemma and then applying Theorem 15.34 to $\|X\|_{H_1}^2 = \tan(\arctan(\|X\|_{H_1}^2))$, we see that if $X \in \mathcal{D}^{1,p}(H_1)$, with $1 \le p \le \infty$, then $\|X\|_{H_1}^2 \in \mathcal{D}^{1,0}$ with $\nabla\|X\|_{H_1}^2 = 2\langle \nabla X, X\rangle_{H_1}$. In particular, if $X \in \mathcal{D}^{1,2p}(H_1)$, then $\|X\|_{H_1}^2 \in \mathcal{D}^{1,p}$. Polarization (see Appendix D) yields that if $X, Y \in \mathcal{D}^{1,2p}$, then $\langle X, Y\rangle_{H_1} \in \mathcal{D}^{1,p}$.

REMARK 15.88. If H_1 instead is a complex Hilbert space, we may regard it as a real Hilbert space with the same norm and inner product $\operatorname{Re}\langle\cdot,\cdot\rangle$; thus Lemma 15.86 holds for this case too, with $\langle \nabla X, X\rangle_{H_1}$ replaced by $\operatorname{Re}\langle \nabla X, X\rangle_{H_1} \in H$. Similarly, if $X \in \mathcal{D}^{1,2p}(H_1)$, then $\|X\|_{H_1}^2 \in \mathcal{D}^{1,p}$ with $\nabla\|X\|_{H_1}^2 = 2\operatorname{Re}\langle \nabla X, X\rangle_{H_1} = \langle \nabla X, X\rangle_{H_1} + \langle X, \nabla X\rangle_{H_1}$.

We continue this section with some further results. Recall that the operators π_n and M_r extend to Hilbert-space-valued variables, see Examples 12.4 and 12.5. We begin with a simpler version of Lemma 15.71 for $p > 1$.

LEMMA 15.89. *Let $1 < p \le \infty$ and $\xi \in H$, and suppose that $X, Y \in L^p$. Then $X \in \mathcal{D}^{\xi,p}$ and $\partial_\xi X = Y$ if and only if, for every real t,*

$$\rho_{t\xi}(X) - X = \int_0^t \rho_{s\xi}(Y)\,ds,$$

where the integral converges in L^1.

PROOF. The integral converges as a Bochner integral in L^1 because $s \mapsto \rho_{s\xi}(Y)$ is a continuous function into L^1 by Theorem 14.1(viii). The result follows by evaluating this integral pointwise as in the proof of Lemma 15.71. □

THEOREM 15.90. *Let $1 < p \le \infty$, $\xi \in H$, $k \ge 0$ and $-1 \le r \le 1$.*
(i) *If $X \in \mathcal{D}^{\xi,p}$, then $M_r X \in \mathcal{D}^{\xi,p}$, and $\partial_\xi(M_r X) = rM_r(\partial_\xi X)$.*
(ii) *If $X \in \mathcal{D}^{k,p}$, then $M_r X \in \mathcal{D}^{k,p}$, and $\nabla^j(M_r X) = r^j M_r(\nabla^j X)$, $0 \le j \le k$.*

PROOF. Let $X \in \mathcal{D}^{\xi,p}$ and $Y = \partial_\xi X$. By Lemma 15.89, for any real t,

$$\rho_{rt\xi}(X) - X = \int_0^{rt} \rho_{s\xi}(Y)\,ds.$$

Applying M_r, which is allowed because the integral converges in L^1 and M_r is a continuous linear operator in L^1, we obtain

$$M_r\big(\rho_{rt\xi}(X)\big) - M_r X = \int_0^{rt} M_r\big(\rho_{s\xi}(Y)\big)\,ds = r\int_0^t M_r\big(\rho_{ru\xi}(Y)\big)\,du,$$

which by Theorem 14.10 yields

$$\rho_{t\xi}(M_rX) - M_rX = r\int_0^t \rho_{u\xi}(M_rY)\,du = \int_0^t \rho_{u\xi}(rM_rY)\,du.$$

Since M_r is bounded on L^p too, we have $M_rX \in L^p$ and $rM_rY \in L^p$, and thus (i) follows by Lemma 15.89.

For (ii), note that (i) yields, for $X \in \mathcal{D}^{1,p}$ and each $\xi \in H$,

$$\partial_\xi(M_rX) = rM_r(\partial_\xi X) = rM_r(\langle \nabla X, \xi\rangle_{H_{\mathbb{C}}}) = \langle rM_r(\nabla X), \xi\rangle_{H_{\mathbb{C}}},$$

which proves the result for $k = 1$ by Theorem 15.64(viii).

Similarly, the result extends to vector-valued variables in $\mathcal{D}^{1,p}(H_0)$ by taking inner products with $h_0 \in H_0$, for every Hilbert space H_0. The result for $k \geq 2$ now follows by induction, using Theorem 15.64(vii). □

REMARK 15.91. The results of Theorem 15.90 are valid also for $p = 1$. A simple proof uses the case $p = 2$ and the facts proved in Section 7 below that $\mathcal{D}^{\xi,2}$ is dense in $\mathcal{D}^{\xi,1}$ and $\mathcal{D}^{k,2}$ in $\mathcal{D}^{k,1}$. We omit the details.

Note also that it follows easily from Theorem 15.90 and Example 12.5 that if $p < \infty$ and $X \in \mathcal{D}^{k,p}$, then $M_rX \to X$ in $\mathcal{D}^{k,p}$ as $r \to 1$.

The hypercontractive properties of M_r also have counterparts for the Sobolev spaces.

THEOREM 15.92. *Let $1 < p \leq q < \infty$, $\xi \in H$ and $k \geq 0$.*

(i) *If $0 \leq r^2 \leq (p-1)/(q-1)$, then M_r maps $\mathcal{D}^{k,p}$ into $\mathcal{D}^{k,q}$ and $\mathcal{D}^{\xi,p}$ into $\mathcal{D}^{\xi,q}$, and is in both cases a contraction.*

(ii) *Let D_{pq} be the set of complex numbers given by Theorem 5.28. If $z \in D_{pq}$, then M_z maps $\mathcal{D}^{k,p}_{\mathbb{C}}$ into $\mathcal{D}^{k,q}_{\mathbb{C}}$ and $\mathcal{D}^{\xi,p}_{\mathbb{C}}$ into $\mathcal{D}^{\xi,q}_{\mathbb{C}}$, and is in both cases a contraction. Moreover, if $X \in \mathcal{D}^{k,p}_{\mathbb{C}}$ $(\mathcal{D}^{\xi,p}_{\mathbb{C}})$ then $z \mapsto M_zX$ is a continuous function $D_{pq} \to \mathcal{D}^{k,q}_{\mathbb{C}}$ $(\mathcal{D}^{\xi,q}_{\mathbb{C}})$ that is analytic in the interior D°_{pq}.*

PROOF. The properties in (i) follow by Theorem 15.64(iv),(v) from Theorem 15.90 and the hypercontractive properties of M_r in Theorem 5.8 and Example 12.5.

For (ii), suppose that $X \in \mathcal{D}^{k,p}_{\mathbb{C}}$. (The case $X \in \mathcal{D}^{\xi,p}_{\mathbb{C}}$ is similar.) The function $F(z) = (M_zX, zM_z(\nabla X), \dots, z^kM_z(\nabla^kX))$ is then, by Theorem 5.28 and Example 12.5, continuous from D_{pq} into $\bigoplus_0^k L^q_{\mathbb{C}}(H_{\mathbb{C}}^{\otimes j})$ and analytic in D°_{pq}.

If $r > 0$ is small enough, then (i) and Theorem 15.90 show that $F(r)$ belongs to the closed subspace $\tilde{\mathcal{D}}^{k,q}_{\mathbb{C}}$ defined in Corollary 15.75. By Theorem G.1, we thus have $F(z) \in \tilde{\mathcal{D}}^{k,q}_{\mathbb{C}}$ for every $z \in D_{pq}$, and the function F is continuous from D_{pq} into $\tilde{\mathcal{D}}^{k,q}_{\mathbb{C}}$ and analytic in D°_{pq}. Moreover, $\|F(z)\|_{\tilde{\mathcal{D}}^{k,q}_{\mathbb{C}}} \leq \|X\|_{k,p}$. Since $\tilde{\mathcal{D}}^{k,q}_{\mathbb{C}}$ is isometric to $\mathcal{D}^{k,q}_{\mathbb{C}}$ by an isometry mapping $F(z)$ to its first component M_zX, see Remark 15.77, the results follow. □

THEOREM 15.93. *Let $1 < p \leq \infty$.*

(i) *If* $\xi \in H$ *and* $X \in \mathcal{D}^{\xi,p}$, *then* $\partial_\xi(\pi_n X) = \pi_{n-1}(\partial_\xi X)$, $n \geq 0$.

(ii) *If* H_0 *is a Hilbert space and* $X \in \mathcal{D}^{1,p}(H)$, *then* $\nabla(\pi_n X) = \pi_{n-1}(\nabla X)$, $n \geq 0$.

PROOF. We use complex scalars, and may assume that $p < \infty$ (by Theorem 15.64(i),(ii)).

Suppose that $X \in \mathcal{D}^{\xi,p}$. By Theorem 5.28, the function $z \mapsto F(z) = (M_z X, zM_z(\partial_\xi X))$ is an analytic function into $L^p \times L^p$, defined on a neighbourhood D°_{pp} of 0. As in the proof of Theorem 15.92, $F(z) \in \tilde{\mathcal{D}}^{\xi,p}_{\mathbb{C}}$ for real $z \in (-1,1)$ by Theorem 15.90, and thus for all $z \in D^\circ_{pp}$ by Theorem G.1.

For small positive z, by Theorem 5.23,

$$F(z) = \Bigl(\sum_0^\infty z^n \pi_n(X), z\sum_0^\infty z^n \pi_n(\partial_\xi X)\Bigr) = \Bigl(\sum_0^\infty z^n \pi_n(X), \sum_1^\infty z^n \pi_{n-1}(\partial_\xi X)\Bigr)$$

$$= \sum_0^\infty z^n \bigl(\pi_n(X), \pi_{n-1}(\partial_\xi X)\bigr),$$

where the sums converge in L^p and $L^p \times L^p$. This Taylor expansion together with Theorem G.2 yields $(\pi_n(X), \pi_{n-1}(\partial_\xi X)) \in \tilde{\mathcal{D}}^{\xi,p}_{\mathbb{C}}$, which proves (i).

For (ii), we observe that (i) and Proposition 15.45 show that for every $\xi \in H$ and $h \in H_0$,

$$\partial_\xi \langle \pi_n(X), h\rangle_{H_0} = \partial_\xi \pi_n \langle X, h\rangle_{H_0} = \pi_{n-1}\partial_\xi \langle X, h\rangle_{H_0} = \pi_{n-1}\langle \nabla X, \xi \otimes h\rangle_{H\otimes H_0};$$

the result follows by another application of Proposition 15.45. □

We generally regard the Gaussian Hilbert space H as fixed in this chapter, but when considering a random variable that is measurable with respect to the σ-field generated by a closed subspace H', it makes sense to consider derivatives and Sobolev spaces defined for the Gaussian Hilbert space H' as well as the corresponding notions for the original space H. We have the following simple consistency result.

LEMMA 15.94. *Suppose that* H' *is a closed subspace of* H, *and let* $\xi \in H'$. *If* X *is* $\mathcal{F}(H')$*-measurable, then* $\partial_\xi X$ *exists for* H *if and only if it exists for* H', *and then the values coincide; moreover,* X *is* ξ*-a.c. for* H *if and only if it is* ξ*-a.c. for* H'.

PROOF. It follows by (14.11) that $\rho_\xi X$ is the same for H and H', and the result follows. □

LEMMA 15.95. *Suppose that* H' *is a closed subspace of* H *and let* Q *denote the orthogonal projection* $H \to H'$ (*or* $H_{\mathbb{C}} \to H'_{\mathbb{C}}$). *Let* $X \in L^p = L^p(\mathcal{F}(H))$, *with* $1 \leq p \leq \infty$, *and let* $X' = \mathrm{E}(X \mid \mathcal{F}(H')) \in L^p(\mathcal{F}(H'))$.

(i) *If* $\xi \in H'$ *and* $X \in \mathcal{D}^{\xi,p}$, *then* $X' \in \mathcal{D}^{\xi,p}$ *and* $\partial_\xi X' = \mathrm{E}(\partial_\xi X \mid \mathcal{F}(H'))$.

(ii) *If* $k \geq 0$ *and* $X \in \mathcal{D}^{k,p}$, *then* $X' \in \mathcal{D}^{k,p}$ (*both for* H' *and for* H) *and* $\nabla^j X' = Q^{\otimes j}(\mathrm{E}(\nabla^j X \mid \mathcal{F}(H')))$, $j \leq k$.

PROOF. Recall that the conditional expectation $\mathrm{E}(\cdot \mid \mathcal{F}(H'))$ is a continuous linear operator in every L^p, $1 \leq p \leq \infty$.

Observe first that for every $\xi \in H'$, $t \in \mathbb{R}$ and $Z \in L^1$,

$$\mathrm{E}\big(e^{-\xi^2}\rho_{t\xi}(Z) \mid \mathcal{F}(H')\big) = e^{-\xi^2}\rho_{t\xi}\big(\mathrm{E}(Z \mid \mathcal{F}(H'))\big), \tag{15.15}$$

since this formula holds by a simple computation when $Z = Y'Y''$, where Y' is a polynomial in $\xi_1', \dots, \xi_n' \in H'$ and Y'' is a polynomial in $\xi_1'', \dots, \xi_n'' \in H'^{\perp}$, such variables are total in L^1 and both sides of (15.15) define continuous linear maps of L^1 into itself, cf. Lemma 15.70. Part (i) now follows by Lemma 15.71, applying the conditional expectation to both sides of (15.12) and again using the continuity in L^1.

If $X \in \mathcal{D}^{1,p}$, then (i) yields for every $\xi \in H'$,

$$\partial_\xi X' = \mathrm{E}(\partial_\xi X \mid \mathcal{F}(H')) = \mathrm{E}(\langle \nabla X, \xi \rangle_{H_\mathbb{C}} \mid \mathcal{F}(H')) = \langle \mathrm{E}(\nabla X \mid \mathcal{F}(H')), \xi \rangle_{H_\mathbb{C}},$$

and it follows that $X' \in \mathcal{D}^{1,p}$ for H', with $\nabla X' = Q\big(\mathrm{E}(\nabla X \mid \mathcal{F}(H'))\big)$. Since furthermore $\partial_\xi X' = 0$ for $\xi \perp H'$, by (14.11), this holds also for H. This proves (ii) when $k = 1$.

The case $k \geq 2$ follows by induction: Suppose that the result holds for $k-1$ and that $u \in H_\mathbb{C}^{\otimes k-1}$. Let $Y = \langle \nabla^{k-1} X', u \rangle_{H_\mathbb{C}^{\otimes k-1}}$. Then, by the induction hypothesis,

$$\begin{aligned} Y &= \langle Q^{\otimes(k-1)}\,\mathrm{E}(\nabla^{k-1} X \mid \mathcal{F}(H')), u \rangle_{H_\mathbb{C}^{\otimes k-1}} \\ &= \mathrm{E}(\langle \nabla^{k-1} X, Q^{\otimes(k-1)} u \rangle_{H_\mathbb{C}^{\otimes k-1}} \mid \mathcal{F}(H')) \end{aligned}$$

and the just proved case yields

$$\begin{aligned} \nabla Y &= Q\big(\mathrm{E}(\nabla \langle \nabla^{k-1} X, Q^{\otimes(k-1)} u \rangle_{H_\mathbb{C}^{\otimes k-1}} \mid \mathcal{F}(H'))\big) \\ &= \langle Q^{\otimes k}\,\mathrm{E}(\nabla^k X \mid \mathcal{F}(H')), u \rangle_{H_\mathbb{C}^{\otimes k-1}}, \end{aligned}$$

which verifies the induction step by Definition 15.41. □

THEOREM 15.96. *Let $1 \leq p < \infty$ and $k \geq 0$. Suppose that $(H_n)_1^\infty$ is an increasing sequence of finite-dimensional subspaces of H and that $X \in \mathcal{D}^{k,p}$ is $\mathcal{F}(\bigcup_1^\infty H_n)$-measurable. Let $X_n = \mathrm{E}(X \mid \mathcal{F}(H_n))$. Then $X_n \to X$ in $\mathcal{D}^{k,p}$.*

PROOF. Lemma 15.95 shows that each $X_n \in \mathcal{D}^{k,p}$. Moreover, $X_n \to X$ in L^p by the martingale convergence theorem; similarly, it follows easily from the formula in Lemma 15.95 that $\nabla^j X_n \to \nabla^j X$ in $L^p(H_\mathbb{C}^{\otimes j})$ for $j \leq k$. □

We end this section with a counterexample, showing that even if $X \in \mathcal{D}^{\xi,2}$ for every $\xi \in H$, ∇X does not have to exist.

EXAMPLE 15.97. Let H be an infinite-dimensional Gaussian Hilbert space with an orthonormal basis $\{\xi_k\}_1^\infty$, and let

$$Y_n = -\big(\cos(n\xi_n) - \mathrm{E}\cos(n\xi_n)\big) = -\cos(n\xi_n) + e^{-n^2/2} \in L^\infty.$$

By Theorem 15.35, Y_n is r.a.c. and $\nabla Y_n = n\sin(n\xi_n) \otimes \xi_n \in L^\infty(H)$. Hence $Y_n \in \mathcal{D}^{1,\infty} \subset \mathcal{D}^{1,2}$.

Let $X_N = \sum_1^N \frac{1}{n} Y_n$. Note that the variables Y_n satisfy $\operatorname{E} Y_n = 0$ and $\operatorname{E} Y_n^2 = \operatorname{Var}(\cos(n\xi_n)) \leq 1$, and that they are independent. Hence the infinite sum $X = \sum_1^\infty \frac{1}{n} Y_n$ converges in L^2, i.e. $X_N \to X$ in L^2 as $N \to \infty$. Moreover, for every $\xi \in H$, similarly,

$$\partial_\xi X_N = \sum_1^N \tfrac{1}{n} \langle \nabla Y_n, \xi \rangle_H = \sum_1^N \langle \xi, \xi_n \rangle_H \sin(n\xi_n) \to \sum_1^\infty \langle \xi, \xi_n \rangle_H \sin(n\xi_n)$$

in L^2, because the variables $\sin(n\xi_n)$ are independent with $\operatorname{E}\sin(n\xi_n) = 0$ (by symmetry), $\operatorname{Var}(\sin(n\xi_n)) \leq 1$ and $\sum_1^\infty \langle \xi, \xi_n \rangle_H^2 < \infty$.

Consequently, by Theorem 15.72, $X \in \mathcal{D}^{\xi,2}$ for every $\xi \in H$, with $\partial_\xi X = \sum_1^\infty \langle \xi, \xi_n \rangle_H \sin(n\xi_n)$; in particular X is r.a.c.

Suppose now that $\nabla X = Z \in L^0(H)$ exists; then $\langle Z, \xi_n \rangle_H = \partial_{\xi_n} X = \sin(n\xi_n)$ for every n, and thus

$$\|Z\|_H^2 = \sum_1^\infty \langle Z, \xi_n \rangle_H^2 = \sum_1^\infty \sin^2(n\xi_n).$$

The latter series diverges a.s., however, because the events $\{\sin^2(n\xi_n) > \frac{1}{2}\}$ are independent and $\operatorname{P}(\sin^2(n\xi_n) > \frac{1}{2}) \to \frac{1}{2}$ as $n \to \infty$, and thus the Borel–Cantelli lemma shows that a.s. $\sin^2(n\xi_n) > \frac{1}{2}$ for infinitely many n. This contradiction shows that ∇X does not exist. In particular, $X \notin \mathcal{D}^{1,2}$.

6. The L^2 case

The Sobolev spaces $\mathcal{D}^{k,2}$ have a simple description in terms of the chaos decomposition. We begin with a result connecting ∂_ξ and the annihilation operator $A(\xi)$ studied in Chapter 13.

THEOREM 15.98. *Let $\xi \in H$. Then $\mathcal{D}^{\xi,2} = \operatorname{Dom}(A(\xi))$, and $\partial_\xi = A(\xi)$ on $\mathcal{D}^{\xi,2}$.*

PROOF. Observe first that if $X \in \mathcal{P}$, then $X \in \mathcal{D}^{\xi,2}$ by Example 15.66; moreover, as observed in Example 15.9, $\partial_\xi X = A(\xi)X$.

Next, consider $X \in \overline{\mathcal{P}}_m$, for some $m \geq 0$. By definition, there exist polynomial variables $X_n \in \mathcal{P}_m$ such that $X_n \to X$ in L^2 as $n \to \infty$. By Theorem 13.7(iv), $A(\xi)$ is continuous on $\overline{\mathcal{P}}_m$, and thus also $\partial_\xi X_n = A(\xi)X_n \to A(\xi)X$ in L^2. Consequently, Theorem 15.72 yields $X \in \mathcal{D}^{\xi,2}$ and $\partial_\xi X = A(\xi)X$.

If now $X \in \operatorname{Dom}(A(\xi))$, then Theorem 13.7(v) implies that $\pi_{\leq m} X \to X$ and $A(\xi)(\pi_{\leq m} X) \to A(\xi)X$ in L^2 as $m \to \infty$. Since $\pi_{\leq m} X \in \overline{\mathcal{P}}_m$, the case just treated shows that $\pi_{\leq m} X \in \mathcal{D}^{\xi,2}$ and $A(\xi)(\pi_{\leq m} X) = \partial_\xi(\pi_{\leq m} X)$, and Theorem 15.72 again shows that $X \in \mathcal{D}^{\xi,2}$ and $\partial_\xi X = A(\xi)X$.

Conversely, if $X \in \mathcal{D}^{\xi,2}$, let $Y = \partial_\xi X$ and $X_n = \pi_n(X)$, $n \geq 0$. By Theorem 15.93, $\partial_\xi(X_n) = \pi_{n-1}Y$; moreover $\partial_\xi(X_n) = A(\xi)X_n$, since $X_n \in \overline{\mathcal{P}}_n$. Hence,

$$\sum_{n=0}^{\infty}(\|X_n\|_2^2 + \|A(\xi)X_n\|_2^2) = \sum_{n=0}^{\infty}(\|X_n\|_2^2 + \|\partial_\xi X_n\|_2^2)$$
$$= \sum_{n=0}^{\infty}(\|X_n\|_2^2 + \|\pi_{n-1}Y\|_2^2) = \|X\|_2^2 + \|Y\|_2^2 < \infty,$$

and Theorem 13.7(v) shows that $X \in \operatorname{Dom}(A(\xi))$. □

The proof above also yields the following important fact. (Cf. Theorem 13.7(vi).)

COROLLARY 15.99. *Let $\xi \in H$. Then $\mathcal{P}$ is dense in $\mathcal{D}^{\xi,2}$.*

PROOF. Let $\mathcal{D}_0^{\xi,2}$ be the closed hull of $\mathcal{P}$ in $\mathcal{D}^{\xi,2}$. The proof of Theorem 15.98 shows first that $\overline{\mathcal{P}}_* \subseteq \mathcal{D}_0^{\xi,2}$, and then that $\operatorname{Dom}(A(\xi)) \subseteq \mathcal{D}_0^{\xi,2}$. Consequently, $\mathcal{D}^{\xi,2} = \operatorname{Dom}(A(\xi)) = \mathcal{D}_0^{\xi,2}$. □

THEOREM 15.100. *The following are equivalent, for any orthonormal basis $\{\xi\}_{i\in\mathcal{I}}$ in H.*
(i) $X \in \mathcal{D}^{1,2}$.
(ii) $X \in \mathcal{D}^{\xi_i,2}$ *for each* ξ_i, *and* $\sum_i \|\partial_{\xi_i} X\|_2^2 < \infty$.
(iii) $X \in L^2$ *and* $\mathcal{N}^{1/2}X \in L^2$, *i.e.* $\sum_1^\infty n\|\pi_n(X)\|_2^2 < \infty$.
Furthermore, when these conditions hold,

$$\|\nabla X\|_{L^2(H_{\mathbf{C}})}^2 = \sum_i \|\partial_{\xi_i} X\|_2^2 = \|\mathcal{N}^{1/2}X\|_2^2 = \sum_1^\infty n\|\pi_n(X)\|_2^2 \qquad (15.16)$$

and

$$\|X\|_{1,2} = \|(\mathcal{N}+1)^{1/2}X\|_2 = \Big(\sum_0^\infty (n+1)\|\pi_n(X)\|_2^2\Big)^{1/2}. \qquad (15.17)$$

PROOF. (i) ⇒ (ii) If $X \in \mathcal{D}^{1,2}$, then $X \in \mathcal{D}^{\xi,2}$ for every $\xi \in H$, and

$$\|\nabla X\|_{L^2(H_{\mathbf{C}})}^2 = \mathrm{E}\,\|\nabla X\|_{H_{\mathbf{C}}}^2 = \mathrm{E}\sum_i |\langle \nabla X, \xi_i\rangle_{H_{\mathbf{C}}}|^2 = \mathrm{E}\sum_i |\partial_{\xi_i} X|^2$$
$$= \sum_i \|\partial_{\xi_i} X\|_2^2. \qquad (15.18)$$

Hence (ii) holds. (Even if the basis is uncountable, only countably many terms in the sums are non-zero.)

(ii) $\Rightarrow$ (iii) Suppose first that $X \in \overline{\mathcal{P}}_m$ for some finite m. Then $X \in \mathrm{Dom}(\mathcal{N}^{1/2}) \cap \mathrm{Dom}(\mathcal{N})$, and by Theorems 13.15 and 15.98

$$\sum_1^\infty n\|\pi_n(X)\|_2^2 = \langle \mathcal{N}X, X\rangle = \sum_i \langle A^*(\xi_i)A(\xi_i)X, X\rangle$$
$$= \sum_i \|A(\xi_i)X\|_2^2 = \sum_i \|\partial_{\xi_i} X\|_2^2. \tag{15.19}$$

Now suppose that $X \in L^2$ is such that (ii) holds. It follows from Theorem 15.93 that, for each i, $\|\partial_{\xi_i}(\pi_{\le m}X)\|_2 = \|\pi_{\le m-1}(\partial_{\xi_i}X)\|_2 \nearrow \|\partial_{\xi_i}X\|_2$ as $m \to \infty$. We apply (15.19) to $\pi_{\le m}X$ and let $m \to \infty$; monotone convergence implies that (15.19) also holds for X, and thus (iii) holds.

(iii) $\Rightarrow$ (i) Suppose first that $X \in \mathcal{D}^{1,2}$. Then (15.16) holds by (15.18) and (15.19), and (15.17) follows.

Next, if $n \ge 0$ and $X \in \mathcal{P}_n$, then $X \in \mathcal{D}^{1,2}$ by Example 15.66 and $\|X\|_{1,2} \le \sqrt{n+1}\|X\|_2$ by (15.17). It follows from Theorem 15.72 that $\overline{\mathcal{P}}_n \subset \mathcal{D}^{1,2}$.

Finally, if X is such that (iii) holds, let $X_n = \pi_{\le n}(X) \in \overline{\mathcal{P}}_n \subset \mathcal{D}^{1,2}$. By (15.16), if $1 \le m < n$,

$$\|\nabla X_n - \nabla X_m\|_{L^2(H_{\mathrm{C}})}^2 = \|\mathcal{N}^{1/2}(X_n - X_m)\|_2^2 = \sum_{m+1}^n k\|\pi_k(X)\|_2^2;$$

hence $(\nabla X_n)_1^\infty$ is a Cauchy sequence in $L^2(H_{\mathrm{C}})$, and thus it converges. Since also $X_n \to X$ in L^2, Theorem 15.72 yields $X \in \mathcal{D}^{1,2}$. □

COROLLARY 15.101. *If $X, Y \in \mathcal{D}^{1,2}$, then*

$$\mathrm{E}\langle \nabla X, \nabla Y\rangle_{H_{\mathrm{C}}} = \langle \mathcal{N}^{1/2}X, \mathcal{N}^{1/2}Y\rangle_{L^2} = \sum_1^\infty n\langle \pi_n(X), \pi_n(Y)\rangle.$$

PROOF. By (15.16) and polarization, cf. Appendix D. □

The characterization in Theorem 15.100 is easily extended to higher derivatives; we state the result for general Hilbert-space-valued variables. (Note that $\mathcal{N} = \sum_0^\infty n\pi_n$ also extends to such variables.) We use $\asymp$ to denote equivalence within constants depending on k only, and let (as in Chapter 11) $(n)_k = n(n-1)\cdots(n-k+1)$.

THEOREM 15.102. *Let H_0 be a Hilbert space and $k \ge 0$. Then $X \in \mathcal{D}^{k,2}(H_0)$ if and only if $X \in L^2(H_0)$ and $\mathcal{N}^{k/2}X \in L^2(H_0)$, and then*

$$\|X\|_{k,2,H_0} \asymp \|(\mathcal{N}+1)^{k/2}X\|_{L^2(H_0)} = \Big(\sum_{n=0}^\infty (n+1)^k \|\pi_n(X)\|_{L^2(H_0)}^2\Big)^{1/2}; \tag{15.20}$$

moreover

$$\|\nabla^k X\|_{L^2(H^{\otimes k}\otimes H_0)} = \Big(\sum_{n=k}^\infty (n)_k \|\pi_n(X)\|_{L^2(H_0)}^2\Big)^{1/2}. \tag{15.21}$$

PROOF. Consider first $X \in \mathcal{D}^{1,2}(H_0)$. Let $(h_i)_{i\in\mathcal{I}}$ be an orthonormal basis in H_0 (regarding H_0 as a real Hilbert space in the complex case too). Then $\langle X, h_i\rangle_{H_0} \in \mathcal{D}^{1,2}$ for each i, and $\nabla\langle X, h_i\rangle_{H_0} = \langle \nabla X, h_i\rangle_{H_0}$. Thus, using Theorem 15.100 on each $\langle \nabla X, h_i\rangle_{H_0}$,

$$\begin{aligned}\|\nabla X\|^2_{L^2(H\otimes H_0)} &= \mathrm{E}\,\|\nabla X\|^2_{H\otimes H_0} = \mathrm{E}\sum_i \|\nabla\langle X, h_i\rangle_{H_0}\|^2_H \\ &= \sum_i \|\nabla\langle X, h_i\rangle_{H_0}\|^2_{L^2(H)} = \sum_i\sum_n n\|\pi_n\langle X, h_i\rangle_{H_0}\|^2_2 \\ &= \sum_n\sum_i n\,\mathrm{E}\,|\langle \pi_n(X), h_i\rangle_{H_0}|^2 = \sum_1^n n\,\mathrm{E}\,\|\pi_n(X)\|^2_{H_0}. \quad (15.22)\end{aligned}$$

(The calculation is valid even if the basis $\{h_i\}$ is uncountable, since the random variables are a.s. separably valued and thus only a countable number of terms in the sums are non-zero.)

This verifies (15.21) for $k = 1$. We prove the case $k \geq 2$ by induction. (The case $k = 0$ is trivial.)

Thus, suppose that (15.21) holds for some fixed k and all Hilbert spaces H_0 and variables $X \in \mathcal{D}^{k,2}(H_0)$. If $X \in \mathcal{D}^{k+1,2}(H_0)$, then $\nabla X \in \mathcal{D}^{k,2}(H \otimes H_0)$, and thus, by the induction hypothesis,

$$\begin{aligned}\|\nabla^{k+1} X\|^2_{L^2(H^{\otimes(k+1)}\otimes H_0)} &= \|\nabla^k(\nabla X)\|^2_{L^2(H^{\otimes k}\otimes H\otimes H_0)} \\ &= \sum_{n=k}^{\infty}(n)_k\|\pi_n(\nabla X)\|_{L^2(H\otimes H_0)}.\end{aligned}$$

Moreover, by Theorem 15.93 and (15.22),

$$\|\pi_n(\nabla X)\|^2_{L^2(H\otimes H_0)} = \|\nabla(\pi_{n+1}(X))\|^2_{L^2(H\otimes H_0)} = (n+1)\|\pi_{n+1}(X)\|^2_{L^2(H_0)},$$

and thus

$$\|\nabla^{k+1} X\|^2_{L^2(H^{\otimes(k+1)}\otimes H_0)} = \sum_{n=k}^{\infty}(n+1)_{k+1}\|\pi_{n+1}(X)\|^2_{L^2(H_0)},$$

which verifies the induction step; this completes the proof of (15.21).

Consequently, for $X \in \mathcal{D}^{k,2}(H_0)$,

$$\begin{aligned}\|X\|^2_{k,2} &= \sum_{j=0}^{k}\|\nabla^j X\|^2_{L^2(H^{\otimes j}\otimes H_0)} = \sum_{n=0}^{\infty}\sum_{j=0}^{k}(n)_j\|\pi_n(X)\|^2_{L^2(H_0)} \\ &\asymp \sum_{n=0}^{\infty}(n+1)^k\|\pi_n(X)\|^2_{L^2(H_0)} = \|(\mathcal{N}+1)^{k/2}X\|^2_{L^2(H_0)},\end{aligned}$$

which verifies (15.20), and shows that X, $\mathcal{N}^{k/2}X \in L^2(H_0)$.

In order to show the converse, we first show that for every $X \in L^2(H_0)$ and $n \geq 0$, $\pi_n(X) \in \mathcal{D}^{k,2}(H_0)$. Note first that, by Example 15.69, $\mathcal{P} \otimes$

$H_0 \subset \mathcal{D}^{k,2}(H_0)$. Moreover, if $X \in \mathcal{P} \otimes H_0$, then $\pi_n(X) \in \mathcal{P} \otimes H_0$ too by Corollary 3.25, and thus $\pi_n(X) \in \mathcal{D}^{k,2}(H_0)$ and (15.20) yields

$$\|\pi_n(X)\|_{k,2,H_0} \asymp (n+1)^{k/2}\|\pi_n(X)\|_{L^2(H_0)} \leq (n+1)^{k/2}\|X\|_{L^2(H_0)}. \quad (15.23)$$

Now let $X \in L^2(H_0)$ and take a sequence $(X_j)_1^\infty$ in $\mathcal{P} \otimes H_0$ such that $X_j \to X$ in $L^2(H_0)$ as $j \to 0$. (This is possible since $\mathcal{P} \otimes H_0$ is dense.) It follows from (15.23) that $(\pi_n(X_j))_{j=1}^\infty$ is a Cauchy sequence in $\mathcal{D}^{k,2}(H_0)$; since $\mathcal{D}^{k,2}(H_0)$ is complete by Corollary 15.73 and $\pi_n(X_j) \to \pi_n(X)$ in $L^2(H_0)$, this implies $\pi_n(X) \in \mathcal{D}^{k,2}(H_0)$.

Finally, suppose that $X \in L^2(H_0)$ with $\mathcal{N}^{k/2}X \in L^2(H_0)$. By the result just proved, $\pi_{\leq n}X = \sum_0^n \pi_m(X) \in \mathcal{D}^{k,2}(H_0)$ for every $n \geq 0$. Furthermore, if $n > m \geq 0$, (15.20) implies that

$$\|\pi_{\leq n}(X) - \pi_{\leq m}(X)\|_{k,2,H_0}^2 \asymp \sum_{j=m+1}^{n} j^k \|\pi_j(X)\|_{L^2(H_0)}^2.$$

Since $\sum_0^\infty j^k\|\pi_j(X)\|_{L^2(H_0)}^2 = \|\mathcal{N}^{k/2}X\|_{L^2(H_0)}^2 < \infty$, it follows that $(\pi_{\leq n}(X))_1^\infty$ is a Cauchy sequence in $\mathcal{D}^{k,2}(H_0)$, and the completeness of $\mathcal{D}^{k,2}(H_0)$ yields $X \in \mathcal{D}^{k,2}(H_0)$. □

As for Theorem 15.98, the proof yields an important corollary.

COROLLARY 15.103. *Let H_0 be a Hilbert space and $k \geq 0$. Then $\mathcal{P} \otimes H_0$ is dense in $\mathcal{D}^{k,2}(H_0)$. In particular, $\mathcal{P}$ is dense in $\mathcal{D}^{k,2}$.*

PROOF. Let $\mathcal{D}_0^{k,2}(H_0)$ be the closed hull of $\mathcal{P} \otimes H_0$ in $\mathcal{D}^{k,2}(H_0)$. If $X \in \mathcal{D}^{k,2}(H_0)$, then the proof of Theorem 15.102 shows first that $\pi_n(X) \in \mathcal{D}_0^{k,2}(H_0)$ and then that $X = \lim_{n\to\infty} \pi_n(X) \in \mathcal{D}_0^{k,2}(H_0)$. □

7. Dense subspaces

We next want to show that if $p < \infty$, then the polynomial variables are dense in $\mathcal{D}^{\xi,p}$ and $\mathcal{D}^{k,p}$. (For $p = 2$, this was shown in the preceding section.) We begin with the case $p > 1$.

THEOREM 15.104. *Suppose that $1 < p < \infty$. Then $\mathcal{P}$ is dense in $\mathcal{D}^{\xi,p}$ for every $\xi \in H$, and in $\mathcal{D}^{k,p}$ for every $k \geq 0$.*

PROOF. We treat $\mathcal{D}^{k,p}$; the proof for $\mathcal{D}^{\xi,p}$ is similar. Moreover, we use complex scalars; the real case follows immediately.

Let $\mathcal{D}_0^{k,p}$ denote the closed hull of $\mathcal{P}$ in $\mathcal{D}^{k,p}$; we thus want to show that $\mathcal{D}_0^{k,p} = \mathcal{D}^{k,p}$. Note that for $p = 2$, this has already been shown in Corollary 15.103.

Now assume $p \neq 2$, and let $X \in \mathcal{D}^{k,p}$. We use the Mehler transform, and first show that $M_r X \in \mathcal{D}_0^{k,p}$ if $r > 0$ is small enough, treating the cases $p < 2$ and $p > 2$ separately.

If $p < 2$, then $\mathcal{D}^{k,2}$ is continuously included in $\mathcal{D}^{k,p}$, and thus $\mathcal{D}_0^{k,2} \subset \mathcal{D}_0^{k,p}$. If $0 \le r \le (p-1)^{1/2}$, then Theorem 15.92(i) shows that $M_r X \in \mathcal{D}^{k,2} = \mathcal{D}_0^{k,2} \subset \mathcal{D}_0^{k,p}$.

If $p > 2$, we instead first observe that $X \in \mathcal{D}^{k,p} \subset \mathcal{D}^{k,2} = \mathcal{D}_0^{k,2}$, and thus there exists a sequence $(X_n)_1^\infty$ in $\mathcal{P}$ such that $X_n \to X$ in $\mathcal{D}^{k,2}$ as $n \to \infty$. Theorem 15.92(i) now shows that if $0 \le r \le (p-1)^{-1/2}$, then M_r maps $\mathcal{D}^{k,2}$ continuously into $\mathcal{D}^{k,p}$, and thus $M_r X_n \to M_r X$ in $\mathcal{D}^{k,p}$. Since $M_r X_n \in \mathcal{P}$, this yields $M_r X \in \mathcal{D}_0^{k,p}$.

Now consider the function $z \mapsto M_z X$, which by Theorem 15.92(ii) is an analytic function $D_{pp}^\circ \to \mathcal{D}^{k,p}$ with a continuous extension to D_{pp}. Theorem G.1 applies, and shows that $M_z X \in \mathcal{D}_0^{k,p}$ for every $z \in D_{pp}$; in particular, $X = M_1 X \in \mathcal{D}_0^{k,p}$. □

In order to extend this result to $p = 1$, we need other methods. We begin with $\mathcal{D}^{\xi,1}$.

LEMMA 15.105. *Let $\xi \in H$. If $X \in L^1$ is ξ-a.c. and $\partial_\xi X - \xi X \in L^1$, then* $\mathrm{E}(\partial_\xi X - \xi X) = 0$.

PROOF. By assumption and Theorem 15.21, for a.e. $\omega \in \Omega$, the function $t \mapsto \rho_{t\xi} X$ is absolutely continuous and $\frac{d}{dt}\rho_{t\xi} X = \rho_{t\xi}(\partial_\xi X)$ for a.e. t. (We use an absolutely continuous version of $\rho_{t\xi}(X)$ and any jointly measurable version of $\rho_{t\xi}(\partial_\xi X)$.) Consequently, for a.e. ω, with $\sigma^2 = \mathrm{E}\,\xi^2$, $t \mapsto {:}e^{-t\xi}{:}\, \rho_{t\xi} X = e^{-t\xi - t^2\sigma^2/2} \rho_{t\xi} X$ is absolutely continuous and for a.e. t,

$$\begin{aligned} \frac{d}{dt}({:}e^{-t\xi}{:}\, \rho_{t\xi} X) &= e^{-t\xi - t^2\sigma^2/2} \rho_{t\xi}(\partial_\xi X) - (\xi + t\sigma^2) e^{-t\xi - t^2\sigma^2/2} \rho_{t\xi} X \\ &= {:}e^{-t\xi}{:}\, \rho_{t\xi}(\partial_\xi X - \xi X). \end{aligned}$$

Consequently, for any real t,

$$ {:}e^{-t\xi}{:}\, \rho_{t\xi} X - X = \int_0^t {:}e^{-s\xi}{:}\, \rho_{s\xi}(\partial_\xi X - \xi X)\, ds, \quad \text{a.s.} \tag{15.24}$$

Since Theorem 14.1(v) yields ${:}e^{-s\xi}{:}\, \rho_{s\xi}(\partial_\xi X - \xi X) \in L^1$ with constant norm (viz. $\|\partial_\xi X - \xi X\|_1$), the integral in (15.24) converges also as a Bochner integral in L^1. Fubini's theorem yields, using Theorem 14.1(v) again,

$$\begin{aligned} 0 = \mathrm{E}({:}e^{-t\xi}{:}\, \rho_{t\xi} X) - \mathrm{E}\, X &= \int_0^t \mathrm{E}\big({:}e^{-s\xi}{:}\, \rho_{s\xi}(\partial_\xi X - \xi X)\big)\, ds \\ &= \int_0^t \mathrm{E}(\partial_\xi X - \xi X)\, ds = t\, \mathrm{E}(\partial_\xi X - \xi X). \end{aligned}$$

The result follows by taking $t = 1$. □

THEOREM 15.106. *Let $1 \le p < \infty$ and $\xi \in H$. Then $\mathcal{P}$ is dense in $\mathcal{D}^{\xi,p}$.*

PROOF. By Theorem 15.104, it remains only to consider $p = 1$.

Since the mapping $X \mapsto (X, \partial_\xi X)$ is an isomorphism of $\mathcal{D}^{\xi,1}$ onto the closed subspace $\tilde{\mathcal{D}}^{\xi,1}$ of $L^1 \times L^1$, cf. Remark 15.77, the Hahn–Banach theorem shows that every continuous linear functional on $\mathcal{D}^{\xi,1}$ can be written as $X \mapsto \mathrm{E}(UX) + \mathrm{E}(V\partial_\xi X)$ for some $U, V \in L^\infty$. By the Hahn–Banach theorem again, it suffices to show that if such a linear functional vanishes on $\mathcal{P}$, then it vanishes identically on $\mathcal{D}^{\xi,1}$.

Thus, let $(U, V) \in L^\infty \times L^\infty$ with $\mathrm{E}(UZ) + \mathrm{E}(V\partial_\xi Z) = 0$ for every $Z \in \mathcal{P}$. Since $U, V \in L^2$, this condition can be written

$$\langle U, Z\rangle_{L^2} = -\langle V, \partial_\xi Z\rangle_{L^2} = -\langle V, A(\xi)Z\rangle_{L^2}, \qquad Z \in \mathcal{P}.$$

By Theorem 13.7(i),(vi),(v), $V \in \mathrm{Dom}(A^*(\xi)) = \mathrm{Dom}(A(\xi))$ and $A^*(\xi)V = -U$. Hence Theorem 15.98 shows that $V \in \mathcal{D}^{\xi,2}$, and, using Theorem 13.7(vii),

$$\partial_\xi V = A(\xi)V = \xi V - A^*(\xi)V = \xi V + U.$$

In particular, V is ξ-a.c.

Let $X \in \mathcal{D}^{\xi,1}$. By Theorem 15.22, XV is ξ-a.c., and by Theorem 15.7,

$$\partial_\xi(XV) = X\partial_\xi V + V\partial_\xi X = X(\xi V + U) + V\partial_\xi X;$$

thus

$$\partial_\xi(XV) - \xi(XV) = XU + V\partial_\xi X.$$

Since $X, \partial_\xi X \in L^1$, and $U, V \in L^\infty$, XV and $\partial_\xi(XV) - \xi(XV)$ belong to L^1, and Lemma 15.105 yields

$$0 = \mathrm{E}\big(\partial_\xi(XV) - \xi(XV)\big) = \mathrm{E}(XU + V\partial_\xi X) = \mathrm{E}(UX) + \mathrm{E}(V\partial_\xi X),$$

as required. □

In order to treat $\mathcal{D}^{k,1}$, we first study the finite-dimensional case.

EXAMPLE 15.107. Consider again the d-dimensional Gaussian Hilbert space consisting of the linear functions on $(\mathbb{R}^d, \gamma^d)$, where γ is the standard Gaussian measure on $\mathbb{R}$ and $d \geq 1$. Let $\{\xi_i\}_1^d$ be the standard orthonormal basis in H consisting of the coordinate functions. The $\mathcal{F}(H)$-measurable functions are just the measurable functions on $\mathbb{R}^d$, and $L^p = L^p(\mathbb{R}^d, \gamma^d)$. (Note that a random variable f may also be written $f(\xi_1, \dots, \xi_d)$.)

Let $1 \leq p < \infty$ and $\xi = \sum_1^d a_i\xi_i \in H$. If f is a polynomial, then, by Example 15.5,

$$\partial_\xi f = \sum_{i=1}^d a_i \partial_i f. \tag{15.25}$$

More generally, if $f \in \mathcal{D}^{\xi,p}$, then $\partial_\xi f = g$ for some function $g \in L^p(\mathbb{R}^d, \gamma^d)$, and Theorem 15.106 shows that there exist polynomials p_n on $\mathbb{R}^d$ such that $p_n \to f$ in $\mathcal{D}^{\xi,p}$. Thus, using (15.25), $p_n \to f$ and $\partial_\xi p_n = \sum_i a_i\partial_i p_n \to g$ in $L^p(\mathbb{R}^d, \gamma^d)$. Since convergence in $L^p(\mathbb{R}^d, \gamma^d)$ implies convergence as distributions in $\mathcal{D}'(\mathbb{R}^d)$, and each partial derivative ∂_i is a continuous operator on $\mathcal{D}'(\mathbb{R}^d)$, it follows that $g = \sum_i a_i \partial_i f$ as distributions. In other words, (15.25) holds for every $f \in \mathcal{D}^{\xi,p}$ with the derivatives interpreted as distribution derivatives. (See e.g. Rudin (1991) or Treves (1967) for the definition of distributions and distribution derivatives.)

Since $\mathcal{D}^{1,p} \subset \mathcal{D}^{\xi,p}$ for every $\xi \in H$, it follows that if $f \in \mathcal{D}^{1,p}$, then all distribution derivatives $\partial_i f \in L^p(\mathbb{R}^d, \gamma^d)$.

Furthermore, if $f \in \mathcal{D}^{k,p}$ with $k \geq 2$, then $\partial_i f = \langle \nabla f, \xi_i \rangle_{H_\mathbb{C}} \in \mathcal{D}^{k-1,p}$, and, by induction, $\partial^\alpha f \in L^p$ for every multi-index α with $|\alpha| \leq k$.

Conversely, let $1 \leq p < \infty$, and suppose that $f \in L^p(\mathbb{R}^d, \gamma^d)$ is such that the distributional derivatives $\partial_i f \in L^p(\mathbb{R}^d, \gamma^d)$. Choose some $\psi \in C_0^\infty(\mathbb{R}^d)$ with $\psi(0) = 1$ and $\int_{\mathbb{R}^d} \psi(x)\,dx = 1$, and let $f_N(x) = f(x)\psi(x/N)$, $N \geq 1$. Then each f_N has compact support and thus $f_N \in L^p(\mathbb{R}^d, dx)$. Furthermore, define, for $\varepsilon > 0$, $\psi_\varepsilon(x) = \varepsilon^{-d}\psi(x/\varepsilon)$, and let $f_{N,\varepsilon} = f_N * \psi_\varepsilon \in C_0^\infty(\mathbb{R}^d)$. By Theorem 15.35 and Example 15.67, $f_{N,\varepsilon} \in \mathcal{D}^{1,p}$ with $\nabla f_{N,\varepsilon} = \sum_1^d \partial_i f_{N,\varepsilon} \otimes \xi_i$.

First, let $\varepsilon \to 0$. By a standard result on regularization, $f_{N,\varepsilon} \to f_N$ in $L^p(\mathbb{R}^d, dx)$, and thus also in $L^p(\mathbb{R}^d, \gamma^d)$. Moreover,

$$\partial_i f_N(x) = \partial_i f(x)\psi(x/N) + N^{-1} f(x) \partial_i \psi(x/N) \in L^p(\mathbb{R}^d, dx),$$

and thus, as $\varepsilon \to 0$, $\partial_i f_{N,\varepsilon} = \partial_i f_N * \psi_\varepsilon \to \partial_i f_N$ in $L^p(\mathbb{R}^d, dx) \subset L^p(\mathbb{R}^d, \gamma^d)$. Hence $\nabla f_{N,\varepsilon} \to \sum_1^d \partial_i f_N \otimes \xi_i$ in $L^p(H_\mathbb{C})$.

By Theorem 15.72, $f_N \in \mathcal{D}^{1,p}$ with $\nabla f_N = \sum_1^d \partial_i f_N \otimes \xi_i$.

Next, let $N \to \infty$. Then, by dominated convergence, $f_N \to f$ and $\partial_i f_N \to \partial_i f$ in $L^p(\mathbb{R}^d, \gamma^d)$, and it follows, again by Theorem 15.72, that $f \in \mathcal{D}^{1,p}$.

Consequently, if $1 \leq p < \infty$,

$$\mathcal{D}^{1,p} = \{f \in L^p(\mathbb{R}^d, \gamma^d) : \partial_i f \in L^p(\mathbb{R}^d, \gamma^d) \text{ for } i = 1, \dots, d\},$$

with ∂_i interpreted as a distributional derivative. In fact, this extends to $p = \infty$, by the case $p < \infty$ and Theorem 15.64(v). The same argument shows also that if $\xi = \sum_i a_i \xi_i \in H$ and $1 \leq p \leq \infty$, then

$$\mathcal{D}^{\xi,p} = \{f \in L^p(\mathbb{R}^d, \gamma^d) : \sum_i a_i \partial_i f \in L^p(\mathbb{R}^d, \gamma^d)\}$$

and, if $k \geq 1$ and $1 \leq p \leq \infty$,

$$\mathcal{D}^{k,p} = \{f \in L^p(\mathbb{R}^d, \gamma^d) : \partial^\alpha f \in L^p(\mathbb{R}^d, \gamma^d) \text{ for every } \alpha \text{ with } |\alpha| \leq k\}.$$

We observe also that, since $f_{N,\varepsilon}$ above belongs to $C_0^\infty(\mathbb{R}^d)$, the argument above shows that $C_0^\infty(\mathbb{R}^d)$ is dense in $\mathcal{D}^{k,p}$ when $k \geq 0$ and $1 \leq p < \infty$.

Finally, we can now prove the main density theorem; we include several alternatives.

THEOREM 15.108. *Let $1 \leq p < \infty$ and $k \geq 0$. Then each of the following is a dense subspace of $\mathcal{D}^{k,p}$.*

(i) *The set $\mathcal{P}$ of polynomial variables.*

(ii) *The set* $\mathcal{S}_0$ *of variables of the form* $f(\xi_1, \dots, \xi_n)$, *with* $n \geq 1$, $f \in C_0^\infty(\mathbb{R}^n)$ *and* $\xi_1, \dots, \xi_n \in H$.
(iii) $\mathcal{D}^{l,q}$, *for every* l *and* q *with* $l \geq k$ *and* $p \leq q \leq \infty$.
(iv) $\mathcal{D}^{\infty-}$.

PROOF. Observe first that all four sets are linear spaces and that they are contained in $\mathcal{D}^{k,p}$ by Examples 15.66 and 15.67 and Theorem 15.64(i).

Consider first $\mathcal{S}_0$. Note first that we have already shown in Example 15.107 that (a subset of) $\mathcal{S}_0$ is dense for the special d-dimensional Gaussian Hilbert space considered there. Since every d-dimensional Gaussian Hilbert space is isomorphic to this space (and the case $d = 0$ is trivial), it follows that $\mathcal{S}_0$ is dense in $\mathcal{D}^{k,p}$ for every finite-dimensional Gaussian Hilbert space.

Now suppose that our Gaussian Hilbert space H has infinite dimension. Let $X \in \mathcal{D}^{k,p}$, and let $(\xi_k)_1^\infty$ be a sequence in H such that X is measurable with respect to the σ-field generated by $\{\xi_k\}_1^\infty$. (If H is separable, we may for example take any orthonormal basis.) Let H_n be the subspace of H spanned by $\{\xi_k\}_1^n$, and $X_n = \mathrm{E}(X \mid \mathcal{F}(H_n))$. Then $X_n \in \mathcal{D}^{k,p}$ for H_n by Lemma 15.95, and thus, since the already proved finite-dimensional case applies to H_n, X_n belongs to the closed hull $\overline{\mathcal{S}_0}$ of $\mathcal{S}_0$. Furthermore, Theorem 15.96 shows that $X_n \to X$ in $\mathcal{D}^{k,p}$, and thus X too belongs to $\overline{\mathcal{S}_0}$.

This completes the proof for $\mathcal{S}_0$. Since $\mathcal{S}_0 \subset \mathcal{D}^{l,q}$ and $\mathcal{S}_0 \subset \mathcal{D}^{\infty-}$ by Example 15.67, these spaces too are dense in $\mathcal{D}^{k,p}$.

Finally, we consider $\mathcal{P}$. We have already shown in Theorem 15.104 that $\mathcal{P}$ is dense in $\mathcal{D}^{k,p}$ when $1 < p < \infty$; thus we now only have to treat $p = 1$. Let $\mathcal{D}_0^{k,1}$ be the closed hull of $\mathcal{P}$ in $\mathcal{D}^{k,1}$. Since $\mathcal{P}$ is dense in $\mathcal{D}^{k,2}$ and the inclusion $\mathcal{D}^{k,2} \subset \mathcal{D}^{k,1}$ is continuous, $\mathcal{D}_0^{k,1} \supset \mathcal{D}^{k,2}$; moreover, we have just shown that $\mathcal{D}^{k,2}$ is dense in $\mathcal{D}^{k,1}$, and thus $\mathcal{D}_0^{k,1} = \mathcal{D}^{k,1}$. □

COROLLARY 15.109. *$\mathcal{P}$ and $\mathcal{S}_0$ are dense in $\mathcal{D}^{\infty-}$.* □

THEOREM 15.110. *Let $1 \leq p < \infty$ and $k \geq 0$. The families $\{e^\xi \mid \xi \in H\}$ and $\{:e^\xi: \mid \xi \in H\}$ are total in $\mathcal{D}^{k,p}$. In the complex case, the same is true for the families $\{e^{i\xi} \mid \xi \in H\}$, $\{:e^{i\xi}: \mid \xi \in H\}$, $\{e^\zeta \mid \zeta \in H_{\mathbb{C}}\}$ and $\{:e^\zeta: \mid \zeta \in H_{\mathbb{C}}\}$.*

PROOF. Since a Wick exponential $:e^\zeta:$ is a constant times the exponential e^ζ, see Theorem 3.33, it suffices to consider the families with Wick exponentials. Note that the families are subsets of $\mathcal{D}^{k,p}$, see Example 15.68.

We use complex scalars. Let A denote one of the families in the statement, and let M be the closed linear hull of A in $\mathcal{D}^{k,p}$.

Let $\xi \in H$. Then $z \mapsto F(z) = :e^{z\xi}:$ is an entire analytic function $\mathbb{C} \to \mathcal{D}^{k,p}$, as is easily shown directly or by applying Theorem 15.92(ii) to $:e^{z\xi}: = M_{z/R} :e^{R\xi}:$ for large R.

For the family $\{:e^{i\xi}:\}$ we have $F(iz) \in A \subset M$ for every real z, and for the other two families $F(z) \in A \subset M$ when $z \in \mathbb{R}$. Thus Theorem G.1 (applied to $F(iz)$ or $F(z)$) shows that in all three cases, $F(z) \in M$ for every complex z.

Moreover, since $F(z)$ has the Taylor expansion $\sum_0^\infty \frac{1}{n!} :\xi^n : z^n$, Theorem G.2 yields $:\xi^n: \in M$ for every $n \geq 0$. Theorem 3.29 now yields $\mathcal{P} \subset M$, and thus Theorem 15.108 yields $M = \mathcal{D}^{k,p}$. □

REMARK 15.111. It follows that these families are total in $\mathcal{D}^{\infty,p}$, $1 \leq p < \infty$, and in $\mathcal{D}^{\infty-}$ too.

COROLLARY 15.112. *If* $1 \leq p < \infty$ *and* $\xi \in H$, *then the families in Theorem* 15.110 *are total in* $\mathcal{D}^{\xi,p}$ *too.*

PROOF. This follows from Theorem 15.110 since $\mathcal{D}^{1,p}$ is a dense subspace of $\mathcal{D}^{\xi,p}$ (by Theorem 15.106) and the inclusion is continuous. □

REMARK 15.113. The density results above fail for $p = \infty$. First, it is obvious that $\mathcal{D}^{k,\infty}$ does not even contain any variables in $\mathcal{P}$ or $\{e^\xi\}$ except constants, since non-constant such variables are unbounded. Secondly, although $\mathcal{S}_0$ and $\{e^{i\xi}\}$ are contained in $\mathcal{D}^{k,\infty}$ by Examples 15.67 and 15.68, they are not dense or total even in the one-dimensional case, since their closed linear hulls in $\mathcal{D}^{k,\infty}$ in this case contain only variables $f(\xi)$ where $\xi \in H$ and f has k continuous derivatives.

However, the proof above (projection to finite dimension, truncation and regularization) shows that $\mathcal{S}_0$ is dense in $\mathcal{D}^{k,\infty}$ in the weak topology defined by the linear functionals $X \mapsto \sum_0^k \mathrm{E}\langle \nabla^j X, Y_j \rangle_{H_{\mathbb{C}}^{\otimes j}}$ with $Y_j \in L^1(H_{\mathbb{C}}^{\otimes j})$. Note that if we identify $\mathcal{D}^{k,\infty}$ and $\tilde{\mathcal{D}}^{k,\infty}$, this weak topology is the restriction to $\tilde{\mathcal{D}}^{k,\infty}$ of the natural weak*-topology on $\bigoplus_0^k L^\infty(H_{\mathbb{C}}^{\otimes j})$, so it is reasonable to call it *the weak*-topology* on $\mathcal{D}^{k,\infty}$. (The reader interested in functional analysis may further observe that $\tilde{\mathcal{D}}^{k,\infty}$ is a closed subspace of $\bigoplus_0^k L^\infty(H_{\mathbb{C}}^{\otimes j})$ in the weak*-topology, which is easily shown using $\tilde{\mathcal{D}}^{k,\infty} = \tilde{\mathcal{D}}^{k,2} \cap \bigoplus_0^k L^\infty(H_{\mathbb{C}}^{\otimes j})$; it follows that $\tilde{\mathcal{D}}^{k,\infty}$, and thus also $\mathcal{D}^{k,\infty}$, may be regarded as the dual space of a quotient space of $\bigoplus_0^k L^1(H_{\mathbb{C}}^{\otimes j})$, and the just defined weak*-topology on $\mathcal{D}^{k,\infty}$ is the weak*-topology for this duality.)

Moreover, $\{e^{i\xi}\}_{\xi \in H}$ and $\{:e^{i\xi}:\}_{\xi \in H}$ are total subsets of $\mathcal{D}^{k,\infty}$ in the weak*-topology; this can by shown by a Fourier expansion of elements in $\mathcal{S}_0$ as in the special case $k = 1$ in the proof of Theorem 15.131 below.

REMARK 15.114. The density results above extend to the Hilbert-space-valued case. In fact, if $X \in \mathcal{D}^{k,p}(H_0)$ with $k \geq 0$ and $1 \leq p < \infty$, let $H_1, H_2, \dots$ be an increasing sequence of finite-dimensional subspaces of H_0 such that $X \in \overline{\bigcup_1^\infty H_n}$ a.s., and let Q_n be the orthogonal projection onto H_n. Then it is easily seen, using Theorem 15.64(xi), that $Q_n X \in \mathcal{D}^{k,p}(H_n)$ and $Q_n X \to X$ in $\mathcal{D}^{k,p}(H_0)$. It follows that the set $\{X \otimes h : X \in \mathcal{D}^{k,p},\ h \in H_0\}$ is total in $\mathcal{D}^{k,p}(H_0)$; thus by Theorem 15.108 e.g. $\mathcal{P} \otimes H_0$ is dense in $\mathcal{D}^{k,p}(H_0)$.

8. The Meyer inequalities

The results in Section 6 for the case $p = 2$ have important extensions to the general case $1 < p < \infty$.

We let c_p and C_p in this section denote various positive constants that depend on p only; the meaning may shift from one occurrence to the next. Similarly, $C_{p,k}$ denotes various constants that depend on p and k. We use $\asymp$ to denote equivalence within constants of the type $C_{p,k}$.

We proceed in several steps; the main part of the proof is the following partial result.

LEMMA 15.115. *If $1 < p < \infty$ and $X \in \mathcal{P}$, then*

$$\|\nabla X\|_{L^p(H_{\mathbf{C}})} \leq C_p \|\mathcal{N}^{1/2} X\|_p.$$

PROOF. For notational convenience we consider only real variables; the complex case follows by taking the real and imaginary parts separately. Moreover, we assume that the Gaussian Hilbert space H is indexed by a Hilbert space H_1; i.e. that there exists an isometry $h \mapsto \xi_h$ of a Hilbert space H_1 onto H. (This is no loss of generality; we may for example take $H_1 = H$.) Let H' be a second Gaussian Hilbert space indexed by H_1, such that the variables in H and H' are independent; this is another way of saying that $H \oplus H'$ is a Gaussian Hilbert space indexed by $H_1 \oplus H_1$. Denote the defining isometry $H_1 \to H'$ by $h \mapsto \xi'_h$.

Define, for $-\infty < \theta < \infty$, a linear operator ρ_θ in $H \oplus H'$ by

$$\rho_\theta(\xi_h) = \cos\theta \cdot \xi_h + \sin\theta \cdot \xi'_h,$$
$$\rho_\theta(\xi'_h) = -\sin\theta \cdot \xi_h + \cos\theta \cdot \xi'_h,$$

extended by linearity. It is easily seen that ρ_θ is an isometry of $H \oplus H'$ onto itself with inverse $\rho_{-\theta}$; moreover $\rho_\theta \rho_\tau = \rho_{\theta+\tau}$ for all real θ and τ. (ρ_θ can be interpreted as a rotation.)

The operator $R_\theta = \Gamma(\rho_\theta)$ maps $\mathcal{P}(H \oplus H')$ into itself, and preserves distributions by Theorem 4.11. In particular, $\|R_\theta X\|_p = \|X\|_p$ for every $X \in \mathcal{P}(H \oplus H')$.

Let π'_1 denote the orthogonal projection $\mathcal{P}(H') \to H'$ (or $H'_{\mathbf{C}}$), and extend it to a linear operator, also denoted π'_1, in $\mathcal{P}(H \oplus H')$ by

$$\pi'_1(XY) = X\pi'_1(Y) \qquad \text{for } X \in \mathcal{P}(H),\ Y \in \mathcal{P}(H').$$

(This is well-defined; in fact, $\mathcal{P}(H \oplus H')$ is the algebraic tensor product of $\mathcal{P}(H)$ and $\mathcal{P}(H')$, and this operator is the tensor product $I \otimes \pi'_1$, cf. Appendix E, in particular Example E.6.)

Consider now the gradient operator ∇ on $\mathcal{P}(H)$. By our standard definition above, it maps into $\mathcal{P}(H) \otimes H \subset L^0(H)$, but in this proof we use the isometry $\xi_h \mapsto \xi'_h$ of H onto H' and regard ∇ as a map into $\mathcal{P}(H) \otimes H' \subset \mathcal{P}(H \oplus H')$. Thus, for any $\xi_h \in H$, cf. Example 15.31,

$$\nabla(:\xi_h^n:) = n:\xi_h^{n-1}:\xi'_h.$$

Furthermore, by the definition of $R_\theta = \Gamma(\rho_\theta)$, using also Theorem 3.20,

$$\begin{aligned} R_\theta(:\xi_h^n:) &= :(\rho_\theta \xi_h)^n: = :(\cos\theta \cdot \xi_h + \sin\theta \cdot \xi_h')^n: \\ &= \sum_{k=0}^n \binom{n}{k} \cos^{n-k}\theta \sin^k\theta \, :\xi_h^{n-k}: \, :\xi_h'^{\,k}: \end{aligned}$$

and thus

$$\pi_1'\big(R_\theta(:\xi_h^n:)\big) = n\cos^{n-1}\theta\sin\theta \, :\xi_h^{n-1}: \xi_h' = \cos^{n-1}\theta\sin\theta \, \nabla(:\xi_h^n:).$$

By polarization, cf. Theorem D.1, it follows that

$$\pi_1' R_\theta X = \cos^{n-1}\theta \sin\theta \, \nabla X \tag{15.26}$$

for every $X \in \mathcal{P}(H) \cap H^{:n:}$.

Let

$$\varphi(\theta) = \begin{cases} (4\pi)^{-1/2} |\ln\cos\theta|^{-1/2} \operatorname{sign}\theta, & 0 < |\theta| < \pi/2, \\ 0, & \pi/2 \le |\theta| \le \pi. \end{cases}$$

The function φ is singular at 0; it is easily seen that $\varphi(\theta) = (2\pi)^{-1/2}\theta^{-1} + O(\theta)$. Hence φ is not integrable on $(-\pi, \pi)$, but if g is, for example, a differentiable function on $[-\pi, \pi]$, then the principal value integral

$$\text{p.v.} \int_{-\pi}^{\pi} \varphi(\theta) g(\theta)\, d\theta = \lim_{\varepsilon \to 0} \int_{\varepsilon < |\theta| < \pi} \varphi(\theta) g(\theta)\, d\theta = \int_0^\pi \varphi(\theta)\big(g(\theta) - g(-\theta)\big)\, d\theta$$

exists (note that φ is odd). We can write

$$\varphi(\theta) = c \cot\tfrac{\theta}{2} + \varphi_1(\theta),$$

for some constant c $(= (8\pi)^{-1/2})$ and a bounded function φ_1. Since $\varphi_1 \in L^\infty(\mathbb{T}) \subset L^1(\mathbb{T})$, convolution with φ_1 is a bounded operator on $L^p(\mathbb{T})$; moreover, convolution with $\cot\frac{\theta}{2}$, in the principal value sense, is also bounded on $L^p(\mathbb{T})$ by Riesz' theorem (Riesz 1926, Zygmund 1959, VII.(2.4)). Hence, if g is smooth, the function

$$\varphi * g(\tau) = \text{p.v.} \int_{-\pi}^{\pi} \varphi(\theta) g(\tau - \theta)\, d\theta = \int_0^\pi \varphi(\theta)\big(g(\tau - \theta) - g(\tau + \theta)\big)\, d\theta$$

is well-defined and $\|\varphi * g\|_{L^p(\mathbb{T})} \le C_p \|g\|_{L^p(\mathbb{T})}$.

Now define, for $X \in \mathcal{P}(H) \subset \mathcal{P}(H \oplus H')$,

$$F(X) = \int_0^\pi \varphi(\theta)\big(R_\theta(X) - R_{-\theta}(X)\big)\, d\theta.$$

(The integral converges in L^p and $F(X) \in \mathcal{P}(H \oplus H')$, because $R_\theta(X)$ is a trigonometric polynomial in θ.) Thus, letting $R.(X)$ denote $R_\theta(X)$ as a function of θ,

$$R_\tau F(X) = \int_0^\pi \varphi(\theta)\big(R_{\tau+\theta}(X) - R_{\tau-\theta}(X)\big)\, d\theta = -\varphi * R.(X)(\tau)$$

and hence

$$\int_0^{2\pi} |R_\tau F(X)|^p \, d\tau \le C_p \int_0^{2\pi} |R_\theta(X)|^p \, d\theta. \tag{15.27}$$

We take expectations; since R_τ is an isometry in L^p,

$$\mathrm{E} \int_0^{2\pi} |R_\tau F(X)|^p \, d\tau = \int_0^{2\pi} \|R_\tau F(X)\|_p^p \, d\tau = 2\pi \|F(X)\|_p^p$$

and similarly for the right hand side of (15.27). Hence, (15.27) implies

$$\|F(X)\|_p \le C_p \|X\|_p, \tag{15.28}$$

i.e. the operator F is bounded in L^p.

If $X \in \mathcal{P}(H) \cap H^{:n:}$, with $n \ge 1$, then, using the change of variables $\cos\theta = e^{-t^2}$ and (15.26),

$$\begin{aligned}
\pi_1' F(X) &= \int_0^{\pi} \varphi(\theta)\big(\pi_1' R_\theta(X) - \pi_1' R_{-\theta}(X)\big) \, d\theta \\
&= \int_0^{\pi/2} 2\varphi(\theta) \cos^{n-1}\theta \sin\theta \, \nabla X \, d\theta \\
&= \frac{2}{\sqrt{\pi}} \int_0^{\infty} e^{-nt^2} dt \, \nabla X \\
&= n^{-1/2} \nabla X = \nabla \mathcal{N}^{-1/2} X.
\end{aligned}$$

By linearity, for every $X \in \mathcal{P}(H)$ with $\mathrm{E}\, X = \pi_0(X) = 0$,

$$\pi_1' F(X) = \nabla \mathcal{N}^{-1/2} X.$$

For the final step in the proof, it is convenient to assume that H is defined on one probability space $(\Omega, \mathcal{F}, \mathrm{P})$, H' on another probability space $(\Omega', \mathcal{F}', \mathrm{P}')$, and $H \oplus H'$ on the product space $\Omega \times \Omega'$. Since the norm on H' is proportional to the L^p-norm by (1.1), and the projection π_1' is bounded on $L^p(\Omega', \mathcal{F}', \mathrm{P}')$ by Theorem 5.14, we then obtain, for any $X \in \mathcal{P}(H)$ with $\mathrm{E}\, X = 0$ and a.e. $\omega \in \Omega$,

$$\|\nabla \mathcal{N}^{-1/2} X\|_{H'} = \|\pi_1' F(X)\|_{H'} = C_p \|\pi_1' F(X)\|_{L^p(\Omega')} \le C_p \|F(X)\|_{L^p(\Omega')},$$

and thus by Fubini's theorem

$$\|\nabla \mathcal{N}^{-1/2} X\|_{L^p(H')} \le C_p \big\| \|F(X)\|_{L^p(\Omega')} \big\|_{L^p(\Omega)} = C_p \|F(X)\|_p. \tag{15.29}$$

The result follows by combining (15.29) and (15.28), and replacing X by $\mathcal{N}^{1/2} X$. □

The converse to Lemma 15.115 follows by duality.

LEMMA 15.116. *If $1 < p < \infty$ and $X \in \mathcal{P}$, then*

$$\|\mathcal{N}^{1/2} X\|_p \le C_p \|\nabla X\|_{L^p(H_{\mathrm{C}})}.$$

PROOF. Let $Y \in \mathcal{P}$. Then by Theorem 15.100 and polarization,

$$\langle \mathcal{N}^{1/2}X, \mathcal{N}^{1/2}Y\rangle = \langle \nabla X, \nabla Y\rangle_{L^2(H_{\mathbb{C}})} = \mathrm{E}\langle \nabla X, \nabla Y\rangle_{H_{\mathbb{C}}}.$$

The Cauchy–Schwarz and Hölder inequalities and Lemma 15.115 yield, with q equal to the conjugate exponent $p/(p-1)$,

$$|\langle \mathcal{N}^{1/2}X, \mathcal{N}^{1/2}Y\rangle| \le \|\nabla X\|_{L^p(H_{\mathbb{C}})}\|\nabla Y\|_{L^q(H_{\mathbb{C}})} \le C_q\|\nabla X\|_{L^p(H_{\mathbb{C}})}\|\mathcal{N}^{1/2}Y\|_q.$$

If $Z \in \mathcal{P}$, we may take $Y = \mathcal{N}^{-1/2}(Z - \mathrm{E}\,Z)$ and thus $\mathcal{N}^{1/2}Y = Z - \mathrm{E}\,Z$. Then $\langle \mathcal{N}^{1/2}X, \mathcal{N}^{1/2}Y\rangle = \langle \mathcal{N}^{1/2}X, Z - \mathrm{E}\,Z\rangle = \langle \mathcal{N}^{1/2}X, Z\rangle$ and $\|\mathcal{N}^{1/2}Y\|_q = \|Z - \mathrm{E}\,Z\|_q \le 2\|Z\|_q$, and we thus obtain

$$|\langle \mathcal{N}^{1/2}X, Z\rangle| \le C_p\|\nabla X\|_{L^p(H_{\mathbb{C}})}\|Z\|_q, \qquad Z \in \mathcal{P},$$

and the result follows. □

We next extend these result to Hilbert-space-valued random variables.

Combining Lemmas 15.115 and 15.116 and Theorem 12.1 (for the operator $\nabla\mathcal{N}^{-1/2}$), we obtain the following.

LEMMA 15.117. *If $1 < p < \infty$, H_0 is a Hilbert space and $X \in \mathcal{P} \otimes H_0$, then*

$$\|\nabla X\|_{L^p(H\otimes H_0)} \asymp \|\mathcal{N}^{1/2}X\|_{L^p(H_0)}.$$

□

This is easily extended to higher derivatives. We write $(\mathcal{N})_k = \mathcal{N}(\mathcal{N} - 1)\cdots(\mathcal{N} - k + 1)$.

LEMMA 15.118. *If $1 < p < \infty$, $k \ge 0$, H_0 is a Hilbert space and $X \in \mathcal{P} \otimes H_0$, then*

$$\|\nabla^k X\|_{L^p(H^{\otimes k}\otimes H_0)} \asymp \|(\mathcal{N})_k^{1/2}X\|_{L^p(H_0)}.$$

PROOF. This follows from Lemma 15.117 by induction as in the proof of Theorem 15.102. □

REMARK 15.119. The result for higher derivatives may also be obtained by taking higher projections π_k in the proof of Lemma 15.115, see Pisier (1988).

The final step is to extend this equivalence to other variables than polynomial ones. We also use Theorem 5.47 to replace the term $(\mathcal{N})_k^{1/2}$ by simpler expressions.

THEOREM 15.120. *Suppose that $1 < p < \infty$ and $k \ge 0$. If $X \in L^p$, then $X \in \mathcal{D}^{k,p}$ if and only if $\mathcal{N}^{k/2}X$ exists in L^p (or, equivalently, $(\mathcal{N}+1)^{k/2}X$ exists in L^p). Moreover, we then have*

$$\|X\|_{k,p} \asymp \|(\mathcal{N}+1)^{k/2}X\|_p \asymp \|X\|_p + \|\mathcal{N}^{k/2}X\|_p \asymp |\mathrm{E}\,X| + \|\mathcal{N}^{k/2}X\|_p, \tag{15.30}$$

$$\|X\|_{k,p} \asymp \|\nabla^k X\|_{L^p(H_{\mathbb{C}}^{\otimes k})} + \|X\|_p \tag{15.31}$$

and

$$\|\nabla^k X\|_{L^p(H_{\mathbb{C}}^{\otimes k})} \asymp \|(I - \pi_{\leq k-1})\mathcal{N}^{k/2}X\|_p \asymp \|(I - \pi_{\leq k-1})(\mathcal{N}+1)^{k/2}X\|_p. \tag{15.32}$$

PROOF. We begin by proving the equivalences (15.30)–(15.32) for polynomial variables. Thus, assume for the time being $X \in \mathcal{P}$.

We apply Theorem 5.47 repeatedly. First, let $f(n) = n^{k/2}\mathbf{1}[n \geq k]$ and $g(n) = (n)_k^{1/2}$; then $g(n) = 0 \iff n < k \iff f(n) = 0$, and $g(n)/f(n) = \prod_1^{k-1}(1 - j/n)^{1/2} = \varphi(1/n)$ for $n \geq k$, where $\varphi(z) = \prod_1^{k-1}(1 - jz)^{1/2}$ is analytic for $|z| < 1/(k-1)$. Since further $\varphi(0) \neq 0$, Theorem 5.47(iii) shows that

$$\|(\mathcal{N})_k^{1/2}X\|_p \asymp \|(I - \pi_{\leq k-1})\mathcal{N}^{k/2}X\|_p. \tag{15.33}$$

Similarly, Theorem 5.47(iii), with the same $f(n)$ but $g(n) = (n+1)^{k/2}\mathbf{1}[n \geq k]$, shows that

$$\|(I - \pi_{\leq k-1})\mathcal{N}^{k/2}X\|_p \asymp \|(I - \pi_{\leq k-1})(\mathcal{N}+1)^{k/2}X\|_p. \tag{15.34}$$

Lemma 15.118, (15.33) and (15.34) show that (15.32) holds for polynomial variables X.

In particular, since $\pi_{\leq k-1}$ is bounded on L^p,

$$\|\nabla^k X\|_{L^p(H_{\mathbb{C}}^{\otimes k})} \leq C_{p,k}\|(\mathcal{N}+1)^{k/2}X\|_p.$$

Furthermore, Theorem 5.39 implies that if $j \leq k$, then

$$\|(\mathcal{N}+1)^{j/2}X\|_p = \|(\mathcal{N}+1)^{-(k-j)/2}(\mathcal{N}+1)^{k/2}X\|_p \leq C_{p,k-j}\|(\mathcal{N}+1)^{k/2}X\|_p.$$

Consequently,

$$\begin{aligned} \|X\|_{k,p} &\leq \sum_{j=0}^{k} \|\nabla^j X\|_{L^p(H_{\mathbb{C}}^{\otimes j})} \leq \sum_{j=0}^{k} C_{p,j}\|(\mathcal{N}+1)^{j/2}X\|_p \\ &\leq C_{p,k}\|(\mathcal{N}+1)^{k/2}X\|_p. \end{aligned} \tag{15.35}$$

Conversely,

$$\begin{aligned} \|(\mathcal{N}+1)^{k/2}X\|_p &\leq \|(I - \pi_{\leq k-1})(\mathcal{N}+1)^{k/2}X\|_p + \sum_{n=0}^{k-1} \|\pi_n(\mathcal{N}+1)^{k/2}X\|_p \\ &\leq C_{p,k}\|\nabla^k X\|_{L^p(H_{\mathbb{C}}^{\otimes k})} + \sum_{n=0}^{k-1}(n+1)^{k/2}\|\pi_n X\|_p \\ &\leq C_{p,k}\|\nabla^k X\|_{L^p(H_{\mathbb{C}}^{\otimes k})} + C_{p,k}\|X\|_p \end{aligned}$$

and

$$\|\nabla^k X\|_{L^p(H_{\mathbb{C}}^{\otimes k})} + \|X\|_p \leq 2\|X\|_{k,p}. \tag{15.36}$$

Combining (15.35)–(15.36) we obtain the first equivalence in (15.30) and (15.31), still for polynomial variables. Moreover, by Theorem 5.47 again, $\|\mathcal{N}^{k/2}X\|_p \leq C_{p,k}\|(\mathcal{N}+1)^{k/2}X\|_p$, and

$$\begin{aligned}\|(\mathcal{N}+1)^{k/2}X\|_p &\leq \|(I-\pi_0)(\mathcal{N}+1)^{k/2}X\|_p + \|\pi_0 X\|_p\\ &\leq C_{p,k}\|\mathcal{N}^{k/2}X\|_p + |\operatorname{E}X|,\end{aligned}$$

and by Theorem 5.39

$$|\operatorname{E}X| \leq \|X\|_p \leq \|(\mathcal{N}+1)^{k/2}X\|_p.$$

Hence the remaining equivalences in (15.30) follow.

Since $\mathcal{D}^{k,p}$ is complete by Corollary 15.73 and the polynomial variables are dense by Theorem 15.104, $X \in \mathcal{D}^{k,p}$ if and only there exists a sequence X_n of polynomial variables which is a Cauchy sequence in the norm $\|\cdot\|_{k,p}$ and converges to X in L^p; it follows easily from the equivalences above for polynomial variables and Theorem 5.45 that this happens if and only if $\mathcal{N}^{k/2}X$ exists in L^p, cf. Definition 5.46. By Theorem 5.47, this space also equals the set of X such that $(\mathcal{N}+1)^{k/2}X$ exists in L^p.

The equivalences (15.30)–(15.32) hold for all $X \in \mathcal{D}^{k,p}$ by continuity. □

This result is due to Meyer (1984), and the equivalences are known as the *Meyer inequalities.* The proof above is due to Pisier (1988).

REMARK 15.121. The theorem extends to Hilbert-space-valued variables. The proof is the same, using Theorem 12.1 to extend the multiplier theorems.

REMARK 15.122. We have defined $\mathcal{D}^{k,p}$ for non-negative integers k, but we may now extend the definition to any real parameter, provided $1 < p < \infty$.

For $s \geq 0$ and $1 < p < \infty$ we define $\mathcal{D}^{s,p} = \{X \in L^p : (1+\mathcal{N})^{s/2}X \in L^p\}$, equipped with the norm $\|X\|'_{s,p} = \|(1+\mathcal{N})^{s/2}X\|_p$; this is a Banach space and when s is an integer, it is by Theorem 15.120 the same space as defined earlier (with an equivalent norm). This space can also be defined as the completion of $\mathcal{P}$ in the norm $\|\cdot\|'_{s,p}$, cf. Definition 5.46 and Theorem 5.45.

For $s < 0$, $\mathcal{D}^{s,p}$ can still be defined as the completion of $\mathcal{P}$ in the norm $\|(1+\mathcal{N})^{s/2}X\|_p$, but this is no longer a space of random variables. (Its elements can be interpreted as 'generalized random variables' or 'distributions'.) We then have the natural dualities $(\mathcal{D}^{s,p})^* = \mathcal{D}^{-s,p'}$, $-\infty < s < \infty$ and $1 < p < \infty$, and $(\mathcal{D}^{\infty-})^* = \bigcup_{-\infty<s<\infty,\,1<p<\infty} \mathcal{D}^{s,p}$.

For further results and applications, see Watanabe (1984); for other classes of generalized random variables, see Kuo (1996).

COROLLARY 15.123. $\overline{\mathcal{P}}_* \subset \mathcal{D}^{\infty-}$.

PROOF. If $X \in \overline{\mathcal{P}}_*$, then by Theorem 3.50, $\mathcal{N}^{k/2}X \in \overline{\mathcal{P}}_* \subset L^p$ for every $k \geq 0$ and $p < \infty$, and thus $X \in \mathcal{D}^{k,p}$. □

COROLLARY 15.124. *If $1 < p < \infty$ and $k \geq 2$, then $\mathcal{N}$ maps $\mathcal{D}^{k,p}$ continuously into $\mathcal{D}^{k-2,p}$.* □

COROLLARY 15.125. *If $X \in \mathcal{D}^{\infty-}$, then $\mathcal{N}X \in \mathcal{D}^{\infty-}$.* □

We saw in Section 5 that the Mehler transform preserves the Sobolev spaces. We can now show that it actually increases the smoothness dramatically.

COROLLARY 15.126. *If $1 < p < \infty$, $X \in L^p$ and $-1 < r < 1$, then the Mehler transform $M_r X \in \mathcal{D}^{\infty,p}$.*

PROOF. By Theorem 5.41, $\mathcal{N}^k r^{\mathcal{N}} X \in L^p$ for every $k \geq 0$, and thus $M_r X = r^{\mathcal{N}} X \in \mathcal{D}^{k,p}$ by Theorem 15.120. □

Although the norm $\|X\|_{k,p}$ was defined using $\nabla^j X$ for all $j \leq k$, (15.31) shows that (at least when $1 < p < \infty$) it suffices to consider $j = 0$ and $j = k$ and that the intermediate terms (for $k \geq 2$) may be ignored. The following results show further aspects of this.

COROLLARY 15.127. *Let $1 < p < \infty$ and $k \geq 0$. If $(X_n)_1^\infty$ is a sequence in $\mathcal{D}^{k,p}$ such that $X_n \to X$ in L^p and $\nabla^k X_n \to Y$ in $L^p(H_{\mathbb{C}}^{\otimes k})$ as $n \to \infty$, for some variables $X \in L^p$ and $Y \in L^p(H_{\mathbb{C}}^{\otimes k})$, then $X \in \mathcal{D}^{k,p}$ and $\nabla^k X = Y$. In other words, ∇^k, defined on the subspace $\mathcal{D}^{k,p}$ of L^p, is a closed densely defined operator $L^p \to L^p(H_{\mathbb{C}}^{\otimes k})$.*

PROOF. It follows from (15.31) that $(X_n)_1^\infty$ is a Cauchy sequence in $\mathcal{D}^{k,p}$. Hence $X_n \to Z$ in $\mathcal{D}^{k,p}$ for some $Z \in \mathcal{D}^{k,p}$, but then $X_n \to Z$ in L^p and $\nabla^k X_n \to \nabla^k Z$ in $L^p(H_{\mathbb{C}}^{\otimes k})$. Consequently, $X = Z \in \mathcal{D}^{k,p}$ and $\nabla^k X = \nabla^k Z = Y$. □

This corollary shows that $\mathcal{D}^{k,p}$ is (for $1 < p < \infty$ at least) the natural domain of ∇^k in L^p, for $k \geq 2$ too.

THEOREM 15.128. *Let $1 < p < \infty$ and $k \geq 0$. Then the following are equivalent.*

(i) $X \in \mathcal{D}^{k,p}$.
(ii) $X \in \mathcal{D}^{k,1}$, $X \in L^p$ *and* $\nabla^k X \in L^p(H_{\mathbb{C}}^{\otimes k})$.

PROOF. It is clear from the definitions that (i) $\Rightarrow$ (ii).

Conversely, suppose that (ii) holds. Let $0 < r < 1$. By Corollary 15.126, $M_r X \in \mathcal{D}^{k,p}$. On the other hand, Theorem 15.90 with $p = 1$, see Remark 15.91, shows that $\nabla^k(M_r X) = r^k M_r(\nabla^k X)$, which converges in $L^p(H_{\mathbb{C}}^{\otimes k})$ to $\nabla^k X$ as $r \to 1$ by the assumption $\nabla^k X \in L^p(H_{\mathbb{C}}^{\otimes k})$, see Example 12.5. Since also $M_r X \to X$ in L^p as $r \to 1$, Corollary 15.127 yields $X \in \mathcal{D}^{k,p}$. □

THEOREM 15.129. *Let $1 < p < \infty$ and $k \geq 0$. Then the following are equivalent.*

(i) $X \in \mathcal{D}^{k,p}$.

(ii) $X \in L^p$ *and there exists* $Y \in L^p(H_{\mathbb{C}}^{\otimes k})$ *such that* $\pi_n(Y) = \nabla^k(\pi_{n+k}X)$ *for each* $n \geq 0$.

Moreover, we then have $Y = \nabla^k X$.

PROOF. Suppose that (i) holds; then (ii) holds with $Y = \nabla^k X$ because $\pi_n \nabla^k = \nabla^k \pi_{n+k}$ on $\mathcal{P}$, and both $\pi_n \nabla^k$ and $\nabla^k \pi_{n+k}$ define continuous operators $\mathcal{D}^{k,p} \to L^p(H_{\mathbb{C}}^{\otimes k})$.

Conversely, suppose that (ii) holds. Let $0 < r < 1$. By Corollary 15.126, $M_r X \in \mathcal{D}^{k,p}$, and thus, using the already proved implication (i) $\Rightarrow$ (ii),

$$\begin{aligned}\pi_n(\nabla^k M_r X) &= \nabla^k(\pi_{n+k}(M_r X)) = r^{n+k}\nabla^k(\pi_{n+k}X) = r^{n+k}\pi_n Y \\ &= r^k \pi_n(M_r Y).\end{aligned}$$

Hence $\nabla^k M_r X = r^k M_r Y$. Consequently, as $r \to 1$, using Example 12.5, $M_r X \to X$ in L^p and $\nabla^k M_r X \to Y$ in $L^p(H_{\mathbb{C}}^{\otimes k})$, and the result follows by Corollary 15.127. □

9. The divergence operator

The operator ∇ restricted to $\mathcal{D}^{1,p}$, $1 < p < \infty$, is a densely defined operator $L^p_{\mathbb{C}} \to L^p(H_{\mathbb{C}})$; it thus has an adjoint which is a closed densely defined operator $L^{p'}(H_{\mathbb{C}}) \to L^{p'}_{\mathbb{C}}$, where, as throughout this section, p' is the conjugate exponent defined by $1/p + 1/p' = 1$. This adjoint is known as the *divergence operator*, and is denoted by δ.

In order to cover all p at once, and achieve greater generality, we make the following definition, which soon will be shown to include the case $1 < p < \infty$ just mentioned.

DEFINITION 15.130. Let $X \in L^1(H_{\mathbb{C}})$. Then δX *exists* (*in* L^1), and $\delta X = Y$, if there exists a random variable $Y \in L^1_{\mathbb{C}}$ such that $\mathrm{E}\langle X, \nabla Z\rangle_{H_{\mathbb{C}}} = \mathrm{E}(Y\overline{Z})$ for every $Z \in \mathcal{D}^{1,\infty}_{\mathbb{C}}$ (or, equivalently, $\mathrm{E}\langle X, \nabla Z\rangle_{H_{\mathbb{C}}} = \mathrm{E}(YZ)$ for every $Z \in \mathcal{D}^{1,\infty}_{\mathbb{R}}$). If further $X \in L^p(H_{\mathbb{C}})$ and $\delta X \in L^p_{\mathbb{C}}$, we say that δX *exists in* L^p.

Note that δX is uniquely defined, when it exists, by Lemma 2.7 and Example 15.68. Clearly, δ is a closed linear operator $L^1(H_{\mathbb{C}}) \to L^1_{\mathbb{C}}$. Moreover, if $X \in L^1(H)$ and δX exists, then $\delta X \in L^1_{\mathbb{R}}$. It is easy to see that δX exists in L^p for every finite p when $X \in \mathcal{P} \otimes H_{\mathbb{C}}$, see Example 15.136 below. Hence δ restricts to a closed densely defined operator $L^p(H) \to L^p_{\mathbb{R}}$ and $L^p(H_{\mathbb{C}}) \to L^p_{\mathbb{C}}$ for $1 \leq p < \infty$.

Observe also, by taking $Z = 1$ in the definition, that $\mathrm{E}\,\delta X = 0$ whenever δX exists.

We obtain some equivalent conditions by restricting Z above to suitable subsets of $\mathcal{D}^{1,\infty}$. Let $\mathcal{S}_0$ be as in Example 15.67.

THEOREM 15.131. *Let* $X \in L^1(H_{\mathbb{C}})$ *and* $Y \in L^1$. *Then the following are equivalent.*

(i) $\delta X = Y$.

(ii) $\mathrm{E}\langle X, \nabla Z\rangle_{H_{\mathbb{C}}} = \mathrm{E}(Y\overline{Z})$ *for every* $Z \in \mathcal{S}_0$.
(iii) $\mathrm{E}\big(i\langle X, \xi\rangle_{H_{\mathbb{C}}} e^{i\xi}\big) = \mathrm{E}(Ye^{i\xi})$ *for every* $\xi \in H$.

PROOF. (i) implies (ii) by the definition, since $\mathcal{S}_0 \subset \mathcal{D}^{1,\infty}$ by Example 15.67. Conversely, if (ii) holds, then the equality extends from $Z \in \mathcal{S}_0$ to general $Z \in \mathcal{D}^{1,\infty}$, and thus (i) holds, by the arguments in Example 15.107 and the proof of Theorem 15.108; this is equivalent to the claim in Remark 15.113 that $\mathcal{S}_0$ is weak*-dense in $\mathcal{D}^{1,\infty}$.

Similarly, (i) implies (iii) because $e^{-i\xi} \in \mathcal{D}^{1,\infty}_{\mathbb{C}}$ and $\nabla e^{-i\xi} = -ie^{-i\xi} \otimes \xi$ by Examples 15.68 and 15.32.

Finally, we show that (iii) $\Rightarrow$ (ii). Assume (iii), and let $Z \in \mathcal{S}_0$; thus $Z = \varphi(\xi_1, \dots, \xi_n)$ where $\varphi \in C_0^\infty(\mathbb{R}^n) \subset S(\mathbb{R}^n)$ and $\xi_1, \dots, \xi_n \in H$. Since the Fourier transform maps $S(\mathbb{R}^n)$ onto itself (Rudin 1991, Theorem 7.7), $\varphi(x) = \int_{\mathbb{R}^n} \psi(t) e^{it\cdot x}\, dt$ for some $\psi \in S(\mathbb{R}^n)$. It follows that

$$\partial_k \varphi(x) = \int_{\mathbb{R}^n} it_k \psi(t) e^{it\cdot x}\, dt, \qquad 1 \le k \le n.$$

Consequently,

$$Z = \int_{\mathbb{R}^n} \psi(t) e^{i\sum_1^n t_k \xi_k}\, dt$$

and, using Theorem 15.35,

$$\nabla Z = \sum_1^n \partial_k \varphi(\xi_1, \dots, \xi_n) \otimes \xi_k = \sum_1^n \int_{\mathbb{R}^n} it_k \psi(t) e^{i\sum_1^n t_k\xi_k}\, dt \otimes \xi_k.$$

Thus, using Fubini's theorem and the assumption,

$$\begin{aligned}
\mathrm{E}\langle X, \nabla\overline{Z}\rangle_{H_{\mathbb{C}}} &= \mathrm{E}\Big(\sum_1^n \int_{\mathbb{R}^n} it_k \psi(t) e^{i\sum_1^n t_k\xi_k} \langle X, \xi_k\rangle_{H_{\mathbb{C}}}\, dt\Big) \\
&= \int_{\mathbb{R}^n} \psi(t)\, \mathrm{E}\Big(ie^{i\sum_1^n t_k\xi_k} \langle X, \sum_1^n t_k\xi_k\rangle_{H_{\mathbb{C}}}\Big)\, dt \\
&= \int_{\mathbb{R}^n} \psi(t)\, \mathrm{E}\big(Ye^{i\sum_1^n t_k\xi_k}\big)\, dt = \mathrm{E}(YZ).
\end{aligned}$$

Hence (ii) holds. □

The next theorem gives further alternative characterizations of the divergence for the case $p > 1$, and shows that $\delta\colon L^p(H_{\mathbb{C}}) \to L^p_{\mathbb{C}}$ is indeed the adjoint of $\nabla\colon L^{p'}_{\mathbb{C}} \to L^{p'}(H_{\mathbb{C}})$. (For the real case, $\delta\colon L^p(H) \to L^p_{\mathbb{R}}$ is the adjoint of $\nabla\colon L^{p'}_{\mathbb{R}} \to L^{p'}(H)$.)

THEOREM 15.132. *Let* $1 < p \le \infty$ *and let* $X \in L^p(H_{\mathbb{C}})$, $Y \in L^p$. *Then the following are equivalent.*
(i) $\delta X = Y$.
(ii) $\mathrm{E}\langle X, \nabla Z\rangle_{H_{\mathbb{C}}} = \mathrm{E}(Y\overline{Z})$ *for every* $Z \in \mathcal{D}^{1,p'}$.

(iii) $\mathrm{E}\langle X, \nabla Z\rangle_{H_{\mathbb{C}}} = \mathrm{E}(YZ)$ *for every* $Z \in \mathcal{P}_{\mathbb{R}}$.
(iv) $\delta\pi_{n-1}(X) = \pi_n(Y)$ *for every* $n \geq 0$.

PROOF. (i) $\Leftrightarrow$ (ii) and (iii) $\Leftrightarrow$ (ii) are immediate consequences of the definition and the facts that $\mathcal{D}^{1,\infty}$ and $\mathcal{P}$ are dense subsets of $\mathcal{D}^{1,p'}$, cf. Theorem 15.108.

Finally, if $n \geq 0$ and $Z \in \mathcal{P}_{\mathbb{R}}$, $\mathrm{E}\langle \pi_{n-1}(X), \nabla Z\rangle_{H_{\mathbb{C}}} = \mathrm{E}\langle X, \pi_{n-1}(\nabla Z)\rangle_{H_{\mathbb{C}}} = \mathrm{E}\langle X, \nabla\pi_n(Z)\rangle_{H_{\mathbb{C}}}$ and $\mathrm{E}(\pi_n(Y)Z) = \mathrm{E}(Y\pi_n(Z))$. Hence, by (i) $\Leftrightarrow$ (iii) applied to $\pi_{n-1}(X)$ and $\pi_n(Y)$, (iv) holds if and only if $\mathrm{E}\langle X, \nabla\pi_n(Z)\rangle_{H_{\mathbb{C}}} = \mathrm{E}(Y\pi_n(Z))$ for every $Z \in \mathcal{P}_{\mathbb{R}}$ and every n. Consequently (iii) $\Rightarrow$ (iv), because $Z \in \mathcal{P}_{\mathbb{R}}$ implies $\pi_n(Z) \in \mathcal{P}_{\mathbb{R}}$, and (iv) $\Rightarrow$ (iii) because $Z = \sum_n \pi_n(Z)$ for $Z \in \mathcal{P}_{\mathbb{R}}$. $\square$

COROLLARY 15.133. *Let* $1 < p \leq \infty$. *If* $X \in L^p(H_{\mathbb{C}})$, *then* δX *exists in* L^p *if and only if* $|\,\mathrm{E}\langle \nabla Z, X\rangle_{H_{\mathbb{C}}}| \leq C\|Z\|_{p'}$, *for some* $C < \infty$ *and all real polynomial variables* $Z \in \mathcal{P}_{\mathbb{R}}$. *Moreover, we then have* $\|\delta X\|_p = \sup\{|\,\mathrm{E}\langle X, \nabla Z\rangle_{H_{\mathbb{C}}}| : Z \in \mathcal{P}_{\mathbb{C}}, \|Z\|_{p'} \leq 1\}$. $\square$

REMARK 15.134. The divergence operator as defined above really corresponds to $-\operatorname{Div}$ in the Euclidean case, where $\partial/\partial x_i$ is anti-self-adjoint, but the sign is of no particular importance to us.

We next study some special cases where the divergence exists.

THEOREM 15.135. *If* $X \in L^2$ *and* $\xi \in H$, *then* $\delta(X \otimes \xi)$ *exists in* L^2 *if and only if* $A^*(\xi)X$ *exists, and then* $\delta(X \otimes \xi) = A^*(\xi)X$.

PROOF. By Theorems 15.132 and 15.98, $\delta(X \otimes \xi) = Y \in L^2$ if and only if

$$\langle Z, Y\rangle_{L^2} = \mathrm{E}\langle \nabla Z, X \otimes \xi\rangle_{H_{\mathbb{C}}} = \mathrm{E}(\langle \nabla Z, \xi\rangle_{H_{\mathbb{C}}}\overline{X}) = \mathrm{E}((\partial_\xi Z)\overline{X}) = \langle A(\xi)Z, X\rangle_{L^2}$$

for all $Z \in \mathcal{P}$, which by Theorem 13.7(i),(vi) is equivalent to $Y = A^*(\xi)X$. $\square$

EXAMPLE 15.136. If $X = \sum_1^m X_i \otimes \xi_i \in \mathcal{P} \otimes H$, then $\delta X = \sum_1^m A^*(\xi_i)X_i \in \mathcal{P}$. In particular, in this case δX exists in L^p for every $p < \infty$.

Theorem 15.135 shows that δX also exists in L^p if $X = \sum_1^m X_i \otimes \xi_i$ with $X_i \in \overline{\mathcal{P}}_*$ and $\xi_i \in H$. In particular, this holds if all $X_i \in H^{:n:}$ for some $n \geq 0$. This result extends to the completed Hilbert space tensor product $H^{:n:} \otimes H_{\mathbb{C}}$, which is a closed subspace of $L^2(H_{\mathbb{C}})$ by Example E.12. Note that $H^{:n:} \otimes H_{\mathbb{C}} \subset L^p(H_{\mathbb{C}})$ for every $p < \infty$ by Example 12.4.

THEOREM 15.137. *Let* $n \geq 0$. *If* $X \in H^{:n:} \otimes H_{\mathbb{C}}$, *then* δX *exists in* L^p *for every* $p < \infty$. *Moreover,* $\delta X \in H^{:n+1:}$ *and* $\|\delta X\|_2 \leq \sqrt{n+1}\|X\|_{H^{:n:}\otimes H}$.

PROOF. If $Z \in \mathcal{P}_{\mathbb{C}}$, then $\pi_n \nabla Z = \nabla\pi_{n+1}Z$ and thus, by Theorem 15.100,

$$\begin{aligned}|\,\mathrm{E}\langle \nabla Z, X\rangle_{H_{\mathbb{C}}}| &= |\,\mathrm{E}\langle \nabla\pi_{n+1}Z, X\rangle_{H_{\mathbb{C}}}| \leq \|\nabla\pi_{n+1}Z\|_{L^2(H_{\mathbb{C}})}\|X\|_{L^2(H_{\mathbb{C}})}\\ &= \sqrt{n+1}\|\pi_{n+1}Z\|_2\|X\|_{L^2(H_{\mathbb{C}})} \leq \sqrt{n+1}\|Z\|_2\|X\|_{L^2(H_{\mathbb{C}})}.\end{aligned}$$

It follows by Corollary 15.133 that δX exists in L^2, with the estimate $\|\delta X\|_2 \leq \sqrt{n+1}\|X\|_{L^2(H_{\mathbb{C}})}$. Moreover, $\langle \delta X, Z\rangle_{L^2} = \mathrm{E}\langle X, \nabla Z\rangle_{H_{\mathbb{C}}} = 0$ if $Z \in \mathcal{P} \cap H^{:m:}$ for some $m \neq n+1$; hence $\delta X \in H^{:n+1:}$ (cf. Corollary 3.26). This further implies that $\delta X \in L^p_{\mathbb{C}}$, $p < \infty$. □

COROLLARY 15.138. *If $X \in \mathcal{D}^{1,2}(H_{\mathbb{C}})$, then δX exists in L^2 and we have $\|\delta X\|_2 \leq \|X\|_{1,2,H_{\mathbb{C}}}$.*

PROOF. $X = \sum_0^\infty \pi_n(X)$ and, by Theorems 15.137 and 15.102, the sum $\sum_0^\infty \delta(\pi_n(X))$ converges in L^2 with

$$\|\sum_0^\infty \delta(\pi_n(X))\|_2^2 = \sum_0^\infty \|\delta(\pi_n(X))\|_2^2 \leq \sum_0^\infty (n+1)\|\pi_n(X)\|_2^2$$
$$= \|X\|^2_{L^2(H_{\mathbb{C}})} + \|\nabla X\|^2_{L^2(H\otimes H_{\mathbb{C}})} = \|X\|^2_{1,2,H_{\mathbb{C}}}.$$

The result follows since δ is a closed operator $L^2(H_{\mathbb{C}}) \to L^2$. □

Theorem 15.137 extends to $p \neq 2$ as follows.

THEOREM 15.139. *Let $1 < p < \infty$. If $X \in \mathcal{D}^{1,p}(H_{\mathbb{C}})$, then δX exists in L^p, and δ is a bounded linear operator $\mathcal{D}^{1,p}(H_{\mathbb{C}}) \to L^p_{\mathbb{C}}$.*

PROOF. Let $X \in \mathcal{D}^{1,p}(H_{\mathbb{C}})$. By Theorem 15.120 and Remark 15.121, $(\mathcal{N}+1)^{1/2}X \in L^p(H_{\mathbb{C}})$. If $Z \in \mathcal{P}$ with $\mathrm{E}\,Z = 0$, then $(\mathcal{N}+1)^{-1/2}\nabla Z = \nabla\mathcal{N}^{-1/2}Z$ and thus, using Hölder's inequality and Lemma 15.115,

$$|\,\mathrm{E}\langle X, \nabla Z\rangle_{H_{\mathbb{C}}}| = |\,\mathrm{E}\langle(\mathcal{N}+1)^{1/2}X, (\mathcal{N}+1)^{-1/2}\nabla Z\rangle_{H_{\mathbb{C}}}|$$
$$\leq \|(\mathcal{N}+1)^{1/2}X\|_{L^p(H_{\mathbb{C}})}\|\nabla\mathcal{N}^{-1/2}Z\|_{L^{p'}(H_{\mathbb{C}})} \leq C_p\|X\|_{1,p,H_{\mathbb{C}}}\|Z\|_{p'}.$$

The result follows by Corollary 15.133. □

COROLLARY 15.140. *Let $1 < p < \infty$ and $k \geq 1$. If $X \in \mathcal{D}^{k,p}(H_{\mathbb{C}})$, then $\delta X \in \mathcal{D}^{k-1,p}_{\mathbb{C}}$.*

PROOF. An immediate consequence of Theorems 15.139 and 15.120 (with Remark 15.121) and the identity $\mathcal{N}^\alpha\delta X = \delta(\mathcal{N}+1)^\alpha X$, with $\alpha = (k-1)/2$. □

COROLLARY 15.141. *If $X \in \mathcal{D}^{\infty-}(H_{\mathbb{C}})$, then $\delta X \in \mathcal{D}^{\infty-}_{\mathbb{C}}$.* □

We also have a product rule. (The rule differs from the corresponding rule in ordinary (Euclidean) vector calculus by the minus sign, cf. Remark 15.134.)

THEOREM 15.142. *Let $1 < p < \infty$. If $X \in \mathcal{D}^{1,p}(H_{\mathbb{C}})$ and $Y \in \mathcal{D}^{1,p'}$, then $\delta(YX)$ exists and*

$$\delta(YX) = Y\delta X - \langle\nabla Y, \overline{X}\rangle_{H_{\mathbb{C}}}. \tag{15.37}$$

PROOF. Note first that by Theorem 15.139, δX exists in L^p, and thus, by Hölder's inequality, the right hand side of (15.37) belongs to $L^1_{\mathbb{C}}$; furthermore, $YX \in L^1(H_{\mathbb{C}})$.

Next, let $Z \in \mathcal{D}^{1,\infty}_{\mathbb{R}}$. Then $YZ \in \mathcal{D}^{1,p'}$ by Corollary 15.80, and $\nabla(YZ) = Z\nabla Y + Y\nabla Z$. Consequently, by Theorem 15.132,

$$\begin{aligned} \mathrm{E}(ZY\delta(X)) &= \mathrm{E}\langle \nabla(YZ), \overline{X}\rangle_{H_{\mathbb{C}}} = \mathrm{E}\langle Y\nabla(Z), \overline{X}\rangle_{H_{\mathbb{C}}} + \mathrm{E}\langle Z\nabla(Y), \overline{X}\rangle_{H_{\mathbb{C}}} \\ &= \mathrm{E}\langle YX, \nabla(Z)\rangle_{H_{\mathbb{C}}} + \mathrm{E}(Z\langle \nabla(Y), \overline{X}\rangle_{H_{\mathbb{C}}}). \end{aligned}$$

Hence,

$$\mathrm{E}\langle YX, \nabla(Z)\rangle_{H_{\mathbb{C}}} = \mathrm{E}\big(Z\big(Y\delta(X) - \langle \nabla(Y), \overline{X}\rangle_{H_{\mathbb{C}}}\big)\big),$$

and the result follows by Definition 15.130. □

Our next goal is to prove that δ is a local operator, at least on $\mathcal{D}^{1,p}(H_{\mathbb{C}})$ with $p > 1$. We begin with a lemma.

LEMMA 15.143. *Let $1 < p < \infty$. If $X \in \mathcal{D}^{1,p}(H)$ and $\varphi\colon \mathbb{R} \to \mathbb{R}$ is continuously differentiable with compact support, then*

$$\delta\big(\varphi(\|X\|_H^2)X\big) = \varphi(\|X\|_H^2)\delta X - 2\varphi'(\|X\|_H^2)\langle \nabla X, X \otimes X\rangle_{H\otimes H}. \qquad (15.38)$$

PROOF. Suppose first that $p \geq 2$, so that $p' \leq p$. Then the result follows by Lemma 15.86 and Theorem 15.142.

Suppose now $1 < p < 2$. By Remark 15.114 there exists a sequence $X_n \in \mathcal{P} \otimes H$ such that $X_n \to X$ in $\mathcal{D}^{1,p}(H)$, and thus $X_n \to X$ in $L^p(H)$, $\nabla X_n \to \nabla X$ in $L^p(H \otimes H)$ and, by Theorem 15.139, $\delta X_n \to \delta X$ in L^p; by selecting a subsequence we may further assume that these three limits hold a.s. too. By the already proved case $p = 2$, (15.38) holds for each X_n.

Now let $n \to \infty$. Then $\varphi(\|X_n\|_H^2)X_n \to \varphi(\|X\|_H^2)X$ a.s.; since φ is bounded and X_n converges in $L^1(H)$, the sequence $\{\varphi(\|X_n\|_H^2)X_n\}$ is uniformly integrable and thus $\varphi(\|X_n\|_H^2)X_n \to \varphi(\|X\|_H^2)X$ in $L^1(H)$.

Similarly, if $T(X)$ denotes the right hand side of (15.38), $C_1 = \sup|\varphi(x)|$ and $C_2 = \sup|x^2\varphi'(x)|$, then $|T(X_n)| \leq C_1|\delta X_n| + 2C_2\|\nabla X_n\|_{H\otimes H}$, and thus $\{T(X_n)\}$ is uniformly integrable and $T(X_n) \to T(X)$ in L^1 as $n \to \infty$. Since δ is a closed operator, (15.38) holds. □

THEOREM 15.144. *If $X, Y \in \mathcal{D}^{1,p}(H_{\mathbb{C}})$ for some $p > 1$, and $A \in \mathcal{F}$ is an event such that $X = Y$ a.s. on A, then $\delta X = \delta Y$ a.s. on A.*

PROOF. It suffices to consider the case $Y = 0$; moreover, by writing $X = X_1 + iX_2$ with $X_1, X_2 \in \mathcal{D}^{1,p}(H)$, we may assume that $X \in \mathcal{D}^{1,p}(H)$.

Let $\varphi\colon \mathbb{R} \to \mathbb{R}$ be a continuously differentiable function with compact support and $\varphi(0) = 1$, and apply Lemma 15.143 to $\varphi(nx)$ for $n \geq 1$. Thus

$$\delta\big(\varphi(n\|X\|_H^2)X\big) = \varphi(n\|X\|_H^2)\delta X - 2n\varphi'(n\|X\|_H^2)\langle \nabla X, X \otimes X\rangle_{H\otimes H}. \qquad (15.39)$$

Let $n \to \infty$. Then $\varphi(n\|X\|_H^2) \to \mathbf{1}[X{=}0]$ a.s., and thus, by dominated convergence, $\varphi(n\|X\|_H^2)X \to \mathbf{1}[X{=}0]X = 0$ in $L^p(H)$ and $\varphi(n\|X\|_H^2)\delta X \to \mathbf{1}[X{=}0]\delta X$ in L^p. Similarly, $2n\varphi'(n\|X\|_H^2)\langle \nabla X, X \otimes X\rangle_{H\otimes H} \to 0$ a.s., and by dominated convergence also in L^1 (and L^p), because

$$\begin{aligned}|2n\varphi'(n\|X\|_H^2)\langle \nabla X, X \otimes X\rangle_{H\otimes H}| &\le 2(n\|X\|_H^2)|\varphi'(n\|X\|_H^2)|\,\|\nabla X\|_{H\otimes H} \\ &\le C\|\nabla X\|_{H\otimes H} \in L^p.\end{aligned}$$

Since δ is a closed operator, we obtain by taking these limits in (15.39),

$$0 = \delta(0) = \mathbf{1}[X{=}0]\delta X,$$

and thus $\delta X = \mathbf{1}[X{=}0]\delta X = 0$ a.s. on A. □

We next show that the number or Ornstein–Uhlenbeck operator $\mathcal{N}$ equals the divergence of the gradient; hence it corresponds to the Laplacian.

THEOREM 15.145. *Let $1 < p < \infty$ and $X \in L^p$. Then ∇X and $\delta(\nabla X)$ exist in L^p if and only if $\mathcal{N}X$ exists in L^p, and then $\mathcal{N}X = \delta(\nabla X)$.*

PROOF. Observe first that if $X \in \mathcal{D}^{1,p}$ and $Z \in \mathcal{P}_{\mathbb{R}}$, then

$$\mathrm{E}\langle \nabla X, \nabla Z\rangle_{H_{\mathrm{C}}} = \mathrm{E}(\mathcal{N}^{1/2}X\mathcal{N}^{1/2}Z) = \mathrm{E}(X\mathcal{N}Z).$$

In fact, if $X \in \mathcal{P}$, then the result holds by Corollary 15.101, and the general case follows by continuity (keeping $Z \in \mathcal{P}_{\mathbb{R}}$ fixed).

Consequently, if ∇X exists in L^p and $\delta(\nabla X) = Y \in L^p$, then $\mathrm{E}(YZ) = \mathrm{E}\langle \nabla X, \nabla Z\rangle_{H_{\mathrm{C}}} = \mathrm{E}(X\mathcal{N}Z)$ for every real polynomial variable $Z \in \mathcal{P}_{\mathbb{R}}$, which is equivalent to $\pi_n(Y) = n\pi_n(X)$ for each n, i.e. $Y = \mathcal{N}X$. (Cf. Definition 5.46.)

Conversely, if $\mathcal{N}X$ exists in L^p, then $X \in \mathcal{D}^{2,p} \subset \mathcal{D}^{1,p}$ by Theorem 15.120, so ∇X exists in L^p; moreover, for every $Z \in \mathcal{P}_{\mathbb{R}}$, $\mathrm{E}\langle \nabla X, \nabla Z\rangle_{H_{\mathrm{C}}} = \mathrm{E}(X\mathcal{N}Z) = \mathrm{E}((\mathcal{N}X)Z)$. By Theorem 15.132, $\delta(\nabla X)$ exists in L^p and equals $\mathcal{N}X$. □

This result leads to simple proofs of some properties of $\mathcal{N}$. We first show that $\mathcal{N}$ is a local operator. Recall that $\mathcal{N}X$ exists in L^p if and only if $X \in \mathcal{D}^{2,p}$, for $1 < p < \infty$, by Theorem 15.120.

THEOREM 15.146. *If $X, Y \in \mathcal{D}^{2,p}$ for some $p > 1$, and $A \in \mathcal{F}$ is an event such that $X = Y$ a.s. on A, then $\mathcal{N}X = \mathcal{N}Y$ a.s. on A.*

PROOF. By Theorems 15.39 and 15.144, $\nabla X = \nabla Y$ and $\delta(\nabla X) = \delta(\nabla Y)$ a.s. on A, and the result follows by Theorem 15.145. □

We next show a chain rule for $\mathcal{N}$, which behaves like a second order differential operator.

THEOREM 15.147. *Let $1 < p < \infty$. Suppose that $X_1, \dots, X_m \in \mathcal{D}^{2,p}_{\mathbb{R}}$ and that $\varphi\colon \mathbb{R}^m \to \mathbb{R}$ is twice continuously differentiable and such that, with $X = (X_1, \dots, X_m)$ and for every i and j in $1, \dots, m$, $\varphi(X) \in L^p$, $\partial_i\varphi(X) \in L^{p'}$, $\partial_i\varphi(X)\nabla X_i \in L^p(H)$, $\partial_i\varphi(X)\mathcal{N}X_i \in L^p$, $\partial_i\partial_j\varphi(X)\nabla X_j \in L^{p'}(H)$ and $\partial_i\partial_j\varphi(X)\langle\nabla X_i, \nabla X_j\rangle_H \in L^p$. Then $\varphi(X) \in \mathcal{D}^{2,p}_{\mathbb{R}}$ and*

$$\mathcal{N}(\varphi(X)) = \sum_{i=1}^{m} \partial_i\varphi(X)\mathcal{N}X_i - \sum_{i,j=1}^{m} \partial_i\partial_j\varphi(X)\langle\nabla X_i, \nabla X_j\rangle_H.$$

PROOF. By Theorem 15.78, $\varphi(X) \in \mathcal{D}^{1,p}$ and

$$\nabla\varphi(X) = \sum_{i=1}^{m} \partial_i\varphi(X)\nabla X_i;$$

similarly each $\partial_i\varphi(X) \in \mathcal{D}^{1,p'}$ and $\nabla(\partial_i\varphi(X)) = \sum_{j=1}^{m} \partial_i\partial_j\varphi(X)\nabla X_j$. Since also $\nabla X_i \in \mathcal{D}^{1,p}(H)$, Theorem 15.142 yields

$$\begin{aligned}\delta(\nabla\varphi(X)) &= \sum_{i=1}^{m} \delta(\partial_i\varphi(X)\nabla X_i) = \sum_{i=1}^{m} \bigl(\partial_i\varphi(X)\delta(\nabla X_i) - \langle\nabla\partial_i\varphi(X), \nabla X_i\rangle_H\bigr)\\ &= \sum_{i=1}^{m} \partial_i\varphi(X)\delta(\nabla X_i) - \sum_{i,j=1}^{m} \partial_i\partial_j\varphi(X)\langle\nabla X_j, \nabla X_i\rangle_H,\end{aligned}$$

and the result follows from Theorem 15.145, since the right hand side by assumption belongs to L^p. □

COROLLARY 15.148. *Let $2 \le p, q, r < \infty$ with $1/p = 1/q + 1/r$ and assume that $X \in \mathcal{D}^{2,q}_{\mathbb{R}}$ and $Y \in \mathcal{D}^{2,r}_{\mathbb{R}}$. Then $XY \in \mathcal{D}^{2,p}$ and*

$$\mathcal{N}(XY) = X\mathcal{N}(Y) + Y\mathcal{N}(X) - 2\langle\nabla X, \nabla Y\rangle_H. \tag{15.40}$$

PROOF. Note that $X, Y \in \mathcal{D}^{2,p}_{\mathbb{R}} \subseteq \mathcal{D}^{2,p'}_{\mathbb{R}}$, since $q, r \ge p \ge p'$. It follows, using Hölder's inequality, that Theorem 15.147 applies with $\varphi(x, y) = xy$. □

This corollary extends by a simple continuity argument to the full range $1 < p, q, r < \infty$; we omit the details.

COROLLARY 15.149. *If $X, Y \in \mathcal{D}^{\infty-}$, then $\langle\nabla X, \nabla Y\rangle_{H_{\mathbb{C}}} \in \mathcal{D}^{\infty-}$.*

PROOF. If X and Y are real, then (15.40) yields

$$\langle\nabla X, \nabla Y\rangle_H = \tfrac{1}{2}\bigl(X\mathcal{N}(Y) + Y\mathcal{N}(X) - \mathcal{N}(XY)\bigr),$$

which belongs to $\mathcal{D}^{\infty-}$ by Corollaries 15.125 and 15.84. The complex case follows by linearity. (Of course, it is also possible to give a direct proof, similarly to Theorem 15.83.) □

EXAMPLE 15.150. Let $2 \le p < \infty$. If $\xi_1, \dots, \xi_n$ is an orthonormal family in H, and $f\colon \mathbb{R}^n \to \mathbb{C}$ is twice continuously differentiable and such that $f(\xi_1, \dots, \xi_n) \in L^p$, $\partial_i f(\xi_1, \dots, \xi_n) \in L^q$ for some $q > p$ and $\partial_i \partial_j f(\xi_1, \dots, \xi_n) \in L^p$ for for all $i, j = 1, \dots, n$, then Theorem 15.147 shows that $f(\xi_1, \dots, \xi_n) \in \mathcal{D}^{2,p}$ and

$$\mathcal{N} f(\xi_1, \dots, \xi_n) = -\sum_{i=1}^n \frac{\partial^2 f}{\partial x_i^2}(\xi_1, \dots, \xi_n) + \sum_{i=1}^n \xi_i \frac{\partial f}{\partial x_i}(\xi_1, \dots, \xi_n).$$

In particular, this applies to every polynomial f.

10. Smoothness

The central smoothness result, which is the basis of many applications of the Malliavin calculus, is as follows. (Cf. the related Theorem 15.54 on existence of density.) Recall that $L^{\infty-} = \bigcap_{p<\infty} L^p$.

THEOREM 15.151. *Suppose that $X_1, \dots, X_n \in \mathcal{D}_{\mathbb{R}}^{\infty-}$, and let W be the determinant of the Malliavin matrix $(\langle \nabla X_i, \nabla X_j \rangle_H)_{i,j=1}^n$. If $W \neq 0$ a.s. and $W^{-1} \in L^{\infty-}$, then $X = (X_1, \dots, X_n)$ has an infinitely differentiable density in $\mathbb{R}^n$; moreover, the density belongs to the class $S(\mathbb{R}^n)$ of rapidly decreasing functions.*

PROOF. We prove a series of lemmas. The first two are versions of the chain rule for the function $1/x$, where the singularity at the origin requires special attention.

LEMMA 15.152. *Let $1 \le p < \infty$. Suppose that $X \in \mathcal{D}_{\mathbb{R}}^{1,0}$, $X \neq 0$ a.s., $X^{-1} \in L^p$ and $X^{-2}\nabla X \in L^p(H)$. Then $X^{-1} \in \mathcal{D}^{1,p}$ and $\nabla X^{-1} = -X^{-2}\nabla X$.*

PROOF. Let, for $\varepsilon > 0$, $Y_\varepsilon = X/(X^2 + \varepsilon)$. By Theorem 15.34, $Y_\varepsilon \in \mathcal{D}^{1,0}$ with

$$\nabla Y_\varepsilon = -\frac{X^2 - \varepsilon}{(X^2 + \varepsilon)^2} \nabla X;$$

since further $|Y_\varepsilon| \le |X|^{-1} \in L^p$ and $\|\nabla Y_\varepsilon\|_H \le X^{-2} \|\nabla X\|_H \in L^p$, $Y_\varepsilon \in \mathcal{D}^{1,p}$. Moreover, dominated convergence yields $Y_\varepsilon \to X^{-1}$ in L^p and $\nabla Y_\varepsilon \to -X^{-2}\nabla X$ in $L^p(H)$ as $\varepsilon \to 0$, and the result follows by Theorem 15.72. □

LEMMA 15.153. *Suppose that $X \in \mathcal{D}_{\mathbb{R}}^{\infty-}$, $X \neq 0$ a.s., and $X^{-1} \in L^{\infty-}$. Then $X^{-1} \in \mathcal{D}^{\infty-}$.*

PROOF. We prove by induction on k that if $k \ge 0$, then $X^{-1} \in \mathcal{D}^{k,p}$ for every $p < \infty$.

Suppose that this induction hypothesis holds for some $k \ge 0$. Given $p < \infty$, we thus have $X^{-1} \in \mathcal{D}^{k,3p}$; since further $\nabla X \in \mathcal{D}^{k,3p}(H)$ by assumption, Theorem 15.83 (twice) shows that $X^{-2}\nabla X \in \mathcal{D}^{k,p}(H) \subset L^p(H)$. By Lemma 15.152, $X^{-1} \in \mathcal{D}^{1,p}$ and $\nabla X^{-1} \in \mathcal{D}^{k,p}(H)$, and thus $X^{-1} \in \mathcal{D}^{k+1,p}$

by Theorem 15.64(vii). This verifies the induction step, and the proof is completed by observing that the case $k = 0$ follows by the assumptions. □

For the final lemma we define $C_b^\infty(\mathbb{R}^n) = \{\varphi \in C^\infty(\mathbb{R}^n) : \partial^\alpha \varphi$ is bounded on $\mathbb{R}^n$ for every multi-index $\alpha\}$.

LEMMA 15.154. *Suppose that the assumptions of Theorem 15.151 hold. If $Y \in \mathcal{D}^{\infty-}$ and α is a multi-index, then there exists $Y_\alpha \in \mathcal{D}^{\infty-}$ such that*

$$\mathrm{E}\big(\partial^\alpha \varphi(X) Y\big) = \mathrm{E}\big(\varphi(X) Y_\alpha\big)$$

for every $\varphi \in C_b^\infty(\mathbb{R}^n)$.

PROOF. Let $M_{ij} = \langle \nabla X_i, \nabla X_j \rangle_H$ and let $(L_{ij})_{i,j=1}^n$ be the inverse of the Malliavin matrix $(M_{ij})_{i,j=1}^n$. We begin by proving $L_{ij} \in \mathcal{D}^{\infty-}$.

By Corollary 15.149, each $M_{ij} \in \mathcal{D}^{\infty-}$. Since the determinant W is a polynomial in M_{ij}, it follows from Corollary 15.84 that $W \in \mathcal{D}^{\infty-}$. Consequently, by Lemma 15.153 and the assumptions, $W^{-1} \in \mathcal{D}^{\infty-}$. Finally, each L_{ij} equals W^{-1} times a polynomial in M_{kl}, and thus $L_{ij} \in \mathcal{D}^{\infty-}$ by Corollary 15.84 again.

Now assume $\varphi \in C_b^\infty(\mathbb{R}^n)$. By Theorem 15.78, $\varphi(X) \in \mathcal{D}^{1,2}$ and $\nabla\varphi(X) = \sum_i \partial_i \varphi(X) \nabla X_i$. Hence, for $j = 1, \dots, n$,

$$\langle \nabla\varphi(X), \nabla X_j \rangle_{H_{\mathbb{C}}} = \sum_i \partial_i \varphi(X) M_{ij}$$

and thus, multiplying by the inverse matrix,

$$\partial_k \varphi(X) = \sum_j \langle \nabla\varphi(X), \nabla X_j \rangle_{H_{\mathbb{C}}} L_{jk} = \langle \nabla\varphi(X), \sum_j L_{jk} \nabla X_j \rangle_{H_{\mathbb{C}}}.$$

Consequently,

$$\mathrm{E}(\partial_k \varphi(X) Y) = \mathrm{E}\langle \nabla\varphi(X), \overline{Y} \sum_j L_{jk} \nabla X_j \rangle_{H_{\mathbb{C}}}. \tag{15.41}$$

Let $Z_k = Y \sum_j L_{jk} \nabla X_j$. Since $Y \in \mathcal{D}^{\infty-}$ and $\nabla X_j \in \mathcal{D}^{\infty-}(H)$ by assumption, and $L_{jk} \in \mathcal{D}^{\infty-}$ was proved above, Corollary 15.85 shows that $Z_k \in \mathcal{D}^{\infty-}(H_{\mathbb{C}})$. Thus Corollary 15.141 shows that $\delta Z_k \in \mathcal{D}^{\infty-}$, and consequently it follows form (15.41) and Theorem 15.132 that

$$\mathrm{E}\big(\partial_k \varphi(X) Y\big) = \mathrm{E}\langle \nabla\varphi(X), \overline{Z}_k \rangle_{H_{\mathbb{C}}} = \mathrm{E}\big(\varphi(X) \delta Z_k\big) = \mathrm{E}\big(\varphi(X) Y_k\big),$$

with $Y_k = \delta Z_k$. This proves the result when $|\alpha| = 1$, and the general case follows by induction on $|\alpha|$. □

In order to complete the proof of Theorem 15.151, we apply Lemma 15.154 with $Y = 1$ and $\varphi(x) = e^{it\cdot x}$, $t \in \mathbb{R}^n$; since $\partial^\alpha \varphi(x) = i^{|\alpha|} t^\alpha e^{it\cdot x}$, this yields

$$|t^\alpha \,\mathrm{E}\, e^{it\cdot X}| = |\,\mathrm{E}\, \delta^\alpha \varphi(X)| = |\,\mathrm{E}(\varphi(X) Y_\alpha)| \le \mathrm{E}\,|Y_\alpha|.$$

It follows that the characteristic function $\psi(t) = \mathrm{E}\, e^{it\cdot X}$ satisfies

$$|\psi(t)| \le C_k |t|^{-k}, \qquad t \in \mathbb{R}^n,$$

for every k and some $C_k < \infty$; as is well-known, this implies by the Fourier inversion formula that X has an infinitely differentiable density given by $f(x) = (2\pi)^{-n} \int_{\mathbb{R}^n} \psi(t)e^{-it\cdot x}\, dt$. Moreover, f and all derivatives $\partial^\alpha f$ are bounded.

Finally, for every multi-index β, the same argument with $Y = X^\beta \in \mathcal{D}^{\infty-}$ shows that $|\operatorname{E}(e^{it\cdot X}X^\beta)| \le C_{k,\beta}|t|^{-k}$. Since $\operatorname{E}(e^{it\cdot X}X^\beta)$ is the Fourier transform of $x^\beta f(x)$, it follows that $\partial^\alpha(x^\beta f(x))$ is bounded for all multi-indices α and β, and thus $f \in S(\mathbb{R}^n)$. □

11. A new look at the Skorohod integral

If H is a Gaussian Hilbert space indexed by a real Hilbert space H_1, then, as stated in Remark 15.28, it is convenient to regard ∇ as a densely defined operator $L^0_{\mathbb{R}} \to L^0(H_1)$ or $L^0_{\mathbb{C}} \to L^0(H_{1\mathbb{C}})$; similarly, δ may be regarded as a densely defined operator $L^p(H_1) \to L^p_{\mathbb{R}}$ or $L^p(H_{1\mathbb{C}}) \to L^p_{\mathbb{C}}$, $1 \le p < \infty$.

In this section we consider the special case where $H_1 = L^2_{\mathbb{R}}(M, \mathcal{M}, \mu)$ for some σ-finite measure space $(M, \mathcal{M}, \mu)$. Then, as discussed in detail in Section 7.2, the defining isometry $I\colon H_1 = L^2_{\mathbb{R}}(M, \mathcal{M}, \mu) \to H$ yields a stochastic integral. We shall prove that the divergence operator δ coincides with the corresponding Skorohod integral (Gaveau and Trauber 1982); for further results see Nualart (1995, 1997+) and Accardi and Pikovski (1996). We use the notation of Section 7.2, in particular the Skorohod integral of a stochastic process X_t on M is denoted by $\int_M X_t\, dZ(t)$ and if $f \in L^2_{\mathbb{R}}(M, \mathcal{M}, \mu)$, then $\int_M f\, dZ = I(f)$.

Note first that $L^0(H_{1\mathbb{C}}) = L^0(\Omega, \mathcal{F}(H), \operatorname{P}; L^2_{\mathbb{C}}(M, \mathcal{M}, \mu))$ may be identified with the space of all complex measurable functions X on $\Omega \times M$ such that $\|X(\omega, t)\|_{L^2(M)} < \infty$ a.s., see Appendix C. Moreover,

$$L^p(H_{1\mathbb{C}}) = \{X \in L^0_{\mathbb{C}}(\Omega \times M) : \| \|X(\omega, t)\|_{L^2(M)} \|_{L^p(\Omega)} < \infty\}.$$

In particular, $L^2(H_{1\mathbb{C}}) = L^2_{\mathbb{C}}(\Omega \times M) = L^2(M; L^2_{\mathbb{C}})$. (Similarly, $L^2(H_1) = L^2_{\mathbb{R}}(\Omega \times M) = L^2(M; L^2_{\mathbb{R}})$.) Hence, an element $X \in L^2(H_{1\mathbb{C}})$ may be regarded as the (a.e. defined) stochastic process $t \mapsto X_t = X(\omega, t) \in L^2_{\mathbb{C}}$, and $\int_M \|X_t\|_2^2\, d\mu = \|X\|^2_{L^2(H_{1\mathbb{C}})} < \infty$. Conversely, every stochastic process X_t on M such that $\int_M \|X_t\|_2^2\, d\mu < \infty$ arises in this way.

Consequently, if $X \in L^0$ and ∇X exists, we may regard $\nabla X \in L^0(H_{\mathbb{C}})$ as a stochastic process indexed by M; we denote this process by $(\partial_t X)_{t \in M}$. (Note, however, that this process is defined only as an a.e. defined function of t, i.e. modulo null functions; hence we can in general not give a meaning to $\partial_t X$ for an individual t.) The defining relation for ∇X may then be written

$$\begin{aligned} \partial_{I(f)} X(\omega) &= \langle \nabla X(\omega), I(f) \rangle_{H_{\mathbb{C}}} = \langle \partial_t X(\omega), f(t) \rangle_{L^2(M,\mu)} \\ &= \int_M f(t) \partial_t X(\omega)\, d\mu(t), \end{aligned}$$

for $f \in L^2_{\mathbb{R}}(M, \mathcal{M}, \mu)$.

By Theorem 15.34, the chain rule holds for ∂_t: if $X_1, \dots, X_n$ are real, $\partial_t X_1, \dots, \partial_t X_n$ exist and f is continuously differentiable on $\mathbb{R}^n$, then

$$\partial_t f(X_1, \dots, X_n) = \sum_{i=1}^{n} \frac{\partial f}{\partial x_i}(X_1, \dots, X_n)\partial_t X_i, \qquad \text{for a.e. } t.$$

(Recall that this makes sense only as a.e. defined functions of $t \in M$, and not for individual t.)

EXAMPLE 15.155. If $f \in L^2(M, \mathcal{M}, \mu)$, then $\nabla I(f) = 1 \otimes I(f)$, see Example 15.31, and thus

$$\partial_t\Big(\int f\, dZ\Big) = \partial_t(I(f)) = f(t).$$

From Example 15.31 then follows, for every $n \geq 1$,

$$\partial_t(\,{:}I(f)^n{:}\,) = nf(t)\,{:}I(f)^{n-1}{:}.$$

Substituting f_1 for f, and using Theorem 7.26, we see that this yields

$$\partial_t \int_{M^n} f(t_1, \dots, t_n)\, dZ^n = n \int_{M^{n-1}} f(t, t_1, \dots, t_{n-1})\, dZ^{n-1} \tag{15.42}$$

when $f(t_1, \dots, t_n) = f_1(t_1) \cdots f_1(t_n)$ for some $f_1 \in L^2(M, \mathcal{M}, \mu)$. This can also be written

$$\partial_t \int_{M^n} f(t_1, \dots, t_n)\, dZ^n = \sum_{i=1}^{n} \int_{M^{n-1}} f(t_1, \dots, t_{i-1}, t, t_{i+1}, \dots, t_n)\, dZ^{n-1}. \tag{15.43}$$

By polarization, see Theorem D.2, (15.43) holds also when $f(t_1, \dots, t_n) = f_1(t_1) \cdots f_n(t_n)$ with $f_1, \dots, f_n \in L^2(M, \mathcal{M}, \mu)$. Such functions form a total subset of $L^2(M^n, \mu^n)$, cf. Example E.10. Moreover, it follows from Theorems 7.26 and 15.100 that both sides of (15.43) define continuous linear maps $L^2(M^n) \to L^2(\Omega \times M) = L^2(M; L^2)$, and thus, by continuity, (15.43) holds for every $f \in L^2(M^n, \mu^n)$.

As a consequence, (15.42) holds for every symmetric $f \in L^2(M^n, \mu^n)$.

Dually, we can regard δ as an operator mapping stochastic processes on M to random variables. In the L^2 case, this turns out to give another definition of the Skorohod integral.

THEOREM 15.156. *Let $X \in L^2(H_{1\mathbb{C}}) = L^2(M; L^2_{\mathbb{C}})$. Then δX exists in L^2 if and only if the Skorohod integral $\int_M X_t\, dZ(t)$ exists, and then $\delta X = \int_M X_t\, dZ(t)$.*

PROOF. Consider first the case $X = Y \otimes h$, with $Y \in \overline{\mathcal{P}}_*$ and $h \in H_1$. Then $X_t = h(t)Y$, $t \in M$, and thus Theorem 7.40 yields

$$\int_M X_t\, dZ(t) = \int_M h(t)Y\, dZ(t) = Y \odot \int_M h(t)\, dZ(t) = Y \odot I(h),$$

while Theorem 15.135 yields

$$\delta X = \delta(Y \otimes I(h)) = A^*(I(h))Y = I(h) \odot Y.$$

Hence the result holds in this case.

By Theorems 7.39(ii) and 15.137, both $X \mapsto \int_M X_t\, dZ(t)$ and δ are continuous linear operators $H^{:n:} \otimes H_{1\mathbb{C}} = \pi_n(L^2(H_{1\mathbb{C}})) \to L^2_{\mathbb{C}}$, and since we have just shown that they coincide on a total subset, the result holds when $X = \pi_n(X)$ for some $n \geq 0$. The general case follows from the definition (7.38) and Theorem 15.132. □

In other words, the Skorohod integral is the adjoint of the map $\nabla = \partial_t \colon \mathcal{D}^{1,2} \subset L^2 \to L^2(M; L^2)$.

EXAMPLE 15.157. By Theorem 15.139, δ maps $\mathcal{D}^{1,2}(H_{1\mathbb{C}})$ into $L^2_{\mathbb{C}}$. It is easily seen that $\mathcal{D}^{1,2}(H_{1\mathbb{C}}) = L^2(M, \mathcal{D}^{1,2}_{\mathbb{C}})$, and that the norm $\|\cdot\|_{1,2,H_{1\mathbb{C}}}$ in this space equals the norm $|||\cdot|||$ considered in Theorem 7.39; hence this result corresponds to Theorem 7.39(ii).

REMARK 15.158. We may define the Skorohod integral of X_t as δX as long as the latter exists in L^1; this gives by Theorem 15.156 an extension of the original definition which allows for Skorohod integrals outside L^2.

A further extension will be given in Chapter 16.

XVI

Transforms

In this chapter we define some transforms that are useful tools for studying Gaussian Hilbert spaces and operators on them. Many of the operators defined in this book yield simple expressions for the transforms; in fact, such operators are sometimes defined by their actions on the transform side, and some of the results proved by other methods earlier in the book may be proved using transforms.

The two basic transforms are defined and studied in Section 1. We use the transforms to extend the definition of the Wick product in Section 2. In the remaining sections we study relations between the transforms and some of the operations defined in previous chapters.

We continue to assume that H is a Gaussian Hilbert space. For convenience we let L^p denote $L^p(\Omega, \mathcal{F}(H), \mathrm{P})$ (or assume that $\mathcal{F}(H) = \mathcal{F}$) in this chapter too. Recall that $L^{1+} = \bigcup_{p>1} L^p$.

1. The T- and S-transforms

We will define two transforms, each mapping random variables in L^1 or L^{1+} to continuous complex-valued functions on the Hilbert space H or its complexification $H_{\mathbf{C}}$.

DEFINITION 16.1.

(i) If either $X \in L^1$ and $\zeta \in H$, or $X \in L^{1+}$ and $\zeta \in H_{\mathbf{C}}$, then

$$TX(\zeta) = \mathrm{E}(Xe^{i\zeta}).$$

(ii) If $X \in L^{1+}$ and $\zeta \in H_{\mathbf{C}}$, then

$$SX(\zeta) = \mathrm{E}(X :e^{\zeta}:).$$

REMARK 16.2. If $X \in L^{1+}$, then TX and SX are thus functions on $H_{\mathbf{C}}$. By considering the restriction to H, they may also be regarded as functions on H (as is always TX for general $X \in L^1$). We do not distinguish notationally between these interpretations; in the sequel many statements are valid for both interpretations, and otherwise it should be clear from the context which interpretation is intended.

These transforms are closely related. In fact, $:e^{\zeta}: = e^{-\mathrm{E}\zeta^2/2}e^{\zeta}$ by Theorem 3.33, which gives the following.

THEOREM 16.3. *The S- and T-transforms are related by the relations*

$$SX(\zeta) = e^{-\mathrm{E}\zeta^2/2}TX(-i\zeta),$$
$$TX(\zeta) = e^{-\mathrm{E}\zeta^2/2}SX(i\zeta)$$

for every $X \in L^{1+}$ and $\zeta \in H_{\mathbf{C}}$. □

These relations make it easy to go between the two transforms. For many purposes it is thus trivial to replace one of the transforms by the other, and the choice of transform to use is only a matter of convenience. It turns out that often the S-transform is more convenient, for example because of its simple rule for Wick products in Theorem 16.19 below; hence the S-transform is more widely used. Note however that the T-transform has the advantage of being defined on all of L^1; for problems involving L^1 the T-transform is thus often the only choice. (An alternative would be to consider the S-transform for imaginary arguments in iH only.)

EXAMPLE 16.4. Taking X as the constant 1 we find, for all $\zeta \in H_{\mathbf{C}}$,

$$T1(\zeta) = \mathrm{E}\, e^{i\zeta} = e^{-\mathrm{E}\zeta^2/2},$$
$$S1(\zeta) = \mathrm{E}(\,{:}e^{i\zeta}{:}\,) = 1.$$

EXAMPLE 16.5. Suppose that $\xi_1, \dots, \xi_n \in H$ (or $H_{\mathbf{C}}$); then ${:}\xi_1 \cdots \xi_n{:} \in H^{:n:}$ and, for $\zeta \in H_{\mathbf{C}}$, by Theorem 3.9,

$$S(\,{:}\xi_1 \cdots \xi_n{:}\,)(\zeta) = \mathrm{E}(\,{:}\xi_1 \cdots \xi_n{:}\ {:}e^{\zeta}{:}\,) = \mathrm{E}(\,{:}\xi_1 \cdots \xi_n{:} \sum_0^\infty \frac{1}{k!} {:}\zeta^k{:}\,)$$
$$= \sum_0^\infty \frac{1}{k!} \mathrm{E}(\,{:}\xi_1 \cdots \xi_n{:}\ {:}\zeta^k{:}\,) = \frac{1}{n!} \mathrm{E}(\,{:}\xi_1 \cdots \xi_n{:}\ {:}\zeta^n{:}\,)$$
$$= \prod_{j=1}^n \mathrm{E}(\xi_j \zeta).$$

Theorem 16.3 now gives

$$T(\,{:}\xi_1 \cdots \xi_n{:}\,)(\zeta) = i^n e^{-\mathrm{E}\zeta^2/2} \prod_{j=1}^n \mathrm{E}(\xi_j \zeta).$$

In particular for $\xi, \zeta \in H_{\mathbf{C}}$,

$$S\xi(\zeta) = \mathrm{E}(\xi\zeta),$$
$$T\xi(\zeta) = i e^{-\mathrm{E}\zeta^2/2}\, \mathrm{E}(\xi\zeta).$$

EXAMPLE 16.6. Let $\xi, \zeta \in H_{\mathbf{C}}$. Then by Corollary 3.37,

$$S(\,{:}e^{\xi}{:}\,)(\zeta) = \mathrm{E}(\,{:}e^{\xi}{:}\ {:}e^{\zeta}{:}\,) = e^{\mathrm{E}\,\xi\zeta},$$

and thus

$$T(\,{:}e^{\xi}{:}\,)(\zeta) = e^{i\,\mathrm{E}\,\xi\zeta - \mathrm{E}\zeta^2/2}.$$

By Theorem 3.33 we also have

$$S(e^{\xi})(\zeta) = e^{\mathrm{E}\,\xi^2/2+\mathrm{E}\,\xi\zeta},$$
$$T(e^{\xi})(\zeta) = e^{\mathrm{E}\,\xi^2/2+i\,\mathrm{E}\,\xi\zeta-\mathrm{E}\,\zeta^2/2} = e^{\mathrm{E}(\xi+i\zeta)^2/2}.$$

EXAMPLE 16.7. Let $X = e^{a\xi^2}$ with ξ standard normal and $\operatorname{Re} a < 1/2$ (so that $X \in L^{1+}$). By the calculations in Example 3.41, for every real or complex t,

$$SX(t\xi) = \mathrm{E}\big(e^{a\xi^2} :e^{t\xi}:\big) = (1-2a)^{-1/2} e^{\frac{a}{1-2a}t^2};$$

thus, by Theorem 16.3,

$$TX(t\xi) = (1-2a)^{-1/2} e^{-t^2/2(1-2a)}.$$

EXAMPLE 16.8. Consider a one-dimensional Gaussian Hilbert space $\{t\xi : t \in \mathbb{R}\}$, where ξ is standard normal. Any $\mathcal{F}(H)$-measurable random variable is of the form $f(\xi)$, and if $X = f(\xi) \in L^1$, then

$$TX(t\xi) = \mathrm{E}\big(f(\xi)e^{it\xi}\big) = \int_{-\infty}^{\infty} f(x)e^{itx}\tfrac{1}{\sqrt{2\pi}}e^{-x^2/2}\,dx,$$

which is the Fourier transform $\hat{f}_1(t)$ of the function $f_1(x) = f(x)\frac{1}{\sqrt{2\pi}}e^{-x^2/2}$; note that $\|f_1\|_{L^1(\mathbb{R},dx)} = \|f(\xi)\|_{L^1}$.

REMARK 16.9. We define the S- and T-transforms as functions on the Gaussian Hilbert space itself or its complexification. This is natural in our abstract setting, but in concrete applications it may be advantageous to change the definitions a little to make the transforms defined on some other spaces. (This is the traditional approach, cf. for example Hida, Kuo, Potthoff and Streit (1993).)

Thus if H is a Gaussian Hilbert space indexed by some other (real) Hilbert space H_1, which by Definition 1.18 means that we are given an isometry $h \mapsto \xi_h$ of H_1 onto H, then it is natural and convenient to use this isometry to transfer the S- and T-transforms to functions on H_1 or $H_{1\mathbb{C}}$; thus we let for example $TX(h) = \mathrm{E}(Xe^{i\xi_h})$, $h \in H_1$. Moreover, it may be convenient to consider only the restriction to some dense subspace of H_1.

A typical case is when H_1 is $L^2_{\mathbb{R}}(\mathbb{R}^d)$; then the T-transform of a random variable in L^1 may be regarded as a continuous (non-linear) functional on $L^2_{\mathbb{R}}(\mathbb{R}^d)$, and the S- and T-transforms of a random variable in L^{1+} may be regarded as analytic functions on $L^2_{\mathbb{C}}(\mathbb{R}^d)$. Alternatively, they may (by taking the restriction) be regarded as functions on the Schwartz space $\mathcal{S}_{\mathbb{R}}(\mathbb{R}^d)$ or $\mathcal{S}_{\mathbb{C}}(\mathbb{R}^d)$, or on some other convenient space of test functions.

It should be clear that the difference between these different definitions is only notational.

The basic properties of the T- and S-transforms are given in the following theorems. We refer to Appendix G and Hervé (1989) for the definition of analytic functions on Banach spaces.

THEOREM 16.10.

(i) *If $X \in L^1$, then TX is a bounded continuous function on H. In fact,*

$$|TX(\xi)| \leq \|X\|_1, \qquad \xi \in H. \tag{16.1}$$

Moreover, $TX = 0$ on H if and only if $X = 0$ a.s. Hence T is an injective continuous linear map of L^1 into $C_b(H)$, the Banach space of (complex) bounded continuous functions on H.

X is real if and only if $\overline{TX(\xi)} = TX(-\xi)$ on H.

(ii) *If $X \in L^{1+}$, then TX is an analytic function on $H_{\mathbf{C}}$. Hence T is an injective linear map of L^{1+} into the space of analytic functions on $H_{\mathbf{C}}$.*

For every $p > 1$,

$$|TX(\zeta)| \leq \exp\Bigl(\frac{p}{2(p-1)}\operatorname{E}(\operatorname{Im}\zeta)^2\Bigr)\|X\|_p$$
$$\leq \exp\Bigl(\frac{p}{2(p-1)}\|\zeta\|_2^2\Bigr)\|X\|_p, \qquad \zeta \in H_{\mathbf{C}}. \tag{16.2}$$

Hence T is a continuous mapping $L^p \mapsto C(H_{\mathbf{C}})$ for every $p > 1$, if the space $C(H_{\mathbf{C}})$ of (complex) continuous functions of $H_{\mathbf{C}}$ is given the pointwise topology.

PROOF. (i) Linearity and (16.1) follow at once from the definition, since $|e^{i\xi}| \leq 1$. Consequently, if $X_n \to X$ in L^1, then $TX_n \to TX$ uniformly on H. Since Example 16.6 and Theorem 2.12 (or Example 16.5 and Theorem 3.29) show that TX is continuous when X belongs to a dense subset of L^1, this implies that TX is continuous for every $X \in L^1$. Finally, injectivity holds by Lemma 2.7.

(ii) Hölder's inequality and

$$\|e^{i\zeta}\|_q^q = \operatorname{E} e^{-q\operatorname{Im}\zeta} = e^{q^2\operatorname{E}(\operatorname{Im}\zeta)^2/2}$$

(with $q = p/(p-1)$, the conjugate exponent) yield (16.2).

The transforms in Example 16.6 or 16.5 are analytic on $H_{\mathbf{C}}$, so Theorem 2.12 or Theorem 3.29 (respectively) implies that if $X \in L^p$ with $1 < p < \infty$, then there exist X_n such that $X_n \to X$ in L^p and TX_n is analytic on $H_{\mathbf{C}}$. Since (16.2) shows that then $TX_n(\zeta) \to TX(\zeta)$ uniformly on each ball $\{\|\zeta\| < r\}$, TX is analytic.

The remaining assertions follow immediately; note that continuity into $C(H_{\mathbf{C}})$ with the pointwise topology means just that the evaluation at each point $\zeta \in H_{\mathbf{C}}$ is continuous, which follows by (16.2). □

THEOREM 16.11. *If $X \in L^{1+}$, then SX is an analytic function on $H_{\mathbf{C}}$; in particular, SX is a continuous function on H. Moreover, $SX = 0$ on H (or $H_{\mathbf{C}}$) if and only if $X = 0$. Hence S is an injective linear map of L^{1+} into the space of analytic functions on $H_{\mathbf{C}}$.*

For every $p > 1$,

$$|SX(\zeta)| \le \exp\Big(\tfrac{1}{2(p-1)}\,\mathrm{E}(\mathrm{Re}\,\zeta)^2 + \tfrac{1}{2}\,\mathrm{E}(\mathrm{Im}\,\zeta)^2\Big)\|X\|_p$$
$$\le \exp\big(\tfrac{1}{2}(\tfrac{1}{p-1} \vee 1)\|\zeta\|_2^2\big)\|X\|_p, \qquad \zeta \in H_{\mathbf{C}}.$$

Hence S is a continuous mapping $L^p \mapsto C(H_{\mathbf{C}})$ for every $p > 1$, if the space $C(H_{\mathbf{C}})$ of (complex) continuous functions of $H_{\mathbf{C}}$ is given the pointwise topology.

Furthermore,

$$S(\overline{X})(\zeta) = \overline{SX(\overline{\zeta})};$$

hence X is real if and only if SX is real on H.

PROOF. As for Theorem 16.10, using Corollary 3.38, or by Theorems 16.3 and 16.10. □

REMARK 16.12. Equip L^{1+} with the natural inductive topology, see Remark 14.3. Then the continuity statements in Theorems 16.10(ii) and 16.11 can be restated as: The T- and S-transforms are continuous linear maps of L^{1+} into $C(H_{\mathbf{C}})$. Moreover, the proof shows that we may here replace the pointwise topology on $C(H_{\mathbf{C}})$ by the stronger topology of uniform convergence on compact sets (the compact-open topology) or the even stronger topology of uniform convergence on bounded sets. The latter topology is not a vector space topology if $\dim H = \infty$, but if we let $C_\beta(H_{\mathbf{C}})$ be the space of all complex continuous functions on $H_{\mathbf{C}}$ that are bounded on each bounded set, equipped with the topology given by the semi-norms $\|f\|_r = \sup\{|f(\xi)| : \|\xi\| \le r\}$, we obtain a Fréchet space such that S and T are continuous linear maps of L^{1+} into $C_\beta(H_{\mathbf{C}})$.

REMARK 16.13. If $X \in L^1$, then actually $TX(\xi) \to 0$ as $\|\xi\|_H \to \infty$. (This may be regarded as an extension of the Riemann–Lebesgue lemma, cf. Example 16.8.) In fact, this holds for the transforms in Example 16.6 (and 16.5), and the continuity argument in the proof above then shows that it holds for any $X \in L^1$.

REMARK 16.14. Example 16.8 shows that the T-transform of a random variable in L^1 in general does not have any smoothness beyond continuity.

In particular, the T-transform TX on H of some $X \in L^1$ does not in general have an extension to an analytic function on $H_{\mathbf{C}}$, or even on a neighbourhood of the origin.

REMARK 16.15. If $X \in L^{1+}$ and TX vanishes in a neighbourhood of 0 in H or $H_{\mathbf{C}}$, then by analytic continuation, TX vanishes everywhere and thus $X = 0$. The same holds for SX. Consequently, for any $\delta > 0$, the S- and T-transforms are injective continuous linear maps of L^{1+} into the space of bounded continuous functions on $\{\xi \in H : \|\xi\|_2 < \delta\}$ or the space of bounded analytic functions on $\{\zeta \in H_{\mathbf{C}} : \|\zeta\|_2 < \delta\}$.

Note that Example 16.8 shows that this does not extend to the T-transform of L^1; it is well-known that any C^∞-function on $\mathbb{R}$ with compact support is the Fourier transform of some function in $L^1(\mathbb{R}, dx)$, and in particular there are Fourier transforms that vanish in a neighbourhood of 0 but not identically. Hence, for any $\delta > 0$, there exists, even for a one-dimensional Gaussian Hilbert space, a non-zero $X \in L^1$ such that $TX = 0$ on $\{\xi \in H : \|\xi\| < \delta\}$.

THEOREM 16.16.
(i) *If $X \in H^{:n:}$, then $S(X)$ is homogeneous of degree n on $H_{\mathbb{C}}$.*
(ii) *If $X \in L^{1+}$, then*

$$SX(\zeta) = \sum_0^\infty S(\pi_n(X))(\zeta), \qquad \zeta \in H_{\mathbb{C}},$$

where the sum converges absolutely for each ζ.
(iii) *Conversely, if $X \in L^{1+}$ and $\psi_0, \psi_1, \dots$ is a sequence of functions on H such that ψ_n is homogeneous of degree n and $\sum_0^\infty \psi_n(\zeta) = SX(\zeta)$ (with conditionally convergent sum) for every $\zeta \in H$, then $\psi_n = S(\pi_n(X))$.*

PROOF. (i) Let $\zeta \in H_{\mathbb{C}}$ and $t \in \mathbb{C}$. If $X = {:}\xi_1 \cdots \xi_n{:}$ for some $\xi_1, \dots, \xi_n \in H$, it follows from Example 16.5 that

$$SX(t\zeta) = t^n SX(\zeta); \tag{16.3}$$

such X are total in $H^{:n:}$ by Corollary 3.27, and both sides of (16.3) are continuous linear functionals of $X \in H^{:n:}$; hence the equality holds for all $X \in H^{:n:}$.

(ii) Suppose $X \in L^p$ with $p > 1$ and let $p' = p/(p-1)$ be the conjugate exponent. Since $\sum_0^\infty \|\frac{1}{n!}{:}\zeta^n{:}\|_{p'} < \infty$, as a consequence of Theorem 5.19,

$$SX(\zeta) = \sum_0^\infty \tfrac{1}{n!} \operatorname{E}(X {:}\zeta^n{:}),$$

where the sum converges absolutely by Hölder's inequality. Moreover,

$$\tfrac{1}{n!} \operatorname{E}(X {:}\zeta^n{:}) = \langle \pi_n(X), \tfrac{1}{n!} {:}\bar{\zeta}^n{:} \rangle = \langle \pi_n(X), {:}e^{\bar{\zeta}}{:} \rangle = S(\pi_n(X))(\zeta).$$

(iii) Let $\zeta \in H$. Then, for every real t, by (ii) and (i),

$$SX(t\zeta) = \sum_{n=0}^\infty S(\pi_n(X))(t\zeta) = \sum_{n=0}^\infty t^n S(\pi_n(X))(\zeta),$$

and by assumption,

$$SX(t\zeta) = \sum_{n=0}^\infty \psi_n(t\zeta) = \sum_{n=0}^\infty t^n \psi_n(\zeta).$$

Hence, if we let $a_n = S(\pi_n(X))(\zeta) - \psi_n(\zeta)$, then $\sum_0^\infty a_n t^n = 0$ for every $t \in \mathbb{R}$. This implies that $a_n = 0$ for each n. □

EXAMPLE 16.17. Suppose that $\dim H = d < \infty$. Let γ be the standard Gaussian measure on $H_{\mathbb{C}}$; if we identify $H_{\mathbb{C}}$ with $\mathbb{C}^d$, using any orthonormal basis $\{\xi_k\}$ in H, and let λ denote the Lebesgue measure, $d\gamma = \pi^{-d}e^{-|z|^2}\,d\lambda$.

Example 16.5 shows that, for any multi-index α, $S(:\xi^\alpha:)$ is the function z^α on $\mathbb{C}^d$; moreover, the coordinate functions z_k are independent standard complex Gaussian variables on $(\mathbb{C}^d, \gamma)$ and by direct computation, Example 3.32 or Theorem 1.36, the functions z^α are orthogonal and $\|z^\alpha\|^2_{L^2(\gamma)} = \alpha!$.

Let $A(\mathbb{C}^d, \gamma)$ be the *Segal–Bargmann space* defined in Remark 4.4, i.e. the subspace of $L^2(\mathbb{C}^d, \gamma)$ consisting of analytic functions. This is a closed subspace so $A(\mathbb{C}^d, \gamma)$ is a Hilbert space; moreover, $\{z^\alpha\}_\alpha$ is an orthogonal basis. Theorem 3.21 now shows that S is an isometry of $L^2_{\mathbb{C}}$ onto $A(\mathbb{C}^d, \gamma)$; in other words, an analytic function f on $\mathbb{C}^d$ is the S-transform of a variable in $L^2_{\mathbb{C}}$ if and only if $\int_{\mathbb{C}^d} |f|^2\,d\gamma < \infty$.

It follows by Theorem 16.16 (or directly) that S maps $H^{:n:}$ onto the space of homogeneous analytic polynomials of degree n on $\mathbb{C}^d$.

If H and V_σ are as in Example 13.5, and the natural identification of $H_{\mathbb{C}}$ and $\mathbb{C}^d$ is used, the isometry $SV_\sigma^{-1}\colon L^2(\mathbb{R}^d, dx) \to A(\mathbb{C}^d, \gamma)$ is called the *Bargmann transform* (Folland 1989, Section 1.6).

REMARK 16.18. If H is finite-dimensional, then every homogeneous analytic function on $H_{\mathbb{C}}$ is the S-transform of some $X \in L^2$ (with $X \in H_{\mathbb{C}}^{:n:}$ for some n.) This fails when $\dim H = \infty$; for example, $\Phi(\zeta) = \mathrm{E}\,\zeta^2$ is analytic and homogeneous of degree 2, but not the S-transform of any $X \in H^{:2:}$

THEOREM 16.19. *If $X, Y \in \overline{\mathcal{P}}_*$, then*

$$S(X \odot Y) = S(X)S(Y).$$

Hence the S-transform is an algebra isomorphism of $\overline{\mathcal{P}}_*$ onto an algebra of analytic functions on $H_{\mathbb{C}}$.

PROOF. By (bi)linearity, it suffices to verify this for $X \in H^{:n:}$ and $Y \in H^{:m:}$ for some $n, m \geq 0$. By Corollary 3.27 and continuity (Theorems 3.47 and 16.11), we may furthermore suppose $X = :\xi_1 \cdots \xi_n:$ and $Y = :\eta_1 \cdots \eta_m:$; but this case follows from Example 16.5. □

COROLLARY 16.20. *If $X, Y \in \overline{\mathcal{P}}_*$, then*

$$T(X \odot Y)(\zeta) = e^{\mathrm{E}\zeta^2/2}TX(\zeta)TY(\zeta), \qquad \zeta \in H_{\mathbb{C}}.$$

PROOF. As for Theorem 16.19, using Example 16.5; or by Theorems 16.19 and 16.3. □

There are special results for the transforms acting on L^2. Since S and T are injective on L^2, their ranges $S(L^2)$ and $T(L^2)$ may be given Hilbert space structures so that S and T are isometries. Theorem F.5 and Example 16.6 yield the following theorems.

THEOREM 16.21. *S is an isometry of L^2 onto $S(L^2)$. $S(L^2_{\mathbb{C}})$ is a complex reproducing Hilbert space on H or $H_{\mathbb{C}}$ with reproducing kernel*

$$K(\xi,\eta) = e^{\langle \xi,\eta\rangle} = e^{\mathrm{E}(\xi\overline{\eta})}.$$

$S(L^2_{\mathbb{R}})$ is a real reproducing Hilbert space on H, with the same reproducing kernel. □

THEOREM 16.22. *T is an isometry of L^2 onto $T(L^2)$. $T(L^2_{\mathbb{C}})$ is a complex reproducing Hilbert space on H or $H_{\mathbb{C}}$ with reproducing kernel*

$$K(\xi,\eta) = e^{-\mathrm{E}(\xi-\overline{\eta})^2/2}.$$

□

REMARK 16.23. $T(L^2_{\mathbb{R}})$ is *not* a space of real functions on H.

REMARK 16.24. A closely related transform is the *Hermite transform* (Holden, Øksendal, Ubøe and Zhang 1996). In order to define it, we fix an orthonormal basis $\{\xi_i\}_{i\in\mathcal{I}}$ in H, let $\Phi\colon \ell^2_{\mathbb{C}}(\mathcal{I}) \to H_{\mathbb{C}}$ be the isometry $(a_i)_{i\in\mathcal{I}} \mapsto \sum_{\mathcal{I}} a_i\xi_i$, and define

$$\mathcal{H}X(a) = SX(\Phi(a)) = SX(\sum_{\mathcal{I}} a_i\xi_i), \qquad a = (a_i) \in \ell^2_{\mathbb{C}}(\mathcal{I}).$$

Note that the transform depends on the choice of the basis in H.

By taking a restriction, we can also regard $\mathcal{H}X$ as defined on some given subspace of $\ell^2_{\mathbb{C}}(\mathcal{I})$.

The isometry Φ defines H as a Gaussian Hilbert space indexed by $\ell^2_{\mathbb{R}}(\mathcal{I})$, and the Hermite transform is the corresponding version of the S-transform, as discussed in Remark 16.9. Thus the difference between the Hermite transform and the S-transform is mainly notational, and results for the S-transform transfer immediately to the Hermite transform. For example, $\mathcal{H}X$ is an analytic function on $\ell^2_{\mathbb{C}}(\mathcal{I})$ for every $X \in L^{1+}$, and $\mathcal{H}X = 0$ if and only if $X = 0$ a.s.

We will not consider the Hermite transform further, and leave the formulation of additional results to the reader.

2. More general Wick products

We defined in Chapter 3 the Wick product of two random variables with finite chaos decompositions. The S-transform enables us to define the Wick product more generally. For related definitions and applications, see Hida and Ikeda (1967), Gjessing, Holden, Lindstrøm, Øksendal, Ubøe and Zhang (1993) and Holden, Øksendal, Ubøe and Zhang (1996).

DEFINITION 16.25. Suppose that $X, Y \in L^{1+}$ and $p > 1$. We say that the *Wick product of X and Y exists in L^p* if there exists a $Z \in L^p$ such that

$$SX(\zeta)SY(\zeta) = SZ(\zeta) \tag{16.4}$$

for every $\zeta \in H$; this Z is the Wick product and is denoted by $X \odot Y$.

The Wick product *exists in* L^{1+} if it exists in some L^p, $p > 1$.

Note that $X \odot Y$ is uniquely defined (when it exists) by Theorem 16.11.

REMARK 16.26. In Definition 16.25 we require (16.4) for $\zeta \in H$; since the S-transforms are analytic by Theorem 16.11, it is equivalent to require (16.4) for all complex $\zeta \in H_{\mathbf{C}}$, or for all imaginary ζ such that $i\zeta \in H$.

It follows from Theorem 16.19 that Definition 16.25 extends the previous definitions of Wick product in Chapter 3. Using the T-transform and Corollary 16.20, we can extend it even further.

DEFINITION 16.27. Suppose that $X, Y \in L^1$. The *Wick product of* X *and* Y *exists in* L^1 if there exists a $Z \in L^1$ such that

$$e^{\mathrm{E}\zeta^2/2} TX(\zeta) TY(\zeta) = TZ(\zeta)$$

for every $\zeta \in H$; this Z is the Wick product and is denoted by $X \odot Y$.

It follows from Theorem 16.3 and Remark 16.26 that these definitions are compatible: if $X, Y \in L^{1+}$ and $p > 1$, then the Wick product exists in L^p according to the first definition if and only if it exists in L^1 according to the second definition and further belongs to L^p, and the definitions yield the same product.

REMARK 16.28. It follows easily that the compatibility stated in Theorem 3.6 extends: if H_1 is a closed subspace of a Gaussian Hilbert space H and $X, Y \in L^1(\Omega, \mathcal{F}(H_1), \mathrm{P})$, then the Wick product of X and Y exists for H if and only if it exists for H_1, and the two products coincide. We omit the details.

EXAMPLE 16.29. If $\xi, \eta \in H_{\mathbf{C}}$, then by Example 16.6, for every $\zeta \in H_{\mathbf{C}}$,

$$S(:e^{\xi}:)(\zeta) S(:e^{\eta}:)(\zeta) = e^{\mathrm{E}\xi\zeta} e^{\mathrm{E}\eta\zeta} = e^{\mathrm{E}(\xi+\eta)\zeta} = S(:e^{\xi+\eta}:)(\zeta).$$

Thus $:e^{\xi}: \odot :e^{\eta}:$ exists in L^p for every $p < \infty$, and

$$:e^{\xi}: \odot :e^{\eta}: = :e^{\xi+\eta}:.$$

EXAMPLE 16.30. Since $e^{\xi} = e^{\mathrm{E}\xi^2/2} :e^{\xi}:$ by Theorem 3.33, it follows from Example 16.29 that

$$e^{\xi} \odot e^{\eta} = e^{-\mathrm{E}(\xi\eta)} e^{\xi+\eta}, \qquad \xi, \eta \in H_{\mathbf{C}}.$$

In particular, if $\xi, \eta \in H$, then

$$e^{i\xi} \odot e^{i\eta} = e^{\mathrm{E}(\xi\eta)} e^{i(\xi+\eta)}.$$

Since $e^{i\xi}$, $e^{i\eta}$ and $e^{i(\xi+\eta)}$ have norm 1 in every L^p, $0 < p \leq \infty$, but $e^{\mathrm{E}\xi\eta}$ can be arbitrarily large, this example shows that there is no Wick version of Hölder's inequality. In conjunction with the closed graph theorem, it further shows that if $0 \neq \xi \in H$, then there exists $X \in L^{\infty}$ such that $e^{i\xi} \odot X$ does *not* exist in L^1.

EXAMPLE 16.31. Let $\xi \in H$ be standard normal. By the preceding example, $e^{s\xi} \odot e^{it\xi} = e^{-ist+(s+it)\xi}$ for every complex s and t, and by a careful differentiation with respect to s we obtain

$$\xi \odot e^{it\xi} = (\xi - it)e^{it\xi}.$$

(This formula follows also, more simply, from Theorem 16.36 below and Example 13.10.)

In particular, for every $p \geq 1$, $\|\xi \odot e^{it\xi}\|_p = t + O(1) \to \infty$ as $t \to \infty$, which by the closed graph theorem implies that there exists $X \in L^\infty$ such that $\xi \odot X$ does *not* exist in L^p. This can also be seen from the formulae in Example 13.11, which by Theorem 16.36 show that $\xi\odot$ acts as a differential operator rather than a multiplication.

EXAMPLE 16.32. If ξ is a standard normal variable, $\operatorname{Re} a$, $\operatorname{Re} b$, $\operatorname{Re} c < 1/2$ and $\frac{a}{1-2a} + \frac{b}{1-2b} = \frac{c}{1-2c}$, then by Example 16.7 (and Remark 16.28),

$$e^{a\xi^2} \odot e^{b\xi^2} = (1-2a)^{-1/2}(1-2b)^{-1/2}(1-2c)^{1/2}e^{c\xi^2}.$$

Taking for example $1/6 < a < 1/4$, $b = a$ and $c = 2a/(1+2a)$, we obtain an example with $X = Y \in L^2$ where $X \odot Y$ exists in L^{1+} but not in L^2.

Examples 16.30 and 16.31 show that the definitions above do not define the Wick product for all X and Y in L^2. An explicit example is the following.

EXAMPLE 16.33. Let $\xi \in H$ be standard normal, and let $X = \mathbf{1}[\xi > 0]$; thus $X \in L^\infty$. We claim that the Wick product $X \odot X$ does not exist in L^1. For simplicity we assume that H is the one-dimensional Gaussian Hilbert space spanned by ξ. (The general case follows easily, cf. Remark 16.28.)

Suppose that $X \odot X$ exists in L^1. Then $X \odot X = h(\xi)$ for some measurable function h, and by Example 16.8, with $g_1(x) = \frac{1}{\sqrt{2\pi}}e^{-x^2/2}\mathbf{1}[x > 0]$ and $h_1(x) = \frac{1}{\sqrt{2\pi}}e^{-x^2/2}h(x)$,

$$TX(t\xi) = \hat{g}_1(t),$$
$$T(X \odot X)(t\xi) = \hat{h}_1(t),$$

and thus by Definition 16.27,

$$\hat{h}_1(t) = e^{t^2/2}\big(\hat{g}_1(t)\big)^2, \qquad -\infty < t < \infty. \tag{16.5}$$

Since $\hat{h}_1(t) = T(X \odot X)(t\xi)$ is bounded by Theorem 16.10, (16.5) yields $e^{t^2/2}|\hat{g}_1(t)|^2 \leq C$ for some $C < \infty$, and thus

$$|\hat{g}_1(t)| \leq C^{1/2}e^{-t^2/4}.$$

Consequently, $\hat{g}_1 \in L^1(\mathbb{R})$, but this implies that g_1 is continuous, a contradiction.

The following theorem shows that for variables in L^{1+}, the extended Wick product can alternatively be defined using the chaos decomposition and termwise Wick products.

THEOREM 16.34. *If $X, Y \in L^{1+}$ and $X \odot Y$ exists in L^{1+}, then*

$$\pi_n(X \odot Y) = \sum_{k=0}^{n} \pi_k(X) \odot \pi_{n-k}(Y), \qquad n \geq 0.$$

Conversely, if $X, Y, Z \in L^{1+}$ and $\pi_n(Z) = \sum_{k=0}^{n} \pi_k(X) \odot \pi_{n-k}(Y)$ for every $n \geq 0$, then $X \odot Y = Z$.

PROOF. By Theorem 16.16(ii), for every $\zeta \in H_{\mathbf{C}}$,

$$SX(\zeta)SY(\zeta) = \sum_{k=0}^{\infty}\sum_{l=0}^{\infty} S(\pi_k(X))(\zeta)S(\pi_l(Y))(\zeta),$$

where the double sum converges absolutely. Rearranging the double sum according to $k + l$, we obtain $SX(\zeta)SY(\zeta) = \sum_0^\infty \psi_n(\zeta)$, where, using Theorem 16.19,

$$\psi_n(\zeta) = \sum_{k=0}^{n} S(\pi_k(X))(\zeta)S(\pi_{n-k}(Y))(\zeta) = S\Big(\sum_{k=0}^{n} \pi_k(X) \odot \pi_{n-k}(Y)\Big)(\zeta);$$

note that ψ_n is homogeneous of degree n.

If $Z = X \odot Y \in L^{1+}$, then, by Theorem 16.16(iii), $\psi_n = S(\pi_n(Z))$ for every $n \geq 0$, and the result follows since S is injective.

Conversely, if $\pi_n(Z) = \sum_{k=0}^{n} \pi_k(X) \odot \pi_{n-k}(Y)$, then $S(\pi_n(Z)) = \psi_n$ and Theorem 16.16(ii) again implies $SZ(\zeta) = SX(\zeta)SY(\zeta)$. $\square$

REMARK 16.35. Another possible way to extend the Wick product is by continuity. For example, fixing $p \in (0, \infty)$, we may say that $X \odot Y = Z$ for two given random variables $X, Y \in L^p$ if there exist sequences $(X_n)_1^\infty$ and $(Y_n)_1^\infty$ of polynomial variables in $\mathcal{P}$ such that $X_n \to X$, $Y_n \to Y$ and $X_n \odot Y_n \to Z$ in L^p. If $p \geq 1$, it follows by continuity of the T-transform, see Theorem 16.10, that then $Z = X \odot Y$ according to Definition 16.27 (and if $p > 1$ thus also according to Definition 16.25); in particular, Z is unique if it exists.

We do not know whether this definition is equivalent to our definitions above, i.e. whether such sequences $(X_n)_1^\infty$ and $(Y_n)_1^\infty$ always exist when $X \odot Y$ is defined by Definition 16.25 or Definition 16.27.

The creation operator $A^*(\xi)$ was defined in Chapter 13 by $A^*(\xi)(X) = \xi \odot X$ for $X \in \overline{\mathcal{P}}_*$, and then extended to a larger domain. This extension is given by the extended Wick product as follows.

THEOREM 16.36. *Let $X \in L^2$ and $\xi \in H$. Then $X \in \operatorname{Dom}(A^*(\xi))$ if and only if the Wick product $\xi \odot X$ exists in L^2, and then $A^*(\xi)X = \xi \odot X$. In particular, for $\zeta \in H_{\mathbf{C}}$,*

$$S\big(A^*(\xi)X\big)(\zeta) = S\xi(\zeta)SX(\zeta) = \mathrm{E}(\zeta\xi)SX(\zeta).$$

PROOF. If $X \in \mathrm{Dom}(A^*(\xi))$ and $\zeta \in H$, then by Examples 13.10 and 16.5,

$$\begin{aligned} S(A^*(\xi)X)(\zeta) &= \langle A^*(\xi)X, :e^{\zeta}:\rangle = \langle X, A(\xi):e^{\zeta}:\rangle \\ &= \langle X, \mathrm{E}(\xi\zeta):e^{\zeta}:\rangle = \mathrm{E}(\xi\zeta)SX(\zeta) = S\xi(\zeta)SX(\zeta). \end{aligned}$$

(Alternatively, this follows from the case $X \in \overline{\mathcal{P}}_*$ by continuity.) Hence $\xi \odot X$ exists in L^2 with $\xi \odot X = A^*(\xi)X$.

Conversely, if $\xi \odot X$ exists in L^2, then for $\zeta \in H$,

$$\langle \xi \odot X, :e^{\zeta}:\rangle = S(\xi \odot X)(\zeta) = \mathrm{E}(\xi\zeta)SX(\zeta) = \langle X, \mathrm{E}(\xi\zeta):e^{\zeta}:\rangle = \langle X, \partial_\xi :e^{\zeta}:\rangle.$$

Since the family $\{:e^{\zeta}:\}_{\zeta\in H}$ is total in $\mathcal{D}^{\xi,2}$ by Corollary 15.112, it follows by continuity and Theorem 15.98 that

$$\langle \xi \odot X, Y\rangle = \langle X, \partial_\xi Y\rangle = \langle X, A(\xi)Y\rangle$$

for every $Y \in \mathcal{D}^{\xi,2} = \mathrm{Dom}(A(\xi))$. Hence $A^*(\xi)X = \xi \odot X$. □

The creation operator, which is a special case of the Wick product by Theorem 16.36, is a local operator by Theorems 15.135 and 15.144. The next example shows that, in contrast, the Wick product is *not* local.

EXAMPLE 16.37. Consider as in Example 16.8 a one-dimensional Gaussian Hilbert space $\{t\xi\}$ with ξ standard normal. Let $X = f(\xi)$ and $Y = g(\xi)$ where $f(x) = e^{ax^2/2}$ and $g(x) = \mathbf{1}[0 < x < 1]$, with $0 < a < 1$. Let further $f_1(x) = f(x)\frac{1}{\sqrt{2\pi}}e^{-x^2/2} = \frac{1}{\sqrt{2\pi}}e^{-(1-a)x^2/2}$ and $g_1(x) = g(x)\frac{1}{\sqrt{2\pi}}e^{-x^2/2}$. Then, by Example 16.8, $TX(t\xi) = \hat{f}_1(t) = (1-a)^{-1/2}e^{-t^2/2(1-a)}$ and $TY(t\xi) = \hat{g}_1(t)$. Consequently,

$$e^{t^2/2}TX(t\xi)TY(t\xi) = (1-a)^{-1/2}e^{-at^2/2(1-a)}\hat{g}_1(t) = \hat{\varphi}(t)\hat{g}_1(t),$$

with $\varphi(x) = (2\pi a)^{-1/2}e^{-(1-a)x^2/2a}$.

Let $h_1 = \varphi * g_1 \in L^1(dx)$, $h(x) = h_1(x)\sqrt{2\pi}e^{x^2/2}$ and $Z = h(\xi) \in L^1$. Then

$$TZ(t\xi) = \hat{h}_1(t) = \hat{\varphi}(t)\hat{g}_1(t) = e^{t^2/2}TX(t\xi)TY(t\xi)$$

for every $t \in \mathbb{R}$, and thus $X \odot Y = Z$ by Definition 16.25.

Since $\varphi > 0$ and $g_1 \geq 0$ on $\mathbb{R}$, $h_1(x) > 0$ for every x and thus $Z > 0$ everywhere, although $Y = 0$ on $\{\xi < 0\}$. Hence Wick multiplication by X is not local.

Taking $a < 1/2$ we further have, as is easily shown, $X \in L^2$, $Y \in L^\infty$ and $Z \in L^\infty$; in particular $X \odot Y$ then exists in L^2.

3. Fock space operators

Let H_1 and H_2 be two Gaussian Hilbert spaces, defined on probability spaces $(\Omega_1, \mathcal{F}_1, \mathrm{P}_1)$ and $(\Omega_2, \mathcal{F}_2, \mathrm{P}_2)$ respectively, and let $A\colon H_1 \to H_2$ be a bounded linear operator. Then A has a unique extension to a complex linear operator $H_{1\mathbb{C}} \to H_{2\mathbb{C}}$, which we also denote by A. Moreover, the adjoint of this complex linear version of A is the extension of the adjoint $A^*\colon H_2 \to H_1$ of A; hence we can use the notation A^* for both the real and the complex versions.

THEOREM 16.38. *If* $A\colon H_1 \to H_2$ *has norm* $\|A\| \le 1$, *then*

$$S(\Gamma(A)X)(\zeta) = SX(A^*\zeta), \qquad X \in L^{1+}(\Omega_1, \mathcal{F}_1, \mathrm{P}_1),\ \zeta \in H_{2\mathbb{C}},$$
$$T(\Gamma(A)X)(\zeta) = e^{\mathrm{E}(A^*\zeta)^2/2 - \mathrm{E}\zeta^2/2} TX(A^*\zeta), \quad X \in L^{1}(\Omega_1, \mathcal{F}_1, \mathrm{P}_1),\ \zeta \in H_2.$$

If $X \in L^{1+}(\Omega_1, \mathcal{F}_1, \mathrm{P}_1)$, *then the latter equation holds for all* $\zeta \in H_{2\mathbb{C}}$.

PROOF. It follows from Theorems 4.12, 16.10 and 16.11 that for each fixed $\zeta \in H_{2\mathbb{C}}$, both sides of both equalities are continuous linear functionals of $X \in L^p(\Omega_1)$, $p > 1$; if $\zeta \in H_2$ is real, then the two sides of the second equation are even continuous linear functionals on $L^1(\Omega_1)$. Hence it suffices to verify the equalities for a set of X that is total in each $L^p(\Omega_1)$, $1 \le p < \infty$; we choose to verify them on the set of Wick exponentials, which is total by Corollary 3.40.

Thus, let $X = {:}e^{\xi}{:}$, with $\xi \in H_1$ and let $\zeta \in H_{2\mathbb{C}}$. Then, by Example 4.8, $\Gamma(A)X = {:}e^{A\xi}{:}$ and thus, using Example 16.6 twice,

$$S(\Gamma(A)X)(\zeta) = e^{\mathrm{E}((A\xi)\zeta)} = e^{\mathrm{E}(\xi A^*\zeta)} = SX(A^*\zeta).$$

Furthermore, by Theorem 16.3,

$$\begin{aligned} T(\Gamma(A)X)(\zeta) &= e^{-\mathrm{E}\zeta^2/2} S(\Gamma(A)X)(i\zeta) = e^{-\mathrm{E}\zeta^2/2} SX(iA^*\zeta) \\ &= e^{-\mathrm{E}\zeta^2/2 + \mathrm{E}(A^*\zeta)^2/2} TX(A^*\zeta). \end{aligned}$$

□

Taking $H_1 = H_2$ and $A = rI$, we obtain as a special case results for the Mehler transform.

COROLLARY 16.39. *If* H *is a Gaussian Hilbert space and* $-1 \le r \le 1$, *then*

$$S(M_r X)(\zeta) = SX(r\zeta), \qquad X \in L^{1+},\ \zeta \in H_{\mathbb{C}},$$
$$T(M_r X)(\zeta) = e^{(r^2-1)\mathrm{E}\zeta^2/2} TX(r\zeta), \qquad X \in L^1,\ \zeta \in H.$$

□

Another application of Theorem 16.38 is the following result for conditional expectations.

THEOREM 16.40. *Let H be a Gaussian Hilbert space and H_1 a closed subspace of H. Suppose that $X \in L^1$ and let $X_1 = \mathrm{E}(X \mid \mathcal{F}(H_1))$. Suppose further that $\zeta_1 \in H_{1\mathbf{C}}$ and $\zeta_2 \in H_{1\mathbf{C}}^{\perp} \subset H_{\mathbf{C}}$.*

(i) *If $X \in L^{1+}$, then $SX_1(\zeta_1+\zeta_2) = SX(\zeta_1)$.*

(ii) *If $X \in L^{1+}$ or $\zeta_1, \zeta_2 \in H$, then $TX_1(\zeta_1+\zeta_2) = e^{-\mathrm{E}\zeta_2^2/2}TX(\zeta_1)$.*

PROOF. By Theorem 4.9, $X_1 = \Gamma(P)X$, where $P\colon H \to H_1$ is the orthogonal projection, and the result follows by Theorem 16.38. □

COROLLARY 16.41. *Let H be a Gaussian Hilbert space and H_1 a closed subspace of H. If $X \in L^{1+}$, then X is $\mathcal{F}(H_1)$-measurable if and only if*

$$SX(\zeta) = SX(\zeta + \zeta_1)$$

for every $\zeta \in H$ and $\zeta_1 \in H_1^{\perp}$. (Then the formula holds for $\zeta \in H_{\mathbf{C}}$ and $\zeta_1 \in H_{1\mathbf{C}}^{\perp}$ too.)

PROOF. Let $X_1 = \mathrm{E}(X \mid \mathcal{F}(H_1))$. Then X is $\mathcal{F}(H_1)$-measurable if and only if $X = X_1$, and thus if and only if $SX = SX_1$, and the result follows by Theorem 16.40. □

If we want to extend Theorem 16.38 to the case of a complex linear map $A\colon H_{1\mathbf{C}} \to H_{2\mathbf{C}}$, we have to be careful with the definition of the adjoint operator. Recall that $A^*\colon H_{2\mathbf{C}} \to H_{1\mathbf{C}}$ is defined by

$$\langle \zeta_1, A^*\zeta_2 \rangle = \langle A\zeta_1, \zeta_2 \rangle$$

and that the map $A \mapsto A^*$ is anti-linear. We define another linear operator $A^t\colon H_{2\mathbf{C}} \to H_{1\mathbf{C}}$ by $A^t(\zeta) = \overline{A^*(\bar{\zeta})}$, or, equivalently,

$$\mathrm{E}(\zeta_1 A^t(\zeta_2)) = \mathrm{E}(A(\zeta_1)\zeta_2).$$

Note that if $A\colon H_1 \to H_2$ is a real operator, then $A^t = A^*$.

REMARK 16.42. If we express the operators as matrices, using some orthonormal bases in H_1 and H_2, then the matrix of A^t is the transpose of the matrix of A, while the matrix of A^* is the Hermitian conjugate.

THEOREM 16.43. *If $A\colon H_{1\mathbf{C}} \to H_{2\mathbf{C}}$ has norm $\|A\| \le 1$, then*

$$S(\Gamma(A)X)(\zeta) = SX(A^t\zeta), \qquad X \in L^2(\Omega_1, \mathcal{F}_1, \mathrm{P}_1),\ \zeta \in H_{2\mathbf{C}},$$

$$T(\Gamma(A)X)(\zeta) = e^{\mathrm{E}(A^t\zeta)^2/2 - \mathrm{E}\zeta^2/2}TX(A^t\zeta), \qquad X \in L^2(\Omega_1, \mathcal{F}_1, \mathrm{P}_1),\ \zeta \in H_{2\mathbf{C}}.$$

PROOF. As for Theorem 16.38. □

Of course, if we know that $\Gamma(A)$ extends continuously to an operator $L^p(\Omega_1) \to L^q(\Omega_2)$ with $p, q > 1$, then these formulae hold for every $X \in L^p$. For example, this yields the following extension of Corollary 16.39.

COROLLARY 16.44. *If $1 < p < \infty$ and D_{pp} is as in Theorem 5.28, then for every $z \in D_{pp}$,*

$$S(M_z X)(\zeta) = SX(z\zeta), \qquad X \in L^p,\ \zeta \in H_{\mathbb{C}},$$
$$T(M_z X)(\zeta) = e^{(z^2-1)\operatorname{E}\zeta^2/2} TX(z\zeta), \qquad X \in L^p,\ \zeta \in H_{\mathbb{C}}.$$

□

4. Stochastic integration

In this section, I is a Gaussian stochastic integral on a measure space $(M, \mathcal{M}, \mu)$, and Z is the corresponding Gaussian stochastic measure. We assume for simplicity that $I\colon L^2_{\mathbb{R}}(M, \mathcal{M}, \mu) \to H$ is onto. As in Remark 16.9, we regard the S-transform of a random variable as a function on $L^2_{\mathbb{R}}(M, \mathcal{M}, \mu)$ or $L^2_{\mathbb{C}}(M, \mathcal{M}, \mu)$; if we slightly abuse the notation by using S for both versions,

$$SX(\varphi) = SX(I(\varphi)) = \operatorname{E}(X :e^{I(\varphi)}:), \qquad X \in L^{1+},\ \varphi \in L^2(M, \mathcal{M}, \mu).$$

In particular, the transform of a Gaussian variable $I(f) \in H$ is given by, cf. Example 16.5,

$$S(I(f))(\varphi) = S(I(f))(I(\varphi)) = \operatorname{E}(I(f)I(\varphi)) = \int_M f\varphi \, d\mu. \tag{16.6}$$

For multiple stochastic integrals there is a similar formula.

THEOREM 16.45. *If $f \in L^2(M^n, \mathcal{M}^n, \mu^n)$, then $\int_{M^n} f \, dZ^n \in H^{:n:}$ has the S-transform, for $\varphi \in L^2(M, \mathcal{M}, \mu)$,*

$$\begin{aligned} S\Big(\int_{M^n} f \, dZ^n\Big)(\varphi) &= \langle f, \bar{\varphi}^{\otimes n} \rangle_{L^2(M^n)} \\ &= \int_{M^n} f(t_1, \dots, t_n)\varphi(t_1) \cdots \varphi(t_n) \, d\mu(t_1) \cdots d\mu(t_n). \end{aligned} \tag{16.7}$$

PROOF. Suppose first that $f = f_1 \otimes \cdots \otimes f_n$. Then by Theorem 7.26, $\int_{M^n} f \, dZ^n = \,:\!\int_M f_1 \, dZ \cdots \int_M f_n \, dZ\!:$ and thus, by Theorem 16.19 and (16.6),

$$\begin{aligned} S\Big(\int_{M^n} f \, dZ^n\Big)(\varphi) &= \prod_{i=1}^n S\Big(\int_M f_i \, dZ\Big)(\varphi) = \prod_{i=1}^n \int_M f_i \varphi \, d\mu \\ &= \int_{M^n} f_1(t_1) \cdots f_n(t_n)\varphi(t_1) \cdots \varphi(t_n) \, d\mu(t_1) \cdots d\mu(t_n), \end{aligned}$$

as claimed. Since such functions f are total in $L^2(M^n)$ and, for fixed φ, the expressions in (16.7) are continuous linear functions of f, cf. Theorems 7.26 and 16.11, the result holds for every f. □

Recall the definition of the Skorohod integral in Section 7.3.

THEOREM 16.46. *If $X_t \in L^2(M, \mathcal{M}, \mu; L^2(\Omega, \mathcal{F}(H), \mathrm{P}))$ is a random function such that the Skorohod integral $\int_M X_t \, dZ(t)$ exists, then*

$$S\Big(\int_M X_t \, dZ(t)\Big)(\varphi) = \int_M S(X_t)(\varphi) \, \varphi(t) \, d\mu(t), \qquad \varphi \in L^2(M, \mathcal{M}, \mu).$$

REMARK 16.47. The corresponding formula for the T-transform is, by Theorem 16.3,

$$T\Big(\int_M X_t \, dZ(t)\Big)(\varphi) = i \int_M T(X_t)(\varphi) \, \varphi(t) \, d\mu(t), \qquad \varphi \in L^2(M, \mathcal{M}, \mu).$$

We begin the proof of the theorem with a lemma.

LEMMA 16.48. *If the random function $X_t \in L^2(M, \mathcal{M}, \mu; L^2(\Omega, \mathcal{F}(H), \mathrm{P}))$ and $\varphi \in L^2(M, \mathcal{M}, \mu)$, then the integral $\int_M S(X_t)(\varphi)\varphi(t) \, d\mu(t)$ converges absolutely and*

$$\int_M S(X_t)(\varphi)\varphi(t) \, d\mu(t) = \sum_{n=0}^{\infty} \int_M S(\pi_n(X_t))(\varphi)\varphi(t) \, d\mu(t).$$

PROOF. By Theorem 16.11, $X \mapsto S(X)(\varphi)$ is a continuous linear functional on $L^2(\Omega, \mathcal{F}, \mathrm{P})$. Thus, for fixed φ, $X_t \mapsto S(X_t)(\varphi)$ is a bounded linear map $\widetilde{L}^2 = L^2(M; L^2(\Omega)) \to L^2(M, \mathcal{M}, \mu)$, and the Cauchy–Schwarz inequality implies that the mapping $X_t \mapsto \psi(X_t) = \int_M S(X_t)(\varphi)\varphi(t) \, d\mu(t)$ defines a continuous linear functional ψ on $\widetilde{L}^2$, with the integral absolutely convergent.

Moreover, since $X_t = \sum_0^\infty \pi_n(X_t)$ in $\widetilde{L}^2$ by dominated convergence, it follows that $\psi(X_t) = \sum_0^\infty \psi(\pi_n(X_t))$. □

PROOF OF THEOREM 16.46. By the definition (7.38), $\int_M X_t \, dZ(t) = \sum_{n=0}^\infty \int_M \pi_n(X_t) \, dZ(t)$ and thus

$$S\Big(\int_M X_t \, dZ(t)\Big)(\varphi) = \sum_{n=0}^{\infty} S\Big(\int_M \pi_n(X_t) \, dZ(t)\Big).$$

This and Lemma 16.48 show that it suffices to prove the claim for each $\pi_n(X)$ separately; equivalently, we may assume that $X_t \in H^{:n:}$ for some fixed n.

Thus, let as in (7.33) $X_t = \int_{M^n} F_t \, dZ^n$; then by (7.36) $\int_M X_t \, dZ(t) = \int F \, dZ^{n+1}$, with F defined by (7.34). Theorem 16.45 now yields

$$S(X_t)(\varphi) = \int_{M^n} F_t(t_1, \dots, t_n)\varphi(t_1) \cdots \varphi(t_n) \, d\mu(t_1) \cdots d\mu(t_n)$$

and

$$\begin{aligned} S\Big(\int_M X_t \, dZ(t)\Big)(\varphi) &= \int_{M^{n+1}} F(t_1, \dots, t_{n+1})\varphi(t_1) \cdots \varphi(t_{n+1}) \, d\mu(t_1) \cdots d\mu(t_{n+1}) \\ &= \int_M S(X_{t_{n+1}})\varphi(t_{n+1}) \, d\mu(t_{n+1}). \end{aligned}$$

□

Since Itô integrals are special cases of Skorohod integrals by Theorem 7.41, we obtain in particular the following.

COROLLARY 16.49. *Let B_t, $t \geq 0$, be a Brownian motion and let X_t be a square integrable predictable process. Then*

$$S\Big(\int_0^\infty X_t\,dB(t)\Big)(\varphi) = \int_0^\infty S(X_t)(\varphi)\varphi(t)\,dt, \qquad \varphi \in L^2([0,\infty)).$$

□

The converse to Theorem 16.46 also holds.

THEOREM 16.50. *If $X_t \in L^2(M,\mathcal{M},\mu; L^2(\Omega,\mathcal{F}(H),\mathrm{P}))$ and there exists $Y \in L^2(\Omega,\mathcal{F},\mathrm{P})$ such that*

$$S(Y)(\varphi) = \int_M S(X_t)(\varphi)\varphi(t)\,d\mu(t), \qquad \varphi \in L^2(M,\mathcal{M},\mu),$$

then the Skorohod integral $\int_M X_t\,dZ(t)$ exists and equals Y.

PROOF. Let $\psi_n = S\big(\int_M \pi_n(X_t)\,dZ(t)\big)$; by Theorem 16.46,

$$\psi_n(\varphi) = \int_M S(\pi_n(X_t))(\varphi)\varphi(t)\,d\mu(t).$$

Lemma 16.48 and the assumption thus yield, for every $\varphi \in L^2(M,\mathcal{M},\mu)$,

$$\begin{aligned}\sum_{n=0}^\infty \psi_n(\varphi) &= \sum_{n=0}^\infty \int_M S(\pi_n(X_t))(\varphi)\varphi(t)\,d\mu(t)\\ &= \int_M S(X_t)(\varphi)\varphi(t)\,d\mu(t) = S(Y)(\varphi).\end{aligned} \tag{16.8}$$

Moreover, $S(\pi_n(X_t))$ is homogeneous of degree n by Theorem 16.16(i), and thus, for every real λ and $\varphi \in L^2(M,\mathcal{M},\mu)$,

$$\psi_n(\lambda\varphi) = \int_M \lambda^n S(\pi_n(X_t))(\varphi)\lambda\varphi(t)\,d\mu(t) = \lambda^{n+1}\psi_n(\varphi).$$

Consequently, ψ_n is homogeneous of degree $n+1$, so ψ_{n-1} is homogeneous of degree n. Since (16.8) may be written $S(Y)(\varphi) = \sum_0^\infty \psi_{n-1}(\varphi)$, with $\psi_{-1}(\varphi) = 0$, Theorem 16.16(iii) yields

$$S(\pi_n(Y)) = \psi_{n-1} = S\Big(\int_M \pi_{n-1}(X_t)\,dZ(t)\Big)$$

and thus

$$\pi_n(Y) = \int_M \pi_{n-1}(X_t)\,dZ(t), \qquad n \geq 1,$$

while $S(\pi_0(Y)) = \psi_{-1} = 0$ and thus $\pi_0(Y) = 0$. The definition (7.38) of the Skorohod integral now yields

$$\int_M X_t \, dZ(t) = \sum_{n=0}^{\infty} \pi_{n+1}(Y) = Y.$$

□

We can now use the S-transform to prove an extension of Theorem 7.40 with the extended Wick product defined in Section 2.

THEOREM 16.51. *Suppose that $X \in L^2(\Omega, \mathcal{F}, \mathrm{P})$ and $f \in L^2(M, \mathcal{M}, \mu)$. Then*

$$\int_M f(t) X \, dZ(t) = X \odot \int_M f(t) \, dZ(t)$$

in the sense that if one side exists in L^2, then so does the other.

PROOF. We have, using (16.6),

$$\begin{aligned}\int_M S\big(f(t)X\big)(\varphi)\varphi(t) \, d\mu(t) &= \int_M f(t) S(X)(\varphi)\varphi(t) \, d\mu(t) \\ &= S(X)(\varphi) \int_M f(t)\varphi(t) \, d\mu(t) = S(X)(\varphi) S(I(f))(\varphi).\end{aligned}$$

Hence, by Theorems 16.46 and 16.50 and Definition 16.25, the Skorohod integral exists if and only if the Wick product does, and then they coincide.

□

Theorems 16.46 and 16.50 make it possible to define more general Skorohod integrals in L^p, $p > 1$. Using the T-transform analogously, cf. Remark 16.47, we can further extend this to L^1 as follows.

DEFINITION 16.52. If $X = (X_t)_{t \in M} \in L^1(M \times \Omega) = L^1(M; L^1(\Omega)) = L^1(\Omega; L^1(M))$ and $Y \in L^1$ are such that

$$i \int T(X_t)(\varphi) \, \varphi(t) \, d\mu(t) = T(Y)(\varphi) \tag{16.9}$$

for every $\varphi \in L^2_{\mathbb{R}}(M) \cap L^\infty_{\mathbb{R}}(M)$, then we say that the Skorohod integral $\int_M X_t \, dZ(t)$ equals Y.

Note that the integral in the left hand side of (16.9) converges for $\varphi \in L^2_{\mathbb{R}}(M) \cap L^\infty_{\mathbb{R}}(M)$ since $|T(X_t)(\varphi)| \le \|X_t\|_{L^1} \in L^1(M)$, and that $\int_M X_t \, dZ(t)$ is unique, if it exists, by Theorem 16.10 and the fact that $L^2_{\mathbb{R}}(M) \cap L^\infty_{\mathbb{R}}(M)$ is dense in $L^2_{\mathbb{R}}(M)$.

Note further that if $(X_t)_{t \in M} \in L^2(M; L^1(\Omega))$ and $Y = \int_M X_t \, dZ(t)$, then $T(X_t)(\varphi) \in L^2(M)$ and thus the left hand side of (16.9) is defined for every $\varphi \in L^2_{\mathbb{R}}(M)$; moreover, (16.9) then holds for every $\varphi \in L^2_{\mathbb{R}}(M)$, as can be seen by applying (16.9) to truncations of φ and using the continuity of $T(X_t)$ and $T(Y)$ on $L^2_{\mathbb{R}}(M)$ together with dominated convergence.

We will see in Section 6 that this general definition of the Skorohod integral extends the definition given in Remark 15.158.

5. The Cameron–Martin shift

By Theorem 14.1(iv), the S-transform can, on H, alternatively be defined as follows.

THEOREM 16.53. *If $X \in L^{1+}$ and $\xi \in H$, then*

$$SX(\xi) = \mathrm{E}\,\rho_\xi X.$$

□

This leads to the following results.

THEOREM 16.54. *If $X \in L^{1+}$, $\xi \in H$ and $\zeta \in H_{\mathbb{C}}$, then*

$$S(\rho_\xi X)(\zeta) = SX(\zeta + \xi).$$

PROOF. If $\zeta \in H$, then by Theorems 16.53 and 14.1(ii),

$$S(\rho_\xi X)(\zeta) = \mathrm{E}\,\rho_\zeta(\rho_\xi(X)) = \mathrm{E}\,\rho_{\zeta+\xi}(X) = SX(\zeta + \xi).$$

The result for general $\zeta \in H_{\mathbb{C}}$ follows by analytic continuation. □

COROLLARY 16.55. *Let H be a Gaussian Hilbert space and H_1 a closed subspace of H. If $X \in L^0$, then X is $\mathcal{F}(H_1)$-measurable if and only if $\rho_\xi X = X$ for every $\xi \perp H_1$.*

PROOF. Suppose first that $X \in L^{1+}$. If $\xi \in H$, then $\rho_\xi X = X$ if and only if $SX(\zeta) = S(\rho_\xi X)(\zeta)$ for every $\zeta \in H$, which by Theorem 16.54 holds if and only if $SX(\zeta) = SX(\zeta + \xi)$ for every $\zeta \in H$. The result in the case $X \in L^{1+}$ thus follows by Corollary 16.41.

In general, let $f(x) = x/(|x| + 1)$. Since $f\colon \mathbb{C} \to \mathbb{C}$ is invertible, X is $\mathcal{F}(H_1)$-measurable if and only if $f(X)$ is, and, using Theorem 14.1(iii), $\rho_\xi X = X$ if and only if $\rho_\xi(f(X)) = f(X)$. The result thus follows from the special case just treated applied to $f(X) \in L^\infty \subset L^{1+}$. □

In particular, taking $H_1 = \{0\}$, this yields another proof of Theorem 14.9.

6. Malliavin calculus

We begin with the directional derivative. We define, for any function F on H or $H_{\mathbb{C}}$,

$$\partial_\xi F(\zeta) = \frac{d}{dt} F(\zeta + t\xi)\big|_{t=0}, \qquad \xi, \zeta \in H\ (H_{\mathbb{C}})$$

(provided this derivative exists).

THEOREM 16.56. *If $X \in \mathcal{D}^{\xi,p}$ with $p > 1$ and $\xi \in H$, then*

$$S(\partial_\xi X) = \partial_\xi(SX).$$

PROOF. Let $1 < q < p$. By Theorem 14.1(viii), $s \mapsto \rho_{s\xi}(\partial_\xi X)$ is a continuous map into L^q; it follows from Lemma 15.89 that, as $t \to 0$, $(\rho_{t\xi}(X) - X)/t \to \partial_\xi X$ in L^q. Thus, for every $\zeta \in H_{\mathbf{C}}$, since the S-transform is continuous on L^q by Theorem 16.11, using also Theorem 16.54,

$$\frac{SX(\zeta + t\xi) - SX(\zeta)}{t} = S\Big(\frac{\rho_{t\xi}X - X}{t}\Big)(\zeta) \to S(\partial_\xi X)(\zeta).$$

□

The formula for the T-transform is more complicated.

THEOREM 16.57. *If $X \in \mathcal{D}^{\xi,1}$ with $\xi \in H$, then $\partial_\xi(TX)$ exists on H and*

$$T(\partial_\xi X)(\zeta) = -i\partial_\xi TX(\zeta) - i\,\mathrm{E}(\xi\zeta)TX(\zeta), \qquad \zeta \in H.$$

PROOF. Suppose first that $X \in \mathcal{D}^{\xi,2}$. Then Theorems 16.3 and 16.56 yield

$$\begin{aligned}\partial_\xi TX(\zeta) &= \partial_\xi\big(e^{-\mathrm{E}\zeta^2/2}SX(i\zeta)\big)\\ &= -\mathrm{E}(\zeta\xi)e^{-\mathrm{E}\zeta^2/2}SX(i\zeta) + ie^{-\mathrm{E}\zeta^2/2}S(\partial_\xi X)(i\zeta)\\ &= -\mathrm{E}(\zeta\xi)TX(\zeta) + iT(\partial_\xi X)(\zeta).\end{aligned}$$

Hence, for $\zeta \in H$ and $t \in \mathbb{R}$,

$$TX(\zeta + t\xi) = TX(\zeta) + \int_0^t \big(iT(\partial_\xi X)(\zeta + s\xi) - \mathrm{E}(\xi(\zeta + s\xi))TX(\zeta + s\xi)\big)ds. \tag{16.10}$$

For fixed $\zeta \in H$ and $t \in \mathbb{R}$, both sides of (16.10) are, using Theorem 16.10, continuous linear functionals of $X \in \mathcal{D}^{\xi,1}$. Since $\mathcal{D}^{\xi,2}$ is dense in $\mathcal{D}^{\xi,1}$ by Theorem 15.106, this implies that (16.10) holds for every $X \in \mathcal{D}^{\xi,1}$. The integrand in (16.10) is continuous by Theorem 16.10, and it follows that $\partial_\xi(TX)(\zeta) = iT(\partial_\xi X)(\zeta) - \mathrm{E}(\xi\zeta)TX(\zeta)$. □

Converses to these theorems hold too.

THEOREM 16.58. *Let $1 < p \le \infty$ and $\xi \in H$. If $X, Y \in L^p$ with $SY = \partial_\xi SX$, then $X \in \mathcal{D}^{\xi,p}$ and $\partial_\xi X = Y$.*

PROOF. Let $1 < q < p$, and fix $t \in \mathbb{R}$. Define $Z = \int_0^t \rho_{s\xi}(Y)\,ds$; since $s \mapsto \rho_{s\xi}(Y)$ is a continuous function into L^q by Theorem 14.1(ix), the integral converges as a Bochner integral in L^q, and $Z \in L^q$. Since the S-transform is continuous on L^q, we obtain for each fixed $\zeta \in H$, also using Theorem 16.54,

$$\begin{aligned}SZ(\zeta) &= \int_0^t S(\rho_{s\xi}(Y))(\zeta)\,ds = \int_0^t SY(\zeta + s\xi)\,ds = \int_0^t \partial_\xi SX(\zeta + s\xi)\,ds\\ &= SX(\zeta + t\xi) - SX(\zeta) = S(\rho_{t\xi}X)(\zeta) - SX(\zeta),\end{aligned}$$

and thus $Z = \rho_{t\xi}(X) - X$. The result follows by Lemma 15.89. □

LEMMA 16.59. *Let $Z \in L^1$, $t \in \mathbb{R}$ and $\xi, \zeta \in H$. Let further $\alpha = 1/2 + \mathrm{E}\,\xi^2$ and let the probability measure $\nu = \gamma_{\sqrt{2}}$ on $\mathbb{R}$ be the distribution of* $\mathrm{N}(0,2)$. *Then*

$$T\big(e^{-\xi^2}\rho_{t\xi}(Z)\big)(\zeta) = \int_{-\infty}^{\infty} TZ(\zeta + u\xi)e^{-i\alpha ut + \alpha t^2/2 - it\,\mathrm{E}(\xi\zeta)}\,d\nu(u).$$

In particular, $T(e^{-\xi^2}Z)(\zeta) = \int_{-\infty}^{\infty} TZ(\zeta + u\xi)\,d\nu(u)$.

PROOF. Note that $e^{-\xi^2}\rho_{t\xi}(Z) \in L^1$ by Lemma 15.70. We write $\sigma^2 = \mathrm{E}\,\xi^2$ and $\beta = \mathrm{E}(\xi\zeta)$ and have, using Theorem 14.1(v) and the formula $\int e^{ixu}d\nu(u) = e^{-x^2}$, $x \in \mathbb{R}$,

$$\begin{aligned}\mathrm{E}\big(e^{-\xi^2}\rho_{t\xi}(Z)e^{i\zeta}\big) &= \mathrm{E}\big({:}e^{t\xi}{:}\,Z\rho_{-t\xi}(e^{-\xi^2+i\zeta})\big) = \mathrm{E}\,Ze^{t\xi - t^2\sigma^2/2 - (\xi - t\sigma^2)^2 + i(\zeta - t\beta)} \\ &= \mathrm{E}\,Ze^{-(\xi-\alpha t)^2 + \alpha t^2/2 + i\zeta - it\beta} = \mathrm{E}\int_{-\infty}^{\infty} Ze^{iu(\xi - \alpha t) + \alpha t^2/2 + i\zeta - it\beta}d\nu(u).\end{aligned}$$

The result follows by Fubini's theorem. The final formula is obtained by taking $t = 0$. □

THEOREM 16.60. *Let $\xi \in H$. If $X, Y \in L^1$, $\partial_\xi(TX)$ exists and $TY(\zeta) = -i\partial_\xi TX(\zeta) - i\,\mathrm{E}(\xi\zeta)TX(\zeta)$ for $\zeta \in H$, then $X \in \mathcal{D}^{\xi,1}$ and $\partial_\xi X = Y$.*

PROOF. Let $\zeta \in H$, and write $\sigma^2 = \mathrm{E}\,\xi^2$, $\alpha = 1/2 + \sigma^2$ and $\beta = \mathrm{E}(\xi\sigma)$. By the lemma and the assumption, for every real s,

$$\begin{aligned}T\big(e^{-\xi^2}\rho_{s\xi}(Y)\big)(\zeta) &= \int_{-\infty}^{\infty} TY(\zeta + u\xi)e^{-i\alpha us + \alpha s^2/2 - i\beta s}\,d\nu(u) \\ &= -i\int_{-\infty}^{\infty}\frac{d}{du}TX(\zeta + u\xi)e^{-i\alpha us + \alpha s^2/2 - i\beta s}\,d\nu(u) \\ &\qquad - i(\beta + u\sigma^2)\int_{-\infty}^{\infty} TX(\zeta + u\xi)e^{-i\alpha us + \alpha s^2/2 - i\beta s}\,d\nu(u).\end{aligned}$$

The first integral on the right hand side equals by integration by parts, recalling $d\nu(u) = (4\pi)^{-1/2}e^{-u^2/4}\,du$,

$$i\int_{-\infty}^{\infty} TX(\zeta + u\xi)(-i\alpha s - u/2)e^{-i\alpha us + \alpha s^2/2 - i\beta s}\,d\nu(u)$$

and thus we obtain

$$T\big(e^{-\xi^2}\rho_{s\xi}(Y)\big)(\zeta) = \int_{-\infty}^{\infty} TX(\zeta + u\xi)(\alpha s - i\alpha u - i\beta)e^{-i\alpha us + \alpha s^2/2 - i\beta s}\,d\nu(u). \tag{16.11}$$

Let $t \in \mathbb{R}$ and define $W = \int_0^t e^{-\xi^2} \rho_{s\xi}(Y)\, ds$, which exists in L^1 by Lemma 15.70. Then (16.11), Fubini's theorem and Lemma 16.59 yield

$$\begin{aligned} TW(\zeta) &= \int_0^t T(e^{-\xi^2} \rho_{s\xi}(Y))(\zeta)\, ds \\ &= \int_0^t \int_{-\infty}^{\infty} TX(\zeta + u\xi)(-i\alpha u + \alpha s - i\beta) e^{-i\alpha us + \alpha s^2/2 - i\beta s}\, d\nu(u)\, ds \\ &= \int_{-\infty}^{\infty} TX(\zeta + u\xi)\big(e^{-i\alpha ut + \alpha t^2/2 - i\beta t} - 1\big)\, d\nu(u) \\ &= T(e^{-\xi^2} \rho_{t\xi}(X))(\zeta) - T(e^{-\xi^2} X)(\zeta). \end{aligned}$$

Hence $W = e^{-\xi^2} \rho_{t\xi}(X) - e^{-\xi^2} X$, and the result follows by Lemma 15.71. □

REMARK 16.61. Since the annihilation operator $A(\xi) = \partial_\xi$ (with domain $\mathcal{D}^{\xi,2}$) by Theorem 15.98, the theorems above yield characterizations of the annihilation operator by the S- or T-transform; in particular Theorem 16.56 yields

$$S(A(\xi)X) = \partial_\xi SX, \qquad X \in \operatorname{Dom}(A(\xi)).$$

This can be compared to the corresponding result for the creation operator given in Theorem 16.36, viz.

$$S(A^*(\xi)X)(\zeta) = \mathrm{E}(\zeta\xi) SX(\zeta), \qquad X \in \operatorname{Dom}(A^*(\xi)).$$

So far we have considered transforms of real and complex random variables, but it is clear that the definitions extend to Hilbert-space-valued variables; if H_0 is a complex Hilbert space and $X \in L^1(H_0)$, the TX is a bounded continuous function $H \to H_0$, and if $X \in L^{1+}(H_0)$, then TX and SX are analytic functions $H_\mathbb{C} \to H_0$. Recall also, see Appendix G, that if $F\colon H_\mathbb{C} \to \mathbb{C}$ is an analytic function, then F' is an analytic function with values in the dual space $H_\mathbb{C}^*$; note that there is a natural linear isomorphism $\iota\colon H_\mathbb{C} \to H_\mathbb{C}^*$ given by $\iota(\xi)(\xi') = \mathrm{E}(\xi\xi') = \langle \xi, \bar{\xi}' \rangle_{H_\mathbb{C}}$.

The theorems above and Theorem 15.64(viii) yield the following.

THEOREM 16.62. *Let $1 < p \le \infty$ and $X \in L^p$. Then $X \in \mathcal{D}^{1,p}$ if and only if there exists a random variable $Y \in L^p(H_\mathbb{C}^*)$ such that $(SX)' = SY$, and then $Y = \nabla X$ if $H_\mathbb{C}^*$ and $H_\mathbb{C}$ are identified by the natural isomorphism described above; thus*

$$\langle S(\nabla X)(\zeta), \bar{\xi} \rangle_{H_\mathbb{C}} = (SX)'(\zeta)(\xi), \qquad \zeta, \xi \in H_\mathbb{C}.$$

□

THEOREM 16.63. *Let $1 \le p \le \infty$ and $X \in L^p$. Then $X \in \mathcal{D}^{1,p}$ if and only if TX is continuously differentiable on H and there exists a random variable $Y \in L^p(H_\mathbb{C}^*)$ such that*

$$TY(\zeta) = -i(TX)'(\zeta) - iTX(\zeta) \otimes \iota(\zeta), \qquad \zeta \in H,$$

where $\iota\colon H_{\mathbb{C}} \to H_{\mathbb{C}}^*$ *is the natural isomorphism described above; moreover, in this case* $Y = \iota(\nabla X)$. □

For the divergence and number operators we have the following formulae.

THEOREM 16.64. *Let* $1 \le p \le \infty$, $X \in L^p(H_{\mathbb{C}})$ *and* $Y \in L^p$. *Then the following are equivalent.*

(i) δX *exists and* $\delta X = Y$.

(ii) $TY(\zeta) = i\langle TX(\zeta), \zeta\rangle_{H_{\mathbb{C}}}$ *for* $\zeta \in H$.

(iii) (*Provided* $p > 1$) $SY(\zeta) = \langle SX(\zeta), \bar{\zeta}\rangle_{H_{\mathbb{C}}}$ *for* $\zeta \in H$ (*or* $\zeta \in H_{\mathbb{C}}$).

PROOF. Theorem 15.131 yields the equivalence (i) ⇔ (ii), while (ii) ⇔ (iii) follows from Theorem 16.3. □

THEOREM 16.65. *Let* $1 < p \le \infty$ *and* $X \in L^p$. *If* $\mathcal{N}X$ *exists in* L^p, *then*

$$S(\mathcal{N}X)(\zeta) = \partial_\zeta SX(\zeta) = \frac{d}{dt} SX(t\zeta)\Big|_{t=1} = \langle (SX)'(\zeta), \bar{\zeta}\rangle, \qquad \zeta \in H_{\mathbb{C}}.$$

Conversely, if $\partial_\zeta SX(\zeta) = SY(\zeta)$ *for some* $Y \in L^p$ *and every* $\zeta \in H$, *then* $\mathcal{N}X$ *exists in* L^p *and* $\mathcal{N}X = Y$.

PROOF. By Theorem 16.16, $S(\pi_n(X))(\zeta)$ is homogeneous of degree n and $SX(\zeta) = \sum_0^\infty S(\pi_n(X))(\zeta)$; thus $\partial_\zeta SX(\zeta) = \sum_0^\infty nS(\pi_n(X))(\zeta)$. Consequently, using Theorem 16.16 again, $\partial_\zeta SX(\zeta) = SY(\zeta)$ if and only if $\pi_n(Y) = n\pi_n(X)$ for each n, i.e. $Y = \mathcal{N}X$ (cf. Definition 5.46). □

Finally, assume as in Section 4 that $I\colon L^2_{\mathbb{R}}(M, \mathcal{M}, \mu) \to H$ is a Gaussian stochastic integral and that I is onto, and regard the S- and T-transforms as functions on $L^2_{\mathbb{R}}(M, \mathcal{M}, \mu)$. As discussed in Section 15.11, I yields an isomorphism between $L^1(H_{\mathbb{C}})$ and $L^1(\Omega; L^2_{\mathbb{C}}(M))$, and we can regard δ as a mapping from a subset of $L^1(\Omega; L^2_{\mathbb{C}}(M))$ into L^1; moreover this map extends the Skorohod integral.

If $X = (X_t)_{t\in M} \in L^1(\Omega; L^2_{\mathbb{C}}(M))$ and $\xi \in H$, then $TX(\xi) = \mathrm{E}(X\,{:}e^{i\xi}{:}) \in L^2_{\mathbb{C}}(M)$ equals a.e. the function $t \mapsto \mathrm{E}(X_t\,{:}e^{i\xi}{:}) = TX_t(\xi)$, cf. Appendix C. Thus, Theorem 16.64 says in the present setting that $\delta X = Y \in L^1$ if and only if, for every $g \in L^2_{\mathbb{R}}(M)$,

$$TY(g) = i\langle TX(g), g\rangle_{L^2(M)} = i\int_M TX_t(g)g(t)\,d\mu(t).$$

Consequently, if δX exists in L^1, then the Skorohod integral $\int_M X_t\,dZ_t$ defined in Section 4 exists and equals δX. In other words, the definition of the Skorohod integral in Definition 16.52 is more general than the one in Remark 15.158.

Appendices

A. The monotone class theorem

The *monotone class theorem* is a generic name used for several related results. We find it convenient to use the following version, cf. Dellacherie and Meyer (1975, Theorem I.21) and Protter (1990, p. 7).

THEOREM A.1. *Suppose that $\mathcal{A}$ is a vector space of bounded real-valued functions on a set Ω such that $\mathcal{A}$ contains the constant functions, and has the following property: if $(f_n)_1^\infty$ is an increasing sequence of positive, uniformly bounded functions in $\mathcal{A}$, then $\lim_{n\to\infty} f_n \in \mathcal{A}$ (i.e., $\mathcal{A}$ is closed under bounded monotone convergence). Let $\mathcal{C}$ be a subset of $\mathcal{A}$ which is closed under multiplication. Then $\mathcal{A}$ contains every bounded function that is measurable with respect to the σ-field $\mathcal{F}(\mathcal{C})$ generated by $\mathcal{C}$.*

PROOF. Let $\mathcal{C}'$ be the subspace of $\mathcal{A}$ spanned by $\mathcal{C}$ and the constants; then $\mathcal{C}'$ is closed under multiplication and is thus an algebra. Let further $\mathcal{C}'_+ = \{f \in \mathcal{C}' : f \geq 0\}$.

Let $\mathfrak{A}$ be the family of all vector spaces of bounded real functions on Ω that are closed under bounded monotone convergence and contain $\mathcal{C}'$; further, let $\mathcal{A}_0 = \bigcap_{\mathcal{A}_1 \in \mathfrak{A}} \mathcal{A}_1$ be the minimal such space. Since $\mathcal{A} \in \mathfrak{A}$, $\mathcal{A}_0 \subseteq \mathcal{A}$.

Furthermore, if $f \in \mathcal{A}_0$, let $\mathcal{A}_f = \{g \in \mathcal{A}_0 : fg \in \mathcal{A}_0\}$. Then $\mathcal{A}_f$ is a subspace of $\mathcal{A}_0$, and if $f \geq 0$, then $\mathcal{A}_f$ is closed under bounded monotone convergence.

Suppose now that $f \in \mathcal{C}'_+$, then $fg \in \mathcal{C}' \subseteq \mathcal{A}_0$ for every $g \in \mathcal{C}'$ and thus $\mathcal{C}' \subseteq \mathcal{A}_f$; consequently $\mathcal{A}_f \in \mathfrak{A}$ and thus $\mathcal{A}_f \supseteq \mathcal{A}_0$. This shows that if $f \in \mathcal{C}'_+$ and $g \in \mathcal{A}_0$, then $fg \in \mathcal{A}_0$. More generally, if $f \in \mathcal{C}'$ and $g \in \mathcal{A}_0$, let $M = \sup|f|$; then $f + M \in \mathcal{C}'_+$ and thus $fg = (f+M)g - Mg \in \mathcal{A}_0$.

This proves (interchanging f and g) that $\mathcal{C}' \subseteq \mathcal{A}_f$ for every $f \in \mathcal{A}_0$, and the same argument again shows that if $f \in \mathcal{A}_0$ and $f \geq 0$, then $\mathcal{A}_f \supseteq \mathcal{A}_0$ and thus $fg \in \mathcal{A}_0$ for every $g \in \mathcal{A}_0$. Finally we conclude, writing $f = (f+M) - M$ as above, that $fg \in \mathcal{A}_0$ for every $f, g \in \mathcal{A}_0$. In other words, $\mathcal{A}_0$ is an algebra.

Let $\mathcal{F} = \{A \subseteq \Omega : \mathbf{1}_A \in \mathcal{A}_0\}$. Since $\mathcal{A}_0$ is an algebra, $\mathcal{F}$ contains $\varnothing$ and Ω, and is closed under complementation and finite intersections; consequently, $\mathcal{F}$ is a (Boolean) algebra. (The word 'algebra' is used in two different senses here.) Moreover, since $\mathcal{A}_0$ is closed under bounded monotone convergence, $\mathcal{F}$ is a σ-field.

Since $\mathcal{A}_0$ is a linear space, every simple $\mathcal{F}$-measurable function belongs to $\mathcal{A}_0$. Every bounded positive $\mathcal{F}$-measurable function is the limit of an increasing sequence of such functions, and thus belongs to $\mathcal{A}_0$; it follows that every bounded $\mathcal{F}$-measurable function belongs to $\mathcal{A}_0 \subseteq \mathcal{A}$.

It remains to prove that $\mathcal{F} \supseteq \mathcal{F}(\mathcal{C})$, or equivalently that every function in $\mathcal{C}$ is $\mathcal{F}$-measurable. In order to do this we first prove that $\mathcal{A}_0$ is closed under uniform convergence. Suppose that $(f_n)_1^\infty$ is a sequence in $\mathcal{A}_0$ which converges uniformly to f. By selecting a subsequence we may assume that $\sup|f_n - f| \le 2^{-n}$. Let $g_n = f_{n+1} - f_n + 2^{1-n}$; then $0 \le g_n \le 2^{2-n}$, and $\sum_1^n g_k$ converges monotonically to $f - f_1 + 2$. Since each $g_k \in \mathcal{A}_0$ and $\mathcal{A}_0$ is closed under bounded monotone convergence, $f - f_1 + 2 \in \mathcal{A}_0$ and thus $f \in \mathcal{A}_0$.

Now suppose that $f \in \mathcal{A}_0$ and that $\varphi\colon \mathbb{R} \to \mathbb{R}$ is continuous. By the Stone–Weierstrass theorem, there exists a sequence of polynomials $(p_n)_1^\infty$ such that $p_n \to \varphi$ uniformly on $[-\sup|f|, \sup|f|]$ as $n \to \infty$. Consequently, $p_n(f) \to \varphi(f)$ uniformly on Ω, and since we have shown that $\mathcal{A}_0$ is an algebra that is closed under uniform convergence, $\varphi(f) \in \mathcal{A}_0$.

Let $\varphi_n(x) = \max(0, \min(nx, 1))$, and note that $\varphi_n(x)$ increases to $\mathbf{1}[x > 0]$ for every x. Hence, if $f \in \mathcal{A}_0$ and $r \in \mathbb{R}$, then $\varphi_n(f - r)$ increases monotonically to $\mathbf{1}[f > r]$, and it follows that $\mathbf{1}[f > r] \in \mathcal{A}_0$. Consequently, $\{f > r\} \in \mathcal{F}$, which proves that every $f \in \mathcal{A}_0$ is $\mathcal{F}$-measurable. □

COROLLARY A.2. *Suppose that $\mathcal{A}$ is a (real or complex) vector space of bounded functions on a set Ω such that $\mathcal{A}$ contains the constant functions, and has the following property: if $(f_n)_1^\infty$ is an increasing sequence of positive functions in $\mathcal{A}$, and $f = \lim_{n\to\infty} f_n < \infty$ everywhere, then $f \in \mathcal{A}$ (i.e., $\mathcal{A}$ is closed under monotone convergence). Let $\mathcal{C}$ be a set of bounded real functions in $\mathcal{A}$ such that $\mathcal{C}$ is closed under multiplication. Then $\mathcal{A}$ contains every function that is measurable with respect to the σ-field $\mathcal{F}(\mathcal{C})$ generated by $\mathcal{C}$.*

PROOF. Let $\mathcal{A}_1$ be the subspace of bounded real functions in $\mathcal{A}$. Then $\mathcal{A}_1$ is closed under bounded monotone convergence and contains $\mathcal{C}$; thus Theorem A.1 shows that $\mathcal{A}_1$ contains every real bounded $\mathcal{F}(\mathcal{C})$-measurable function. Thus, by monotone convergence, $\mathcal{A}$ contains every positive measurable function, and the result follows since $\mathcal{A}$ is a vector space. □

EXAMPLE A.3. Let $\Omega = \mathbb{R}^{\mathcal{I}}$, where $\mathcal{I}$ is an arbitrary index set, and let $\mathcal{F}$ be the product σ-field $\mathcal{B}^{\mathcal{I}}$ on Ω. Suppose that for every finite subset $\mathcal{J} \subseteq \mathcal{I}$, we are given a probability measure $\mathrm{P}_{\mathcal{J}}$ on $\mathbb{R}^{\mathcal{J}}$; then there is at most one probability measure on $\mathbb{R}^{\mathcal{I}}$ such that the projection onto $\mathbb{R}^{\mathcal{J}}$ induces $\mathrm{P}_{\mathcal{J}}$ for every finite $\mathcal{J}$. This follows from Theorem A.1 by assuming that μ and ν are two such measures, and then taking $\mathcal{A}$ to be the set of all bounded measurable real functions f on Ω such that $\int f\,d\mu = \int f\,d\nu$, and $\mathcal{C}$ as the subset of bounded measurable real functions that depend on a finite number of coordinates only.

In other words, the joint distribution of a family (i.e. stochastic process) $(X_i)_{i\in\mathcal{I}}$ is uniquely determined by the finite-dimensional distributions. (Conversely, by Kolmogorov's theorem (Gihman and Skorohod 1971, Theorem I.4.2), such a measure on $\mathbb{R}^{\mathcal{I}}$ exists provided the measures $\mathrm{P}_{\mathcal{J}}$ are compatible.)

B. Stochastic processes

There are several definitions of stochastic processes that are slightly different but essentially equivalent. For example, a stochastic process may be regarded as an indexed collection of random variables (defined on a common probability space), but also as a random function.

This can be formalized as follows. (For simplicity we consider only real-valued stochastic processes; complex-valued processes are defined similarly.) Let T be an arbitrary index set and let $(\Omega, \mathcal{F}, \mathrm{P})$ be a probability space. We use the standard notation B^A for the set of all functions from A into B. Recall that the set of (real) random variables on $(\Omega, \mathcal{F}, \mathrm{P})$ is the subset of $\mathbb{R}^\Omega$ consisting of the measurable functions. A (real) stochastic process indexed by T, defined on $(\Omega, \mathcal{F}, \mathrm{P})$, may be defined in one of the following three ways.

(i) A collection $\{X_t\}_{t\in T}$ of random variables on $(\Omega, \mathcal{F}, \mathrm{P})$, i.e. a function X from T into $\mathbb{R}^\Omega$ which maps into the set of measurable functions.

(ii) A random function $X(\cdot) \in \mathbb{R}^T$, i.e. a measurable function $X\colon \Omega \to \mathbb{R}^T$ (where $\mathbb{R}^T$ is equipped with the product σ-field).

(iii) A function $X(t,\omega)$ on $T \times \Omega$, such that $X(t,\cdot)$ is measurable for each $t \in T$.

Note that X is defined as an element of $(\mathbb{R}^\Omega)^T$, $(\mathbb{R}^T)^\Omega$ and $\mathbb{R}^{T\times\Omega}$, respectively.

If we first ignore the measurability requirements, the three definitions become obviously equivalent, because a function $X \in \mathbb{R}^{T\times\Omega}$ corresponds uniquely to the functions $t \mapsto (\omega \mapsto X(t,\omega))$ in $(\mathbb{R}^\Omega)^T$ and $\omega \mapsto (t \mapsto X(t,\omega))$ in $(\mathbb{R}^T)^\Omega$. Moreover, it is almost as easy to see that the measurability requirements correspond to each other, and thus the three definitions above are naturally equivalent.

There is, however, a further complication. We have defined two random variables to be equivalent if they are a.s. equal. For stochastic processes, the first two definitions above then lead to two different notions of equivalence of two processes X and Y:

(i) For every t, $X_t = Y_t$ a.s.; equivalently, for each fixed t, $X(t,\omega) = Y(t,\omega)$ except for $\omega \in N_t$, where N_t is a null set that may depend on t.

(ii) A.s., $X(t) = Y(t)$ for all t; equivalently, there exists a null set $N \subset \Omega$ independent of t such that $X(t,\omega) = Y(t,\omega)$ for all t and $\omega \notin N$.

These conditions are equivalent when T is countable, but in general (ii) is stronger. Both notions of equivalence are useful. The second (stronger) is natural for the study of pathwise properties, and it is usually taken as the definition of equivalence in stochastic process literature.

For our purposes, the first (weaker) version is the natural one. Thus we formally define a stochastic process to be an indexed collection of random variables $\{X_t\}_{t\in T}$, and identify two such processes $\{X_t\}$ and $\{Y_t\}$ with $X_t = Y_t$ a.s. for each t.

A *version* of the stochastic process X is a function Y on $T \times \Omega$ such that $Y(t,\cdot) = X_t$ a.s. for every fixed t. We may thus talk about the properties of the paths $t \to Y(t,\omega)$ of a version. For example, if T is a topological space, a *continuous version* is a version Y such that $Y(\cdot,\omega)$ is a continuous function on t for every $\omega \in \Omega$. If T is a measurable space, we similarly say that a version Y is *measurable* if $t \to Y(t,\omega)$ is a measurable function on T for every $\omega \in \Omega$; moreover, we say that Y is *jointly measurable* if $Y(t,\omega)$ is a measurable function on $T \times \Omega$.

Note that although our definition of a stochastic process involves the weak notion of equivalence, it allows us to define the distribution of the process as the induced probability measure on $\mathbb{R}^T$. In fact, if $X(t,\omega)$ and $Y(t,\omega)$ are two versions of the same process, then $\mathrm{P}(X(\cdot,\omega) \in A) = \mathrm{P}(Y(\cdot,\omega) \in A)$ for every measurable subset of $\mathbb{R}^T$, because every such subset is determined by some countable subset of the coordinates (i.e., $A = A_0 \times \mathbb{R}^{T\setminus T_0}$ with $A_0 \subseteq \mathbb{R}^{T_0}$ for some countable subset T_0 of T).

C. Banach-space-valued functions and random variables

We occasionally consider functions or random variables with values in a (real or complex) Banach space B. (We are mainly interested in the Hilbert space case, but the general Banach case is not more difficult.) For a detailed account of integration theory for Banach-space-valued functions, see for example Dunford and Schwartz (1958, Chapter III, in particular Section III.6).

A B-valued random variable (defined on a probability space $(\Omega,\mathcal{F},\mathrm{P})$) is a measurable function $X\colon \Omega \to B$, where the Banach space B is equipped with its Borel σ-field. There are no problems with this definition when the space B is *separable* (i.e., has a countable dense subset), but there are serious difficulties when B is non-separable. (For example, the product σ-field on $B\times B$ may be strictly smaller than the Borel σ-field for the product topology on $B \times B$, and the addition map $B \times B \to B$ may be non-measurable. Cf. Billingsley (1968).) Many authors therefore consider only random variables with values in a separable Banach space (which covers most applications), but we will be slightly more general; we allow the space B to be non-separable but require the random variable to live in a separable subspace. (Cf. also Ledoux and Talagrand (1991, Section 2.1).)

DEFINITION C.1. Let B be a Banach space, and denote the norm in B by $\|\cdot\|_B$.

(i) A B-valued random variable is a.s. separably valued if there exists a separable subspace $B_1 \subseteq B$ such that $X \in B_1$ a.s.

(ii) $L^0(B)$ denotes the space of all a.s. separably valued B-valued random variables, with the topology of convergence in probability, i.e. $X_n \to X$ in $L^0(B)$ if $\|X_n - X\|_B \xrightarrow{\text{p}} 0$.

(iii) If $0 < p < \infty$, then $L^p(B) = \{X \in L^0(B) : \operatorname{E} \|X\|_B^p < \infty\}$.

A measurable simple function is a finite sum $\sum_1^n \mathbf{1}_{E_i} b_i$, with $E_i \subseteq \Omega$ measurable and $b_i \in B$. Some equivalent conditions for measurability are as follows; we omit the proof, cf. Dunford and Schwartz (1958, III.6.10–13).

PROPOSITION C.2. *The following are equivalent.*

(i) *$X : \Omega \to B$ is measurable and a.s. separably valued (i.e., $X \in L^0(B)$).*

(ii) *$X : \Omega \to B$ is a.s. separably valued and $x^*(X)$ is measurable $\Omega \to \mathbb{R}$ or $\mathbb{C}$ for every $x^* \in B^*$, the dual space of B. (Functions satisfying the latter condition are known as* weakly measurable.*)*

(iii) *There exists a sequence X_n of measurable simple functions $\Omega \to B$ such that $X_n \to X$ a.s.*

□

In particular, the measurable simple functions are dense in $L^0(B)$; they are also dense in $L^p(B)$ for every $p < \infty$ (Dunford and Schwartz 1958, III.3.8).

We use the following simple uniqueness result for a.s. separably valued variables.

PROPOSITION C.3. *If $X \in L^0(B)$ is such that $x^*(X) = 0$ a.s. for every $x^* \in B^*$, or more generally for x^* in some total set in B^*, then $X = 0$ a.s.*

PROOF. Note first that the set of all $x^* \in B^*$ such that $x^*(X) = 0$ a.s. is a closed subspace of B^*; hence the weaker assumption implies that $x^*(X) = 0$ a.s. for all x^*.

Let $B_1 \subseteq B$ be a separable subspace such that $X \in B_1$ a.s., and choose a countable dense subset $\{x_i\}_1^\infty$ of B_1. Let $x_i^* \in B^*$ be such that $\|x_i^*\| = 1$ and $\langle x_i^*, x_i \rangle = \|x_i\|$.

If $0 \neq x \in B_1$, then there exists an i such that $\|x - x_i\| < \frac{1}{2}\|x\|$ and thus $|\langle x_i^*, x - x_i \rangle| \leq \|x - x_i\| < \|x_i\| = \langle x_i^*, x_i \rangle$; hence $\langle x_i^*, x \rangle \neq 0$. Consequently,

$$\{\omega : X(\omega) \neq 0\} \subseteq \{\omega : X(\omega) \notin B_1\} \cup \bigcup_1^\infty \{\omega : \langle x_i^*, X(\omega) \rangle \neq 0\}.$$

□

More generally, L^p-spaces of B-valued functions on an arbitrary measure space $(M, \mathcal{M}, \mu)$ are defined similarly as for probability spaces.

If f is a scalar function on M and $b \in B$, we let $f \otimes b$ denote the B-valued function $fb \colon x \mapsto f(x)b$. Obviously, $f \otimes b$ is measurable if f is, and $\|f \otimes b\|_{L^p(B)} = \|f\|_{L^p} \|b\|_B$, $0 < p < \infty$.

Lebesgue integrals of B-valued functions can be defined as for scalar functions; if $f \in L^1(M, \mathcal{M}, \mu; B)$, then $\int_M f\,d\mu \in B$. Furthermore,

$$\|\int_M f\,d\mu\|_B \leq \int_M \|f\|_B\,d\mu. \tag{C.1}$$

(In particular, if X is a random variable in $L^1(B)$, then $\mathrm{E}\,X \in B$ is well-defined.) Such Banach-space-valued integrals are known as *Bochner integrals* and we sometimes use this name for emphasis.

An important special case of B-valued functions is when the Banach space B itself is a Lebesgue space on some other measure space $(M_1, \mathcal{M}_1, \mu_1)$. If $(M, \mathcal{M}, \mu)$ and $(M_1, \mathcal{M}_1, \mu_1)$ are σ-finite and $1 \leq p, q \leq \infty$, then an element of $L^p(M, \mathcal{M}, \mu; L^q(M_1, \mathcal{M}_1, \mu_1))$ may be regarded as a measurable function on $M \times M_1$ (uniquely defined $(\mu \times \mu_1)$-a.e.); moreover, if f is a measurable function on $M \times M_1$ that belongs to $L^1(M, \mathcal{M}, \mu; L^q(M_1, \mathcal{M}_1, \mu_1))$, then the Bochner integral $\int_M f\,d\mu \in L^q(M_1, \mathcal{M}_1, \mu_1)$ equals a.s. the pointwise integral $\int_M f(x, x_1)\,d\mu(x)$ (Dunford and Schwartz 1958, III.11.17). It follows further that an element of $L^0(M, \mathcal{M}, \mu; L^q(M_1, \mathcal{M}_1, \mu_1))$, i.e. a measurable function $M \to L^q(M_1, \mathcal{M}_1, \mu_1)$ $(1 \leq q \leq \infty)$, may be regarded as a measurable function on $M \times M_1$ such that $f(x, \cdot) \in L^q(M_1)$ for a.e. $x \in M$.

In this connection we note the following simple consequence of (the integral version of) Minkowski's inequality.

PROPOSITION C.4. *Let $M_1, \mathcal{M}_1, \mu_1$ and $M_2, \mathcal{M}_2, \mu_2$ be two σ-finite measure spaces. If f is a measurable function on $M_1 \times M_2$ and $0 < p \leq q \leq \infty$, then*

$$\|\,\|f\|_{L^p(M_1,d\mu_1)}\|_{L^q(M_2,d\mu_2)} \leq \|\,\|f\|_{L^q(M_2,d\mu_2)}\|_{L^p(M_1,d\mu_1)}.$$

PROOF. Since $q/p \geq 1$, (C.1) applies in $L^{q/p}$ and yields

$$\begin{aligned}\|\,\|f\|_{L^p(M_1,d\mu_1)}\|^p_{L^q(M_2,d\mu_2)} &= \Big\|\int |f(x,y)|^p d\mu_1(x)\Big\|_{L^{q/p}(M_2,d\mu_2)} \\ &\leq \int \|\,|f(x,y)|^p\|_{L^{q/p}(M_2,d\mu_2)} d\mu_1(x) \\ &= \|\,\|f\|_{L^q(M_2,d\mu_2)}\|^p_{L^p(M_1,d\mu_1)}.\end{aligned}$$

□

D. Polarization

Polarization is the standard name for a procedure to recover a bilinear or sesquilinear form from the corresponding quadratic form, and similar results for multilinear forms and operators.

For example, if B is a symmetric bilinear form in a real or complex vector space, and $Q(x) = B(x, x)$, then

$$B(x, y) = \tfrac{1}{4}\big(Q(x + y) - Q(x - y)\big). \tag{D.1}$$

Similarly, if B is a sesquilinear form in a complex vector space, and $Q(x) = B(x,x)$, then

$$B(x,y) = \tfrac{1}{4}\sum_{k=0}^{3} i^k Q(x + i^k y). \tag{D.2}$$

(In particular, (D.1) and (D.2) are valid if B is the inner product in a real or complex Hilbert space, respectively, and $Q(x) = \|x\|^2$.) These formulae make it possible to transfer various properties of Q to B.

Similar explicit formulae may be given also for symmetric multilinear forms in more than two variables, or symmetric multilinear operators into another vector space; for example the following.

THEOREM D.1. *Let V and W be two real or complex vector spaces, and suppose that T is a symmetric n-linear mapping of V^n into W, where $n \geq 1$. If $Q(x) = T(x,\dots,x)$, then*

$$T(x_1,\dots,x_n) = \frac{1}{2^n n!} \sum_{\varepsilon_1,\dots,\varepsilon_n \in \{-1,1\}} \varepsilon_1 \cdots \varepsilon_n Q(\varepsilon_1 x_1 + \cdots + \varepsilon_n x_n)$$

for every $x_1,\dots,x_n \in V$. □

As a corollary, we obtain the following simple uniqueness result.

THEOREM D.2. *Let V and W be two real or complex vector spaces, and suppose that S and T are two symmetric n-linear mappings of V^n into W such that $S(x,\dots,x) = T(x,\dots,x)$ for every $x \in V$. Then $S(x_1,\dots,x_n) = T(x_1,\dots,x_n)$ for every $x_1,\dots,x_n \in V$.* □

E. Tensor products

E.1. Generalities. The name *tensor product* denotes an idea rather than a specific construction. Moreover, this is true in two different senses.

First, tensor products are defined for many types of objects (mainly in algebra and functional analysis), for example groups, rings, vector spaces, Banach spaces, etc., and although the definitions have a common flavour, the details depend on the context. (A general definition, covering at least most of the different cases, can be given in category theory, but we will not go into that.)

Secondly, even if we stick to a particular category, for example vector spaces or Hilbert spaces, the tensor product of two objects is defined only up to isomorphism. By standard abuse of language, we nevertheless usually refer to any of the (isomorphic) choices as 'the tensor product', but we may also refer to different choices as different *realizations* of the tensor product.

The moral of this introduction is that generally it is not so important what a tensor product *is*, only what it *does* (i.e. what properties it has).

In this book we are concerned with tensor products of Hilbert spaces, but as a background we begin with the (algebraic) tensor product of vector spaces without any topology involved. In this appendix, all vector spaces are either real or complex; there is no difference between the two cases.

DEFINITION E.1. If V and W are two vector spaces, then their (*algebraic*) *tensor product* is a vector space, denoted $V \otimes W$, together with a bilinear map $\tau: V \times W \to V \otimes W$, written as $(v, w) \mapsto v \otimes w \in V \otimes W$, with the following *universal property*:

If U is any vector space and $\varphi: V \times W \to U$ is a bilinear map, then there exists a unique linear map $\psi: V \otimes W \to U$ such that $\varphi = \psi \circ \tau$ (i.e., $\varphi(v, w) = \psi(v \otimes w)$ for $v \in V$, $w \in W$).

REMARK E.2. It is important to note that not every element in $V \otimes W$ is of the form $v \otimes w$, but it is always a finite sum of such elements. Note also that the representation $v \otimes w$ is not unique, even when it exists; for example, $v \otimes 0 = 0 \otimes w = 0$ for any v, w.

Uniqueness up to isomorphism of the tensor product follows easily from the universal property. (We omit the details.) Existence may be shown in several ways, for example by the examples below.

EXAMPLE E.3. A standard construction is as follows: first take $\widetilde{U}$ to be the space of all formal finite linear combinations $\sum \lambda_{vw}(v, w)$ of elements in $V \times W$ (with real or complex coefficients λ_{vw}); then let $V \otimes W = \widetilde{U}/U_0$, where U_0 is the subspace spanned by all elements of the forms $(\lambda v, w) - \lambda(v, w)$, $(v, \lambda w) - \lambda(v, w)$, $(v_1 + v_2, w) - (v_1, w) - (v_2, w)$ and $(v, w_1 + w_2) - (v, w_1) - (v, w_2)$, and let $v \otimes w$ be the image of $(v, w) \in \widetilde{U}$ in $V \otimes W = \widetilde{U}/U_0$. It is a simple exercise to verify the universal property.

REMARK E.4. If the reader feels uncomfortable with 'formal finite linear combinations', note that $\widetilde{U}$ may be interpreted as the set of all scalar functions $\lambda: V \times W \to \mathbb{R}$ (or $\mathbb{C}$) that vanish except on a finite set.

There is also an alternative: Let $\widetilde{U}$ be any vector space whose dimension equals the cardinality of $V \times W$, select a basis and denote its elements by e_{vw}, $v \in V$, $w \in W$. Then the construction above can be repeated with (v, w) replaced by e_{vw}.

EXAMPLE E.5. Another useful construction of the tensor product is to choose bases $\{e_i\}_{i \in I}$ in V and $\{f_j\}_{j \in J}$ in W and then let $V \otimes W$ be the space of all formal finite linear combinations of the elements $(i, j) \in I \times J$ (cf. Remark E.4), defining $e_i \otimes f_j = (i, j)$ and extending this to a bilinear map $V \times W \to V \otimes W$. Again, the universal property is easily verified.

Note that this example shows that if $\{e_i\}_{i \in I}$ and $\{f_j\}_{j \in J}$ are bases in V and W, then $\{e_i \otimes f_j\}_{I \times J}$ is a basis in $V \otimes W$. In particular, $\dim(V \otimes W) = \dim(V) \dim(W)$.

EXAMPLE E.6. If V and W are some spaces of (real or complex) functions on some sets X and Y, a useful concrete realization of $V \otimes W$ is obtained by defining

$$f \otimes g(x,y) = f(x)g(y), \qquad x \in X,\ y \in Y,$$

and letting $V \otimes W$ be the linear space of functions on $X \times Y$ spanned by $\{f \otimes g : f \in V, g \in W\}$. We omit the proof that this indeed satisfies the universal property.

If now V and W are Banach spaces, then their norms can be used to define a norm on their algebraic tensor product $V \otimes W$. This is in general not complete, but taking the completion we obtain a *Banach space tensor product*; often this too is denoted by $V \otimes W$.

There are, however, several different, inequivalent, ways to define a norm on $V \otimes W$, and thus there are several different Banach space tensor products. For general Banach spaces, there are at least two commonly used norms, yielding two different tensor products, see for example Schaefer (1971) or Treves (1967). For Hilbert spaces, the only case we are interested in, there is a third choice which yields a Hilbert space; this tensor product is much more important for us and it is the only one used in this book.

DEFINITION E.7. Let V and W be Hilbert spaces. It is easily seen (for example using the universal property twice) that there exists a unique inner product on the algebraic tensor product $V \otimes W$ such that

$$\langle v \otimes w, v' \otimes w' \rangle_{V \otimes W} = \langle v, v' \rangle_V \langle w, w' \rangle_W. \tag{E.1}$$

The *Hilbert space tensor product* of V and W is the Hilbert space obtained by completing the algebraic tensor product in the norm corresponding to this inner product.

We denote this tensor product too by $V \otimes W$. It is easily seen that this definition is equivalent to the following (where we also change to symbols more common for Hilbert spaces).

DEFINITION E.8. If H_1 and H_2 are two Hilbert spaces, their tensor product is a Hilbert space $H_1 \otimes H_2$ together with a bilinear map $H_1 \times H_2 \to H_1 \otimes H_2$, denoted by $(f_1, f_2) \mapsto f_1 \otimes f_2 \in H_1 \otimes H_2$, such that the range of this map is total in $H_1 \otimes H_2$ and for all $f_1, g_1 \in H_1$, $f_2, g_2 \in H_2$,

$$\langle f_1 \otimes f_2, g_1 \otimes g_2 \rangle_{H_1 \otimes H_2} = \langle f_1, g_1 \rangle_{H_1} \langle f_2, g_2 \rangle_{H_2}. \tag{E.2}$$

It follows from (E.2) that for any $f_1 \in H_1$ and $f_2 \in H_2$,

$$\|f_1 \otimes f_2\|_{H_1 \otimes H_2} = \|f_1\|_{H_1} \|f_2\|_{H_2}. \tag{E.3}$$

The Hilbert space tensor product may be constructed by completing the algebraic tensor product as in Definition E.7, but there are also direct constructions. Some important constructions are given in the following examples; we omit the detailed verifications.

EXAMPLE E.9. It follows easily from the definition that if $\{e_i\}_{i\in I}$ and $\{f_j\}_{j\in J}$ are orthonormal bases in H_1 and H_2, then $\{e_i \otimes f_j\}_{(i,j)\in I\times J}$ is an orthonormal basis in $H_1 \otimes H_2$.

Consequently, given any two such bases, we may construct $H_1 \otimes H_2$ as the space of formal sums $\sum_{ij} \lambda_{ij} e_i \otimes f_j$ with $\sum |\lambda_{ij}|^2 < \infty$; this space can be identified with $\ell^2(I \times J)$.

In particular, if $\dim(H_1) = m$ and $\dim(H_2) = n$ are finite, we may realize $H_1 \otimes H_2$ as the space of $m \times n$ matrices.

EXAMPLE E.10. The tensor product of two L^2-spaces $L^2(X_1, \mu_1)$ and $L^2(X_2, \mu_2)$, where μ_1 and μ_2 are σ-finite measures, has the concrete realization $L^2(X_1 \times X_2, \mu_1 \times \mu_2)$ with $f \otimes g(x_1, x_2) = f(x_1)g(x_2)$; cf. Example E.6.

EXAMPLE E.11. More generally, if $H_1 \subseteq L^2(X_1, \mu_1)$ and $H_2 \subseteq L^2(X_2, \mu_2)$ are two closed subspaces, where μ_1 and μ_2 are σ-finite measures, then $H_1 \otimes H_2$ can be realized as the closed subspace of $L^2(X_1 \times X_2, \mu_1 \times \mu_2)$ spanned by $\{f_1 \otimes f_2 : f_1 \in H_1,\ f_2 \in H_2\}$.

EXAMPLE E.12. The tensor product $L^2(X, \mu) \otimes H$, where (X, μ) is a measure space and H is any Hilbert space, may be realized as $L^2(X, \mu; H)$ with $f \otimes h = fh$ (which coincides with the notation in Appendix C).

More generally, if $H_1 \subseteq L^2(X, \mu)$, then $H_1 \otimes H$ may be identified with the closed subspace of $L^2(X, \mu; H)$ spanned by $\{f \otimes h : f \in H_1,\ h \in H\}$.

EXAMPLE E.13. Let H_1^* and H_2^* denote the dual spaces. (Yes, these are isomorphic to H_1 and H_2, but we prefer to distinguish them here.) If $f \in H_1$ and $g \in H_2$, let $f \otimes g$ be the bilinear functional on $H_1^* \times H_2^*$ given by $f \otimes g(h_1, h_2) = h_1(f)h_2(g)$. Then $H_1 \otimes H_2$ can be realized as the space of all bounded bilinear functionals u on $H_1^* \times H_2^*$ that are *Hilbert–Schmidt*, i.e. satisfy $\sum_{ij} |u(e_i, f_j)|^2 < \infty$ for some (or, equivalently, any) orthonormal bases $\{e_i\}_{i\in\mathcal{I}}$ in H_1 and $\{f_j\}_{j\in\mathcal{J}}$ in H_2 (cf. Appendix H).

We have so far considered only tensor products of two spaces, but the definitions and results are easily extended to several spaces by considering multilinear maps instead of bilinear. Alternatively, we can define tensor products of several spaces by induction from the case of two spaces, since there are natural isomorphisms $H_1 \otimes (H_2 \otimes H_3) \cong (H_1 \otimes H_2) \otimes H_3 \cong H_1 \otimes H_2 \otimes H_3$ (in both the algebraic and the Hilbert space cases).

In particular, we may form the tensor powers $H^{\otimes n}$ of any Hilbert space. (We define $H^{\otimes 1} = H$ for any space, and let $H^{\otimes 0}$ be the one-dimensional space of scalars ($\mathbb{R}$ or $\mathbb{C}$); thus $\dim(H^{\otimes n}) = \dim(H)^n$ for any $n \geq 0$.)

E.2. Symmetric tensor powers. Note that if $f, g \in H$, then in general $f \otimes g \neq g \otimes f$ in $H \otimes H$. (Cf. for example Example E.10.) In many applications it is, however, natural to impose commutativity also, which leads to the introduction of the *symmetric tensor power*.

REMARK E.14. There is also a corresponding notion of *anti-symmetric* tensor power, but that does not occur in this book.

In the algebraic case, the symmetric tensor power can be defined by a universal property as above, but now considering only symmetric multilinear maps (and thus in particular requiring τ to be symmetric). In the Hilbert space case, which is the only one that we will consider in the remainder of this appendix, we will use the following analogue of Definition E.8. (There is also a corresponding version of Definition E.7; these are equivalent.)

DEFINITION E.15. If H is a Hilbert space and $n \geq 1$, the symmetric tensor power is a Hilbert space $H^{\odot n}$ together with a symmetric n-linear map $H \times \cdots \times H \to H^{\odot n}$, denoted by $(f_1, \ldots, f_n) \mapsto f_1 \odot \cdots \odot f_n$, such that the range of this map is total in $H^{\odot n}$ and

$$\langle f_1 \odot \cdots \odot f_n, g_1 \odot \cdots \odot g_n \rangle = \sum_{\pi \in \mathfrak{S}_n} \prod_{i=1}^{n} \langle f_i, g_{\pi(i)} \rangle, \tag{E.4}$$

where $\mathfrak{S}_n$ is the symmetric group of all permutations of $\{1, \ldots, n\}$. For $n = 0$, we define $H^{\odot 0}$ to be the one-dimensional space of scalars.

Note that $H^{\odot 1} = H$. Observe also that (E.4) in particular yields

$$\langle f^{\odot n}, g^{\odot n} \rangle = n! \, \langle f, g \rangle^n$$

and thus

$$\|f^{\odot n}\|_{H^{\odot n}} = \sqrt{n!} \|f\|_H^n, \qquad f \in H, \, n \geq 0. \tag{E.5}$$

Again, the symmetric tensor power is defined only up to equivalence, and there are several constructions. One way to construct the symmetric tensor power is to let $H^{\odot n}$ be the subspace of $H^{\otimes n}$ that is invariant under all permutations of coordinates, i.e. under the mappings defined by

$$f_1 \otimes \cdots \otimes f_n \mapsto f_{\pi(1)} \otimes \cdots \otimes f_{\pi(n)}, \qquad \pi \in \mathfrak{S}_n,$$

and let

$$f_1 \odot \cdots \odot f_n = \frac{1}{\sqrt{n!}} \sum_{\pi \in \mathfrak{S}_n} f_{\pi(1)} \otimes \cdots \otimes f_{\pi(n)}. \tag{E.6}$$

If $H = L^2(M, \mu)$, where μ is σ-finite, we can thus construct the symmetric tensor power as the space of symmetric functions in $L^2(M^n, \mu^n)$, using Example E.10 and (E.6). An equivalent realization, which is more convenient for us, is obtained by changing the measure to accommodate the factorial in (E.6). We state this as a proposition.

PROPOSITION E.16. *Suppose that $(M, \mathcal{M}, \mu)$ is a σ-finite measure space. Let $\mathcal{M}^{\odot n}$ be the σ-field of symmetric measurable subsets of M^n, and let $\mu^{\odot n} =$*

$\frac{1}{n!}\mu^n$. *We may identify* $L^2(M,\mathcal{M},\mu)^{\odot n}$ *with* $L^2(M^n,\mathcal{M}^{\odot n},\mu^{\odot n})$, *which equals the subspace of symmetric functions in* $L^2(M^n,\mathcal{M}^n,\frac{1}{n!}\mu^n)$, *by letting*

$$f_1 \odot \cdots \odot f_n(x_1,\dots,x_n) = \sum_{\pi\in\mathfrak{S}_n}\prod_{i=1}^{n} f_i(x_{\pi(i)}). \tag{E.7}$$

(*We interpret* $(M^0,\mathcal{M}^{\odot 0},\mu^{\odot 0})$ *as a space with one point of measure* 1.) □

The Hilbert space direct sum

$$\bigoplus_{n=0}^{\infty} H^{\odot n} = \{(f_n)_0^\infty : f_n \in H^{\odot n} \text{ with } \sum_0^\infty \|f_n\|^2 < \infty\}$$

is called the (*symmetric*) *Fock space* over H; we denote the Fock space by $\Gamma(H)$.

REMARK E.17. Some authors prefer different normalizations in (E.4), making the factorials, which are inevitable in the theory, appear at other places in the formulae. The definition chosen here seems to be the best for our purposes.

REMARK E.18. More suggestive notations such as $\exp(H)$ are sometimes used for the Fock space.

We also use the following result.

PROPOSITION E.19. *If H is a Hilbert space, then the multiplication*

$$(f_1\odot\cdots\odot f_n)\odot(f_{n+1}\odot\cdots\odot f_{n+m}) = f_1\odot\cdots\odot f_{n+m} \tag{E.8}$$

may be extended to a continuous bilinear operation (also denoted $\odot$) $H^{\odot n}\times H^{\odot m}\to H^{\odot(n+m)}$ for any $n,m\ge 0$. Moreover,

$$\|X\odot Y\| \le \binom{n+m}{m}^{1/2}\|X\|\|Y\|, \qquad X\in H^{\odot m},\ Y\in H^{\odot n}, \tag{E.9}$$

where the constant is best possible (except in the trivial case $\dim H = 0$).

PROOF. Let us first consider the general (non-symmetric) tensor powers $H^{\otimes n}$. There is a (unique) natural isomorphism $\otimes\colon H^{\otimes n}\otimes H^{\otimes m}\to H^{\otimes(n+m)}$ such that

$$(f_1\otimes\cdots\otimes f_n)\otimes(f_{n+1}\otimes\cdots\otimes f_{n+m}) = f_1\otimes\cdots\otimes f_{n+m}$$

for all sequences $f_i\in H$, $i=1,\dots,n+m$; this is easily verified by taking any orthonormal basis in H and using Example E.9 repeatedly.

We now use the realization (E.6) and regard $H^{\odot n}$ as a subspace of $H^{\otimes n}$; let p_n denote the orthogonal projection of $H^{\otimes n}$ onto $H^{\odot n}$ and note that $p_n(f_1\otimes\cdots\otimes f_n) = (n!)^{-1/2} f_1\odot\cdots\odot f_n$. Then

$$\begin{aligned} p_{n+m}\big((f_1\odot\cdots\odot f_n)\otimes(f_{n+1}\odot\cdots\odot f_{n+m})\big) &= (n!\,m!)^{1/2}p_{n+m}(f_1\otimes\cdots\otimes f_{n+m})\\ &= (n!\,m!)^{1/2}(n+m)!^{-1/2} f_1\odot\cdots\odot f_{n+m}. \end{aligned}$$

We may thus define $\odot$ by

$$X \odot Y = \binom{n+m}{n}^{1/2} p_{n+m}(X \otimes Y), \qquad X \in H^{\odot m}, Y \in H^{\odot n},$$

and it follows that (E.8) and (E.9) hold. The constant $\binom{n+m}{n}^{1/2}$ is best possible, as follows by considering $f^{\odot n} \odot f^{\odot m}$, with any $f \neq 0$, and using (E.5). □

By Proposition E.19, the algebraic direct sum $\Gamma_*(H) = \sum_{n=0}^{\infty} H^{\odot n}$ is a graded commutative algebra which is called the *symmetric tensor algebra* of H. It is dense in the Fock space $\Gamma(H)$ defined above. Note, however, that the multiplication (E.8) is not uniformly bounded for varying n and m; hence it is not bounded on $\Gamma_*(H)$ and does not extend to make $\Gamma(H)$ an algebra.

E.3. Operators. If V, V', W, W' are vector spaces and $A\colon V \to V'$, $B\colon W \to W'$ are linear operators, then the mapping $(v, w) \mapsto Av \otimes Bw$ is a bilinear map of $V \times W$ into $V' \otimes W'$; by the universal property it thus induces a (unique) linear map, denoted $A \otimes B\colon V \otimes W \to V' \otimes W'$, such that

$$A \otimes B(v \otimes w) = (Av) \otimes (Bw), \qquad v \in V, w \in W.$$

We next want to show the corresponding result for the Hilbert space tensor product. Recall that the norm on the tensor product is given by the inner product defined by (E.1).

PROPOSITION E.20. *Suppose that $A\colon V \to V'$ and $B\colon W \to W'$ are two bounded linear operators between Hilbert spaces. If u belongs to the algebraic tensor product $V \otimes W$, then*

$$\|A \otimes B(u)\|_{V' \otimes W'} \leq \|A\| \, \|B\| \, \|u\|_{V \otimes W}. \tag{E.10}$$

Consequently, $A \otimes B$ extends to a bounded linear operator, also denoted $A \otimes B$, between the Hilbert space tensor products $V \otimes W$ and $V' \otimes W'$. Furthermore,

$$\|A \otimes B\| = \|A\| \, \|B\|.$$

PROOF. We begin by considering the case $B = I$ on $W = W'$.

The element u can be represented as a finite sum $\sum_1^n v_i \otimes w_i$. Taking any such representation, we may express the elements $w_i \in W$ as linear combinations of some finite set of orthonormal vectors $e_j \in W$, $w_i = \sum_j a_{ij} e_j$. (Simply take any orthonormal basis in the subspace spanned by $\{w_i\}$.) Thus $u = \sum_{ij} a_{ij} v_i \otimes e_j = \sum_j x_j \otimes e_j$, with $x_j = \sum_i a_{ij} v_i \in V$. In other words, there exists a representation $u = \sum_j x_j \otimes e_j$, with $\{e_j\}$ orthonormal.

By (E.1), the elements $x_j \otimes e_j$ are orthogonal in $V \otimes W$, and thus

$$\|u\|_{V \otimes W}^2 = \sum_j \|x_j \otimes e_j\|_{V \otimes W}^2 = \sum_j \|x_j\|_V^2. \tag{E.11}$$

Since $(A \otimes I)u = \sum_j A(x_j) \otimes e_j$, we similarly obtain

$$\|(A \otimes I)u\|^2_{V' \otimes W} = \sum_j \|A(x_j)\|^2_{V'}. \tag{E.12}$$

But $\|A(x_j)\|^2_{V'} \leq \|A\|^2 \|x_j\|^2_V$ for every j, and thus (E.11) and (E.12) yield

$$\|(A \otimes I)u\|^2_{V' \otimes W} \leq \|A\|^2 \|u\|^2_{V \otimes W}.$$

We have proved (E.10) in the special case $B = I$. The same argument, with the spaces interchanged, applies also to the case $A = I$, and the general case follows by these special cases and the factorization $A \otimes B = (A \otimes I)(I \otimes B)$.

This shows $\|A \otimes B\| \leq \|A\| \|B\|$, and the converse inequality follows easily using (E.3). □

By induction, the proposition extends to tensor products of several Hilbert spaces. In particular, every bounded linear operator $A\colon H_1 \to H_2$ between two Hilbert spaces induces a (unique) bounded linear map $A^{\otimes n}\colon H_1^{\otimes n} \to H_2^{\otimes n}$, for every $n \geq 0$, such that

$$A^{\otimes n}(f_1 \otimes \cdots \otimes f_n) = Af_1 \otimes \cdots \otimes Af_n, \qquad f_1, \dots, f_n \in H_1.$$

Moreover, $\|A^{\otimes n}\| = \|A\|^n$. (For $n = 0$, $A^{\otimes 0}$ is the identity operator in the space $H_1^{\otimes 0} = H_2^{\otimes 0}$ of scalars.)

The corresponding result for symmetric tensor powers holds as well.

PROPOSITION E.21. *Any bounded linear operator $A\colon H_1 \to H_2$ between two Hilbert spaces induces a (unique) bounded linear map $A^{\odot n}\colon H_1^{\odot n} \to H_2^{\odot n}$, for every $n \geq 0$, such that*

$$A^{\odot n}(f_1 \odot \cdots \odot f_n) = Af_1 \odot \cdots \odot Af_n, \qquad f_1, \dots, f_n \in H_1. \tag{E.13}$$

Moreover, $\|A^{\odot n}\| = \|A\|^n$.

If furthermore $\|A\| \leq 1$, then these tensor powers of A combine to a bounded linear map

$$\Gamma(A) = \bigoplus_0^\infty A^{\odot n}\colon \Gamma(H_1) \to \Gamma(H_2)$$

such that $\Gamma(A)((f_n)_0^\infty) = (A^{\odot n} f_n)_0^\infty$ (where $f_n \in H_1^{\odot n}$); this map has norm 1.

PROOF. Let us, as above, regard $H_i^{\odot n}$ as subspaces of $H_i^{\otimes n}$ using (E.6). Then $A^{\otimes n}$ maps $H_1^{\odot n} \subset H_1^{\otimes n}$ into $H_2^{\odot n} \subset H_2^{\otimes n}$, and we can define $A^{\odot n}$ as the restriction of $A^{\otimes n}$. It follows that (E.13) holds and that $\|A^{\odot n}\| \leq \|A\|^n$; the converse inequality follows by considering $A^{\odot n}(f^{\odot n}) = (Af)^{\odot n}$, using (E.5). (The case $n = 0$ is trivial, with $A^{\odot 0}$ the identity on $H_1^{\odot 0} = H_2^{\odot 0} = \mathbb{R}$ or $\mathbb{C}$.)

The final part follows directly, and

$$\|\Gamma(A)\| = \sup_{n \geq 0} \|A^{\odot n}\| = \sup_{n \geq 0} \|A\|^n = 1.$$

□

The mapping $A \mapsto \Gamma(A)$ has nice functorial properties.

PROPOSITION E.22. *Suppose that* $A\colon H_1 \to H_2$ *and* $B\colon H_2 \to H_3$ *are linear operators with* $\|A\|, \|B\| \le 1$, *where* H_1, H_2, H_3 *are Hilbert spaces. Then* $\Gamma(BA) = \Gamma(B)\Gamma(A)\colon \Gamma(H_1) \to \Gamma(H_3)$ *and* $\Gamma(A^*) = \Gamma(A)^*\colon \Gamma(H_2) \to \Gamma(H_1)$. *Moreover,* $\Gamma(I) = I$ *on* $\Gamma(H)$ *for any Hilbert space* H.

PROOF. Obvious, using (E.13) and (E.4). □

REMARK E.23. If $\|A\| > 1$, we may still define $\Gamma(A)$ as above, but it now becomes a densely defined *unbounded* operator. Furthermore, the same holds for any unbounded, densely defined A. We do not need these cases.

REMARK E.24. The operator $\Gamma(A)$ is called the *second quantization* of A. This name is also sometimes used for the related (unbounded, closely defined) operator $d\Gamma(A)$ on $\Gamma(H)$ defined by

$$d\Gamma(A)(f_1 \odot \cdots \odot f_n) = \sum_{i=1}^{n} f_1 \odot \cdots \odot A(f_i) \odot \cdots \odot f_n$$

when $A\colon H \to H$. These operators are related by the relation $e^{d\Gamma(A)} = \Gamma(e^A)$ (at least formally; if A is self-adjoint and negative definite, this is an equality between well-defined bounded operators).

An important example is $d\Gamma(I)$, which equals the number operator defined in Example 4.7.

F. Reproducing Hilbert Spaces

The theory of reproducing Hilbert spaces was developed by Aronszajn (1950). We give here a summary of the theory as it is used in this book.

DEFINITION F.1. A reproducing Hilbert space is a Hilbert space H of functions on some set T such that each point evaluation $f \mapsto f(t)$ is a continuous linear functional on H. (We may consider either a real Hilbert space of real functions or a complex Hilbert space of complex functions; the only difference is that the conjugations below are not necessary in the real case.)

EXAMPLE F.2. The sequence space ℓ^2 is a reproducing Hilbert space of functions on $\{1, 2, \dots\}$. On the other hand, $L^2([0,1])$ and $L^2(\mathbb{R}^d)$ are *not* reproducing Hilbert spaces; since the functions in these spaces are defined only a.e., the point evaluations are not even defined; moreover, on the subspaces of continuous functions the point evaluations are defined but not continuous (in the L^2-norm).

Since a continuous linear functional on a Hilbert space is given by the inner product with an element of the space, the definition implies the existence of (unique) functions $K_t \in H$, $t \in T$, such that

$$f(t) = \langle f, K_t \rangle_H, \qquad f \in H,\ t \in T. \tag{F.1}$$

(Conversely, every Hilbert space H of functions on T such that (F.1) holds for some family $K_t \in H$ is a reproducing Hilbert space.)

The function $K(s,t) = K_t(s)$ on $T \times T$ is known as the *reproducing kernel*; we give some of its basic properties as a theorem.

THEOREM F.3. *Let H be a reproducing Hilbert space of functions on some set T. Then there exists a unique function K on $T \times T$ (the reproducing kernel) such that:*

(i) $K(s,t) = \overline{K(t,s)}$. *($K$ is Hermitian.)*
(ii) $\sum_{i,j=1}^n \lambda_i \overline{\lambda}_j K(t_i, t_j) \geq 0$ *for any finite sequences $\{t_i\}_1^n$ of points in T and $\{\lambda_i\}_1^n$ of real or complex numbers. (K is positive semi-definite.)*
(iii) *The function $K(\cdot,t)\colon s \mapsto K(s,t)$ belongs to H for every $t \in T$.*
(iv) $f(t) = \langle f, K(\cdot,t)\rangle_H$ *for every $f \in H$ and $t \in T$.*
(v) $K(s,t) = \langle K(\cdot,t), K(\cdot,s)\rangle_H$ *for every $s,t \in T$.*
(vi) *For every $t \in T$, $\|K(\cdot,t)\|_H = K(t,t)^{1/2}$; this is also the norm of the linear functional $f \mapsto f(t)$ on H.*
(vii) *The family $\{K(\cdot,t)\}_{t\in T}$ is total in H.*

PROOF. Our notation is such that $K(\cdot,t) = K_t$. Hence (iii) and (iv) are the same as the definition (F.1), which also proves uniqueness. (vii) follows because $f \perp K(\cdot,t)$ for all $t \in T$ implies $f = 0$ by (iv).

Specializing (iv) to $f = K(\cdot,s)$ yields

$$K(t,s) = \langle K(\cdot,s), K(\cdot,t)\rangle, \qquad s,t \in T,$$

which is the same as (v).

Writing (v) as $K(s,t) = \langle K_t, K_s\rangle$, we obtain (i) since the inner product is Hermitian. Similarly, (ii) follows by

$$\sum_{i,j} \lambda_i \overline{\lambda}_j K(t_i,t_j) = \langle \sum_j \overline{\lambda}_j K_{t_j}, \sum_i \overline{\lambda}_i K_{t_i}\rangle_H = \| \sum_i \overline{\lambda}_i K_{t_i} \|_H^2 \geq 0.$$

Finally, $\|K_t\|_H = \langle K_t, K_t\rangle_H^{1/2} = K(t,t)^{1/2}$, and this equals the norm of $f \mapsto \langle f, K_t\rangle = f(t)$, which proves (vi). □

REMARK F.4. For a real reproducing Hilbert space, (i) says that K is symmetric. Moreover, in the real case it does not matter whether we allow complex or only real λ_i in (ii). The proof above works for real λ_i, but the complex case follows from this because the symmetry $K(t_i,t_j) = K(t_j,t_i)$ yields

$$\sum_{i,j} \lambda_i \overline{\lambda}_j K(t_i,t_j) = \sum_{i,j} \operatorname{Re}\lambda_i \operatorname{Re}\lambda_j K(t_i,t_j) + \sum_{i,j} \operatorname{Im}\lambda_i \operatorname{Im}\lambda_j K(t_i.t_j).$$

Conversely, as will be shown in Theorem F.7, the reproducing kernel determines the reproducing Hilbert space; moreover, any function K on $T \times T$ that satisfies (i) and (ii) above is the reproducing kernel of a reproducing Hilbert space.

The reproducing Hilbert spaces occurring in this book arise through the following general construction.

THEOREM F.5. *Let H be a Hilbert space and let $\{h_t\}_{t\in T}$ be a total set in H. Define a linear mapping R from H into the space of all functions on T by $R(h)(t) = \langle h, h_t\rangle_H$. Then R is injective, and if we define an inner product on $R(H) = \{R(h) : h \in H\}$ by*

$$\langle f, g\rangle_{R(H)} = \langle R^{-1}(f), R^{-1}(g)\rangle_H, \tag{F.2}$$

which makes $R\colon H \to R(H)$ an isometry, then $R(H)$ becomes a reproducing Hilbert space with reproducing kernel

$$K(s,t) = \langle h_t, h_s\rangle_H.$$

PROOF. If $R(h) = 0$, then $h \perp h_t$ for every $t \in T$, but since $\{h_t\}$ is total, this implies $h = 0$. Thus R is injective and the inner product (F.2) on $R(H)$ is well-defined. This definition makes R a linear isometry. In particular, $R(H)$ is isometric to H and is thus a Hilbert space. Finally, if $f \in R(H)$ and $t \in T$, then

$$f(t) = R(R^{-1}f)(t) = \langle R^{-1}f, h_t\rangle_H = \langle f, R(h_t)\rangle_{R(H)}.$$

This proves that $R(H)$ is a reproducing Hilbert space, and that the reproducing kernel is given by $K_t = R(h_t)$, and thus

$$K(s,t) = K_t(s) = R(h_t)(s) = \langle h_t, h_s\rangle_H.$$

□

In order to show existence of a reproducing Hilbert space with given kernel, we first show a standard result on Hilbert spaces.

LEMMA F.6. *Suppose that T is a set and that K is a function on $T \times T$ such that K is Hermitian and semi-definite. Then there exist a Hilbert space H and elements $h_t \in H$, $t \in T$, such that*

$$\langle h_s, h_t\rangle_H = K(s,t).$$

PROOF. One simple, standard construction is to begin with the vector space V of all functions on T that differ from 0 at only finitely many points. The functions $\delta_t = \mathbf{1}_{\{t\}}$ form a basis of V. Define an inner product in V by

$$\langle f, g\rangle_V = \sum_{s,t\in T} f(s)\overline{g(t)}K(s,t).$$

This is Hermitian and semi-definite, by the conditions on K, but not necessarily strictly definite. Thus let $V_0 = \{f \in V : \langle f, f\rangle_V = 0\}$. The Cauchy–Schwarz inequality implies that V_0 is a subspace of V and that the inner product is well-defined and definite on the quotient space V/V_0. Now let H be the completion of V/V_0, and let h_t be the element in H corresponding to δ_t in V; then $\langle h_s, h_t\rangle_H = \langle \delta_s, \delta_t\rangle_V = K(s,t)$. □

THEOREM F.7. *Suppose that T is a set and that K is a function on $T \times T$ such that K is Hermitian and semi-definite. Then there exists a unique reproducing Hilbert space H with reproducing kernel K.*

PROOF. Since $\overline{K}$ satisfies the same assumptions as K, there exist by Lemma F.6 a Hilbert space E and elements $h_t \in E$ such that

$$\langle h_s, h_t \rangle = \overline{K(s,t)} = K(t,s).$$

Replacing E by the closed linear span of $\{h_t\}$, we may assume that the set $\{h_t\}$ is total in E. Theorem F.5 then produces a reproducing Hilbert space $R(E)$ on T with reproducing kernel $\langle h_t, h_s \rangle_E = K(s,t)$.

In order to show uniqueness, observe that if H_1 and H_2 are two reproducing Hilbert spaces with reproducing kernel K, then, by Theorem F.3(iii),(vii),(v), the linear space H_0 spanned by the functions $K(\cdot, t)$ is a dense subspace of both H_1 and H_2, and the inner products in H_1 and H_2 coincide on H_0. Hence also the norms coincide on H_0. If $f \in H_1$, then there exists a sequence $\{f_n\} \subset H_0$ such that $f_n \to f$ in H_1, and in particular $f_n(t) \to f(t)$ for each t (because H_1 is a reproducing Hilbert space). Since $\{f_n\}$ is a Cauchy sequence in H_1 and $\{f_n\} \subset H_0$, where the norms coincide, it is also a Cauchy sequence in H_2. Thus $f_n \to g$ in H_2, for some g. But this implies $f_n(t) \to g(t)$ for each $t \in T$, and thus $g(t) = f(t)$. Hence $f \in H_2$ so $H_1 \subseteq H_2$, and by symmetry $H_1 = H_2$. It follows also, by continuity, that the inner products in H_1 and H_2 coincide, so $H_1 = H_2$ as Hilbert spaces. □

We end this section with a simple result. Further results on changes of the underlying set or of the space are given in Aronszajn (1950).

THEOREM F.8. *Suppose that H is a reproducing Hilbert space of functions on some set T, with reproducing kernel K, and suppose that T' is a subset of T such that the only function in H that vanishes on all of T' is the zero function. Then the set $H' = \{f|_{T'} : f \in H\}$ of restrictions of functions in H to T' is a reproducing Hilbert space on T' and its reproducing kernel is the restriction of K to $T' \times T'$. (The norm on H' is given by $\|f|_{T'}\|_{H'} = \|f\|_H$.)*

PROOF. It follows from the definitions that H' is a reproducing Hilbert space with reproducing kernel given by $K_t|_{T'}$, $t \in T'$. We leave the details to the reader. □

G. Analytic functions in Banach spaces

Analytic Banach-space-valued functions, defined on a domain in either the complex plane or another complex Banach space, are defined as follows, see Hervé (1989) for further details and results.

First, a function f defined on an open subset D of the complex plane with values in a complex Banach space B is analytic if the derivative $f'(z) = \lim_{w \to z}\big((f(w) - f(z))/(w - z)\big)$ exists (in B) for every $z \in D$. As in the complex-valued case, there are a number of equivalent conditions, for example that f has a power series expansion (with coefficients in B) in every disc contained in D. Moreover, f is analytic if and only if the complex function $z \mapsto \langle x^*, f(z)\rangle_B$ is analytic for every $x^* \in B^*$.

Secondly, a function f defined on an open subset D of a complex Banach space B, with values in the complex plane or in another complex Banach space, is analytic if it is continuous and if for every $x, y \in B$, the function $f(x + zy)$ is an analytic function in the subset $\{z \in \mathbb{C} : x + zy \in D\}$ of the complex plane. (Continuity follows from the second condition in the finite-dimensional case, but not in general.)

If D is open in B and $f \colon D \to B_1$ is analytic, then the derivative f' is an analytic function on D with values in the Banach space $L(B, B_1)$ of bounded linear operators $B \to B_1$. In particular, if $f \colon D \to \mathbb{C}$, then $f' \colon D \to B^*$.

We use the following simple results.

THEOREM G.1. *Suppose that $f \colon D \to B$ is an analytic function, where D is a connected open subset of the complex plane and B is a complex Banach space. Suppose further that $0 \in D$.*

If M is a closed subspace of B, and $f(z) \in M$ when $0 < z < r$ for some $r > 0$, then $f(z) \in M$ for every $z \in D$.

If furthermore f has a continuous extension to the closure $\overline{D}$, then $f(z) \in M$ for every $z \in \overline{D}$.

PROOF. Let $x^* \in B^*$ annihilate M. Then the complex function $z \mapsto \langle x^*, f(z)\rangle_B$ is analytic in D and equals 0 when z is small and positive. By the standard uniqueness theorem for analytic functions, this function is identically 0 in D. This shows that for every $z \in D$, $\langle x^*, f(z)\rangle_B = 0$ whenever $x^* \in M^\perp$, which implies $f(z) \in M$ by a standard application of the Hahn–Banach theorem.

The final statement follows by continuity. □

THEOREM G.2. *Suppose that $f \colon D \to B$ is an analytic function, where D is an open subset of the complex plane and B is a complex Banach space. Suppose further that $0 \in D$ and that f has the Taylor expansion $f(z) = \sum_0^\infty a_n z^n$ for small positive z, with $a_n \in B$.*

If M is a closed subspace of B, and $f(z) \in M$ when $0 < z < r$ for some $r > 0$, then $a_n \in M$ for every $n \geq 0$.

PROOF. Let $x^* \in B^*$ annihilate M. Then the complex function $z \mapsto \langle x^*, f(z)\rangle_B$ is analytic in D, has the Taylor expansion $\sum_0^\infty \langle x^*, a_n\rangle z^n$ for small positive z and equals 0 for such z. Consequently, $\langle x^*, a_n\rangle = 0$ for every n, whenever $x^* \in M^\perp$, which implies $a_n \in M$. □

H. Hilbert–Schmidt operators and singular numbers

Let H_1 and H_2 be two Hilbert spaces (possibly the same).

The *Hilbert–Schmidt norm* of a linear operator $T \colon H_1 \to H_2$ is defined by $\|T\|_{HS}^2 = \sum_i \|Te_i\|^2$, where $\{e_i\}_{i \in \mathcal{I}}$ is an orthonormal basis in H_1; this is independent of the choice of the basis, since for any orthonormal basis $\{f_j\}_{j\in\mathcal{J}}$ in H_2

$$\sum_i \|Te_i\|^2 = \sum_{i,j} |\langle Te_i, f_j\rangle|^2 = \sum_{i,j} |\langle e_i, T^* f_j\rangle|^2 = \sum_j \|T^* f_j\|^2.$$

A Hilbert–Schmidt operator is an operator with finite Hilbert–Schmidt norm.

It is easily seen that $\|T\|_{HS} \geq \|T\|$, the operator norm, and that a Hilbert–Schmidt operator is compact.

DEFINITION H.1. If $T\colon H_1 \to H_2$ is a bounded linear operator, we define its *singular numbers* $s_k(T)$, $k = 1, 2, \ldots$, by

$$s_k(T) = \inf\{\|T - R\| : R \text{ is an operator } H_1 \to H_2 \text{ with finite rank} < k\}.$$

We list some immediate consequences of the definition.

(i) $\{s_k(T)\}_1^\infty$ is a non-increasing sequence of non-negative numbers;

$$s_1(T) \geq s_2(T) \geq \cdots \geq 0.$$

(ii) $s_1(T) = \|T\|$.

(iii) $\lim_{k\to\infty} s_k(T) = 0$ if and only if T lies in the closed hull of the set of finite rank operators, i.e. if and only if T is a compact operator.

(iv) $s_k(T^*) = s_k(T)$, where T^* is the adjoint of T.

When T is compact, the squared singular numbers $s_k(T)^2$ are the eigenvalues (counted with multiplicities) of the compact self-adjoint operator T^*T, possibly ignoring or adding 0's. This yields another, equivalent, definition of singular numbers for compact operators. For details, and many other results on singular numbers, see e.g. Simon (1979b).

In the compact case, the singular numbers allow us a kind of diagonalization of the operator, as decribed in the following theorem. (But note that even if $H_1 = H_2$, we generally need two different bases.)

THEOREM H.2. *If $T\colon H_1 \to H_2$ is compact, and has N non-zero singular numbers $\{s_k(T)\}_1^N$, with $0 \leq N \leq \infty$, then there exist orthonormal bases $\{e_k\}_{k=1}^N \cup \{e'_i\}_{i\in\mathcal{I}}$ in H_1 and $\{f_k\}_{k=1}^N \cup \{f'_j\}_{j\in\mathcal{J}}$ in H_2 such that $Te_k = s_k(T) f_k$, $1 \leq k \leq N$ and $Te'_i = 0$, $i \in \mathcal{I}$.*

PROOF. Since T^*T is a compact self-adjoint operator, the spectral theorem (Conway 1990, Theorem II.5.1) shows that H_1 has an orthonormal basis consisting of eigenvectors of T^*T. Let $\{e'_i\}_\mathcal{I}$ be the elements of this basis with eigenvalue 0, and let $\{e_k\}_1^N$ be the remaining elements, ordered such that e_k has eigenvalue $s_k(T)^2$. Then $\langle Te_k, Te_l\rangle = \langle T^*Te_k, e_l\rangle = s_k(T)^2\delta_{kl}$ for any $k, l \leq N$. Hence the vectors $f_k = s_k(T)^{-1}Te_k$ form an orthonormal set in H_2, which may be completed to an orthonormal basis by adding a set $\{f'_j\}_\mathcal{J}$. If $i \in \mathcal{I}$, then $\|Te'_i\|^2 = \langle T^*Te'_i, e'_i\rangle = 0$ and thus $Te'_i = 0$. □

COROLLARY H.3. *If $T\colon H_1 \to H_2$, then the Hilbert–Schmidt norm is given by $\|T\|_{HS}^2 = \sum_1^\infty s_k(T)^2$. In particular, T is Hilbert–Schmidt if and only if $\sum_1^\infty s_k(T)^2 < \infty$.* □

References

L. ACCARDI & I. PIKOVSKI (1996), Nonadapted stochastic calculus as third quantization. *Random Oper. Stoch. Eq.* **4**, 77–89.

R.A. ADAMS & F.H. CLARKE (1979), Gross' logarithmic Sobolev inequality: a simple proof. *Amer. J. Math.* **101**, 1265–1270.

N. ARONSZAJN (1950), Theory of reproducing kernels. *Trans. Amer. Math. Soc.* **68**, 337–404.

J.C. BAEZ, I.E. SEGAL & Z. ZHOU (1992), *Introduction to Algebraic and Constructive Quantum Field Theory.* Princeton Univ. Press, Princeton, N.J.

D. BAKRY (1994), L'hypercontractivité et son utilisation en théorie des semigroupes. In *Lectures in Probability Theory,* Ecole d'Eté de Probabilités de Saint-Flour XXII-1992. Lect. Notes Math. **1581**, Springer-Verlag, Berlin, 1–114.

A.D. BARBOUR, M. KAROŃSKI & A. RUCIŃSKI (1989), A central limit theorem for decomposable random variables with applications to random graphs. *J. Combin. Th. Ser. B* **47**, 125–145.

W. BECKNER (1975), Inequalities in Fourier analysis. *Ann. Math.* **102**, 159–182.

D.R. BELL (1987), *The Malliavin Calculus.* Longman, Harlow, Essex.

D. BETOUNES & M. REDFERN (1996), The Stratonovich integral via the renormalization operator on Fock space. *Stochastics Stoch. Rep.* **56**, 161–178.

P. BILLINGSLEY (1968), *Convergence of Probability Measures.* Wiley, New York.

V.I. BOGACHEV (1996), Gaussian measures on linear spaces. *Itogi Nauki i Tekhniki, Mat. Anal.* **15** (Russian). English transl. *J. Math. Sci.* **79**, 933–1034.

A. BONAMI (1970), Étude des coefficients de Fourier des fonctions de $L^p(G)$. *Ann. Inst. Fourier (Grenoble)* **20**:2, 335–402.

C. BORELL (1982), Positivity improving operators and hypercontractivity. *Math. Z.* **180**, 225–234.

C. BORELL & S. JANSON (1982), Converse hypercontractivity. In *Séminaire Initiation à l'Analyse,* 21e année, 1981/1982. Publ. Math. Univ. Pierre Marie Curie **54**, No. 4.

N. BOULEAU & F. HIRSCH (1986), Propriétés d'absolue continuité dans les espaces de Dirichlet et applications aux équations différentielles stochastiques. In *Séminaire de Probabilités XX*. Lect. Notes Math. **1204**, Springer-Verlag, Berlin, 131–161.

N. BOULEAU & F. HIRSCH (1991), *Dirichlet Forms and Analysis on Wiener Space.* De Gruyter, Berlin.

H.J. BRASCAMP & E.H. LIEB (1976), Best constants in Young's inequality, its converse, and its generalization to more than three functions. *Adv. Math.* **20**, 151–173.

A. BUDHIRAJA & G. KALLIANPUR (1995), Hilbert space valued traces and multiple Stratonovich integrals with statistical applications. *Probab. Math. Statist.* **15**, 127–163.

R.H. Cameron & W.T. Martin (1944), Transformations of Wiener integrals under translations. *Ann. Math.* **45**, 386–396.

R.H. Cameron & W.T. Martin (1947), Fourier–Wiener transforms of functionals belonging to L_2 over the space C. *Duke Math. J.* **14**, 99–107.

D.L. Cohn (1980), *Measure Theory*. Birkhäuser, Boston, Mass.

J.B. Conway (1990), *A Course in Functional Analysis*. 2nd ed., Springer-Verlag, New York.

C. Dellacherie & P.-A. Meyer (1975), *Probabilités et potentiel*. Hermann, Paris. English transl. *Probabilities and Potential*. North-Holland, Amsterdam (1978).

P. Diaconis & S. Zabell (1991), Closed form summation for classical distributions: variations on a theme of De Moivre. *Statist. Sci.* **6**, 284–302.

R.L. Dobrushin & R.A. Minlos (1977), Polynomials in linear random functions. *Uspekhi Mat. Nauk* **32**:2, 67–122 (Russian). English transl. *Russian Math. Surveys* **32**:2, 71–127.

N. Dunford & J.T. Schwartz (1958), *Linear Operators*, Part I. Interscience, New York.

E.B. Dynkin & A. Mandelbaum (1983), Symmetric statistics, Poisson point processes and multiple Wiener integrals. *Ann. Statist.* **11**, 739–745.

J.B. Epperson (1989), The hypercontractive approach to exactly bounding an operator with complex Gaussian kernel. *J. Funct. Anal.* **87**, 1–30.

H. Federer (1969), *Geometric Measure Theory*. Springer-Verlag, New York.

G.B. Folland (1989), *Harmonic Analysis in Phase Space*. Princeton Univ. Press, Princeton, N.J.

J. Gasch & L. Maligranda (1994), On vector-valued inequalities of the Marcinkiewicz–Zygmund, Herz and Krivine type. *Math. Nachr.* **167**, 95–129.

B. Gaveau & P. Trauber (1982), L'intégrale stochastique comme opérateur de divergence dans l'espace fonctionnel. *J. Funct. Anal.* **46**, 230–238.

I.M. Gelfand & N.Ya. Vilenkin (1961), *Generalized functions*, Vol. 4. Fizmatgiz, Moscow (Russian). English transl. Academic Press, New York (1964).

I.I. Gihman & A.V. Skorohod (1971), *The Theory of Stochastic Processes*, Vol. I. Nauka, Moscow (Russian). English transl. Springer-Verlag, Berlin (1974).

H. Gjessing, H. Holden, T. Lindstrøm, B. Øksendal, J. Ubøe & T. Zhang (1993), The Wick product. In *Proceedings of the Third Finnish–Soviet Symposium on Probability Theory and Mathematical Statistics, Turku, 1991*, Frontiers in Pure and Applied Probability **1**, eds. H. Niemi et al. TVP Publishers, Moscow, 29–67.

J. Glimm & A. Jaffe (1981), *Quantum Physics*. Springer-Verlag, New York.

L. Gross (1967), Abstract Wiener spaces. In *Proceedings of the Fifth Berkeley Symposium on Mathematical Statistics and Probability*, 1965, Vol. II, Part 1. Univ. of California Press, Berkeley, 31–42.

L. Gross (1974), Analytic vectors for representations of the canonical commutation relations and nondegeneracy of ground states. *J. Funct. Anal.* **17**, 104–111.

L. Gross (1975), Logarithmic Sobolev inequalities. *Amer. J. Math.* **97**, 1061–1083.

L. Gross (1993), Logarithmic Sobolev inequalities and contractive properties of semigroups. In *Dirichlet Forms*. Lect. Notes Math. **1563**, Springer-Verlag, Berlin, 54–88.

A. Grothendieck (1956), Résumé de la théorie métrique des produits tensoriels topologiques. *Bol. Soc. Mat. São Paulo* **8**, 1–79.

U. HAAGERUP (1987), A new upper bound for the complex Grothendieck constant. *Israel J. Math.* **60**, 199–224.

P. HALL (1979), On the invariance principle for *U*-statistics. *Stoch. Proc. Appl.* **9**, 163–174.

M. HERVÉ (1989), *Analyticity in Infinite Dimensional Spaces.* De Gruyter, Berlin.

T. HIDA & M. HITSUDA (1976), *Gaussian Processes.* Kinokuniya, Tokyo (Japanese). English transl. Amer. Math. Soc., Providence, R.I. (1991).

T. HIDA & N. IKEDA (1967), Analysis on Hilbert space with reproducing kernel arising from multiple Wiener integral. In *Proceedings of the Fifth Berkeley Symposium on Mathematical Statistics and Probability*, 1965, Vol. II, Part 1. Univ. of California Press, Berkeley, 117–143.

T. HIDA, H.-H. KUO, J. POTTHOFF & L. STREIT (1993), *White Noise.* Kluwer, Dordrecht, Netherlands.

W. HOEFFDING (1948), A class of statistics with asymptotically normal distribution. *Ann. Math. Statist.* **19**, 293–325.

W. HOEFFDING (1961), The strong law of large numbers for *U*-statistics. Institute of Statistics, Univ. of North Carolina, Mimeograph series No. 302.

H. HOLDEN, B. ØKSENDAL, J. UBØE & T. ZHANG (1996), *Stochastic Partial Differential Equations.* Birkhäuser, Boston, Mass.

Y.Z. HU & P.-A. MEYER (1988), Sur les intégrales multiples de Stratonovich. In *Séminaire de Probabilités XXII.* Lect. Notes Math. **1321**, Springer-Verlag, Berlin, 72–81.

I.A. IBRAGIMOV & Y.A. ROZANOV (1970), *Gaussian Random Processes.* Nauka, Moscow (Russian). English transl. Springer-Verlag, New York (1978).

N. IKEDA & S. WATANABE (1984), An introduction to Malliavin's calculus. In *Stochastic Analysis,* Proceedings, Taniguchi Inter. Symp. on Stoch. Anal., Katata and Kyoto, 1982, ed. K. Itô. North-Holland, Amsterdam, 1–52.

K. ITÔ (1951), Multiple Wiener integral. *J. Math. Soc. Japan* **3**, 157–169.

K. ITÔ (1993), An elementary approach to Malliavin fields. In *Asymptotic problems in probability theory: Wiener functionals and asymptotics,* Proceedings, Sanda and Kyoto 1990. Longman, Harlow, Essex, 35–89.

S.R. JAMMALAMADAKA & S. JANSON (1986), Limit theorems for a triangular scheme of *U*-statistics with applications to inter-point distances. *Ann. Probab.* **14**, 1347–1358.

S. JANSON (1983), On hypercontractivity for multipliers on orthogonal polynomials. *Ark. Mat.* **21**, 97–110.

S. JANSON (1990), A functional limit theorem for random graphs with applications to subgraph count statistics. *Random Struct. Alg.* **1**, 15–37.

S. JANSON (1994), *Orthogonal Decompositions and Functional Limit Theorems for Random Graph Statistics.* Memoirs Amer. Math. Soc. **534**, Amer. Math. Soc., Providence, R.I.

S. JANSON (1995), A graph Fourier transform and proportional graphs. *Random Struct. Alg.* **6**, 341–351.

S. JANSON (1997+), On complex hypercontractivity. To appear.

S. JANSON & K. NOWICKI (1991), The asymptotic distributions of generalized *U*-statistics with applications to random graphs. *Probab. Th. Rel. Fields* **90**, 341–375.

S. JANSON & M.J. WICHURA (1983), Invariance principles for stochastic area and related stochastic integrals. *Stoch. Proc. Appl.* **16**, 71–84.

J.-P. KAHANE (1985), *Some random series of functions*. 2nd ed., Cambridge Univ. Press, Cambridge.

J.L. KRIVINE (1979), Constantes de Grothendieck et fonctions de type positif sur les sphères. *Adv. Math.* **31**, 16–30.

H.-H. KUO (1975), *Gaussian Measures in Banach Spaces*. Lect. Notes Math. **463**, Springer-Verlag, Berlin.

H.-H. KUO (1996), *White Noise Distribution Theory*. CRC Press, Boca Raton, Fla.

M. LE BELLAC (1991), *Quantum and Statistical Field Theory*. Oxford Univ. Press, Oxford.

M. LEDOUX & M. TALAGRAND (1991), *Probability in Banach Spaces*. Springer-Verlag, Berlin.

P. LÉVY (1948), *Processus stochastiques et mouvement brownien*. Gauthier–Villars, Paris.

T. LINDSTRØM, B. ØKSENDAL & J. UBØE (1992), Wick multiplication and Itô–Skorohod stochastic differential equations. In *Ideas and Methods in Mathematical Analysis, Stochastics, and Applications*, eds. S. Albeverio et al. Cambridge Univ. Press, Cambridge, 183–206.

T. LINDVALL (1973), Weak convergence of probability measures and random functions in the function space $D[0,\infty)$. *J. Appl. Probab.* **10**, 109–121.

P. MAJOR (1981), *Multiple Wiener–Itô integrals*. Lect. Notes Math. **849**, Springer-Verlag, Berlin.

P. MALLIAVIN (1978), Stochastic calculus of variations and hypoelliptic operators. In *Proceedings of the International Symposium on Stochastic Differential Equations, Kyoto, 1976*, ed. K. Itô. Wiley, New York, 195–263.

P. MALLIAVIN (1993), *Intégration, analyse de Fourier, probabilités, analyse gaussienne*. 2nd ed., Masson, Paris. English transl. *Integration and Probability*. Springer-Verlag, New York (1995).

P. MALLIAVIN (1997), *Stochastic Analysis*. Springer-Verlag, Berlin. To appear.

G. MARUYAMA (1950), Notes on Wiener integrals. *Kōdai Math. Sem. Rep. 1950*, 41–44.

H.P. MCKEAN (1969), *Stochastic Integrals*. Academic Press, New York.

F.G. MEHLER (1866), Ueber die Entwicklung einer Function von beliebig vielen Variablen nach Laplaceschen Functionen höherer Ordnung. *J. Reine Angew. Math.* **66**, 161–176.

P.-A. MEYER (1984), Transformations de Riesz pour les lois gaussiennes. In *Séminaire de Probabilités XVIII*. Lect. Notes Math. **1059**, Springer-Verlag, Berlin, 179–193.

P.-A. MEYER (1993), *Quantum Probability for Probabilists*. Lect. Notes Math. **1538**, Springer-Verlag, Berlin.

R. VON MISES (1947), On the asymptotic distribution of differentiable statistical functions. *Ann. Math. Statist.* **18**, 309–348.

N. NAKANISHI (1971), *Graph Theory and Feynman Integrals*. Gordon and Breach, New York.

E. NELSON (1973), The free Markov field. *J. Funct. Anal.* **12**, 211–227.

J. NEVEU (1968), *Processus aléatoires gaussiens*. Presses de l'Université de Montréal, Montréal.

J. NEVEU (1976), Sur l'espérance conditionelle par rapport à un mouvement brownien. *Ann. Inst. H. Poincaré B* **12**, 105–109.

D. NUALART (1995), *The Malliavin Calculus and Related Topics*. Springer-Verlag, New York.

D. NUALART (1997+), Analysis on Wiener space and anticipating stochastic calculus. In *Ecole d'Eté de Probabilités de Saint-Flour XXV-1995*. Lect. Notes Math., Springer-Verlag, Berlin. To appear.

D. NUALART & E. PARDOUX (1988), Stochastic calculus with anticipating integrands. *Probab. Th. Rel. Fields* **78**, 535–581.

D. NUALART & M. ZAKAI (1986), Generalized stochastic integrals and the Malliavin calculus. *Probab. Th. Rel. Fields* **73**, 255–280.

N. OBATA (1994), *White Noise Calculus and Fock Space*. Lect. Notes Math. **1577**, Springer-Verlag, Berlin.

D. OCONE (1987), A guide to the stochastic calculus of variations. In *Stochastic Analysis and Related Topics*, eds. H. Korezlioglu and A.S. Üstünel. Lect. Notes Math. **1316**, Springer-Verlag, Berlin, 1–79.

K.R. PARTHASARATY (1992), *An Introduction to Quantum Stochastic Calculus*. Birkhäuser, Basel.

J. PEETRE (1980), A class of kernels related to the inequalities of Beckner and Nelson. In *A Tribute to Åke Pleijel*. Uppsala University, Uppsala, 171–210.

G. PETERS (1997+), Anticipating flows on the Wiener space generated by vector fields of low regularity. *J. Funct. Anal.* To appear.

G. PISIER (1986), *Factorization of Linear Operators and Geometry of Banach Spaces*. CBMS Regional Conference Series **60**, Amer. Math. Soc., Providence, R.I.

G. PISIER (1988), Riesz transforms: A simple analytic proof of P.A. Meyer's inequality. In *Séminaire de Probabilités XXIII*. Lect. Notes Math. **1321**, Springer-Verlag, Berlin, 405–501.

G. PISIER (1989), *The Volume of Convex Bodies and Banach Space Geometry*. Cambridge Univ. Press, Cambridge.

P. PROTTER (1990), *Stochastic Integration and Differential Equations*. Springer-Verlag, Berlin.

M. RIESZ (1926), Sur les maxima des formes bilinéaires et sur les fonctionelles linéaires. *Acta Math.* **49**, 465–497.

H. RUBIN & R.A. VITALE (1980), Asymptotic distribution of symmetric statistics. *Ann. Statist.* **8**, 165–170.

W. RUDIN (1991), *Functional Analysis*. 2nd ed., McGraw–Hill, New York.

H.H. SCHAEFER (1971), *Topological Vector Spaces*. 3rd printing, Springer-Verlag, New York.

I.E. SEGAL (1956), Tensor algebras over Hilbert spaces. I. *Trans. Amer. Math. Soc.* **81**, 106–134.

B. SIMON (1974), *The $P(\phi)_2$ Euclidean (Quantum) Field Theory*. Princeton Univ. Press, Princeton, N.J.

B. SIMON (1976), A remark on Nelson's best hypercontractive estimates. *Proc. Amer. Math. Soc.* **55**, 376–378.

B. SIMON (1979a), *Functional Integration and Quantum Physics*. Academic Press, New York.

B. SIMON (1979b), *Trace Ideals and Their Applications*. Cambridge Univ. Press, Cambridge.

A.V. SKOROHOD (1975), On a generalization of a stochastic integral. *Teor. Veroyatnost. i Primenen.* **20**, 223–238 (Russian). English transl. *Th. Probab. Appl.* **20**, 219–233.

A. SLOAN (1974), A nonperturbative approach to nondegeneracy of ground states in quantum field theory: polaron models. *J. Funct. Anal.* **16**, 161–191.

P.M. SOARDI (1994), *Potential Theory on Infinite Networks*. Lect. Notes Math. **1590**, Springer-Verlag, Berlin.

C. STEIN (1986), *Approximate Computation of Expectations*. IMS, Hayward, Calif.

D.W. STROOCK (1981), The Malliavin calculus, a functional analytic approach. *J. Funct. Anal.* **44**, 212–257.

H. SUGITA (1985), On a characterization of the Sobolev spaces over an abstract Wiener space. *J. Math. Kyoto Univ.* **25**, 717–725.

M. TALAGRAND (1983), Mesures Gaussiennes sur un espace localement convexe. *Z. Wahrschein. Verw. Geb.* **64**, 181–209.

F. TREVES (1967), *Topological Vector Spaces, Distributions and Kernels*. Academic Press, New York.

A.S. ÜSTÜNEL (1995), *An Introduction to Analysis on Wiener Space*. Lect. Notes Math. **1610**, Springer-Verlag, Berlin.

S. WATANABE (1984), *Lectures on Stochastic Differential Equations and Malliavin Calculus*. Tata Institute of Fundamental Research, Springer-Verlag, Berlin.

F.B. WEISSLER (1979), Two-point inequalities, the Hermite semigroup and the Gauss–Weierstrass semigroup. *J. Funct. Anal.* **32**, 102–121.

G.C. WICK (1950), The evaluation of the collision matrix. *Phys. Rev.* **80**, 268–272.

N. WIENER (1938), The homogeneous chaos. *Amer. J. Math.* **60**, 897–936.

M. YOR (1988), Remarques sur certaines constructions des mouvements browniens fractionnaires. In *Séminaire de Probabilités XXII*. Lect. Notes Math. **1321**, Springer-Verlag, Berlin, 217–224.

M. ZAKAI (1985), The Malliavin Calculus. *Acta Appl. Math.* **3**, 175–207.

Z. ZHOU (1991), The contractivity of the free Hamiltonian semigroup in the L_p space of entire functions. *J. Funct. Anal.* **96**, 407–425.

A. ZYGMUND (1959), *Trigonometric Series*. 2nd ed., Cambridge Univ. Press, Cambridge.

Index of notation

$\overset{d}{=}$ equality in distribution, 1
$\overset{d}{\to}$ convergence in distribution, 1
$\overset{p}{\to}$ convergence in probability, 2
$\|\cdot\|_p$ L^p-norm, 1
$\|\cdot\|_{p,q}$ operator norm $L^p \to L^q$, 57
$\|\cdot\|_{HS}$ Hilbert–Schmidt norm, 142, 327, 328
$\langle\cdot,\cdot\rangle$ inner product, 2
$\mathbf{1}[\dots]$ indicator function, 2
$\mathbf{1}_A$ indicator function, 2
$:\cdots:$ Wick product, 23
$\odot$ (general) Wick product, 23, 293
$\odot$ symmetric tensor product, 42, 319
$\otimes$ tensor product, 42, 316, 317
$\oplus$ direct sum, 4, 14
∇ gradient, 236
$\wedge$ minimum, 2
$\asymp$ equivalence within constants, 63, 259, 267
!! semi-factorial, $n!! = n(n-2)\cdots$, 12

$A(\xi)$ annihilation operator, 200
α multi-index, 29
α strong mixing coefficient, 143
$A^*(\xi)$ creation operator, 200
$\mathcal{B}$ Borel σ-field, 1
B_t Brownian motion, 86
β absolute regularity coefficient, 144
$\mathbb{C}$ complex numbers, 2
C_γ contraction, 101
Cov covariance, 2
Corr correlation, 144
D_n a domain in $\mathbb{R}^n$, 88, 157
D_{pq} a domain in $\mathbb{C}$, 69
$\mathcal{D}^{k,p}$ Gaussian Sobolev space, 237, 245, 246, 272
$\mathcal{D}^{\xi,p}$ directional Sobolev space, 245, 246
$\mathcal{D}^{\infty-}$ Gaussian Sobolev space, 246
d_{TV} total variation distance, 144
∂_i $\partial/\partial x_i$, 228
∂_ξ directional derivative, 229, 304
δ divergence operator, 274
Dom domain of operator, 197
E expectation, 1
$\exp\odot$ tensor exponential, 44
$\mathcal{F}$ σ-field, 1
$\mathcal{F}(A)$ σ-field generated by A, 1
F_W Fourier–Wiener transform, 197
$G(n,p)$ random graph, 166
$\Gamma_*(H)$ symmetric tensor algebra, 42, 321
$\Gamma(H)$ Fock space, 42, 320
$\Gamma(A)$ Fock space operator, 45
γ standard gaussian measure, 7
γ_σ gaussian measure, 198
$H^{:n:}$ homogeneous chaos, 17
$H^{\odot n}$ symmetric tensor power, 42, 319
$H(B)$ a Gaussian Hilbert space, 6
h_n Hermite polynomial, 20
I identity operator, 45
I_n multiple stochastic integral, 88, 99
$\hat{I}_n$ multiple stochastic integral, 98
$\kappa(p)$ a constant, 5
$\kappa(p)_{\mathbb{C}}$ a constant, 15
L^p Lebesgue space, 2
L^0 space of random variables, 2
L^{p+} $\bigcup_{q>p} L^q$, 216, 221
L^{p-} $\bigcap_{q<p} L^q$, 216, 221
L^2_0 subspace of L^2, 91, 151
$\mathcal{M}_r$ Mehler transform on $\mathbb{R}$, 50
M_r Mehler transform, 51
$\mathcal{M}_\mu$ sets with finite μ-measure, 95
M_A space of functions, 161, 167
M_A^0 space of functions, 161, 167
$\widetilde{M}_A^0$ space of functions, 167
N normal distribution, 3
$\mathcal{N}$ number operator, 45, 72
Ω probability space, 1
P probability, 1
$\mathcal{P}$ polynomial variables, 20
$\mathcal{P}_n$ polynomial variables, 17

$\overline{\mathcal{P}}_n$ closure of $\mathcal{P}_n$, 17
$\overline{\mathcal{P}}_*$ $\bigcup \overline{\mathcal{P}}_n$, 20
P_{HK} projection from H to K, 142
$P(\xi)$ momentum operator, 210
Π^2 square integrable predictable processes, 89
π_n projection onto $H^{:n:}$, 22
$\pi_{\leq n}$ projection onto $\overline{\mathcal{P}}_n$, 22
$Q(\xi)$ position operator, 210
$\mathbb{R}$ real numbers, 2
R map to Cameron–Martin space, 121
$R(H)$ Cameron–Martin space, 121
$r(\gamma)$ rank of Feynman diagram, 15
ρ covariance function, 117
ρ maximal correlation coefficient, 143
ρ_k singular number, 142
ρ_ξ Cameron–Martin shift, 216
S S-transform, 286
$\mathcal{S}_0$ smooth variables, 264
S_n^k U-statistic, 150, 157, 166
$\overline{S}_n^k$ asymmetric statistic, 157, 166
s_k singular number, 328
$\mathfrak{S}_n$ symmetric group, 25
supp support graph, 167, 171
T T-transform, 286
U_n^k U-statistic, 150
Var variance, 2
V_σ an operator, 199
$v(\gamma)$ value of Feynman diagram, 15
Z_n combination of U-statistics, 151

Index

absolute regularity, 144
absolute regularity coefficient, 144, 146
absolutely continuous
  along ξ, 234, 239
  distribution, 53–54, 81, 241–242
  function, 11, 122, 205, 225, 233
  measure, 122, 223–226
  ray, 237, 239
  version, 234
Accardi, L., 283
Adams, R.A., 57
annihilation operator, 200–210, 213, 257, 307
Aronszajn, N., 323, 326

Baez, J.C., 197
Bakry, D., 57
Barbour, A.D., 172
Bargmann space, 44, 292
Bargmann transform, 292
Beckner, W., 57, 68, 70, 71
Bell, D.R., 228
Betounes, D., 105
Billingsley, P., 153, 156, 161, 179, 180, 225, 312
Bochner integral, 314
Bogachev, V.I., 125
Bonami, A., 58
Borell, C., 61
Bouleau, N., 228, 242
Brascamp, H.J., 57
Brownian bridge, 133
Brownian motion, 6, 11, 86–95, 97, 100, 118, 119, 121, 122, 125, 133, 216, 225, 226, 229, 302
  fractal, 119
  multiparameter, 119
Brownian sheet, 97
Budhiraja, A., 105, 155

Cameron, R.H., xi, 120, 122, 197, 216
Cameron–Martin shift, 216–227, 248, 304
Cameron–Martin space, 120–126, 225, 226
Cameron–Martin theorem, 225
canonical model, 226
CCR, *see* commutation relations, canonical
chain rule, 237, 240, 251, 279, 281, 284
chaos, 20
  homogeneous, 20
Clarke, F.H., 57
Cohn, D.L., 122, 125, 225, 233, 234
commutation relations
  canonical, 202, 211, 212, 214
  Weyl, 214, 215
complete regularity, 144
contraction, 101
convergence in distribution, 1
convergence in probability, 2
Conway, J.B., 55, 79, 197, 200, 213, 328
covariance function, 117
creation operator, 200–210, 213, 296, 297, 307
cumulant, 27

Dellacherie, C., 309
density function, 54, 56, 82, 135, 242, 281
  normal, 3, 7
Diaconis, P., 206
directional derivative, 229
distribution, 1
divergence, 274, 283, 308
Dobrushin, R.L., ix, 43
Dunford, N., 2, 213, 312–314
Dynkin, E.B., 151, 153, 156

electrical network, 133–141
Epperson, J.B., 68, 70

Federer, H., 242, 243
Feynman diagram, 15–16, 24–29, 65, 101–105, 155, 178
  complete, 15

value, 15
Wick value, 27
Fock space, 42, 44, 45, 320
Folland, G.B., 214, 292
Fourier transform, 52, 71, 199, 288
eigenfunctions, 52
Fourier–Wiener transform, 198–200
free field, 8

Gasch, J., 184
Gaussian
field, 9, 11, 118
Hilbert space, 4, 6
complex, 15, 31, 44, 72
indexed, 8–11, 124, 236, 283, 288
linear space, 4, 6
measure, 7
standard, 7, 44
random variable, 3, 4
centred, 3
complex, 3, 12–15, 31–32
standard, 3
standard complex, 13
symmetric, 3
symmetric complex, 13
vector-valued, 3, 7
stochastic integral, 95–105
complex, 110–116
stochastic measure, 95–105, 118
complex, 111–116
stochastic process, 6, 111–126, 129, 225
stationary, 111–116
Gaveau, B., 283
Gelfand, I.M., 8, 10
Gihman, I.I., 117, 311
Gjessing, H., 293
Glimm, J., 8, 15, 197
gradient, 236
Gross, L., 10, 53, 57, 62
Grothendieck's constant, 188, 192
Grothendieck's theorem, 188
Grothendieck, A., 188, 192

Haagerup, U., 188
Hall, P., 156
harmonic oscillator, 213
Hausdorff–Young inequality, 71
Hermite polynomials, 20, 28–30, 33, 49, 52, 67
Hermite transform, 293
Hervé, M., 288, 326
Hida, T., ix, 8, 117, 228, 288, 293
Hilbert–Schmidt, 10, 79, 240, 318, 327–328
norm, 142, 327, 328
Hirsch, F., 228, 242
Hitsuda, M., ix, 117
Hoeffding, W., 153, 154
Holden, H., ix, 109, 228, 293
Hu, Y.Z., 105
hypercontractivity, 57–72

Ibragimov, I.A., ix, 117, 146, 149
Ikeda, N., 228, 293
indicator function, 2
isometry, 8
Itô, K., 11, 20, 98, 100
Itô integral, 87–95, 103, 104, 108, 302
multiple, 88

Jaffe, A., 8, 15, 197
Jammalamadaka, S.R., 155
Janson, S., 57, 61, 68, 72, 155, 165, 172, 175, 176
Jensen's inequality, 48
joint normal distribution, 3, 5

Kahane, J.-P., ix, 87, 117, 119, 120
Kallianpur, G., 105, 155
Karoński, M., 172
Krawtchouk polynomials, 60
Krivine, J.L., 188
Kuo, H.-H., ix, 8, 10, 228, 272, 288

ladder operator, 213
Le Bellac, M., 15
Ledoux, M., 120, 124, 312
Lévy, P., 119
Lieb, E.H., 57
Lindstrøm, T., 109, 293
Lindvall, T., 157
local operator, 232, 278, 279, 297
logarithmic Sobolev inequality, 62
L^p-norm, 1, 5
Lyapounov's inequality, 2

Major, P., ix, 112
Maligranda, L., 184
Malliavin calculus, 228
Malliavin derivative, 236
Malliavin matrix, 242, 281
Malliavin, P., ix, 228
Mandelbaum, A., 151, 153, 156

Martin, W.T., xi, 120, 122, 197, 216
Maruyama, G., 120, 225
maximal correlation coefficient, 143, 145
McKean, H.P., 86
measure of dependence, 143
Mehler transform, 50–53, 62, 68–71, 130, 253, 273
Mehler, F.G., 50, 51
Meyer inequalities, 272
Meyer, P.-A., ix, 72, 105, 197, 272, 309
Minlos, R.A., ix, 43
von Mises, R., 155
mixing, 143, 149
momentum operator, 210–215
monotone class theorem, 309
multi-index, 29

Nakanishi, N., 15
Nelson, E., 57
Neveu, J., ix, 57, 94, 117
normal, *see* Gaussian
Nowicki, K., 172, 176
Nualart, D., ix, 105, 109, 228, 283
nuclear space, 10
number operator, 45, 207, 279, 308, 323

Obata, N., ix, 8, 10, 228
Ocone, D., 228
Øksendal, B., ix, 109, 228, 293
Ornstein–Uhlenbeck operator, *see* number operator
Ornstein–Uhlenbeck process, 118, 130
Ornstein–Uhlenbeck semigroup, 45, 52, 130

Pardoux, E., 105, 109
Parthasaraty, K.R., 197
Peetre, J., 72
Peters, G., 228, 240
Pikovski, I., 283
Pisier, G., ix, 188, 192, 270, 272
polarization, 314
polynomial variables, 20–22
position operator, 210–215
positivity improving, 56
Potthoff, J., ix, 8, 228, 288
predictable process, 89
  elementary, 88
  square integrable, 89
prediction, 129
Protter, P., 86, 88, 309

quantum field theory, 7, 11, 53, 197–215

r.a.c., 237, 239
random graph, 165, 172–176
random hypergraph, 166, 176
random potential, 133
random variable, 1
  generalized, 272
Redfern, M., 105
reproducing Hilbert space, 121, 323
reproducing kernel, 324
Riesz, M., 191, 268
Rozanov, Y.A., ix, 117, 146, 149
Rubin, H., 155
Ruciński, A., 172
Rudin, W., 197, 205, 264, 275

S-transform, 286–308
Schaefer, H.H., 10, 39, 221, 317
Schrödinger representation, 212, 213
Schwartz, J.T., 2, 213, 312–314
second quantization, 323
Segal, I.E., 20, 43, 197
Segal–Bargmann space, 44, 292
semi-invariant, 27
Simon, B., ix, 8, 57, 328
singular numbers, 142, 328
Skorohod integral, 105–110, 283–285, 300–304
Skorohod, A.V., 105, 117, 311
Sloan, A., 53
Soardi, P.M., 134
Sobolev space, 8, 62, 245, 272
spectral measure, 112–116
  stochastic, 112
Stein's method, 206
Stein, C., 206
stochastic calculus of variation, 228
stochastic integral, 6, 87–116, 300–304
  multiple, 98
Stratonovich integral, 105, 155
Streit, L., ix, 8, 228, 288
strong mixing, 143
strong mixing coefficient, 143, 146
Stroock, D.W., 228
Sugita, H., 247
support graph, 167
symmetric tensor algebra, 42, 321
symmetric tensor power, 42, 318

T-transform, 286–308
Talagrand, M., 120, 124, 125, 312
tempered distributions, 7

tensor product, 42, 315–323
algebraic, 316
Banach space, 317
Hilbert space, 317
of operators, 321
total, 2
total variation distance, 144
Trauber, P., 283
Treves, F., 10, 264, 317
two-point inequality, 58, 68

U-statistic, 150
multi-sample, 165
Ubøe, J., ix, 109, 228, 293
unitary, 8
universal property, 316, 319
Üstünel, A.S., 228

V-statistic, 155
version, 234, 312
Vilenkin, N.Ya., 8, 10
Vitale, R.A., 155

Watanabe distributions, 272
Watanabe, S., ix, 72, 228, 272
weak*-topology, 266
Weissler, F.B., 68
white noise, 8, 9, 97
Wichura, M., 165
Wick exponential, 32–34, 93
Wick multiplication, 217
Wick ordering, 209
Wick product, 23–32, 43, 93, 100, 105, 107, 109, 209, 293–297
general, 23, 38, 103, 293
Wick's theorem, 11
Wick, G.C., 23, 209
Wiener chaos, *see* chaos
Wiener measure, 125
Wiener process, 126
Wiener space, 226, 228, 229
abstract, 10
Wiener, N., 20

ξ-a.c., 234, 239

Yor, M., 119

Zabell, S., 206
Zakai, M., 228
Zhang, T., ix, 109, 228, 293
Zhou, Z., 72, 197
Zygmund, A., 268